LIE ALGEBRAS
THEORY AND ALGORITHMS

North-Holland Mathematical Library

Board of Honorary Editors:

VOLUME 56

ELSEVIER
Amsterdam – Lausanne – New York – Oxford – Shannon – Singapore – Tokyo

Lie Algebras
Theory and Algorithms

Willem A. de Graaf
University of St Andrews
Scotland

2000
ELSEVIER
Amsterdam – Lausanne – New York – Oxford – Shannon – Singapore – Tokyo

ELSEVIER SCIENCE B.V.
Sara Burgerhartstraat 25
P.O. Box 211, 1000 AE Amsterdam, The Netherlands

First edition 2000

Library of Congress Cataloging in Publication Data
A catalog record from the Library of Congress has been applied for.

ISBN: 0 444 50116 9

⊖The paper used in this publication meets the requirements of ANSI/NISO Z39.48-1992 (Permanence of Paper).

Printed in The Netherlands.

Preface

Lie algebras arise naturally in various areas of mathematics and physics. However, such a Lie algebra is often only known by a presentation such as a multiplication table, a set of generating matrices, or a set of generators and relations. These presentations by themselves do not reveal much of the structure of the Lie algebra. Furthermore, the objects involved (e.g., a multiplication table, a set of generating matrices, an ideal in the free Lie algebra) are often large and complex and it is not easy to see what to do with them. The advent of the computer however, opened up a whole new range of possibilities: it made it possible to work with Lie algebras that are too big to deal with by hand. In the early seventies this moved people to invent and implement algorithms for analyzing the structure of a Lie algebra (see, e.g., [7], [8]). Since then many more algorithms for this purpose have been developed and implemented.

The aim of the present work is two-fold. Firstly it aims at giving an account of many existing algorithms for calculating with finite-dimensional Lie algebras. Secondly, the book provides an introduction into the theory of finite-dimensional Lie algebras. These two subject areas are intimately related. First of all, the algorithmic perspective often invites a different approach to the theoretical material than the one taken in various other monographs (e.g., [42], [48], [77], [86]). Indeed, on various occasions the knowledge of certain algorithms allows us to obtain a straightforward proof of theoretical results (we mention the proof of the Poincaré-Birkhoff-Witt theorem and the proof of Iwasawa's theorem as examples). Also proofs that contain algorithmic constructions are explicitly formulated as algorithms (an example is the isomorphism theorem for semisimple Lie algebras that constructs an isomorphism in case it exists). Secondly, the algorithms can be used to arrive at a better understanding of the theory. Performing the algorithms in concrete examples, calculating with the concepts involved, really brings the theory to life.

The book is roughly organized as follows. Chapter 1 contains a general introduction into the theory of Lie algebras. Many definitions are given that are needed in the rest of the book. Then in Chapters 2 to 5 we explore the structure of Lie algebras. The subject of Chapter 2 is the structure of nilpotent and solvable Lie algebras. Chapter 3 is devoted to Cartan subalgebras. These are immensely powerful tools for investigating the structure of semisimple Lie algebras, which is the subject of Chapters 4 and 5 (which culminate in the classification of the semisimple Lie algebras). Then in Chapter 6 we turn our attention towards universal enveloping algebras. These are of paramount importance in the representation theory of Lie algebras. In Chapter 7 we deal with finite presentations of Lie algebras, which form a very concise way of presenting an often high dimensional Lie algebra. Finally Chapter 8 is devoted to the representation theory of semisimple Lie algebras. Again Cartan subalgebras play a pivotal role, and help to determine the structure of a finite-dimensional module over a semisimple Lie algebra completely. At the end there is an appendix on associative algebras, that contains several facts on associative algebras that are needed in the book.

Along with the theory numerous algorithms are described for calculating with the theoretical concepts. First in Chapter 1 we discuss how to present a Lie algebra on a computer. Of the algorithms that are subsequently given we mention the algorithm for computing a direct sum decomposition of a Lie algebra, algorithms for calculating the nil- and solvable radicals, for calculating a Cartan subalgebra, for calculating a Levi subalgebra, for constructing the simple Lie algebras (in Chapter 5 this is done by directly giving a multiplication table, in Chapter 7 by giving a finite presentation), for calculating Gröbner bases in several settings (in a universal enveloping algebra, and in a free Lie algebra), for calculating a multiplication table of a finitely presented Lie algebra, and several algorithms for calculating combinatorial data concerning representations of semisimple Lie algebras. In Appendix A we briefly discuss several algorithms for associative algebras.

Every chapter ends with a section entitled "Notes", that aims at giving references to places in the literature that are of relevance to the particular chapter. This mainly concerns the algorithms described, and not so much the theoretical results, as there are standard references available for them (e.g., [42], [48], [77], [86]).

I have not carried out any complexity analyses of the algorithms described in this book. The complexity of an algorithm is a function giving an estimate of the number of "primitive operations" (e.g., arithmetical operations) carried out by the algorithm in terms of the size of the input. Now the size of a Lie algebra given by a multiplication table is the sum of the sizes of its structure constants. However, the number of steps performed by an algorithm that operates on a Lie algebra very often depends not only on the size of the input, but also (rather heavily) on certain structural properties of the input Lie algebra (e.g., the length of its derived series). Of course, it is possible to consider only the worst case, i.e., Lie algebras having a structure that poses most difficulties for the algorithm. However, for most algorithms it is far from clear what the worst case is. Secondly, from a practical viewpoint worst case analyses are not very useful since in practice one only very rarely encounters the worst case.

Of the algorithms discussed in this book many have been implemented inside several computer algebra systems. Of the systems that support Lie algebras we mention GAP4 ([31]), LiE ([21]) and Magma ([22]). We refer to the manual of each system for an account of the functions that it contains.

I would like to thank everyone who, directly or indirectly, helped me write this book. In particular I am grateful to Arjeh Cohen, without whose support this book never would have been written, as it was his idea to write it in the first place. I am also grateful to Gábor Ivanyos for his valuable remarks on the appendix. Also I gratefully acknowledge the support of the Dutch Technology Foundation (STW) who financed part of my research.

Willem de Graaf

Contents

Chapter 1

Basic constructions

This chapter serves two purposes. First of all it provides an introduction into the theory of Lie algebras. In the first four sections we define what a Lie algebra is, and we give a number of examples of Lie algebras. In Section 1.5 we discuss some generalities concerning algorithms. Furthermore, we describe our approach to calculating with Lie algebras. We describe how to represent a Lie algebra on a computer (namely by an array of structure constants), and we give two examples of algorithms. In subsequent sections we give several constructions of Lie algebras and objects related to them. In many cases these constructions are accompanied by a an algorithm that performs the construction.

A second purpose of this chapter is to serve as reference for later chapters. This chapter contains most basic constructions used in this book. Therefore it has the nature of a collection of sections, sometimes without clear line of thought connecting them.

1.1 Algebras: associative and Lie

Definition 1.1.1 *An* algebra *is a vector space A over a field F together with a bilinear map $m : A \times A \to A$.*

The bilinear map m of Definition 1.1.1 is called a *multiplication*. If A is an algebra and $x, y \in A$, then we usually write xy instead of $m(x, y)$.

Because an algebra A is a vector space, we can consider subspaces of A. A subspace $B \subset A$ is called a *subalgebra* if $xy \in B$ for all $x, y \in B$. It is called an *ideal* if xy and yx lie in B for all $x \in A$ and $y \in B$. Clearly an ideal is also a subalgebra.

Let A and B be two algebras over the field F. A linear map $\theta : A \to B$

is called a *morphism of algebras* if $\theta(xy) = \theta(x)\theta(y)$ for all $x, y \in A$ (where the product on the left hand side is taken in A and the product on the right hand side in B). The map θ is an *isomorphism of algebras* if θ is bijective.

Definition 1.1.2 *An algebra A is said to be* associative *if for all elements $x, y, z \in A$ we have*

$$(xy)z = x(yz) \quad \text{(associative law)}.$$

Definition 1.1.3 *An algebra L is said to be a* Lie *algebra if its multiplication has the following properties:*

(L_1) $xx = 0$ *for all $x \in L$,*

(L_2) $x(yz) + y(zx) + z(xy) = 0$ *for all $x, y, z \in L$ (Jacobi identity).*

Let L be a Lie algebra and let $x, y \in L$. Then $0 = (x + y)(x + y) = xx + xy + yx + yy = xy + yx$. So condition (L_1) implies

$$xy = -yx \text{ for all } x, y \in L. \tag{1.1}$$

On the other hand (1.1) implies $xx = -xx$, or $2xx = 0$ for all $x \in L$. The conclusion is that if the characteristic of the ground field is not 2, then (L_1) is equivalent to (1.1). Using (1.1) we see that the Jacobi identity is equivalent to $(xy)z + (yz)x + (zx)y = 0$ for all $x, y, z \in L$.

Example 1.1.4 Let V be an n-dimensional vector space over the field F. Here we consider the vector space $\text{End}(V)$ of all linear maps from V to V. If $a, b \in \text{End}(V)$ then their product is defined by

$$ab(v) = a(b(v)) \text{ for all } v \in V.$$

This multiplication makes $\text{End}(V)$ into an associative algebra.

For $a, b \in \text{End}(V)$ we set $[a, b] = ab - ba$ The bilinear map $(a, b) \to [a, b]$ is called the *commutator*, or *Lie bracket*. We verify the requirements (L_1) and (L_2) for the Lie bracket. First we have $[a, a] = aa - aa = 0$ so that (L_1) is satisfied. Secondly,

$$[a, [b, c]] + [b, [c, a]] + [c, [a, b]] =$$
$$a(bc - cb) - (bc - cb)a + b(ca - ac)$$
$$- (ca - ac)b + c(ab - ba) - (ab - ba)c = 0.$$

Hence also (L$_2$) holds for the commutator. It follows that the space of linear maps from V to V together with the commutator is a Lie algebra. We denote it by $\mathfrak{gl}(V)$.

Now fix a basis $\{v_1, \ldots, v_n\}$ of V. Relative to this basis every linear transformation can be represented by a matrix. Let $M_n(F)$ be the vector space of all $n \times n$ matrices over F. The usual matrix multiplication makes $M_n(F)$ into an associative algebra. It is isomorphic to the algebra $\mathrm{End}(V)$, the isomorphism being the map that sends a linear transformation to its matrix with respect to the fixed basis. Analogously we let $\mathfrak{gl}_n(F)$ be the Lie algebra of all $n \times n$ matrices with coefficients in F. It is equipped with the product $(a, b) \to [a, b] = ab - ba$ for $a, b \in \mathfrak{gl}_n(F)$. The map that sends a linear transformation to its matrix relative to a fixed basis is an isomorphism of $\mathfrak{gl}(V)$ onto $\mathfrak{gl}_n(F)$.

Let A be an algebra and let $B \subset A$ be a subalgebra. Then B is an algebra in its own right, inheriting the multiplication from its "parent" A. Furthermore, if A is a Lie algebra then clearly B is also a Lie algebra, and likewise if A is associative. If B happens to be an ideal, then, by the following proposition, we can give the quotient space A/B an algebra structure. The algebra A/B is called the *quotient algebra* of A and B.

Proposition 1.1.5 *Let A be an algebra and let $B \subset A$ be an ideal. Let A/B denote the quotient space. Then the multiplication on A induces a multiplication on A/B by $\bar{x}\bar{y} = \overline{xy}$ (where $\bar{x}$ denotes the coset of $x \in A$ in A/B). Furthermore, if A is a Lie algebra, then so is A/B (and likewise if A is an associative algebra).*

Proof. First of all we check that the multiplication on A/B is well defined. So let $x, y \in A$ and $b_1, b_2 \in B$. Then $\bar{x} = \overline{x + b_1}$ and $\bar{y} = \overline{y + b_2}$ and hence

$$\bar{x}\bar{y} = \overline{(x + b_1)}\ \overline{(y + b_2)} = \overline{(x + b_1)(y + b_2)} = \overline{xy + xb_2 + b_1y + b_1b_2} = \overline{xy}.$$

Consequently the product $\bar{x}\bar{y}$ is independent of the particular representatives of $\bar{x}$ and $\bar{y}$ chosen.

The fact that the Lie (respectively associative) structure is carried over to A/B is immediate.   □

Associative algebras and Lie algebras are intimately related in the sense that given an associative algebra we can construct a related Lie algebra and the other way round. First let A be an associative algebra. The commutator yields a bilinear operation on A, i.e., $[x, y] = xy - yx$ for all $x, y \in A$, where the products on the right are the associative products of A. Let A_{Lie} be the underlying vector space of A together with the product $[\ ,\ ]$. It is straightforward to check that A_{Lie} is a Lie algebra (cf. Example 1.1.4). In Chapter 6 we will show that every Lie algebra occurs as a subalgebra of a Lie algebra of the form A_{Lie} where A is an associative algebra. (This is the content of the theorems of Ado and Iwasawa.) For this reason we will use square brackets to denote the product of any Lie algebra.

From a Lie algebra we can construct an associative algebra. Let L be a Lie algebra over the field F. For $x \in L$ we define a linear map

$$\mathrm{ad}_L x : L \longrightarrow L$$

by $\mathrm{ad}_L x(y) = [x, y]$ for $y \in L$. This map is called the *adjoint map* determined by x. If there can be no confusion about the Lie algebra to which x belongs, we also write $\mathrm{ad} x$ in place of $\mathrm{ad}_L x$. We consider the subalgebra of $\mathrm{End}(L)$ generated by the identity mapping together with $\{\mathrm{ad} x \mid x \in L\}$ (i.e., the smallest subalgebra of $\mathrm{End}(L)$ containing 1 and this set). This associative algebra is denoted by $(\mathrm{ad} L)^*$.

Since $\mathrm{ad} x$ is the left multiplication by x, the adjoint map encodes parts of the multiplicative structure of L. We will often study the structure of a Lie algebra L by investigating its adjoint map. This will allow us to use the tools of linear algebra (matrices, eigenspaces and so on). Furthermore, as will be seen, the associative algebra $(\mathrm{ad} L)^*$ can be used to obtain valuable information about the structure of L (see, e.g., Section 2.2).

1.2 Linear Lie algebras

In Example 1.1.4 we encountered the Lie algebra $\mathfrak{gl}_n(F)$ consisting of all $n \times n$ matrices over the field F. By E_{ij}^n we will denote the $n \times n$ matrix with a 1 on position (i, j) and zeros elsewhere. If it is clear from the context which n we mean, then we will often omit it and write E_{ij} in place of E_{ij}^n. So a basis of $\mathfrak{gl}_n(F)$ is formed by all E_{ij} for $1 \leq i, j \leq n$.

Subalgebras of $\mathfrak{gl}_n(F)$ are called *linear Lie algebras*. In this section we construct several linear Lie algebras.

Example 1.2.1 For a matrix a let $\mathrm{Tr}(a)$ denote the trace of a. Let $a, b \in \mathfrak{gl}_n(F)$, then

$$\mathrm{Tr}([a, b]) = \mathrm{Tr}(ab - ba) = \mathrm{Tr}(ab) - \mathrm{Tr}(ba) = \mathrm{Tr}(ab) - \mathrm{Tr}(ab) = 0. \quad (1.2)$$

Set $\mathfrak{sl}_n(F) = \{a \in \mathfrak{gl}_n(F) \mid \mathrm{Tr}(a) = 0\}$, then, since the trace is a linear function, $\mathfrak{sl}_n(F)$ is a linear subspace of $\mathfrak{gl}_n(F)$. Moreover, by (1.2) we see that $[a,b] \in \mathfrak{sl}_n(F)$ if $a,b \in \mathfrak{sl}_n(F)$. Hence $\mathfrak{sl}_n(F)$ is a subalgebra of $\mathfrak{gl}_n(F)$. It is called the *special linear Lie algebra*. The Lie algebra $\mathfrak{sl}_n(F)$ is spanned by all E_{ij} for $i \neq j$ together with the diagonal matrices $E_{ii} - E_{i+1,i+1}$ for $1 \leq i \leq n-1$. Hence the dimension of $\mathfrak{sl}_n(F)$ is $n^2 - 1$.

Let V be an n-dimensional vector space over F. We recall that a *bilinear form* f on V is a bilinear function $f : V \times V \to F$. It is *symmetric* if $f(v,w) = f(w,v)$ and *skew symmetric* if $f(v,w) = -f(w,v)$ for all $v,w \in V$. Furthermore, f is said to be *non-degenerate* if $f(v,w) = 0$ for all $w \in V$ implies $v = 0$. For a bilinear form f on V we set

$$L_f = \{a \in \mathfrak{gl}(V) \mid f(av,w) = -f(v,aw) \text{ for all } v,w \in V\}, \qquad (1.3)$$

which is a linear subspace of $\mathfrak{gl}(V)$.

Lemma 1.2.2 *Let f be a bilinear form on the n-dimensional vector space V. Then L_f is a subalgebra of $\mathfrak{gl}(V)$.*

Proof. For $a,b \in L_f$ we calculate

$$f([a,b]v,w) = f((ab-ba)v,w) = f(abv,w) - f(bav,w)$$
$$= -f(bv,aw) + f(av,bw) = f(v,baw) - f(v,abw) = -f(v,[a,b]w).$$

So for each pair of elements $a,b \in L_f$ we have that $[a,b] \in L_f$ and hence L_f is a subalgebra of $\mathfrak{gl}(V)$. $\qquad\qquad\qquad\qquad\qquad\qquad\qquad\qquad\qquad\square$

Now we fix a basis $\{v_1, \dots, v_n\}$ of V. This allows us to identify V with the vector space F^n of vectors of length n. Also, as pointed out in Example 1.1.4, we can identify $\mathfrak{gl}(V)$ with $\mathfrak{gl}_n(F)$, the Lie algebra of all $n \times n$ matrices over F. We show how to identify L_f as a subalgebra of $\mathfrak{gl}_n(F)$. Let M_f be the $n \times n$ matrix with on position (i,j) the element $f(v_i,v_j)$. Then a straightforward calculation shows that $f(v,w) = v^t M_f w$ (where v^t denotes the transpose of v). The condition for $a \in \mathfrak{gl}_n(F)$ to be an element of L_f translates to $v^t a^t M_f w = -v^t M_f a w$ which must hold for all $v,w \in F^n$. It follows that $a \in L_f$ if and only if $a^t M_f = -M_f a$.

The next three examples are all of the form of L_f for some non-degenerate bilinear form f.

Example 1.2.3 Let f be a non-degenerate skew symmetric bilinear form with matrix

$$M_f = \begin{pmatrix} 0 & I_l \\ -I_l & 0 \end{pmatrix},$$

where I_l denotes the $l \times l$ identity matrix. We shall indicate a basis of L_f. Let $a \in \mathfrak{gl}_{2l}(F)$, then we decompose a into blocks

$$a = \begin{pmatrix} A & B \\ C & D \end{pmatrix}$$

(where A, B, C, D are $l \times l$ matrices). Expanding the condition $a^t M_f = -M_f a$ we get $B = B^t$, $C = C^t$ and $D = -A^t$. Therefore the following matrices constitute a basis of L_f:

- $A_{ij} = E_{ij} - E_{l+j,l+i}$ for $1 \leq i, j \leq l$ (the possibilities for A and D),

- $B_{ij} = E_{i,l+j} + E_{j,l+i}$ for $1 \leq i \leq j \leq l$ (the possibilities for B),

- $C_{ij} = E_{l+i,j} + E_{l+j,i}$ for $1 \leq i \leq j \leq l$ (the possibilities for C).

Counting these, we find that the dimension of L_f is

$$l^2 + \frac{l(l+1)}{2} + \frac{l(l+1)}{2} = 2l^2 + l.$$

This Lie algebra is called the *symplectic Lie algebra* and denoted by $\mathfrak{sp}_{2l}(F)$.

Example 1.2.4 Now let V be a vector space of dimension $2l + 1$. We take f to be a non-degenerate symmetric bilinear form on V with matrix

$$M_f = \begin{pmatrix} 1 & 0 & 0 \\ 0 & 0 & I_l \\ 0 & I_l & 0 \end{pmatrix}.$$

Accordingly we decompose $a \in \mathfrak{gl}_{2l+1}(F)$ as

$$a = \begin{pmatrix} \alpha & p & q \\ u & A & B \\ v & C & D \end{pmatrix}$$

where $\alpha \in F$, p, q are $1 \times l$ matrices, u, v are $l \times 1$ matrices and A, B, C, D are $l \times l$ matrices. Here the condition $a^t M_f = -M_f a$ boils down to $\alpha = -\alpha$, $v^t = -p$, $u^t = -q$, $C^t = -C$, $B^t = -B$, and $D = -A^t$. So a basis of L_f is formed by the following elements:

- $p_i = E_{1,1+i} - E_{l+1+i,1}$, for $1 \leq i \leq l$ (p and v),

- $q_i = E_{1,l+1+i} - E_{1+i,1}$, for $1 \leq i \leq l$ (q and u),

- $A_{ij} = E_{1+i,1+j} - E_{l+1+j,l+1+i}$, for $1 \le i, j \le l$ (A and D),

- $B_{ij} = E_{1+i,l+1+j} - E_{1+j,l+1+i}$, for $1 \le i < j \le l$ (B),

- $C_{ij} = E_{l+1+i,1+j} - E_{l+1+j,1+i}$, for $1 \le i < j \le l$ (C).

In this case the dimension of L_f is $2l^2 + l$. The Lie algebra of this example is called the *odd orthogonal Lie algebra* and is denoted by $\mathfrak{o}_{2l+1}(F)$.

Example 1.2.5 We now construct even orthogonal Lie algebras. Let V be of even dimension $2l$ and let f be a symmetric bilinear form on V with matrix

$$M_f = \begin{pmatrix} 0 & I_l \\ I_l & 0 \end{pmatrix}.$$

Let $a \in \mathfrak{gl}_{2l}(F)$ be decomposed in the same way as in Example 1.2.3. Then $a \in L_f$ if and only if $A = -D^t$, $B = -B^t$, and $C = -C^t$. A basis of L_f is constituted by

- $A_{ij} = E_{i,j} - E_{l+j,l+i}$ for $1 \le i, j \le l$ (A and D),

- $B_{ij} = E_{i,l+j} - E_{j,l+i}$ for $1 \le i < j \le l$ (B),

- $C_{ij} = E_{l+i,j} - E_{l+j,i}$ for $1 \le i < j \le l$, (C).

In this case the dimension of L_f is $2l^2 - l$. This Lie algebra is called the *even orthogonal algebra* and is denoted by $\mathfrak{o}_{2l}(F)$.

Example 1.2.6 We give two more classes of subalgebras of $\mathfrak{gl}_n(F)$. First let $\mathfrak{n}_n(F)$ be the space of all strictly upper triangular matrices. If $a, b \in \mathfrak{n}_n(F)$, then also $ab \in \mathfrak{n}_n(F)$ and consequently $[a, b] = ab - ba \in \mathfrak{n}_n(F)$. So $\mathfrak{n}_n(F)$ is a subalgebra of $\mathfrak{gl}_n(F)$. It is spanned by E_{ij} for $1 \le i < j \le n$.

Let $\mathfrak{b}_n(F)$ be the subspace of $\mathfrak{gl}_n(F)$ consisting of all upper triangular matrices. Then in the same way as above it is seen that $\mathfrak{b}_n(F)$ is a subalgebra of $\mathfrak{gl}_n(F)$. It is spanned by E_{ij} for $1 \le i \le j \le n$.

1.3 Structure constants

Let A be an n-dimensional algebra over the field F with basis $\{x_1, \dots, x_n\}$. For each pair (x_i, x_j) we can express the product $x_i x_j$ as a linear combination of the basis elements of A, i.e., there are elements $c_{ij}^k \in F$ such that

$$x_i x_j = \sum_{k=1}^n c_{ij}^k x_k.$$

On the other hand, the set of n^3 constants c_{ij}^k determines the multiplication completely. To see this, let $x = \sum_{i=1}^n \alpha_i x_i$ and $y = \sum_{j=1}^n \beta_j x_j$ be two arbitrary elements of A. Then

$$xy = \left(\sum_{i=1}^n \alpha_i x_i\right)\left(\sum_{j=1}^n \beta_j x_j\right) = \sum_{i,j=1}^n \alpha_i \beta_j x_i x_j = \sum_{i,j,k=1}^n \alpha_i \beta_j c_{ij}^k x_k.$$

So the multiplication can be completely specified by giving a set of n^3 constants c_{ij}^k. These constants are called *structure constants* (because they determine the algebra structure of A). The next results provide a useful criterion for deciding whether an algebra is a Lie algebra by looking at its structure constants.

Lemma 1.3.1 *Let A be an algebra with basis $\{x_1, \ldots, x_n\}$. Then the multiplication in A satisfies (L_1) and (L_2) if and only if*

$$x_i x_i = 0 \quad and \quad x_i x_j + x_j x_i = 0,$$

and

$$x_i(x_j x_k) + x_j(x_k x_i) + x_k(x_i x_j) = 0$$

for $1 \le i, j, k \le n$.

Proof. Since the Jacobi identity is trilinear it is straightforward to see that it holds for all elements of A if and only if it holds for all basis elements of A. We prove that the first two equations are equivalent to $xx = 0$ for all $x \in A$. First, let $x = \sum_{i=1}^n \alpha_i x_i$ then

$$xx = \sum_{i,j=1}^n \alpha_i \alpha_j x_i x_j = \sum_{i=1}^n \alpha_i^2 x_i x_i + \sum_{i<j} \alpha_i \alpha_j x_i x_j + \sum_{i<j} \alpha_i \alpha_j x_j x_i = 0.$$

Conversely, suppose that $xx = 0$ for all $x \in A$. Then certainly $x_i x_i = 0$. Furthermore, by (1.1), $x_i x_j + x_j x_i = 0$. $\square$

Proposition 1.3.2 *Let A be an algebra with basis $\{x_1, \ldots, x_n\}$. Let the structure constants relative to this basis be c_{ij}^k for $1 \le i, j, k \le n$. Then A is a Lie algebra if and only if the structure constants satisfy the following relations:*

$$c_{ii}^k = 0 \ and \ c_{ij}^k = -c_{ji}^k, \tag{1.4}$$

$$\sum_{l=1}^{n} c_{il}^{m} c_{jk}^{l} + c_{jl}^{m} c_{ki}^{l} + c_{kl}^{m} c_{ij}^{l} = 0 \tag{1.5}$$

for all $1 \leq i, j, k, m \leq n$.

Proof. We use Lemma 1.3.1. Since the x_k are linearly independent $x_i x_i = \sum_k c_{ii}^{k} x_k = 0$ is equivalent to all c_{ii}^{k} being zero. Also $x_i x_j + x_j x_i = 0$ is equivalent to $\sum_k c_{ij}^{k} x_k + \sum_k c_{ji}^{k} x_k = 0$, which is equivalent to $c_{ij}^{k} + c_{ji}^{k} = 0$ for $1 \leq k \leq n$. Finally, it is straightforward to check that (1.5) is equivalent to $x_i(x_j x_k) + x_j(x_k x_i) + x_k(x_i x_j) = 0$. $\qquad\square$

We can describe a Lie algebra by giving a basis and the set of structure constants relative to this basis. This is called a *multiplication table* of the Lie algebra. However, because of (1.4), we only have to list the c_{ij}^{k} with $i < j$. Also the structure constants that are zero are usually not listed.

Example 1.3.3 Let L be a 2-dimensional Lie algebra with basis $\{x_1, x_2\}$. Then

$$[x_1, x_2] = c_{12}^{1} x_1 + c_{12}^{2} x_2. \tag{1.6}$$

So (1.6) is a multiplication table for L. Now suppose that c_{12}^{1} and c_{12}^{2} are not both zero. If $c_{12}^{1} \neq 0$ set $y_1 = c_{12}^{1} x_1 + c_{12}^{2} x_2$ and $y_2 = (1/c_{12}^{1}) x_2$. Otherwise set $y_1 = -x_2$ and $y_2 = -(1/c_{12}^{2}) x_1$. In both cases $\{y_1, y_2\}$ is a basis of L and

$$[y_1, y_2] = y_1. \tag{1.7}$$

With respect to the basis $\{y_1, y_2\}$, (1.7) is a multiplication table for L.

Example 1.3.4 We calculate the structure constants of the Lie algebra $\mathfrak{gl}_n(F)$ of Example 1.1.4. A basis of this Lie algebra is given by the set of all matrices E_{ij} for $1 \leq i, j \leq n$. We note that $E_{ij} E_{kl} = \delta_{jk} E_{il}$ where $\delta_{jk} = 1$ if $j = k$ and it is 0 otherwise. Now

$$[E_{ij}, E_{kl}] = E_{ij} E_{kl} - E_{kl} E_{ij} = \delta_{jk} E_{il} - \delta_{li} E_{kj}.$$

Example 1.3.5 Here we compute the structure constants of the Lie algebra $\mathfrak{sl}_n(F)$ of Example 1.2.1. As pointed out in Example 1.2.1, this Lie algebra has a basis consisting of all matrices $x_{ij} = E_{ij}$ for $1 \leq i \neq j \leq n$, together with $h_i = E_{ii} - E_{i+1,i+1}$ for $1 \leq i \leq n-1$. First of all, since the h_i are all diagonal matrices, we have that $[h_i, h_j] = 0$. Secondly,

$$[h_i, x_{kl}] = [E_{ii} - E_{i+1,i+1}, E_{kl}] = \delta_{ik} E_{il} - \delta_{i+1,k} E_{i+1,l} - \delta_{l,i} E_{ki} + \delta_{l,i+1} E_{k,i+1}$$
$$= c_{ikl} x_{kl}$$

where $c_{ikl} = 0, 1, 2, -1, -2$ depending on i, j and k. For example $c_{ikl} = 2$ if $i = k$ and $i + 1 = l$. Finally,

$$[x_{ij}, x_{kl}] = [E_{ij}, E_{kl}] = \delta_{jk} E_{il} - \delta_{li} E_{kj},$$

which is x_{il} if $j = k$ and $i \neq l$ and $-x_{kj}$ if $i = l$ and $j \neq k$. Furthermore, if $i = l$ and $j = k$ then

$$[x_{ij}, x_{kl}] = E_{ii} - E_{jj} = \begin{cases} h_i + h_{i+1} + \cdots + h_{j-1} & \text{if } i < j, \\ -h_j - h_{j+1} - \cdots - h_{i-1} & \text{if } j < i. \end{cases}$$

Example 1.3.6 As a special case of the previous example we consider the Lie algebra $\mathfrak{sl}_2(F)$. We set $h = E_{11} - E_{22}$, $x = E_{12}$ and $y = E_{21}$. Then $\{x, y, h\}$ is a basis of $\mathfrak{sl}_2(F)$. A multiplication table of this Lie algebra is given by

$$[h, x] = 2x, \; [h, y] = -2y, \; [x, y] = h.$$

The Lie algebra $\mathfrak{sl}_2(F)$ will play a major role in the structure theory of so-called semisimple Lie algebras (see Chapter 5).

1.4   Lie algebras from p-groups

One of the main motivations for studying Lie algebras is the connection with groups. Starting off with a group a Lie algebra is constructed that reflects (parts of) the structure of the group. In many cases it can be shown that questions about the group carry over to questions about the Lie algebra, which are (usually) easier to solve. Various constructions of this kind have been used, each for a specific type of group. For instance, if G is a Lie group, then it can be shown that the tangent space at the identity is a Lie algebra. This connection is for example used to study symmetry groups of differential equations (see [10], [67]). Also if G is an algebraic group, then the tangent space at the identity has the structure of a Lie algebra (see, e.g., [12]). Here we sketch how a Lie algebra can be attached to a p-group. In this case the Lie algebra is not a tangent space. But it is quite clear how the Lie algebra structure relates to the group structure.

Let G be a group. Then for $g, h \in G$ we define their *commutator* to be the element

$$(g, h) = g^{-1} h^{-1} g h.$$

Furthermore, if H is a subgroup of G then (G, H) is the subgroup generated by all elements (g, h) for $g \in G$ and $h \in H$.

We set $\gamma_1(G) = G$, and for $k \geq 1$, $\gamma_{k+1}(G) = (G, \gamma_k(G))$. Then $\gamma_1(G) \geq \gamma_2(G) \geq \cdots$ is called the *lower central series* of G. If $\gamma_n(G) = 1$ for some $n > 0$, then G is said to be *nilpotent*.

A group G is said to be a finite p-group if G has p^k elements, where $p > 0$ is a prime. For the basic facts on p-groups we refer to [40], [43]. We recall that a p-group is necessarily nilpotent.

For a group H we denote by H^p the subgroup generated by all h^p for $h \in H$. Let G be a p-group. We define a series $G = \kappa_1(G) \geq \kappa_2(G) \geq \cdots$ by

$$\kappa_n(G) = (\kappa_{n-1}(G), G)\kappa_m(G)^p,$$

where m is the smallest integer such that $pm \geq n$. This series is called the *Jennings series* of G. The next theorem is [44], Chapter VIII, Theorem 1.13.

Theorem 1.4.1 *We have*

1. $(\kappa_m(G), \kappa_n(G)) \leq \kappa_{n+m}(G)$, *and*

2. $\kappa_n(G)^p \leq \kappa_{pn}(G)$.

Set $G_m = \kappa_m(G)/\kappa_{m+1}(G)$. Then by Theorem 1.4.1, all elements of G_m have order p, and G_m is Abelian. Now since any Abelian group is the direct product of cyclic groups we see that G_m is the direct product of cyclic groups of order p, i.e., $G_m = H_1 \times \cdots \times H_k$ where $H_k \cong \mathbb{Z}/p\mathbb{Z}$. Now fix a generator h_i of H_i. Then any element in G_m can be written as $h_1^{n_1} \cdots h_k^{n_k}$ where $0 \leq n_i \leq p - 1$. Set $V_m = \mathbb{F}_p^k$. Then we have an isomorphism $\sigma_m : G_m \longrightarrow V_m$ of G_m onto the additive group of V_m. It is given by $\sigma_m(h_1^{n_1} \cdots h_k^{n_k}) = (n_1, \dots, n_k)$. For $j \geq 1$ we let τ_j be the composition

$$\tau_j : \kappa_j(G) \xrightarrow{\pi_j} G_j \xrightarrow{\sigma_j} V_j$$

where π_j is the projection map.

Set $L = V_1 \oplus V_2 \oplus \cdots$. Then L is a vector space over $\mathbb{F}_p$. We fix a basis B of L that is the union of bases of the components V_m. Let $x, y \in B$ be such that $x \in V_i$ and $y \in V_j$. Furthermore, let $g \in \kappa_i(G)$ and $h \in \kappa_j(G)$ be such that $\tau_i(g) = x$ and $\tau_j(h) = y$ respectively. Then we define

$$[x, y] = \tau_{i+j}((g, h)).$$

This product is well-defined, i.e., $[x, y]$ does not depend on the choice of g, h. This follows from the identity $(g, fh) = (g, h)(g, f)((g, f), h)$, which holds for all elements f, g, h in any group.

Lemma 1.4.2 *For $x, y \in B$ we have that $[x, x] = 0$ and $[x, y] + [y, x] = 0$.*

Proof. Let $x \in V_i$ and $y \in V_j$. Let $g \in \kappa_i(G)$ and $h \in \kappa_j(G)$ be pre-images of respectively x under τ_i and y under τ_j. Then

$$[x, x] = \tau_{2i}((g, g)) = \tau_{2i}(1) = 0.$$

And

$$\begin{aligned}
[x, y] + [y, x] &= \tau_{i+j}((g, h)) + \tau_{i+j}((h, g)) \\
&= \tau_{i+j}((g, h)(h, g)) \\
&= \tau_{i+j}(1) = 0.
\end{aligned}$$

$\square$

Lemma 1.4.3 *For $x_1, x_2, x_3 \in B$ we have $[x_1, [x_2, x_3]] + [x_2, [x_3, x_1]] + [x_3, [x_1, x_2]] = 0$.*

Proof. Suppose that $x_i \in V_{m_i}$ for $i = 1, 2, 3$. Let $g_i \in \kappa_{m_i}(G)$ be a pre-image of x_i under τ_i for $i = 1, 2, 3$. Then

$$[x_1, [x_2, x_3]] + [x_2, [x_3, x_1]] + [x_3, [x_1, x_2]] =$$
$$\tau_{m_1+m_2+m_3}((g_1, (g_2, g_3))(g_2, (g_3, g_1))(g_3, (g_1, g_2))).$$

Now the result follows from the fact that

$$(g_1, (g_2, g_3))(g_2, (g_3, g_1))(g_3, (g_1, g_2)) \in \kappa_{m_1+m_2+m_3+1}(G)$$

(see [44], Chapter VIII, Lemma 9.2). $\square$

Now we extend the product $[\,,\,]$ to $L \times L$ by bilinearity.

Corollary 1.4.4 *With the product $[\,,\,] : L \times L \to L$ the vector space L becomes a Lie algebra.*

Proof. This follows immediately from the above lemmas (cf. Lemma 1.3.1).
$\square$

Example 1.4.5 Let G be the group generated by three elements g_1, g_2, g_3 subject to the relations $(g_2, g_1) = g_3$, $(g_3, g_1) = (g_3, g_2) = 1$ and $g_1^2 = g_2^2 = g_3$ and $g_3^2 = 1$. The first relation is the same as $g_2 g_1 = g_1 g_2 g_3$, whereas the second and third relations can be written as $g_3 g_1 = g_1 g_3$ and $g_3 g_2 = g_2 g_3$. These relations allow us to rewrite any word in the generators to an expression of the form

$$g_1^{i_1} g_2^{i_2} g_3^{i_3}. \tag{1.8}$$

Using the remaining relations we can rewrite this to a word of the form (1.8) where $0 \leq i_k \leq 1$. This rewriting process is called *collection*. We do not discuss this process here as it is clear how it works in the example. For a more elaborate treatment of the collection process we refer to, e.g., [40], [80]. Every element of G has a unique representation as a word (1.8) such that $i_k = 0, 1$. Hence G contains $2^3 = 8$ elements. We have $\gamma_1(G) = G$, $\gamma_2(G) = \langle g_3 \rangle$ (where $\langle g_3 \rangle$ denotes the subgroup generated by g_3) and $\gamma_3(G) = 1$. The Jennings series of G is $\kappa_1(G) = G$, $\kappa_2(G) = \langle g_3 \rangle$ and $\kappa_3(G) = 1$. So $G_1 = G/\langle g_3 \rangle = \langle \bar{g}_1, \bar{g}_2 \rangle$, where $\bar{g}_2 \bar{g}_1 = \bar{g}_1 \bar{g}_2$. Therefore $G_1 = \{1, \bar{g}_1, \bar{g}_2, \bar{g}_1 \bar{g}_2\}$. Let V_1 be a 2-dimensional vector space over $\mathbb{F}_2$ spanned by $\{e_1, e_2\}$. Let $\sigma_1 : G_1 \to V_1$ be the morphism given by $\sigma_1(\bar{g}_i) = e_i$ for $i = 1, 2$ (so $\sigma_1(\bar{g}_1 \bar{g}_2) = e_1 + e_2$). Also we have that $G_2 = \langle g_3 \rangle / 1 = \{1, g_3\}$. Let V_2 be a 1-dimensional vector space over $\mathbb{F}_2$ spanned by e_3. Then $\sigma_2 : G_2 \to V_2$ is given by $\sigma_2(g_3) = e_3$. Now set $L = V_1 \oplus V_2$. We calculate the Lie product of e_1 and e_2:

$$[e_1, e_2] = \tau_2((g_1, g_2)) = \tau_2(g_3) = e_3.$$

Similarly it can be seen that $[e_1, e_3] = [e_2, e_3] = 0$.

1.5   On algorithms

Here we will not give a precise definition of the notion of "algorithm" (for this the reader is referred to the literature on this subject, e.g., [52]). For us "algorithm" roughly means "a list of consecutive steps, each of which can be performed effectively, that, given the input, produces the output in finite time". So there is an algorithm known for adding natural numbers, but not for proving mathematical theorems.

We are concerned with algorithms that calculate with Lie algebras. So we need to represent Lie algebras, and subalgebras, ideals, and elements thereof in such a way that they can be dealt with by a computer, e.g., as lists of numbers. For Lie algebras there are two solutions to this problem that immediately come to mind: we can represent a Lie algebra as a linear Lie algebra, or by a table of structure constants. We briefly describe both approaches.

Linear Lie algebras: If L happens to be a linear Lie algebra, then we can present L by a set of matrices that form a basis of it. If a and b are two elements of this space, then their Lie product is formed by the commutator, $[a, b] = a \cdot b - b \cdot a$.

Structure constants: As seen in Section 1.3, an n-dimensional Lie algebra can be presented by giving a list of n^3 structure constants c_{ij}^k for $1 \leq i, j, k \leq n$, that satisfy the relations of Proposition 1.3.2. Then L is viewed as a (abstract) vector space with basis $\{x_1, \ldots, x_n\}$. The Lie product of two elements of this space is completely determined by the structure constants.

Let L be given as a linear Lie algebra, i.e., by a basis $\{a_1, \ldots, a_n\}$, where $a_i \in M_r(F)$ for $1 \leq i \leq n$. Then by expressing the products $[a_i, a_j]$ as linear combinations of the basis elements we can compute a multiplication table for L. The other transition, from a Lie algebra given by a table of structure constants to a linear Lie algebra is much harder (and will be treated in Chapter 6). Also for many algorithms that we will encounter, it is essential to know a table of structure constants for the Lie algebra. For these reasons we will always assume that a Lie algebra over the field F is presented by a table of structure constants $c_{ij}^k \in F$ relative to a basis $\{x_1, \ldots, x_n\}$. An element x of a Lie algebra will be represented by a coefficient vector, i.e., a list of n elements $\alpha_i \in F$ such that $x = \alpha_1 x_1 + \cdots + \alpha_n x_n$. Finally, subspaces, subalgebras and ideals will be presented by a basis (i.e., a list of coefficient vectors).

Remark. There is a third way of presenting a Lie algebra, namely by *generators and relations*. A Lie algebra presented in this way is given by a set of (abstract) generators X that are subject to a set of relations R. The Lie algebra specified by this data is the most "general" Lie algebra generated by X subject to the relations in R. The theoretical notions needed for this are treated in Chapter 7. There algorithms will be given for calculating the structure constants of a Lie algebra given by generators and relations. So we do not lose any generality by supposing that the Lie algebras we deal with are given by a multiplication table.

In our algorithms we often need auxiliary algorithms for performing tasks that are not of a Lie algebraic nature, but without which our algorithms would not work. For example, we assume that there are algorithms available for performing the elementary arithmetical operations (addition and multiplication) for the fields that are input to our algorithms. Also we assume that elements of these fields can be represented on a computer. This does not pose any problems for the field of rational numbers $\mathbb{Q}$, nor for number fields, nor for finite fields.

On some occasions we will use a routine that selects a random element from a finite set (for such routines we refer to [53]). The selection procedure is such that every element of the set has the same probability to be chosen. This allows *randomization* in our algorithms: at certain points the outcome of a random choice will determine the path taken in the rest of the algorithm. In general also the output of a randomized algorithm depends on the random choices made; the output may even be wrong. This is clearly an undesirable situation. Therefore we restrict our attention to a class of randomized algorithms called *Las Vegas* algorithms. An algorithm for computing a function $f(x)$ is called Las Vegas if on input a it either computes $f(a)$ correctly with probability $p > 0$, or stops without producing output. It is also required that calls to a Las Vegas algorithm produce independent results. Hence if a Las Vegas algorithm is repeated then it always produces a correct answer. The expected number of repetitions is $1/p$.

We will frequently obtain solutions to the problems we are dealing with as solutions of sets of linear equations in several variables. So we need a routine for solving these. A straightforward method for doing that is known as Gaussian elimination, which works over all fields for which we can perform the elementary arithmetical operations. The basic procedure is described in many monographs on linear algebra, see, e.g., [60].

The Gaussian elimination algorithm allows us to perform the basic operations of linear algebra: constructing a basis of a subspace, testing whether an element lies in a subspace, constructing a basis of the intersection of subspaces, and so forth.

In some cases we need a routine for factorizing polynomials over the input field. If this field is $\mathbb{Q}$, a number field or a finite field, then there are algorithms for doing that. For an overview of algorithms for factorizing polynomials we refer to [53] and [58].

Sometimes we use the connections between associative algebras and Lie algebras to study the structure of a Lie algebra. Doing so we also need algorithms for calculating various objects related to an associative algebra. We have described some of them in Appendix A.

Finally we describe an efficient procedure due to J. T. Schwartz for finding a non-zero of a multivariate polynomial. It is based on the following lemma; for the proof we refer to [74]. The corollary is an immediate consequence.

Lemma 1.5.1 (Schwartz) *Let F be a field. Let $f \in F[X_1, \ldots, X_n]$ be a polynomial of degree d. Let Ω be a subset of F of size N. Then the number of elements $v = (v_1, \ldots, v_n)$ of Ω^n such that $f(v) = 0$ is at most dN^{n-1}.*

Corollary 1.5.2 *Let f and Ω be the same as in the previous lemma. Let v be an element from Ω^n chosen randomly and uniformly. Then the probability that $f(v) = 0$ is at most d/N.*

Let f be as in Lemma 1.5.1. We choose a subset $\Omega \subset F$ of size $2d$. Then by selecting a random element from Ω^n we find a vector v such that $f(v) \neq 0$ with probability $\frac{1}{2}$. By repeating this we find a non-zero of f, and the expected number of steps is 2. So if the ground field is large enough, then we have a Las Vegas algorithm for finding a non-zero of a polynomial.

We end this section by giving two algorithms: one for constructing quotient algebras and one for constructing the adjoint map. Both are fairly straightforward. They also illustrate the formats that we use for describing algorithms in this book. We either explicitly give a list of steps that are executed by the algorithm, or, in the cases where the algorithm is rather straightforward, we simply discuss the main ingredients, and leave the task of formulating a series of steps to the reader.

The following is an algorithm that calculates an array of structure constants of the quotient algebra L/I, where I is an ideal of L.

Algorithm QuotientAlgebra
Input: a finite-dimensional Lie algebra L together with an ideal $I \subset L$.
Output: an array of structure constants for L/I.

Step 1 Let $\{y_1, \dots, y_s\}$ be a basis of I. Select elements $z_1, \dots, z_t$ from the basis of L such that $B = \{z_1, \dots, z_t, y_1, \dots, y_s\}$ is a basis of L.

Step 2 for $1 \leq i, j \leq t$ express $[z_i, z_j]$ as a linear combination of elements from B. Let d_{ij} be the resulting coefficient vectors.

Step 3 For $1 \leq i, j \leq t$ let e_{ij} be the vector containing the first t entries from d_{ij}. Return the set of all e_{ij}.

Comments: We recall that the ideal I is presented by a basis. So the basis $\{y_1, \dots, y_s\}$ is part of the input. Furthermore, all vectors are represented as coefficient vectors. So the Gaussian elimination procedure allows us to perform Step 1 and Step 2.

Now we consider the problem of constructing the matrix of the linear map $\mathrm{ad}x : L \to L$, relative to the basis $\{x_1, \dots, x_n\}$ of L. Let x be given as $x = \sum_{i=1}^{n} \alpha_i x_i$, then

$$\mathrm{ad}x(x_j) = \sum_{i=1}^{n} \alpha_i [x_i, x_j] = \sum_{k=1}^{n} \left(\sum_{i=1}^{n} \alpha_i c_{ij}^k \right) x_k.$$

It follows that the coefficients of the matrix of $\mathrm{ad}x$ are $\sum_{i=1}^{n} \alpha_i c_{ij}^k$. So we have an algorithm AdjointMatrix that for a Lie algebra L and an element $x \in L$ constructs the matrix of $\mathrm{ad}x$ relative to the input basis of L. We note that the coefficients α_i of x are immediately available since x is input as a coefficient vector.

1.6 Centralizers and normalizers

In this section we construct centralizers and normalizers in a Lie algebra L. We also give algorithms for calculating bases of these spaces. For that we assume throughout that L has basis $\{x_1, \ldots, x_n\}$ and structure constants c_{ij}^k relative to this basis (see Section 1.5).

Let S be a subset of L. Then the set

$$C_L(S) = \{x \in L \mid [x, s] = 0 \text{ for all } s \in S\}$$

is called the *centralizer* of S in L. We prove that $C_L(S)$ is a subalgebra of L. Let $x, y \in C_L(S)$ and $s \in S$. Then by the Jacobi identity

$$[[x, y], s] = -[[y, s], x] - [[s, x], y] = 0, \tag{1.9}$$

so that $[x, y] \in C_L(S)$ and $C_L(S)$ is a subalgebra of L.

It is straightforward to see that $C_L(S)$ is equal to the centralizer of K in L, where K is the subspace spanned by S. Therefore, in the algorithm for constructing the centralizer we assume that the input is a basis $\{y_1, \ldots, y_t\}$ of a subspace K of L, where

$$y_l = \sum_{j=1}^{n} \lambda_{lj} x_j. \tag{1.10}$$

Then $x = \sum_i \alpha_i x_i$ lies in $C_L(K)$ if and only if $[x, y_l] = 0$ for $1 \le l \le t$. This is equivalent to

$$\sum_{i=1}^{n} \left(\sum_{j=1}^{n} \lambda_{lj} c_{ij}^k \right) \alpha_i = 0 \text{ for } 1 \le k \le n \text{ and } 1 \le l \le t.$$

It follows that we have nt equations for the n unknown $\alpha_1, \ldots, \alpha_n$. By a Gaussian elimination we can solve these; and therefore we find an algorithm Centralizer for calculating the centralizer of a subspace K of L.

The subset

$$C(L) = \{x \in L \mid [x, y] = 0 \text{ for all } y \in L\}$$

is called the *centre* of L. We have that the centre of L is the centralizer of L in itself, i.e., $C(L) = C_L(L)$. As $[x, y] = 0$ for all $x \in C(L)$ and $y \in L$, it is immediate that $C(L)$ is an ideal in L. The centre is the kernel of the map $\mathrm{ad} : L \to \mathrm{End}(L)$, i.e.,

$$C(L) = \{x \in L \mid \mathrm{ad} x = 0\}.$$

So if we study the structure of L via its adjoint map, then we lose "sight" of the centre.

If $C(L) = L$, then L is said to be *Abelian* or *commutative*.

The algorithm for calculating the centralizer also yields an algorithm for calculating the centre. In this case the requirement for an element $x = \sum_i \alpha_i x_i$ to belong to $C(L)$ is $[x, x_j] = 0$ for $1 \leq j \leq n$, which boils down to

$$\sum_{i=1}^{n} c_{ij}^k \alpha_i = 0 \quad \text{for } 1 \leq j, k \leq n.$$

So we have n^2 equations for the n unknowns $\alpha_1, \ldots, \alpha_n$, which can be solved by Gaussian elimination. This gives us an algorithm Centre.

Let V be a subspace of L. Then the set

$$N_L(V) = \{x \in L \mid [x, v] \in V \text{ for all } v \in V\}$$

is called the *normalizer* of V in L. In the same way as for the centralizer we can prove that the normalizer of V in L is a subalgebra of L. If V happens to be a subalgebra of L, then V is an ideal in the Lie algebra $N_L(V)$.

Now we describe an algorithm for calculating the normalizer. Let V be the subspace of L spanned $y_1, \ldots, y_t$, where the y_l are as in (1.10). Then $x = \sum_i \alpha_i x_i$ is an element of $N_L(V)$ if and only if there are β_{lm} for $1 \leq l, m \leq t$ such that

$$[x, y_l] = \beta_{l1} y_1 + \cdots + \beta_{lt} y_t \quad \text{for } l = 1, \ldots, t.$$

This amounts to the following linear equations in the variables α_i and β_{lm}:

$$\sum_{i=1}^{n} \left(\sum_{j=1}^{n} \lambda_{lj} c_{ij}^k \right) \alpha_i - \sum_{m=1}^{t} \lambda_{mk} \beta_{lm} = 0 \quad \text{for } 1 \leq k \leq n \text{ and } 1 \leq l \leq t.$$

Again by a Gaussian elimination we can solve these equations. However, we are not interested in the values of the β_{lm}, so we throw the part of the solution that corresponds to these variables away, and we find a basis of $N_L(V)$. As a consequence we have an algorithm Normalizer for calculating the normalizer of a subspace V of L.

Example 1.6.1 Let L be the Lie algebra with basis $\{x_1, \ldots, x_5\}$ and multiplication table

$$[x_1, x_4] = x_1, \; [x_1, x_5] = -x_2, \; [x_2, x_4] = x_2, \; [x_2, x_5] = x_1, \; [x_4, x_5] = x_3.$$

(As usual we only list products $[x_i, x_j]$ for $i < j$; and we omit those that are 0.) We calculate a basis of the centre of L. Let $x = \sum_{i=1}^{5} \alpha_i x_i$ be an arbitrary element of L. Then $x \in C(L)$ if and only if $[x_i, x] = 0$ for $1 \leq i \leq 5$. So $0 = [x_1, x] = \alpha_4 x_1 - \alpha_5 x_2$, from which it follows that $\alpha_4 = \alpha_5 = 0$. Then also $[x_2, x] = 0$. It is easily seen that $[x_3, x] = 0$. From $0 = [x_4, x] = -\alpha_1 x_1 - \alpha_2 x_2 + \alpha_5 x_3$ we infer that $\alpha_1 = \alpha_2 = \alpha_5 = 0$. Finally $[x_5, x] = \alpha_1 x_2 - \alpha_2 x_1 - \alpha_4 x_3$, from which $\alpha_1 = \alpha_2 = \alpha_4 = 0$. It is seen that only α_3 can be non-zero. It follows that $C(L)$ is spanned by x_3.

Let V be the subspace of L spanned by x_1, x_5. Let $x = \sum_{i=1}^{5} \alpha_i x_i$ be an element of L. Then $x \in N_L(V)$ if and only if

$$[x, x_1] = a x_1 + b x_5$$
$$[x, x_5] = c x_1 + d x_5.$$

The first of these requirements produces $\alpha_5 = 0$ and $a + \alpha_4 = 0$. The second boils down to $c - \alpha_2 = 0$ and $\alpha_1 = \alpha_4 = 0$. Hence $N_L(V)$ is spanned by x_2, x_3.

1.7 Chains of ideals

Here we construct chains of ideals of L. First we note that, in order to check that a subspace I of L is an ideal, it suffices to prove that $[x, y] \in I$ for all $x \in L$ and $y \in I$. This follows from (1.1).

Let L be a Lie algebra and let V, W be subspaces of L. Then the linear span of the elements $[v, w]$ for $v \in V$ and $w \in W$ is called the *product space* of V and W. It is denoted by $[V, W]$.

Lemma 1.7.1 *Let I and J be ideals of L. Then also $[I, J]$ is an ideal of L.*

Proof. Let $x \in I$, $y \in J$, and $z \in L$ then by the Jacobi identity

$$[z, [x, y]] = -[x, [y, z]] - [y, [z, x]],$$

which lies in $[I, J]$. Now if $u \in [I, J]$ is a linear combination of elements of the form $[x, y]$ for $x \in I$ and $y \in J$, then also, by linearity of the product, for all $z \in L$ we have that $[z, u]$ is a linear combination of elements of the form $[a, b]$ for $a \in I$ and $b \in J$. So $[I, J]$ is an ideal of L.   □

The following is an algorithm for calculating the product space $[K_1, K_2]$ of two subspaces K_1 and K_2 of L.

Algorithm ProductSpace
Input: a finite-dimensional Lie algebra L together with two subspaces $K_1, K_2 \subset L$, given by bases $\{y_1, \dots, y_s\}$ and $\{z_1, \dots, z_t\}$ respectively.
Output: a basis of the product space $[K_1, K_2]$.

Step 1 Compute the set A of elements $[y_i, z_j]$ for $1 \leq i \leq s$ and $1 \leq j \leq t$.

Step 2 Calculate a maximally linearly independent subset B of A. Return B.

Comment: The set A computed in Step 1 is just a list of coefficient vectors. So by a Gaussian elimination, we can compute a maximally linearly independent subset B of A. This gives us a basis of $[K_1, K_2]$.

Lemma 1.7.1 implies that the subspace $[L, L]$ is an ideal of L. It is called the *derived subalgebra* of L.

Set $L^1 = L$ and inductively $L^{i+1} = [L, L^i]$. Then by Lemma 1.7.1, L^i is an ideal of L for $i \geq 1$. Hence $L^{i+1} \subset L^i$ for $i \geq 1$, i.e., the series L^i is decreasing. The sequence

$$L = L^1 \supset L^2 \supset \cdots \supset L^i \supset \cdots$$

is called the *lower central series* of L. If $L^k = 0$ for some integer $k > 0$ then L is called *nilpotent*. Furthermore, if L is nilpotent then the smallest integer $c > 0$ such that $L^c \neq 0$ but $L^{c+1} = 0$ is called the *nilpotency class* of L. We note that repeated application of ProductSpace gives us an algorithm LowerCentralSeries for calculating the lower central series and hence for deciding whether a given Lie algebra is nilpotent.

Set $L^{(1)} = L$ and for $i \geq 1$, $L^{(i+1)} = [L^{(i)}, L^{(i)}]$. Then by Lemma 1.7.1 the $L^{(i)}$ are ideals of L for $i \geq 1$. Therefore, $L^{(i+1)}$ is contained in $L^{(i)}$. The sequence

$$L = L^{(1)} \supset L^{(2)} \supset \cdots \supset L^{(i)} \supset \cdots$$

is called the *derived series* of L. If there is an integer $d > 0$ such that $L^{(d)} = 0$, then L is said to be *solvable*. Just as for the lower central series we find an algorithm DerivedSeries for calculating the derived series; and hence an algorithm for deciding whether a Lie algebra is solvable or not.

Lemma 1.7.2 *Let L be a nilpotent Lie algebra. Then L is also solvable.*

Proof. By induction on i we have $L^{(i)} \subset L^i$, and the statement follows. $\square$

Now we define a third series of ideals. Unlike the lower central and derived series, this series is not descending but ascending. Put $C_1(L) = C(L)$ (the centre of L) and for $i \geq 1$ we define $C_{i+1}(L)$ by the relation $C_{i+1}(L)/C_i(L) = C(L/C_i(L))$. It follows that $C_{i+1}(L) \supset C_i(L)$, i.e., the series $C_i(L)$ is increasing. The sequence

$$C(L) = C_1(L) \subset C_2(L) \subset \cdots \subset C_i(L) \subset \cdots$$

is called the *upper central series* of L. Suppose that there is an integer $e > 0$ such that $C_e(L) = C_{e+1}(L)$, then the ideal $C_e(L)$ is called the *hypercentre* of L.

Repeated application of the algorithms Centre and QuotientAlgebra give us an algorithm UpperCentralSeries for calculating the upper central series. Here we remark that for finite-dimensional Lie algebras L the series defined above always stabilize at some point (i.e., there are integers $c, d, e > 0$ such that $L^c = L^{c+1}$, $L^{(d)} = L^{(d+1)}$ and $C_e(L) = C_{e+1}(L)$).

Example 1.7.3 Let L be the 5-dimensional Lie algebra of Example 1.6.1. As seen in that example, $C_1(L)$ is spanned by x_3. In the same way it can be shown that $C(L/C(L)) = 0$ so that $C_2(L) = C_1(L)$.

The lower central and derived series of L can be read off from the multiplication table. It is seen that L^2 is spanned by x_1, x_2, x_3 and L^3 by x_1, x_2, and $L^4 = L^3$. Also $L^{(2)} = L^2$ and $L^{(3)} = 0$. It follows that L is solvable, but not nilpotent.

1.8   Morphisms of Lie algebras

In this section we derive useful properties of morphisms of Lie algebras. Recall that a linear map $\theta : A_1 \to A_2$ from an algebra A_1 into an algebra A_2 is a morphism of algebras if $\theta(ab) = \theta(a)\theta(b)$ for all $a, b \in A_1$. If A_1 and A_2 are Lie algebras, this condition is written as $\theta([a, b]) = [\theta(a), \theta(b)]$. If a morphism of Lie algebras is bijective, then it is said to be an isomorphism and the two Lie algebras are called isomorphic. If this is the case, then we write $A_1 \cong A_2$. Furthermore, by $\ker(\theta) = \{x \in A_1 \mid \theta(x) = 0\}$ we denote the kernel of θ. Also $\operatorname{im}(\theta) = \{\theta(x) \mid x \in A_1\}$ is the image of θ.

Lemma 1.8.1 *Let $\theta : L_1 \to L_2$ be a morphism of Lie algebras. Then we have $L_1/\ker(\theta) \cong \operatorname{im}(\theta)$.*

Proof. First note that $\ker(\theta)$ is an ideal in L_1 and that $\mathrm{im}(\theta)$ is a subalgebra of L_2. Define $\bar{\theta} : L_1/\ker(\theta) \to \mathrm{im}(\theta)$ by $\bar{\theta}(x + \ker(\theta)) = \theta(x)$. This is a well defined morphism of Lie algebras and it clearly is bijective. So it is an isomorphism and the two Lie algebras are isomorphic. $\qquad\square$

Proposition 1.8.2 *For a Lie algebra L with ideals I and J, the following statements hold.*

1. *If $I \subset J$ then the quotient Lie algebra J/I is an ideal of the quotient Lie algebra L/I and we have $(L/I)/(J/I) \cong L/J$.*

2. *The quotient Lie algebra $(I + J)/J$ is isomorphic to $I/(I \cap J)$.*

Proof. 1. Let $\phi : L/I \to L/J$ be the morphism of Lie algebras mapping the coset $x + I \in L/I$ to the coset $x + J \in L/J$; this is well defined because $I \subset J$. Let $x + I \in \ker\phi$, then $x \in J$, and hence $x + I \in J/I$. Since it is clear that $J/I \subset \ker\phi$, we find $\ker\phi = J/I$. Now, as ϕ is clearly surjective, 1. follows from Lemma 1.8.1.

2. Define the map $\psi : I + J \to I/(I \cap J)$ by $\psi(x + y) = x + (I \cap J) \in I/(I \cap J)$ for $x \in I, y \in J$. This is well defined because if $x + y = x' + y'$ for $x' \in I, y' \in J$, then $x - x' = y' - y \in I \cap J$, so that $x + I \cap J = x' + I \cap J$, i.e., $\psi(x + y) = \psi(x' + y')$.

If $x \in I$ and $y \in J$ are such that $x + y \in \ker\psi$, then $x \in I \cap J$. So $x \in J$ and hence also $x + y \in J$. Therefore $\ker\psi \subseteq J$, and consequently $\ker\psi = J$. Thus, 2. follows from Lemma 1.8.1. $\qquad\square$

Example 1.8.3 Let L be a finite-dimensional Lie algebra. We consider the linear map $\mathrm{ad} : L \to \mathfrak{gl}(L)$. We show that ad is a morphism of Lie algebras. For $x, y, z \in L$ we have

$$\mathrm{ad}[x,y](z) = [[x,y],z] = -[[y,z],x] - [[z,x],y] \text{ (Jacobi identity)}$$
$$= [x,[y,z]] - [y,[x,z]] = (\mathrm{ad}x \cdot \mathrm{ad}y - \mathrm{ad}y \cdot \mathrm{ad}x)(z) = [\mathrm{ad}x, \mathrm{ad}y](z).$$

The kernel of ad is $C(L)$, the centre of L. So by Lemma 1.8.1, we see that the image of ad is isomorphic to $L/C(L)$.

1.9 Derivations

Let A be an algebra. A *derivation* d of A is a linear map $d : A \to A$ satisfying

$$d(xy) = d(x)y + xd(y) \text{ for all } x, y \in A.$$

Let $\mathrm{Der}(A)$ be the set of all derivations of A. It is straightforward to check that $\mathrm{Der}(A)$ is a vector space. Furthermore, the commutator defines a product on $\mathrm{Der}(A)$, since for $d_1, d_2 \in \mathrm{Der}(A)$,

$$
\begin{aligned}
[d_1, d_2](xy) &= (d_1 d_2 - d_2 d_1)(xy) \\
&= d_1(d_2(x)y + xd_2(y)) - d_2(d_1(x)y + xd_1(y)) \\
&= d_1(d_2(x))y + d_2(x)d_1(y) + d_1(x)d_2(y) + xd_1(d_2(y)) \\
&\quad -d_2(d_1(x))y - d_1(x)d_2(y) - d_2(x)d_1(y) - xd_2(d_1(y)) \\
&= (d_1(d_2(x)) - d_2(d_1(x)))y + x(d_1(d_2(y)) - d_2(d_1(y))) \\
&= [d_1, d_2](x)y + x[d_1, d_2](y).
\end{aligned}
$$

Consequently the space $\mathrm{Der}(A)$ has the structure of a Lie algebra.

For later use we record the so-called Leibniz formula. Let d be a derivation of the algebra A. Then

$$
d^n(xy) = \sum_{k=0}^{n} \binom{n}{k} d^k(x) d^{n-k}(y) \quad \text{(Leibniz formula)}, \tag{1.11}
$$

which is proved by induction on n.

Let A be a finite-dimensional algebra. We consider the problem of computing a basis of the linear Lie algebra $\mathrm{Der}(A)$. Let $\{y_1, \dots, y_m\}$ be a basis of A and let γ_{ij}^k (for $1 \le i, j, k \le m$) be the structure constants of A relative to this basis. Let $d \in \mathrm{End}(A)$ be a linear map given by $d(y_i) = \sum_{j=1}^{m} d_{ij} y_j$. Now $d \in \mathrm{Der}(A)$ if and only if $d(y_i y_j) = d(y_i) y_j + y_i d(y_j)$ for $1 \le i, j \le m$. This is equivalent to

$$
\sum_{l=1}^{m} \sum_{k=1}^{m} \gamma_{ij}^k d_{kl} y_l = \sum_{l=1}^{m} \sum_{k=1}^{m} \gamma_{kj}^l d_{ik} y_l + \sum_{l=1}^{m} \sum_{k=1}^{m} \gamma_{ik}^l d_{jk} y_l.
$$

And since the y_l are linearly independent, the coefficients of y_l on the left and right hand sides must be equal. Hence $d \in \mathrm{Der}(A)$ if and only if

$$
\sum_{k=1}^{m} \gamma_{ij}^k d_{kl} - \gamma_{kj}^l d_{ik} - \gamma_{ik}^l d_{jk} = 0, \quad \text{for } 1 \le i, j, l \le m.
$$

Which is a system of m^3 linear equations for the m^2 variables d_{ij}. This system can be solved by a Gaussian elimination; as a consequence we find an algorithm Derivations which computes a basis of $\mathrm{Der}(A)$ for any algebra A.

Remark. If A happens to be a Lie algebra, then by (1.1), $y_i y_j = -y_j y_i$. Therefore, $d \in \mathrm{End}(A)$ if and only if $d(y_i y_j) = d(y_i) y_j + y_i d(y_j)$ for $1 \le i \le$

$j \leq m$. So in this case we find $m^2(m+1)/2$ equations (instead of m^3).

Example 1.9.1 Let L be a Lie algebra. Then we claim that for $x \in L$, the map $\mathrm{ad}x : L \to L$ is a derivation of L. Indeed, for $y, z \in L$ we have

$$
\begin{aligned}
\mathrm{ad}x([y,z]) &= [x,[y,z]] \\
&= -[z,[x,y]] - [y,[z,x]] \text{ (Jacobi identity)} \\
&= [[x,y],z] + [y,[x,z]] \\
&= [\mathrm{ad}x(y),z] + [y,\mathrm{ad}x(z)],
\end{aligned}
$$

proving our claim. Derivations of the form $\mathrm{ad}x$ are called *inner*. On the other hand, if a derivation d of L is not of this form, then d is said to be an *outer* derivation.

1.10 (Semi)direct sums

Let L_1 and L_2 be Lie algebras over the field F. In this section we describe how we can define a multiplication on the direct sum (of vector spaces) $L_1 \oplus L_2$ extending the multiplications on L_1, L_2 and making $L_1 \oplus L_2$ into a Lie algebra. Let the product on L_1 be denoted by $[\ ,\ ]_1$ and on L_2 by $[\ ,\ ]_2$. Suppose we have a morphism of Lie algebras $\theta : L_1 \to \mathrm{Der}(L_2)$. Then this map allows us to define an algebra structure on the vector space $L_1 \oplus L_2$ by setting

$$[x_1 + x_2, y_1 + y_2] = [x_1, y_1]_1 + \theta(x_1)(y_2) - \theta(y_1)(x_2) + [x_2, y_2]_2 \qquad (1.12)$$

for $x_1, y_1 \in L_1$ and $x_2, y_2 \in L_2$. (So the product $[x, y]$ for $x \in L_1$ and $y \in L_2$ is formed by applying the derivation $\theta(x)$ to y.)

Lemma 1.10.1 *The multiplication defined by (1.12) makes $L_1 \oplus L_2$ into a Lie algebra.*

Proof. Let $x_1 \in L_1$ and $x_2 \in L_2$. Then

$$[x_1 + x_2, x_1 + x_2] = [x_1, x_1]_1 + \theta(x_1)(x_2) - \theta(x_1)(x_2) + [x_2, x_2]_2 = 0.$$

It suffices to check the Jacobi identity for the elements of a basis of the Lie algebra (cf. Lemma 1.3.1) . Construct a basis of $L_1 \oplus L_2$ by first taking a basis of L_1 and adding a basis of L_2. Let x, y, z be three elements from this basis. We have to consider a few cases: all three elements are from

L_1, two elements are from L_1 (the other from L_2), one element is from L_1, no elements are from L_1. The first and the last case are implied by the Jacobi identity on L_1 and L_2 respectively. The second case follows from the fact that θ is a morphism of Lie algebras. And the third case follows from the fact that $\theta(x)$ is a derivation of L_2 for $x \in L_1$. (We leave the precise verifications to the reader).   □

The Lie algebra $L_1 \oplus L_2$, together with the multiplication defined by (1.12), is called the *semidirect sum* of L_1 and L_2 (with respect to θ). It is straightforward to see that it contains L_2 as an ideal and L_1 as a subalgebra.

On the other hand, suppose that a Lie algebra L contains a subalgebra L_1 and an ideal L_2 such that as a vector space $L = L_1 \oplus L_2$. Then for $x \in L_1$ and $y \in L_2$ we have $\mathrm{ad}x(y) = [x, y] \in L_2$ because L_2 is an ideal. As a consequence $\mathrm{ad}x$ maps L_2 into itself. Furthermore, as seen in Example 1.9.1, the map $\mathrm{ad}x$ is a derivation of L_2. Hence we have a map $\mathrm{ad}_{L_2} : L_1 \to \mathrm{Der}(L_2)$. By Example 1.8.3, ad_{L_2} is a morphism of Lie algebras. It is immediate that the product on L satisfies (1.12) with $\theta = \mathrm{ad}_{L_2}$. So L is the semidirect sum of L_1 and L_2.

If there can be no confusion about the map θ, then the semidirect sum of L_1 and L_2 is denoted by $L_1 \ltimes L_2$.

A special case of the construction of the semidirect sum occurs when we take the map θ to be identically 0. Then the multiplication on $L_1 \oplus L_2$ is given by

$$[x_1 + x_2, y_1 + y_2] = [x_1, y_1]_1 + [x_2, y_2]_2 \text{ for all } x_1, y_1 \in L_1 \text{ and } x_2, y_2 \in L_2.$$

The Lie algebra $L_1 \oplus L_2$ together with this product is called the *direct sum* (of Lie algebras) of L_1 and L_2. It is simply denoted by $L_1 \oplus L_2$. From the way in which the product on $L_1 \oplus L_2$ is defined, it follows that L_1 and L_2 are ideals in $L_1 \oplus L_2$. Conversely suppose that L is a Lie algebra that is the direct sum of two subspaces L_1 and L_2. Furthermore, suppose that L_1 and L_2 happen to be ideals of L. Then $[L_1, L_2]$ is contained in both L_1 and L_2. Therefore $[L_1, L_2] = 0$ and it follows that L is the direct sum (of Lie algebras) of L_1 and L_2.

Example 1.10.2 In general a Lie algebra can be a semidirect sum in more than one way. Let L_1 and L_2 be copies of the same Lie algebra L. Let $\phi : L_1 \to L_2$ be the algebra morphism induced by the identity on L. Define $\theta : L_1 \to \mathrm{Der}(L_2)$ by $\theta(x) = \mathrm{ad}_{L_2}\phi(x)$ for $x \in L_1$. Then θ is the composition of two morphisms of Lie algebras. Consequently, θ is a morphism of Lie algebras and hence we can form the semidirect sum of L_1 and L_2 with respect to θ. Denote the resulting Lie algebra by K. Now let K_1 be the subspace of K spanned by all elements of the form $x - \phi(x)$ for $x \in L_1$. And let K_2 be the subspace spanned by all $\phi(x)$ for $x \in L_1$ (i.e., K_2 is equal to L_2). Then for $x, y \in L_1$,

$$[x - \phi(x), y - \phi(y)] = [x, y] - \theta(x)(\phi(y)) + \theta(y)(\phi(x)) + [\phi(x), \phi(y)]$$
$$= [x, y] - [\phi(x), \phi(y)] + [\phi(y), \phi(x)] + [\phi(x), \phi(y)]$$
$$= [x, y] - [\phi(x), \phi(y)] = [x, y] - \phi([x, y]).$$

Hence K_1 is a subalgebra of K. By a similar calculation it can be shown that $[K_1, K_2] = 0$ so that K_1 and K_2 are ideals of K. So K is the direct sum (of Lie algebras) of K_1 and K_2.

In Section 1.15 we will give an algorithm for finding a decomposition of L as a direct sum of ideals.

1.11   Automorphisms of Lie algebras

Let L be a Lie algebra over the field F. An *automorphism* of L is an isomorphism of L onto itself. Since products of automorphisms are automorphisms and inverses of automorphisms are automorphisms, the automorphisms of L form a group. This group is called the *automorphism group* of L; it is denoted by $\mathrm{Aut}(L)$.

Example 1.11.1 Let V be a finite-dimensional vector space. Let L be a subalgebra of $\mathfrak{gl}(V)$ and $g \in \mathrm{End}(V)$ an invertible endomorphism of V. If $gLg^{-1} = L$ then the map $x \mapsto gxg^{-1}$ is an automorphism of L because $[gxg^{-1}, gyg^{-1}] = gxyg^{-1} - gyxg^{-1} = g[x, y]g^{-1}$.

Now let the ground field F be of characteristic 0. If d is a nilpotent derivation of L, i.e., $d^n = 0$ for some integer $n \geq 0$, then we can define its exponential:

$$\exp d = 1 + d + \frac{1}{2!}d^2 + \cdots + \frac{1}{(n-1)!}d^{n-1}.$$

Lemma 1.11.2 *Let d be a nilpotent derivation of L, then $\exp d$ is an automorphism of L.*

Proof. Because d is nilpotent, there is some integer n with $d^n = 0$. Now we calculate

$$
\begin{aligned}
[(\exp d)x, (\exp d)y] &= \sum_{i=0}^{n-1}\sum_{j=0}^{n-1} \frac{1}{i!j!}[d^i x, d^j y] \\
&= \sum_{m=0}^{2n-2}\left(\sum_{i=0}^{m} \frac{1}{i!(m-i)!}[d^i x, d^{m-i}y]\right) \\
&= \sum_{m=0}^{2n-2} \frac{d^m[x,y]}{m!} \ \text{(by (1.11))} \\
&= \sum_{m=0}^{n-1} \frac{d^m[x,y]}{m!} = (\exp d)[x,y].
\end{aligned}
$$

So $\exp d$ is a morphism of Lie algebras.

The inverse of $\exp d$ is the map

$$
\exp(-d) = \sum_{j=0}^{n-1}(-1)^j \frac{1}{j!}d^j.
$$

It follows that $\exp d$ is a bijective morphism of Lie algebras; i.e., it is an automorphism.   □

In particular, if $\mathrm{ad}x$ is nilpotent then $\exp(\mathrm{ad}x)$ is an automorphism of L. An automorphism of this form is called *inner*. The subgroup of $\mathrm{Aut}(L)$ generated by all inner automorphisms is called the *inner automorphism group*. It is denoted by $\mathrm{Int}(L)$.

1.12 Representations of Lie algebras

Definition 1.12.1 *Let L be a Lie algebra over the field F. A representation of L on a vector space V is a morphism of Lie algebras $\rho : L \to \mathfrak{gl}(V)$.*

A representation $\rho : L \to \mathfrak{gl}(V)$ is said to be *faithful* if $\ker \rho = 0$.

If $\rho : L \to \mathfrak{gl}(V)$ is a representation of the Lie algebra L on the vector space V, then V is said to be an L-module. The representation ρ is said to define an *action* of L on V. If from the context it is clear which representation ρ we mean, then we also write $x \cdot v$ instead of $\rho(x)(v)$. Here the condition that ρ be a morphism of Lie algebras becomes

$$[x,y] \cdot v = x \cdot (y \cdot v) - y \cdot (x \cdot v) \ \text{for all } x, y \in L \text{ and } v \in V. \tag{1.13}$$

Let $\rho : L \to \mathfrak{gl}(V)$ and $\rho' : L \to \mathfrak{gl}(W)$ be two representations of the Lie algebra L. A *morphism* of L-modules from V to W is a linear map $f : V \to W$ such that $\rho'(x)(f(v)) = f(\rho(x)(v))$ for all $v \in V$. If a morphism of L-modules $f : V \to W$ is bijective, then the representations ρ and ρ' are said to be *equivalent*.

Example 1.12.2 Let L be a Lie algebra. As seen in Example 1.8.3 the adjoint map $\mathrm{ad} : L \to \mathfrak{gl}(L)$ is a morphism of Lie algebras. Hence ad is a representation of L. It is called the *adjoint representation*. In the next chapters we will heavily utilize the tools of representation theory to study the adjoint representation of a Lie algebra L. This will yield useful insights in the structure of L.

Now we describe several ways of making new L-modules from known ones. Let L be a Lie algebra over the field F and let V and W be two L-modules. Then the direct sum $V \oplus W$ becomes an L-module by setting

$$x \cdot (v + w) = x \cdot v + x \cdot w \text{ for all } x \in L, \, v \in V \text{ and } w \in W.$$

Also the tensor product $V \otimes W$ can be made into an L-module by

$$x \cdot (v \otimes w) = (x \cdot v) \otimes w + v \otimes (x \cdot w).$$

We verify condition (1.13) for $V \otimes W$:

$$
\begin{aligned}
[x, y] \cdot (v \otimes w) =& ([x, y] \cdot v) \otimes w + v \otimes ([x, y] \cdot w) \\
=& (x \cdot y \cdot v) \otimes w - (y \cdot x \cdot v) \otimes w + \\
& v \otimes (x \cdot y \cdot w) - v \otimes (y \cdot x \cdot w) \\
=& x \cdot (y \cdot (v \otimes w)) - y \cdot (x \cdot (v \otimes w)).
\end{aligned}
$$

By $\mathrm{Hom}_F(V, W)$ we denote the space of all linear maps from V into W. This space can also be made into an L-module by setting

$$(x \cdot f)(v) = x \cdot (f(v)) - f(x \cdot v) \text{ for all } x \in L, \, f \in \mathrm{Hom}_F(V, W) \text{ and } v \in V.$$

We leave the verification of the requirement (1.13) to the reader.

Let V be an L-module and let W be a subspace of V that is stable under L, i.e., $x \cdot w \in W$ for all $x \in L$ and $w \in W$. Then W is called a submodule of V. In this case W is an L-module in its own right. Also, if W is a submodule, then the quotient space V/W can be made into an L-module by setting $x \cdot \bar{v} = \overline{x \cdot v}$, where $\bar{v}$ denotes the coset $v + W$. The space V/W is called the *quotient module* of V and W.

The L-module V is called *irreducible* if it has no submodules other than 0 and V itself. It is called *completely reducible* if V is a direct sum of irreducible L-modules.

Definition 1.12.3 *Let V be a finite-dimensional L-module. A series of L-submodules*

$$0 = V_0 \subset V_1 \subset \cdots \subset V_{n+1} = V$$

such that the L-modules V_{i+1}/V_i are irreducible for $0 \le i \le n$ is called a composition series of V with respect to the action of L.

Lemma 1.12.4 *Let V be a finite-dimensional L-module. Then V has a composition series.*

Proof. The proof is by induction on $\dim V$. The statement is trivial if $\dim V = 0$ or if V is irreducible. Now suppose V is not irreducible and $\dim V > 0$, and assume that the result holds for all L-modules of dimension less than $\dim V$. Let W be a maximal proper submodule of V (such a submodule exists because V is not irreducible). Then by induction W has a composition series. To this series we add V and we obtain a composition series of V (V/W is irreducible since W is a maximal proper submodule). $\square$

1.13   Restricted Lie algebras

In this section all algebras are defined over the field F of characteristic $p > 0$.

Let A be an algebra and let d be a derivation of A. We use the Leibniz formula (1.11), with $n = p$. Since all coefficients $\binom{p}{k}$ are 0 for $1 \le k \le p-1$ we see that

$$d^p(xy) = d^p(x)y + xd^p(y).$$

The conclusion is that d^p is again a derivation of A.

Now let L be a Lie algebra over F. Then for $x \in L$ we have that $\mathrm{ad}x$ is a derivation of L. So $(\mathrm{ad}x)^p$ is also a derivation of L, and it may happen that this derivation is again of the form $\mathrm{ad}y$ for a $y \in L$. Lie algebras L with the property that this happens for all elements x of L are called *restricted*.

If L is a restricted Lie algebra, then there is a map $f_p : L \to L$ such that $(\mathrm{ad}x)^p = \mathrm{ad}f_p(x)$. Now let $c : L \to C(L)$ be *any* map from L into its centre. Then $g_p = f_p + c$ is also a map such that $(\mathrm{ad}x)^p = \mathrm{ad}g_p(x)$. So if $C(L) \ne 0$, there will be a huge number of such maps g_p. Therefore we want to restrict our attention to a "nice" class of maps. In order to see what such a class might look like we study the case where the above phenomenon does not occur, namely when $C(L) = 0$.

First, let A be an associative algebra over F and let $a, b \in A$. By $\mathrm{ad}a : A \to A$ we denote the map defined by $\mathrm{ad}a(b) = ab - ba$. Now

$$(\mathrm{ad}a)^n(b) = \sum_{k=0}^{n} \binom{n}{k}(-1)^k a^k b a^{n-k}, \tag{1.14}$$

which is easily proved by induction. If in (1.14) we set $n = p - 1$, we get (as $\binom{p-1}{k} = (-1)^k$),

$$(\mathrm{ad}a)^{p-1}(b) = \sum_{k=0}^{p-1} a^k b a^{p-1-k}. \tag{1.15}$$

Let T be an indeterminate and consider the polynomial

$$(Ta + b)^p = T^p a^p + b^p + \sum_{i=1}^{p-1} s_i(a, b) T^i, \tag{1.16}$$

where the $s_i(a, b)$ are certain non-commutative polynomial functions of a, b. We differentiate (1.16) with respect to T to obtain

$$\sum_{i=0}^{p-1} (Ta + b)^i a (Ta + b)^{p-1-i} = \sum_{i=1}^{p-1} i s_i(a, b) T^{i-1}.$$

By (1.15), the left-hand side of this equation is equal to $\mathrm{ad}(Ta + b)^{p-1}(a)$. It follows that $s_i(a, b)$ is equal to $\frac{1}{i}$ times the coefficient of T^{i-1} in $\mathrm{ad}(Ta + b)^{p-1}(a)$. In particular, $s_i(a, b)$ is a nested commutator in a and b. Hence also for x, y in a Lie algebra L the expression $s_i(x, y)$ makes sense. Furthermore, $s_i(\mathrm{ad}x, \mathrm{ad}y) = \mathrm{ad}s_i(x, y)$ for $x, y \in L$, since ad is a morphism of Lie algebras (Example 1.8.3).

Example 1.13.1 First let $p = 2$, then $\mathrm{ad}(Ta + b)(a) = [b, a]$. So in this case $s_1(a, b) = [b, a]$. Now let $p = 3$, then

$$\mathrm{ad}(Ta+b)^2(a) = [Ta+b, [Ta+b, a]] = [Ta+b, [b, a]] = T[a, [b, a]] + [b, [b, a]].$$

Hence $s_1(a, b) = [b, [b, a]]$ and $s_2(a, b) = 2[a, [b, a]]$.

Now let $x, y \in L$, where L is a restricted Lie algebra such that $C(L) = 0$. Then by setting $T = 1$ in (1.16) and taking $a = \mathrm{ad}x$ and $b = \mathrm{ad}y$ we get

$$(\mathrm{ad}x + \mathrm{ad}y)^p = (\mathrm{ad}x)^p + (\mathrm{ad}y)^p + \sum_{i=1}^{p-1} s_i(\mathrm{ad}x, \mathrm{ad}y)$$

$$= \mathrm{ad}f_p(x) + \mathrm{ad}f_p(y) + \sum_{i=1}^{p-1} \mathrm{ad}s_i(x, y)$$

$$= \mathrm{ad}\Big(f_p(x) + f_p(y) + \sum_{i=1}^{p-1} s_i(x, y)\Big).$$

So since $C(L) = 0$ we see that $f_p(x+y) = f_p(x) + f_p(y) + \sum_{i=1}^{p-1} s_i(x, y)$. Furthermore, for $\alpha \in F$ and $x \in L$ we have $(\mathrm{ad}\alpha x)^p = \alpha^p \mathrm{ad}f_p(x) = \mathrm{ad}\alpha^p f_p(x)$. Hence $f_p(\alpha x) = \alpha^p f_p(x)$. Now in a restricted Lie algebra we consider maps f_p that have the same properties as the map f_p in the case where L has no centre.

Definition 1.13.2 *Let L be a restricted Lie algebra over F. Then a map $f_p : L \to L$ is called a p-th power mapping, or p-map if*

1. $(\mathrm{ad}x)^p = \mathrm{ad}f_p(x)$ for all $x \in L$,

2. $f_p(x + y) = f_p(x) + f_p(y) + \sum_{i=1}^{p-1} s_i(x, y)$ for $x, y \in L$,

3. $f_p(\alpha x) = \alpha^p f_p(x)$.

Let L be a restricted Lie algebra over F. If it is clear from the context which p-th power map f_p we mean, then we also write x^p instead of $f_p(x)$.

Example 1.13.3 Let A be an associative algebra over the field F of characteristic $p > 0$. Then since A is associative, we can raise its elements to the p-th power. Set $L = A_{Lie}$, the associated Lie algebra. Then for $a \in L$ we have $a^p \in L$. Furthermore, by (1.14) it follows that $(\mathrm{ad}a)^p(b) = \mathrm{ad}(a^p)(b)$. So L is restricted. Also by (1.16), where we set $T = 1$, we have that $a \mapsto a^p$ is a p-th power mapping. As a consequence, any subalgebra K of L that is closed under the map $a \mapsto a^p$ is automatically restricted; and the restriction of $a \mapsto a^p$ to K is a p-th power mapping.

Now let A be an algebra over F and $\mathrm{Der}(A)$ the Lie algebra of all derivations of L. As seen at the start of this section, $d^p \in \mathrm{Der}(A)$ for all $d \in \mathrm{Der}(A)$. Hence $\mathrm{Der}(A)$ is a subalgebra of $\mathrm{End}(A)_{Lie}$ closed under the p-th power map. So $\mathrm{Der}(A)$ is a restricted Lie algebra.

Let f be a bilinear form on the vector space V. Let L_f be the Lie algebra defined by (1.3). Let $a \in L_f$, then $f(a^p v, w) = (-1)^p f(v, a^p w) = -f(v, a^p w)$ so that $a^p \in L_f$. As a consequence L_f is a restricted Lie algebra relative to the map $a \mapsto a^p$.

Suppose we are given a Lie algebra L over F. In principle, when checking whether L is restricted or not, we would have to check whether $(\operatorname{ad}x)^p$ is an inner derivation of L for all $x \in L$. However, by the next result it is enough to check this only for the elements of a basis of L. Furthermore, a-priori it is not clear whether any restricted Lie algebra has a p-th power mapping. Also this is settled by the next proposition.

Proposition 1.13.4 *Let L be a Lie algebra over the field F of characteristic $p > 0$. Let $\{x_1, x_2, \ldots\}$ be a (possibly infinite) basis of L. Suppose that there are $y_i \in L$ for $i \geq 1$ such that $(\operatorname{ad}x_i)^p = \operatorname{ad}y_i$. Then L has a unique p-th power mapping $x \mapsto x^p$ such that $x_i^p = y_i$.*

Proof. We prove uniqueness. Let $f_p : L \to L$ and $g_p : L \to L$ be two p-th power mappings such that $f_p(x_i) = g_p(x_i) = y_i$ for $i \geq 1$. Set $h = f_p - g_p$. Then from the definition of p-th power mapping we have $h(x + y) = h(x) + h(y)$ and $h(\alpha x) = \alpha^p h(x)$, for $x, y \in L$ and $\alpha \in F$. Hence $\ker h$ is a subspace of L. But $x_i \in \ker h$ for all $i \geq 1$. It follows that $h = 0$ and $f_p = g_p$.

The proof of existence is more complicated; it needs the concept of universal enveloping algebras. We defer it to Section 6.3. ☐

On the basis of the last proposition we formulate an algorithm for checking whether a given finite-dimensional Lie algebra is restricted.

Algorithm IsRestricted
Input: a finite-dimensional Lie algebra L defined over a field of characteristic $p > 0$.
Output: true if L is restricted, false otherwise.

Step 1 Select a basis $\{x_1, \ldots, x_n\}$ of L and compute the matrices of $\operatorname{ad}x_i$ for $1 \leq i \leq n$. Let V be the vector space spanned by these matrices.

Step 2 for $1 \leq i \leq n$ determine whether $(\operatorname{ad}x_i)^p \in V$. If one of these is not in V then return false. Otherwise return true.

Let L be a restricted Lie algebra. In order to calculate with p-th power mappings, we have to decide how to represent such a mapping on a computer. Proposition 1.13.4 gives a straightforward solution to this problem.

We can describe a p-th power mapping by giving the images of the elements of a basis of L. Then the image of an arbitrary element of L can be calculated using conditions 2 and 3 from Definition 1.13.2.

Example 1.13.5 Let G be a finite p-group, and L be the corresponding Lie algebra over $F = \mathbb{F}_p$ as constructed in Section 1.4. Then $L = V_1 \oplus \cdots \oplus V_s$, where V_i is isomorphic to the factor group G_i via the isomorphism $\sigma_i : G_i \to V_i$. Now let $x \in V_i$ and let $g \in \kappa_i(G)$ be such that $\tau_i(g) = x$ (where τ_i is as in Section 1.4). Then by Theorem 1.4.1, $g^p \in \kappa_{pi}(G)$. Now we set $x^p = \tau_{pi}(g^p)$. It can be shown that x^p does not depend on the particular pre-image g chosen. Also it can be shown that for $x \in V_i$ we have $(\mathrm{ad}x)^p = \mathrm{ad}x^p$. This together with Proposition 1.13.4 shows that L is a restricted Lie algebra.

Now let G be the group generated by g_1, g_2, g_3 of Example 1.4.5. Let L be the corresponding 3-dimensional Lie algebra over $\mathbb{F}_2$ with basis $\{e_1, e_2, e_3\}$. Then $e_1^2 = \tau_2(g_1^2) = e_3$, and similarly $e_2^2 = e_3$ and $e_3^2 = 0$. Furthermore, using the fact that $s_1(a, b) = [b, a]$ (see Example 1.13.1) we see that

$$(e_1 + e_2)^2 = e_1^2 + e_2^2 + s_1(e_1, e_2) = e_3 + e_3 + e_3 = e_3.$$

1.14 Extension of the ground field

Let L be a Lie algebra over the field F, and let $\tilde{F}$ be an extension field of F. Then the structure constants of L are also elements of $\tilde{F}$, so we can define a second Lie algebra $\tilde{L}$ that has the same structure constants as L, but is defined over $\tilde{F}$. Formally we set $\tilde{L} = L \otimes_F \tilde{F}$. This space consists of finite sums $\sum_i y_i \otimes \lambda_i$, where $y_i \in L$ and $\lambda_i \in \tilde{F}$. For $\otimes$ we have the following rules:

$$(x + y) \otimes \lambda = x \otimes \lambda + y \otimes \lambda,$$
$$x \otimes (\lambda + \mu) = x \otimes \lambda + x \otimes \mu,$$
$$(\alpha x) \otimes \lambda = x \otimes \alpha\lambda \quad \text{for } \alpha \in F,$$
$$\lambda(x \otimes \mu) = x \otimes \lambda\mu.$$

Furthermore, the product in $\tilde{L}$ is defined by $[x \otimes \lambda, y \otimes \mu] = [x, y] \otimes \lambda\mu$. Let $\{x_1, \ldots, x_n\}$ be a basis of L, then the rules for $\otimes$ imply that every $y \in \tilde{L}$ can be written as

$$y = \sum_{i=1}^{n} \lambda_i(x_i \otimes 1) \quad \text{where } \lambda_i \in \tilde{F} \text{ for } 1 \leq i \leq n.$$

It follows that the elements $x_1 \otimes 1, \dots, x_n \otimes 1$ form a basis of $\tilde{L}$. And the structure constants of $\tilde{L}$ relative to this basis equal the structure constants of L relative to the basis $\{x_1, \dots, x_n\}$.

Now let V, W be two subspaces of L; then $\tilde{V} = V \otimes_F \tilde{F}$ and $\tilde{W} = W \otimes_F \tilde{F}$ are subspaces of $\tilde{L}$. Let $\{v_1, \dots, v_s\}$ and $\{w_1, \dots, w_t\}$ be bases of V and W respectively. If we input $\tilde{L}$, $\tilde{V}$ and $\tilde{W}$ (with respective bases $\{x_1 \otimes 1, \dots, x_n \otimes 1\}$, $\{v_1 \otimes 1, \dots, v_s \otimes 1\}$ and $\{w_1 \otimes 1, \dots, w_t \otimes 1\}$) into the algorithm ProductSpace then, because the structure constants of $\tilde{L}$ are equal to those of L, we see that this algorithm performs exactly the same operations as on the input L, V, W. (Solving linear equations with coefficients in F gives an answer with coefficients in F.) Hence

$$[V \otimes_F \tilde{F}, W \otimes_F \tilde{F}] = [V, W] \otimes_F \tilde{F}.$$

Let $L^{(k)}$ denote the k-th term of the derived series of L, then by the above discussion we have that $\tilde{L}^{(k)} = L^{(k)} \otimes_F \tilde{F}$. And the same holds for the terms of the lower central series.

Also, the algorithm Centre solves exactly the same set of linear equations on input L as on input $\tilde{L}$. Hence we see that $C(\tilde{L}) = C(L) \otimes_F \tilde{F}$. And generally, $C_k(\tilde{L}) = C_k(L) \otimes_F \tilde{F}$, where $C_k(L)$ denotes the k-th term of the upper central series of L.

1.15 Finding a direct sum decomposition

Let L be a finite-dimensional Lie algebra. Then L may or may not be the direct sum of two or more ideals. If this happens to be the case then the structure of the direct summands of L may be studied independently. Together they determine the structure of L. So the direct sum decomposition can be a very valuable piece of information. In this section we describe an algorithm to compute a decomposition of L as a direct sum of indecomposable ideals (i.e., ideals that are not direct sums of ideals themselves).

Suppose that $L = I_1 \oplus I_2$ is the direct sum of two ideals I_1, I_2 and I_1 is contained in the centre of L. Then I_1 is called a *central component* of L. First we give a method for finding such a central component if it exists.

Let J_1 be a complementary subspace in $C(L)$ to $C(L) \cap [L, L]$. Then as J_1 is contained in the centre of L, it is an ideal of L. Let J_2 be a complementary subspace in L to J_1 containing $[L, L]$. Then

$$[L, J_2] \subset [L, L] \subset J_2$$

so that J_2 is an ideal of L. Furthermore $L = J_1 \oplus J_2$ and J_1 is central and

J_2 does not contain a central component. The conclusion is that J_1 is a maximal central component.

Now we suppose that $C(L) \subset [L, L]$ (i.e., that L does not have a central component) and we try to decompose L as a direct sum of ideals.

We recall that $M_n(F)$ is the associative matrix algebra consisting of all $n \times n$ matrices over F (cf. Example 1.1.4). From Appendix A we recall that a non-zero element $e \in M_n(F)$ is called an *idempotent* if $e^2 = e$. Two idempotents e_1 and e_2 are called *orthogonal* if $e_1 e_2 = e_2 e_1 = 0$. An idempotent is said to be *primitive* if it is not the sum of two orthogonal idempotents.

Proposition 1.15.1 *Put $n = \dim L$. The Lie algebra L is the direct sum of k (non-zero) ideals $I_1, \dots, I_k$ if and only if the centralizer*

$$C_{M_n(F)}(\mathrm{ad}L) = \{a \in M_n(F) \mid a \cdot \mathrm{ad}x = \mathrm{ad}x \cdot a \text{ for all } x \in L\}$$

contains k orthogonal idempotents $e_1, \dots, e_k$ such that $e_1 + \cdots + e_k$ is the identity on L and $I_r = e_r I_r$ for $r = 1, \dots, k$. Furthermore, the ideal I_r is indecomposable if and only if the corresponding idempotent e_r is primitive.

Proof. First we suppose that

$$L = I_1 \oplus \cdots \oplus I_k,$$

where I_r is a non-zero ideal of L for $1 \leq r \leq k$. For an element $x \in L$ we write $x = x_1 + \cdots + x_k$, where $x_r \in I_r$. Let $e_r : L \to L$ be defined by $e_r(x) = x_r$ (i.e., e_r is the projection onto I_r). Then for $y, z \in L$ we have $e_r \mathrm{ad}y(z) = e_r[y, z] = [y, z]_r = [y, z_r] = \mathrm{ad}y e_r(z)$. Hence $e_r \in C_{M_n(F)}(\mathrm{ad}L)$. Also it is clear that $e_r e_s = 0$ for $r \neq s$, and that $e_1 + \cdots + e_k = 1$.

Now let $e_1, \dots, e_k$ be k orthogonal idempotents in $C_{M_n(F)}(\mathrm{ad}L)$ such that $e_1 + \cdots + e_k = 1$. Then set $I_r = e_r L$ for $1 \leq r \leq k$. Since $e_r^2 = e_r$ we have for $x \in I_r$ and $y \in L$ that $[y, x] = [y, e_r x] = \mathrm{ad}y e_r(x) = e_r \mathrm{ad}y(x) = e_r[y, x]$ so that I_r is an ideal of L. Also

$$L = 1 \cdot L = (e_1 + \cdots + e_k)L = I_1 + \cdots + I_k.$$

Now let $x \in I_r \cap I_s$. Then $x = e_r x$, and also $e_s x = x$ so that $x = e_s e_r x = 0$. Hence L is the direct sum of the I_r.

If e_r is not primitive then, as above, we see that I_r is the direct sum of ideals. Furthermore, if I_r is the direct sum of ideals of I_r (that are then automatically ideals of L), then e_r is not primitive. $\square$

Theorem 1.15.2 *Let A be the associative algebra $C_{M_n(F)}(\mathrm{ad}L)$ and put $Q = A/\mathrm{Rad}(A)$. Suppose that L has no central component. Then Q is commutative.*

Proof. We may suppose that the ground field F is algebraically closed. Indeed, if $\overline{F}$ denotes the algebraic closure of F, then as seen in Section 1.14 we have $C(L) \otimes_F \overline{F} = C(L \otimes_F \overline{F})$ and $[L \otimes_F \overline{F}, L \otimes_F \overline{F}] = [L, L] \otimes_F \overline{F}$. Therefore also $L \otimes_F \overline{F}$ has no central component. Also since a basis of $C_{M_n(F)}(\mathrm{ad}L)$ can be calculated by solving a set of linear equations over F we have $C_{M_n(F)}(\mathrm{ad}L) \otimes_F \overline{F} = C_{M_n(F) \otimes_F \overline{F}}(\mathrm{ad}L \otimes_F \overline{F})$. And finally $\mathrm{Rad}(A \otimes_F \overline{F}) = \mathrm{Rad}(A) \otimes_F \overline{F}$ by Proposition A.1.3. Hence the result for $L \otimes_F \overline{F}$ will imply the result for L.

By Theorem A.1.5 we may write $A = S \oplus \mathrm{Rad}(A)$, where S is a semisimple subalgebra of A isomorphic to Q. We suppose that S is not commutative. By Theorem A.1.4, S decomposes as a direct sum of simple associative algebras that are full matrix algebras over division algebras over F. As F is algebraically closed, the only division algebra over F is F itself (for a proof of this we refer to [69]). Hence from our assumption that S is not commutative it follows that S contains a full matrix algebra of degree at least 2. Now every full matrix algebra of degree ≥ 2 contains a full matrix algebra of degree 2. It follows that S contains elements e_{ij} for $1 \leq i, j \leq 2$ such that $e_{ij}e_{kl} = \delta_{jk}e_{il}$. Then $1 \in A$ decomposes as a sum of three orthogonal idempotents $1 = e_{11} + e_{22} + (1 - e_{11} - e_{22})$ and the first two of these are certainly non-trivial. So by Proposition 1.15.1, L decomposes as a direct sum of ideals

$$L = I_1 \oplus I_2 \oplus I_3$$

where $I_1 = e_{11}L$ and $I_2 = e_{22}L$. Now we have maps

$$f_{21} : I_1 \longrightarrow I_2, \text{ and } f_{12} : I_2 \longrightarrow I_1$$

defined by $f_{21}(x) = e_{21}x$ and $f_{12}(y) = e_{12}y$ for $x \in I_1$ and $y \in I_2$. Since $e_{21}x = e_{22}e_{21}x$ we see that f_{21} maps I_1 into I_2, and likewise f_{12} maps I_2 into I_1. Furthermore $f_{12}(f_{21}(x)) = e_{12}e_{21}x = e_{11}x = x$ for $x \in I_1$. So f_{12} and f_{21} are each others inverses. In particular they are non-singular linear maps.

Because L does not have a central component we have that I_1 and I_2 are not commutative Lie algebras. So there exist $x, y \in I_1$ such that $[x, y] \neq 0$. We calculate

$$f_{21}([x, y]) = e_{21}[x, y] = e_{21}\mathrm{ad}x(y) = \mathrm{ad}x(e_{21}y) = [x, e_{21}y] = 0$$

where last equality follows from $[I_1, I_2] = 0$. But since f_{21} is non-singular this implies that $[x, y] = 0$ and we have reached a contradiction. The conclusion is that S (and hence Q) is commutative.   □

Algorithm DirectSumDecomposition
Input: a finite-dimensional Lie algebra L.
Output: a list of indecomposable ideals of L such that L is their direct sum.

Step 1 Compute the centre $C(L)$ and the derived subalgebra $[L, L]$. Compute a basis of a complement J in $C(L)$ to $C(L) \cap [L, L]$. Let K be a complement to J in L such that K contains $[L, L]$.

Step 2 Set $n := \dim(K)$ and compute the centralizer $A = C_{M_n(F)}(\mathrm{ad}_K K)$.

Step 3 Compute $\mathrm{Rad}(A)$ and set $Q := A/\mathrm{Rad}(A)$. Compute the set of primitive orthogonal idempotents $\bar{e}_1, \dots, \bar{e}_k$ in Q such that $\bar{e}_1 + \cdots + \bar{e}_k = 1 \in Q$.

Step 4 Lift the idempotents $\bar{e}_1, \dots, \bar{e}_k$ to a set of orthogonal idempotents $e_1, \dots, e_k \in A$ such that $e_1 + \cdots + e_r = 1 \in A$. For $1 \le r \le k$ set $I_r = e_r K$.

Step 5 Let $\{y_1, \dots, y_m\}$ be a basis of J and for $1 \le i \le m$ let J_i be the ideal of L spanned by y_i. Return $\{J_1, \dots, J_m\} \cup \{I_1, \dots, I_k\}$.

Comments: If $x_1, \dots, x_n$ is a basis of K, then

$$C_{M_n(F)}(\mathrm{ad}K) = \{a \in M_n(F) \mid a \cdot \mathrm{ad}_K x_i = \mathrm{ad}_K x_i \cdot a \text{ for } 1 \le i \le n\},$$

and hence $C_{M_n(F)}(\mathrm{ad}K)$ can be calculated by solving a system of linear equations. For the calculations in the associative algebra A (calculation of the radical, of central idempotents and lifting them modulo the radical), we refer to Appendix A. We note that since Q is commutative, all idempotents in Q are central. Hence a set of primitive orthogonal idempotents with sum 1 can be found by the algorithm CentralIdempotents (as described in Appendix A).
Remark. A computationally difficult part of the algorithm DirectSumDecomposition is the calculation of the centralizer $C_{M_n(F)}(\mathrm{ad}K)$. The dimension of this algebra may be substantially bigger than the dimension of K. So it would be desirable to have an algorithm that uses Lie algebra methods only. To the best of our knowledge no such algorithm exists for the general case. However, for the special case of *semisimple* Lie algebras we do have

an algorithm that does not "leave" the Lie algebra (see Section 4.12).

1.16   Notes

The algorithms QuotientAlgebra, AdjointMatrix, Centralizer, Centre, Normalizer, ProductSpace, Derivations can be found in [7]. We have followed [71] for the algorithm DirectSumDecomposition.

Usually (cf., [48], [88]), a Lie algebra of characteristic $p > 0$ is called restricted if it has a p-th power mapping. However, due to Proposition 1.13.4 (which is taken from [48]), this is equivalent to our definition of restrictedness.

Chapter 2

On nilpotency and solvability

In this chapter we start our study of the structure of a finite-dimensional Lie algebra. We examine solvable and nilpotent Lie algebras. Furthermore, we show that any finite-dimensional Lie algebra contains a unique maximal nilpotent ideal (called the nilradical), and a unique maximal solvable ideal (called the solvable radical). Section 2.2 is devoted to the nilradical. We give several characterizations of the nilradical, on the basis of which we formulate an algorithm for calculating the nilradical of a finite-dimensional Lie algebra. Section 2.3 is devoted to the solvable radical. We prove its existence and derive several important properties of the solvable radical. Using some of these in Section 2.4 we give a proof of Lie's theorem, which states that a linear "split" solvable Lie algebra of characteristic 0 consists of upper triangular matrices.

In Section 2.5 we prove Cartan's criterion for a Lie algebra of characteristic 0 to be solvable. In Section 2.6 we prove a converse of Cartan's criterion, which yields an algorithm for calculating the solvable radical of a finite-dimensional Lie algebra of characteristic 0. Finally, in Section 2.7 we derive an algorithm for finding a non-nilpotent element in a Lie algebra. This algorithm will be of major importance in Chapter 3, where we give algorithms for calculating a Cartan subalgebra.

2.1 Engel's theorem

Let L be a nilpotent Lie algebra. Then it is straightforward to see that $\mathrm{ad}x$ is a nilpotent endomorphism for all $x \in L$. This section is devoted to proving the converse: if $\mathrm{ad}x$ is nilpotent for all $x \in L$, then L is a nilpotent Lie algebra.

Lemma 2.1.1 *Let V be a finite-dimensional vector space over the field F. Suppose $a \in \mathfrak{gl}(V)$ is nilpotent (i.e., $a^k = 0$ for a $k > 0$). Then the endomorphism $\mathrm{ad}a$ of $\mathfrak{gl}(V)$ is nilpotent.*

Proof. We must prove that there is an $m > 0$ such that $(\mathrm{ad}a)^m(b) = 0$ for all $b \in \mathfrak{gl}(V)$. Because $\mathrm{ad}a(b) = ab - ba$, we have that $(\mathrm{ad}a)^m(b)$ is a sum of elements of the form $c_{ij}a^i b a^j$, where $c_{ij} \in F$ and $i + j = m$. It follows that we can take $m = 2k - 1$. $\qquad\square$

Proposition 2.1.2 *Let L be a Lie subalgebra of $\mathfrak{gl}(V)$. Suppose that all elements $x \in L$ are nilpotent endomorphisms; then there is a non-zero $v \in V$ such that $x \cdot v = 0$ for all $x \in L$.*

Proof. The proof is by induction on $\dim L$. If $\dim L = 0$, then the statement is trivial. So suppose that $\dim L \geq 1$. By induction we may suppose that the statement holds for all Lie algebras of dimension less than $\dim L$. Let K be a maximal proper subalgebra of L. We consider the adjoint representation of K on L, $\mathrm{ad}_L : K \to \mathfrak{gl}(L)$. Let $x \in K$, then $\mathrm{ad}_L x(K) \subset K$. So K is a submodule of L. We form the quotient module and get a representation $\sigma : K \to \mathfrak{gl}(L/K)$. By Lemma 2.1.1, $\sigma(x)$ is nilpotent for all $x \in K$. Therefore we can apply the induction hypothesis and we find that there is a $\bar{y} \in L/K$ such that $\sigma(x) \cdot \bar{y} = 0$ for all $x \in K$. Let y be a pre-image of $\bar{y}$ in L. Then $[x, y] \in K$ for all $x \in K$. Hence the space spanned by K and y is a subalgebra of L and it properly contains K. Since K is maximal, we conclude that L is spanned by K and y.

Now let W be the space of all $v \in V$ such that $x \cdot v = 0$ for all $x \in K$. By induction W is non-zero. Also if $v \in W$, then for $x \in K$,

$$x \cdot (y \cdot v) = y \cdot (x \cdot v) + [x, y] \cdot v = 0$$

because $[x, y] \in K$. Hence y leaves W invariant. But y is nilpotent, and therefore there is a non-zero $w \in W$ such that $y \cdot w = 0$. It follows that $x \cdot w = 0$ for all $x \in L$. $\qquad\square$

Lemma 2.1.3 *Let V be a finite-dimensional vector space. Let $\rho : L \to \mathfrak{gl}(V)$ be an irreducible representation of the Lie algebra L. Suppose that I is an ideal of L such that $\rho(x)$ is nilpotent for all $x \in I$. Then $\rho(x) = 0$ for all $x \in I$.*

Proof. Let W be the subspace of V consisting of all elements $v \in V$ such that $\rho(x)v = 0$ for all $x \in I$. Then $W \neq 0$ by Proposition 2.1.2. Also, for $w \in W$ and $x \in I$ and $y \in L$, we have

$$\rho(x)\rho(y)w = \rho(y)\rho(x)w + \rho([x,y])w = 0$$

so that W is a non-zero L-submodule of V. Since V is irreducible, it follows that $W = V$. $\qquad\qquad\square$

Proposition 2.1.4 *Let* $\rho : L \to \mathfrak{gl}(V)$ *be a finite-dimensional representation of the Lie algebra* L*. Let* $0 = V_0 \subset V_1 \subset \cdots \subset V_{n+1} = V$ *be a composition series of* V *with respect to* L*. Let* I *be an ideal of* L*. Then the following are equivalent:*

- $\rho(x)$ *is nilpotent for all* $x \in I$,

- $\rho(x)V_{i+1} \subset V_i$ *for all* $x \in I$ *and* $0 \leq i \leq n$.

Proof. Suppose that $\rho(x)$ is nilpotent for all $x \in I$. Consider the induced representation $\sigma_{i+1} : L \to \mathfrak{gl}(V_{i+1}/V_i)$. Then also $\sigma_{i+1}(x)$ is nilpotent for all $x \in I$. Hence by Lemma 2.1.3, $\sigma_{i+1}(x) = 0$, implying that $\rho(x)V_{i+1} \subset V_i$. The other implication is trivial. $\qquad\qquad\square$

Theorem 2.1.5 (Engel) *Let* L *be a finite-dimensional Lie algebra. Then* L *is nilpotent if and only if* $\mathrm{ad}x$ *is a nilpotent endomorphism for all* $x \in L$.

Proof. Let $0 = L_0 \subset L_1 \subset \cdots \subset L_{n+1} = L$ be a composition series of L relative to the adjoint representation of L. Suppose that $\mathrm{ad}x$ is nilpotent for all $x \in L$. Then by Proposition 2.1.4 we have that $\mathrm{ad}x(L_{i+1}) \subset L_i$, for $x \in L$ and $0 \leq i \leq n$. Hence $L^k \subset L_{n-k+2}$ (where L^k is the k-th term of the lower central series). It follows that L is nilpotent.

If L is nilpotent, then $\mathrm{ad}x(L^k) \subset L^{k+1}$ for all $x \in L$ and $k \geq 1$. Hence $\mathrm{ad}x$ is nilpotent for all $x \in L$. $\qquad\qquad\square$

Example 2.1.6 Theorem 2.1.5 yields an ad-hoc method to show that a given Lie algebra L is *not* nilpotent, without calculating the lower central series. The only thing we have to do is to find an $x \in L$ such that $\mathrm{ad}x$ is not nilpotent. As an example we consider the Lie algebra with basis $\{x_1, \ldots, x_5\}$ and multiplication table

$$[x_1, x_4] = 2x_1, \quad [x_2, x_3] = x_1, \quad [x_2, x_4] = x_2, \quad [x_2, x_5] = -x_3,$$
$$[x_3, x_4] = x_3, \quad [x_3, x_5] = x_2.$$

Then we see that $\mathrm{ad}x_4$ has an eigenvalue -2 (corresponding to the eigenvector x_1). Hence $\mathrm{ad}x_4$ is not nilpotent, so that L is not a nilpotent Lie algebra. In Section 2.7 we will give an algorithm for finding a non-nilpotent element in a Lie algebra L, provided that L is not nilpotent. This automatically provides a way of testing whether an arbitrary Lie algebra is nilpotent.

2.2   The nilradical

Here we show that an arbitrary Lie algebra contains a unique maximal nilpotent ideal, called the nilradical. We give two characterizations of the nilradical.

Lemma 2.2.1 *Let L be a Lie algebra. If I and J are nilpotent ideals of L, then so is $I + J$.*

Proof. Let (I^k) and (J^k) be the lower central series of I and J respectively. Then because I and J are nilpotent, there is an integer t such that $I^t = J^t = 0$. Hence all m-fold brackets

$$[x_1, [x_2, [\ldots [x_{m-1}, x_m] \ldots]]]$$

of elements of L with at least t of them from I or at least t of them from J are zero. Now set $m = 2t$ and let $x_i = y_i + z_i$, where $y_i \in I$ and $z_i \in J$ for $1 \leq i \leq m$. Then $[x_1, [x_2, \cdots [x_{m-1}, x_m] \cdots]]$ is a linear combination of m-fold brackets containing at least t elements from I or at least t elements from J. Hence $(I + J)^{2t} = 0$. Thus $I + J$ is nilpotent.   □

From Lemma 2.2.1 it follows that a finite-dimensional Lie algebra L contains a unique maximal nilpotent ideal. Indeed, suppose that I and J both are maximal nilpotent ideals. Then $I + J$ is a nilpotent ideal containing both I and J. It follows that $I + J = I$ and $I + J = J$, or $I = J$. The maximal nilpotent ideal of L is called the *nilradical* and denoted by $\mathrm{NR}(L)$.

Proposition 2.2.2 *Let $0 = L_0 \subset L_1 \subset \cdots \subset L_{n+1} = L$ be a composition series of L with respect to the adjoint representation of L. Then $\mathrm{NR}(L)$ is the set of all $x \in L$ such that $\mathrm{ad}x(L_{i+1}) \subset L_i$ for $0 \leq i \leq n$.*

Proof. Let I be the set of all $x \in L$ such that $\mathrm{ad}x(L_{i+1}) \subset L_i$ for $0 \leq i \leq n$. Then I is an ideal of L and $\mathrm{ad}x$ is nilpotent for all $x \in I$. By Theorem 2.1.5 this implies that I is nilpotent and hence $I \subset \mathrm{NR}(L)$. If $x \in \mathrm{NR}(L)$ then, by Theorem 2.1.5, $\mathrm{ad}_{\mathrm{NR}(L)}x$ is nilpotent. Also, since $\mathrm{NR}(L)$ is an ideal,

$\mathrm{ad}x(L) \subset \mathrm{NR}(L)$. Hence $\mathrm{ad}_L x$ is nilpotent. Now by Proposition 2.1.4 we see that $\mathrm{NR}(L) \subset I$. □

From Section 1.1 we recall that $(\mathrm{ad}L)^*$ denotes the associative algebra generated by the identity on L together with all $\mathrm{ad}x$ for $x \in L$. Furthermore, from Appendix A we recall that the radical $\mathrm{Rad}(A)$ of an associative algebra A is its unique maximal nilpotent ideal. The next proposition relates the radical of $(\mathrm{ad}L)^*$ to the nilradical of L.

Proposition 2.2.3 *The nilradical of L is the set of $x \in L$ such that $\mathrm{ad}x \in \mathrm{Rad}((\mathrm{ad}L)^*)$.*

Proof. Let $0 = L_0 \subset L_1 \subset \cdots \subset L_{n+1} = L$ be a composition series of L with respect to to the adjoint representation. Let N be the space of all $a \in (\mathrm{ad}L)^*$ such that $aL_{i+1} \subset L_i$ for $0 \le i \le n$. Then N is a nilpotent ideal in $(\mathrm{ad}L)^*$, and consequently, $N \subset \mathrm{Rad}((\mathrm{ad}L)^*)$. Let $x \in \mathrm{NR}(L)$, then by Proposition 2.2.2, $\mathrm{ad}x \in N$. So $\mathrm{ad}x$ lies in the radical of $(\mathrm{ad}L)^*$.

Let I be the set of $y \in L$ such that $\mathrm{ad}y \in \mathrm{Rad}((\mathrm{ad}L)^*)$. Let $x \in L$ and $y \in I$, then because $\mathrm{ad}[x,y] = \mathrm{ad}x\mathrm{ad}y - \mathrm{ad}y\mathrm{ad}x$ (see Example 1.12.2) we have $\mathrm{ad}[x,y] \in \mathrm{Rad}((\mathrm{ad}L)^*)$. Therefore I is an ideal of L. Also for all $x \in I$ we have that $\mathrm{ad}_L x$ is nilpotent, and as a consequence $\mathrm{ad}_I x$ is nilpotent. Hence by Engel's theorem (Theorem 2.1.5), I is nilpotent so that $I \subset \mathrm{NR}(L)$. □

Using Proposition 2.2.3 we give an algorithm for calculating the nilradical of a Lie algebra L.

Algorithm NilRadical
Input: a finite-dimensional Lie algebra L.
Output: the nilradical of L.

Step 1 Compute a basis of the associative algebra $A = (\mathrm{ad}L)^*$.

Step 2 Compute $R = \mathrm{Rad}(A)$ (see Appendix A).

Step 3 Compute a basis of the space of all $x \in L$ such that $\mathrm{ad}x \in R$. Return this basis.

Comments: The algorithm is justified by Proposition 2.2.3. We remark that the computation in Step 3 can be done by solving a system of linear equations. Furthermore, the algorithm works for Lie algebras defined over fields of characteristic 0 and finite fields (because over these fields we have an algorithm for calculating the radical of an associative algebra, see Appendix A).

Example 2.2.4 Let L be the Lie algebra over $\mathbb{Q}$ with basis $\{x_1, x_2, x_3\}$ and multiplication table

$$[x_1, x_3] = x_1 + x_2, \ [x_2, x_3] = x_2,$$

(as usual we do not list products that are zero). Denote by a_i the matrix of $\mathrm{ad}x_i$, then

$$a_1 = \begin{pmatrix} 0 & 0 & 1 \\ 0 & 0 & 1 \\ 0 & 0 & 0 \end{pmatrix}, \ a_2 = \begin{pmatrix} 0 & 0 & 0 \\ 0 & 0 & 1 \\ 0 & 0 & 0 \end{pmatrix}, \ a_3 = \begin{pmatrix} -1 & 0 & 0 \\ -1 & -1 & 0 \\ 0 & 0 & 0 \end{pmatrix}.$$

Denote the 3×3-identity matrix by a_0. It is straightforward to see that the associative algebra A generated by a_0, a_1, a_2, a_3 is spanned by these matrices together with

$$a_4 = \begin{pmatrix} 0 & 0 & 0 \\ 1 & 0 & 0 \\ 0 & 0 & 0 \end{pmatrix}.$$

We use the algorithm Radical to calculate the radical of A (see Appendix A). Set $a = \sum_{i=0}^{4} \lambda_i a_i$, then $a \in \mathrm{Rad}(A)$ if and only if $\mathrm{Tr}(aa_j) = 0$ for $j = 0, \dots, 4$. We have that

$$a = \begin{pmatrix} \lambda_0 - \lambda_3 & 0 & \lambda_1 \\ \lambda_4 - \lambda_3 & \lambda_0 - \lambda_3 & \lambda_1 + \lambda_2 \\ 0 & 0 & \lambda_0 \end{pmatrix}.$$

Then $\mathrm{Tr}(aa_i) = 0$ if $i = 1, 2, 4$, so we do not get any equations from there. However, $\mathrm{Tr}(aa_0) = -2\lambda_3 + 3\lambda_0$, and $\mathrm{Tr}(aa_3) = 2\lambda_3 - 2\lambda_0$; so $0 = \mathrm{Tr}(aa_0) = \mathrm{Tr}(aa_3)$ implies $\lambda_0 = \lambda_3 = 0$. Hence $\mathrm{Rad}(A)$ is spanned by a_1, a_2, a_4. We see that $\mathrm{NR}(L)$ is spanned by x_1, x_2.

There exist more algorithms for calculating the nilradical of a Lie algebra. However, because for the exposition of some algorithms we need theoretical tools introduced in Chapter 3, we defer a complete discussion to that chapter.

2.3 The solvable radical

We show that a finite-dimensional Lie algebra contains a unique maximal solvable ideal, called the solvable radical. We study the structure of this ideal for linear Lie algebras of characteristic 0, such that the underlying module is irreducible. This leads to a third characterization of the nilradical. We recall that $L^{(k)}$ denotes the k-th term of the derived series of the Lie algebra L.

Proposition 2.3.1 *Let L be a Lie algebra.*

1. *If L is solvable then so are all Lie subalgebras and homomorphic images of L.*

2. *If I is a solvable ideal of L such that L/I is solvable, then L itself is solvable.*

3. *If I and J are solvable ideals of L then so is $I + J$.*

Proof. 1. If K is a subalgebra of L, then the i-th term of the derived series of K is contained in the i-th term of the derived series of L. So, if L is solvable, then the derived series of L ends at 0 and hence the same is true for the derived series of K.

Let $\phi : L \to L'$ be a morphism of Lie algebras, then $[\phi(L), \phi(L)] = \phi([L, L])$. By induction, the i-th term $\phi(L)^{(i)}$ of the derived series of $\phi(L)$ is contained in $\phi(L^{(i)})$. Hence, there is an $i \geq 1$ such that $\phi(L)^{(i)} = 0$, proving that $\phi(L)$ is solvable.

2. As L/I is solvable, $L^{(m)} \subset I$ for some $m \geq 1$. Solvability of I means that $I^{(n)} = 0$ for some n. It follows that $L^{(m+n)} = 0$.

3. Consider the quotient $(I + J)/J$ of $I + J$. According to Proposition 1.8.2 it is isomorphic to $I/(I \cap J)$, which is a quotient of the solvable subalgebra I, and therefore solvable by 1. Since J is also solvable, we can apply 2. to conclude that $I + J$ is solvable. $\qquad\square$

From Proposition 2.3.1 it follows that a finite-dimensional Lie algebra L contains a unique maximal solvable ideal. It is called the *solvable radical* of L; we denote it by $\mathrm{SR}(L)$. Since any nilpotent ideal is also solvable (see Lemma 1.7.2), we immediately see that $\mathrm{SR}(L)$ contains the nilradical $\mathrm{NR}(L)$.

Remark. Let I be a solvable ideal of the quotient algebra $L/\mathrm{SR}(L)$. Let J be the pre-image of I in L. Then by Proposition 2.3.1, J is solvable. Furthermore, J contains $\mathrm{SR}(L)$ so that $J = \mathrm{SR}(L)$. Therefore, $I = 0$. So we see that the solvable radical of $L/\mathrm{SR}(L)$ is trivial. The analogous property does not hold for the nilradical.

Lemma 2.3.2 *Let V be a finite-dimensional vector space over a field of characteristic 0. Let $c \in \text{End}(V)$ be given by*

$$c = \sum_{i=1}^{m} [a_i, b_i]$$

where $a_i, b_i \in \text{End}(V)$. Suppose that $[c, b_i] = 0$ for $1 \leq i \leq m$, then c is nilpotent.

Proof. Here we use that fact that, since V is of characteristic 0, $\text{Tr}(c^k) = 0$ for $k \geq 1$ implies that c is nilpotent. We have

$$c^k = c^{k-1} \sum_{i=1}^{m} a_i b_i - b_i a_i = \sum_{i=1}^{m} (c^{k-1} a_i) b_i - b_i (c^{k-1} a_i) = \sum_{i=1}^{m} [c^{k-1} a_i, b_i].$$

Since the trace of a commutator is always zero, it follows that $\text{Tr}(c^k) = 0$. □

Proposition 2.3.3 *Let V be a finite-dimensional vector space over a field of characteristic 0. Let $L \subset \mathfrak{gl}(V)$ be a linear Lie algebra and suppose that V is an irreducible L-module. Then the solvable radical of L is equal to the centre $C(L)$ of L.*

Proof. Because $C(L)$ is a solvable ideal of L it is contained in $\text{SR}(L)$. So we only have to prove the reverse inclusion. This is equivalent to proving that the ideal $I = [L, \text{SR}(L)]$ is zero.

Suppose that $I \neq 0$, then I is a non-zero subalgebra of $\text{SR}(L)$. Hence by Proposition 2.3.1, I is solvable. So there is an $m \geq 1$ such that $I^{(m)} \neq 0$ and $I^{(m+1)} = 0$. Set $J = [L, I^{(m)}]$. Note that both I and J are ideals of L by Lemma 1.7.1. If $c \in J$, then $c = \sum_i [a_i, b_i]$, where $a_i \in L$ and $b_i \in I^{(m)}$. Since $c \in I^{(m)}$ and $I^{(m)}$ is commutative, we have $[c, b_i] = 0$. Consequently, by Lemma 2.3.2, c is nilpotent. So by Lemma 2.1.3, $J = 0$, and it follows that $I^{(m)} \subset C(L)$. But also $I^{(m)} \subset [L, \text{SR}(L)]$ and hence for $c \in I^{(m)}$, there are $a_i \in L$ and $b_i \in \text{SR}(L)$ such that $c = \sum_i [a_i, b_i]$. Now because $c \in C(L)$, we have $[c, b_i] = 0$. Again by Lemma 2.3.2 it follows that c is nilpotent. Hence Lemma 2.1.3 implies that $I^{(m)} = 0$, and from the assumption that $I \neq 0$ we have derived a contradiction.   □

Corollary 2.3.4 *Let V and L be as in Proposition 2.3.3. Let x be a nilpotent endomorphism of V contained in $\text{SR}(L)$, then $x = 0$.*

Proof. By Proposition 2.3.3, $\mathrm{SR}(L) = C(L)$. Hence x spans an ideal of L. Lemma 2.1.3 now implies that $x = 0$. $\qquad\square$

Corollary 2.3.5 *Let L be a finite-dimensional Lie algebra of characteristic 0. Let R be its solvable radical. Then $[L, R] \subset \mathrm{NR}(L)$.*

Proof. Let $0 = L_0 \subset L_1 \subset \cdots L_{s+1} = L$ be a composition series of L with respect to the adjoint representation. Let $\rho_i : L \to \mathfrak{gl}(L_{i+1}/L_i)$ denote the quotient representation for $0 \le i \le s$. Then $\rho_i(R)$ is a solvable ideal of $\rho_i(L)$. So by Proposition 2.3.3, $\rho_i(R)$ is contained in the centre of $\rho_i(L)$. But this means that $\rho_i([L, R]) = 0$, i.e., $[L, R] \cdot L_{i+1} \subset L_i$. So $\mathrm{ad}_L x$ is nilpotent for all $x \in [L, R]$. In particular $\mathrm{ad}_{[L,R]} x$ is nilpotent for all $x \in [L, R]$. Hence, by Engel's theorem (Theorem 2.1.5) $[L, R]$ is a nilpotent Lie algebra. Now since $[L, R]$ is an ideal of L we conclude that $[L, R] \subset \mathrm{NR}(L)$. $\qquad\square$

We end this section with one more characterization of the nilradical.

Proposition 2.3.6 *Let L be a finite-dimensional Lie algebra of characteristic 0. Let A be the set of all $x \in \mathrm{SR}(L)$ such that $\mathrm{ad}_{\mathrm{SR}(L)} x$ is nilpotent. Let B be the set of all $x \in \mathrm{SR}(L)$ such that $\mathrm{ad}_L x$ is nilpotent. Then $\mathrm{NR}(L) = A = B$.*

Proof. It is straightforward to see that $\mathrm{NR}(L) \subset A$. Also if $x \in A$, then $[x, L] \subset \mathrm{SR}(L)$ and hence $\mathrm{ad}_L x$ is nilpotent. So $A \subset B$. We prove that $B \subset \mathrm{NR}(L)$. Let $0 = L_0 \subset L_1 \subset \cdots \subset L_{s+1} = L$ be a composition series of L with respect to the adjoint representation of L. Let $x \in B$. Let $\sigma_i : L \to \mathfrak{gl}(L_{i+1}/L_i)$ be the quotient representation. Then $\sigma_i(x)$ is nilpotent. Hence by Corollary 2.3.4, $\sigma_i(x) = 0$, i.e., $\mathrm{ad} x(L_{i+1}) \subset L_i$ for $0 \le i \le s$. Now by Proposition 2.2.2, $x \in \mathrm{NR}(L)$. $\qquad\square$

2.4 Lie's theorems

In this section we study the structure of representations of solvable Lie algebras that are "split". It turns out that, in characteristic 0, every such representation is by upper triangular matrices, i.e., a split linear solvable Lie algebra is a subalgebra of $\mathfrak{b}_n(F)$.

Definition 2.4.1 *Let $L \subset \mathfrak{gl}(V)$ be a linear Lie algebra defined over the field F. Then L is called* split *if F contains the eigenvalues of all elements of L.*

Proposition 2.4.2 *Let L be a finite-dimensional solvable Lie algebra over the field F of characteristic 0. Let $\rho : L \to \mathfrak{gl}(V)$ be a finite-dimensional representation of L. Suppose that $\rho(L)$ is split. If V is irreducible and $\dim V > 0$ then V is 1-dimensional.*

Proof. Since L is solvable, the linear Lie algebra $\rho(L)$ is also solvable (Proposition 2.3.1). Furthermore, V is irreducible, so by Proposition 2.3.3, $\rho(L)$ is commutative.

Fix an element $x \in L$. Since $\rho(L)$ is split, there is a $\lambda \in F$ and a vector $v_\lambda \in V$ such that $\rho(x)v_\lambda = \lambda v_\lambda$. Set $W = \{v \in V \mid \rho(x)v = \lambda v\}$, then W is non-zero. Also, if $w \in W$, then for $y \in L$, we have $\rho(x)\rho(y)w = \rho(y)\rho(x)w = \lambda\rho(y)w$. Hence W is invariant under $\rho(L)$ and because V is irreducible, $W = V$. It follows that $\rho(x)$ is λ times the identity on V. So any $x \in L$ acts as multiplication by a scalar on V. Hence any 1-dimensional subspace of V is an L-submodule. So $\dim V = 1$. $\qquad\square$

Corollary 2.4.3 *Let L be a finite-dimensional solvable Lie algebra over the field F of characteristic 0. Let $\rho : L \to \mathfrak{gl}(V)$ be a finite-dimensional representation of L. Suppose $\rho(L)$ is split. Then V contains a common eigenvector for all $\rho(x)$ for $x \in L$.*

Proof. Let $0 = V_0 \subset V_1 \subset \cdots \subset V_n = V$ be a composition series of V relative to the action of L. Then by Proposition 2.4.2, V_1 is 1-dimensional. Hence a basis vector of V_1 will be a common eigenvector. $\qquad\square$

Theorem 2.4.4 (Lie) *Let L be a finite-dimensional solvable Lie algebra of characteristic 0. Let $\rho : L \to \mathfrak{gl}(V)$ be a finite-dimensional representation of L. Suppose that $\rho(L)$ is split. Then there is a basis of V relative to which the matrices of all $\rho(x)$ for $x \in L$ are all upper triangular.*

Proof. Let $0 = V_0 \subset V_1 \subset \cdots \subset V_n = V$ be a composition series of V relative to the action of L. Then the quotient spaces V_i/V_{i-1} are irreducible L-modules. Hence by Proposition 2.4.2, these modules are 1-dimensional. Now let $\{v_1, \ldots, v_n\}$ be a basis of V such that $\bar{v}_i \in V_i/V_{i-1}$ spans V_i/V_{i-1}. Relative to this basis the matrix of $\rho(x)$ is upper triangular for all $x \in L$. $\qquad\square$

Theorem 2.4.5 (Lie) *Let L be an n-dimensional solvable Lie algebra of characteristic 0 such that $\mathrm{ad}_L(L)$ is split. Then there are ideals L_i of L for $0 \leq i \leq n$ such that $\dim L_i = i$ and $0 = L_0 \subset L_1 \subset \cdots \subset L_n = L$.*

Proof. We can take the L_i to be the terms in a composition series of L relative to the adjoint representation. In the same way as in the proof of Theorem 2.4.4 we see that $\dim L_i/L_{i-1} = 1$ and the statement of the theorem follows.   □

2.5   A criterion for solvability

In this section we use the Jordan decomposition of a linear transformation to derive a powerful criterion for a Lie algebra of characteristic 0 to be solvable. Much of the structure theory of semisimple Lie algebras in characteristic 0 is based on it.

Lemma 2.5.1 *Let V be a t-dimensional vector space over the field F. Let $x \in \mathfrak{gl}(V)$ and suppose that there is a basis of V relative to which the matrix of x is diagonal, with entries $d_i \in F$ on the diagonal ($1 \le i \le t$). Then there is a basis of $\mathfrak{gl}(V)$ relative to which $\mathrm{ad}x$ is diagonal with entries $d_i - d_j$ on the diagonal for $1 \le i, j \le t$.*

Proof. Let $\{v_1, \dots, v_t\}$ be a basis of V such that $xv_i = d_i v_i$ for $1 \le i \le t$. Let E_{ij}^t be the endomorphism of V defined by $E_{ij}^t v_k = \delta_{jk} v_i$ (i.e., the matrix of E_{ij}^t has a 1 on position (i, j) and zeros elsewhere). Now $\mathrm{ad}x(E_{ij}^t) = xE_{ij}^t - E_{ij}^t x = (d_i - d_j)E_{ij}^t$ and the lemma follows.   □

Form Appendix A we recall that the Jordan decomposition of a linear transformation a is of the form $a = s + n$, where s is semisimple, n is nilpotent and $[s, n] = 0$.

Lemma 2.5.2 *Let V be a finite-dimensional vector space. Let $a \in \mathfrak{gl}(V)$ and let $a = s + n$ be the Jordan decomposition of a. Then $\mathrm{ad}a = \mathrm{ad}s + \mathrm{ad}n$ is the Jordan decomposition of $\mathrm{ad}a$.*

Proof. We must prove that $\mathrm{ad}s$ is semisimple, $\mathrm{ad}n$ is nilpotent and that $\mathrm{ad}s$ and $\mathrm{ad}n$ commute. The latter follows immediately from the fact that s and n commute (and $[\mathrm{ad}s, \mathrm{ad}n] = \mathrm{ad}[s, n]$). Furthermore, Lemma 2.1.1 states that $\mathrm{ad}n$ is nilpotent. Since s is semisimple, s is diagonalizable. Hence by Lemma 2.5.1 also $\mathrm{ad}s$ is diagonalizable which means that $\mathrm{ad}s$ is semisimple (cf. Proposition A.2.4).   □

Proposition 2.5.3 *Let V be a finite-dimensional vector space over the field F of characteristic 0. Let $M_1 \subset M_2$ be two subspaces of $\mathfrak{gl}(V)$. Set $A = \{x \in \mathfrak{gl}(V) \mid [x, M_2] \subset M_1\}$. Let $x \in A$. If $\mathrm{Tr}(xy) = 0$ for all $y \in A$, then x is nilpotent.*

Proof. We use the fact that x is nilpotent if and only if all its eigenvalues are 0. So let $\lambda_1, \dots, \lambda_n$ be the (not necessarily different) eigenvalues of x. We first prove the statement for the case where F contains all these eigenvalues. Let E be the $\mathbb{Q}$-subspace of F spanned by the λ_i (i.e., $E = \mathbb{Q}\lambda_1 + \cdots + \mathbb{Q}\lambda_n$). We prove that all linear functions $f : E \to \mathbb{Q}$ must be identically 0. From this it follows that $E = 0$ and all eigenvalues of x are 0.

Let $f : E \to \mathbb{Q}$ be an arbitrary linear function. Let $x = s + n$ be the Jordan decomposition of x. Since s is semisimple and $\lambda_1, \dots, \lambda_n$ are also the eigenvalues of s (Proposition A.2.6), there is a basis $\{v_1, \dots, v_n\}$ of V such that $sv_i = \lambda_i v_i$. Let $y \in \mathfrak{gl}(V)$ be the endomorphism defined by $yv_i = f(\lambda_i)v_i$ for $1 \leq i \leq n$ (i.e., the matrix of y is diagonal with $f(\lambda_i)$ on the diagonal). We prove that $\mathrm{ad}y$ can be written as a polynomial in $\mathrm{ad}s$ without constant term.

Let $\{E_{ij}\}$ be the basis of $\mathfrak{gl}(V)$ provided by Lemma 2.5.1. Then we have $\mathrm{ad}s(E_{ij}) = (\lambda_i - \lambda_j)E_{ij}$ and $\mathrm{ad}y(E_{ij}) = (f(\lambda_i) - f(\lambda_j))E_{ij}$. Interpolating we find a polynomial $p \in F[X]$ without constant term and satisfying

$$p(\lambda_i - \lambda_j) = f(\lambda_i - \lambda_j) = f(\lambda_i) - f(\lambda_j) \text{ for } 1 \leq i, j \leq n.$$

Since the matrix of $\mathrm{ad}s$ is diagonal, the matrix of $p(\mathrm{ad}s)$ is also diagonal. Moreover, it has diagonal entries $p(\lambda_i - \lambda_j) = f(\lambda_i) - f(\lambda_j)$. It follows that the matrix of $p(\mathrm{ad}s)$ is exactly the matrix of $\mathrm{ad}y$. Hence $\mathrm{ad}y = p(\mathrm{ad}s)$.

Now since $\mathrm{ad}s$ is the semisimple part of $\mathrm{ad}x$ (Lemma 2.5.2), we have that $\mathrm{ad}s$ is a polynomial in $\mathrm{ad}x$ without constant term (Proposition A.2.6). Therefore, because $\mathrm{ad}x$ maps M_2 into M_1, this also holds for $\mathrm{ad}s$ and since $\mathrm{ad}y$ is a polynomial in $\mathrm{ad}s$ without constant term, also for $\mathrm{ad}y$. Hence $y \in A$. So $0 = \mathrm{Tr}(xy) = \sum_i f(\lambda_i)\lambda_i$. We apply the linear function f to this expression and find $\sum_i f(\lambda_i)^2 = 0$. But $f(\lambda_i) \in \mathbb{Q}$ and hence $f(\lambda_i) = 0$ for $1 \leq i \leq n$. We are done for the case where F contains all eigenvalues of x.

For the general case let $\tilde{F}$ be an extension of F containing all eigenvalues of x. Set $\overline{V} = V \otimes_F \tilde{F}$ and let $\overline{M}_1$ and $\overline{M}_2$ be subspaces of $\mathfrak{gl}(\overline{V})$ spanned by elements $m_1 \otimes 1$ and $m_2 \otimes 1$ for $m_1 \in M_1$ and $m_2 \in M_2$ respectively. Set $\bar{A} = \{x \in \mathfrak{gl}(\overline{V}) \mid [x, \overline{M}_2] \subset \overline{M}_1\}$. Then a basis of $\bar{A}$ is determined by a set of linear equations (analogous to the system of equations determining a basis for the normalizer, Section 1.6). Moreover these equations have coefficients in F and hence $\bar{A} = A \otimes_F \tilde{F}$. It follows that $\mathrm{Tr}((x \otimes 1)\bar{y}) = 0$ for all $\bar{y} \in \bar{A}$. By the proof above it follows that $x \otimes 1$ is nilpotent.   □

Lemma 2.5.4 *Let V be a finite-dimensional vector space. Let $x, y, z \in$ $\mathfrak{gl}(V)$. Then $\mathrm{Tr}([x, y]z) = \mathrm{Tr}(x[y, z])$.*

Proof. We calculate $\mathrm{Tr}([x, y]z) = \mathrm{Tr}(xyz) - \mathrm{Tr}(yxz) = \mathrm{Tr}(xyz) - \mathrm{Tr}(xzy) = \mathrm{Tr}(x[y, z])$.   □

Theorem 2.5.5 (Cartan's criterion for solvability) *Let V be a finite-dimensional vector space of characteristic 0. Let L be a subalgebra of $\mathfrak{gl}(V)$. If $\mathrm{Tr}(xy) = 0$ for all $x \in [L, L]$ and $y \in L$, then L is solvable.*

Proof. Set $A = \{x \in \mathfrak{gl}(V) \mid [x, L] \subset [L, L]\}$. Note that A contains L, and in particular $[L, L]$. Let $u, v \in L$ and $y \in A$. Then according to Lemma 2.5.4,
$$\mathrm{Tr}([u, v]y) = \mathrm{Tr}(u[v, y]) = \mathrm{Tr}([v, y]u).$$
We have $[v, y] \in [L, L]$, so by hypothesis $\mathrm{Tr}([v, y]u) = 0$. Since $[L, L]$ is spanned by elements of the form $[u, v]$ it follows that $\mathrm{Tr}(xy) = 0$ for all $x \in [L, L]$ and $y \in A$. Now we apply Proposition 2.5.3 with $M_1 = [L, L]$ and $M_2 = L$ and conclude that every x in $[L, L]$ is nilpotent. Consequently $\mathrm{ad}x$ is nilpotent for all $x \in [L, L]$ (Lemma 2.1.1). In particular, $\mathrm{ad}_{[L,L]}x$ is nilpotent for all $x \in [L, L]$. Hence by Engel's theorem (Theorem 2.1.5), $[L, L]$ is a nilpotent Lie algebra. In particular $[L, L]$ is solvable (Lemma 1.7.2) and consequently the same is true for L.   □

Corollary 2.5.6 *Let L be a finite-dimensional Lie algebra of characteristic 0, such that $\mathrm{Tr}(\mathrm{ad}x\,\mathrm{ad}y) = 0$ for all $x \in [L, L]$ and $y \in L$. Then L is solvable.*

Proof. We apply Theorem 2.5.5 to the Lie algebra $\mathrm{ad}L \subset \mathfrak{gl}(L)$. It follows that $\mathrm{ad}L$ is solvable. Since the kernel of ad is $C(L)$ which is a solvable ideal, we get that L is solvable (Proposition 2.3.1).   □

2.6   A characterization of the solvable radical

In this section we give a characterization of the solvable radical in characteristic 0. In part this is the converse of Cartan's criterion. This result will enable us to formulate an algorithm for calculating the solvable radical of a Lie algebra of characteristic 0.

Lemma 2.6.1 *Let L be a finite-dimensional Lie algebra of characteristic 0. Let $\rho : L \to \mathfrak{gl}(V)$ be a finite-dimensional irreducible representation of L. Then $\rho([L, L] \cap \mathrm{SR}(L)) = 0$.*

Proof. Set $I = [L, L] \cap \mathrm{SR}(L)$ and let $x \in I$. Then $x = \sum_{i=1}^{m} [y_i, z_i]$ for certain $y_i, z_i \in L$. Hence $\rho(x) = \sum_{i=1}^{m} [\rho(y_i), \rho(z_i)]$. Furthermore, since $x \in \mathrm{SR}(L)$, we know by Proposition 2.3.3, that $\rho(x)$ commutes with $\rho(L)$, in particular $[\rho(x), \rho(z_i)] = 0$. Hence by Lemma 2.3.2, $\rho(x)$ is nilpotent. Now Lemma 2.1.3 finishes the proof. $\square$

Lemma 2.6.2 *Let L be a finite-dimensional Lie algebra of characteristic 0. Let $\rho : L \to \mathfrak{gl}(V)$ be a finite-dimensional representation of L. Then all elements of $\rho([L, L] \cap \mathrm{SR}(L))$ are nilpotent endomorphisms of V.*

Proof. Let $0 = V_0 \subset V_1 \subset \cdots \subset V_{s+1} = V$ be a composition series of V with respect to the action of L. Let $x \in [L, L] \cap \mathrm{SR}(L)$, then by Lemma 2.6.1, the induced actions of x on the quotients V_{i+1}/V_i are all zero. Hence $\rho(x)V_{i+1} \subset V_i$ for $0 \leq i \leq s$. It follows that $\rho(x)$ is nilpotent. $\square$

Corollary 2.6.3 *Let L be a solvable Lie algebra of characteristic 0. Then $[L, L]$ is a nilpotent ideal of L.*

Proof. This follows from Lemma 2.6.2 together with Engel's theorem (Theorem 2.1.5). $\square$

Proposition 2.6.4 *Let L be a finite-dimensional Lie algebra of characteristic 0. Then*

$$\mathrm{SR}(L) = \{x \in L \mid \mathrm{Tr}(\mathrm{ad}x\,\mathrm{ad}y) = 0 \ \text{for all} \ y \in [L, L]\}.$$

Proof. Set $I = \{x \in L \mid \mathrm{Tr}(\mathrm{ad}x\,\mathrm{ad}y) = 0 \ \text{for all} \ y \in [L, L]\}$. Let $0 = L_0 \subset L_1 \subset \cdots \subset L_{s+1} = L$ be a composition series of L relative to the adjoint representation. By Lemma 2.6.2, for all $u \in [L, L] \cap \mathrm{SR}(L)$, the endomorphism $\mathrm{ad}u$ is nilpotent. Hence by Proposition 2.1.4, we see that $\mathrm{ad}u(L_{i+1}) \subset L_i$ for $0 \leq i \leq s$. Consequently, if $x, y \in L$ and $r \in \mathrm{SR}(L)$, then $\mathrm{ad}[y, r](L_{i+1}) \subset L_i$ and hence also $\mathrm{ad}x \cdot \mathrm{ad}[y, r](L_{i+1}) \subset L_i$ for $0 \leq i \leq s$. So $\mathrm{ad}x \cdot \mathrm{ad}[y, r]$ is nilpotent. Using Lemma 2.5.4 we now calculate

$$0 = \mathrm{Tr}(\mathrm{ad}x \cdot \mathrm{ad}[y, r]) = \mathrm{Tr}(\mathrm{ad}[x, y] \cdot \mathrm{ad}r) = \mathrm{Tr}(\mathrm{ad}r \cdot \mathrm{ad}[x, y]).$$

It follows that $SR(L) \subset I$.

For the other inclusion, we first prove that I is an ideal of L. Let $x \in I$ and $y \in [L, L]$ and $z \in L$, then by Lemma 2.5.4,

$$\mathrm{Tr}(\mathrm{ad}[x, z] \cdot \mathrm{ad}y) = \mathrm{Tr}(\mathrm{ad}x \cdot \mathrm{ad}[z, y]) = 0.$$

Hence $[x, z] \in I$ so that I is an ideal of L. Let $x \in I$ and $y \in [I, I]$, then in particular $y \in [L, L]$ and by definition of I, $\mathrm{Tr}(\mathrm{ad}_L x \cdot \mathrm{ad}_L y) = 0$. Now by Cartan's criterion (Theorem 2.5.5) it follows that $\mathrm{ad}_L I$ is solvable. The kernel of $\mathrm{ad}_L : I \to \mathfrak{gl}(L)$ is an Abelian ideal of I. Hence by Proposition 2.3.1, I is solvable which implies $I \subset SR(L)$. □

Corollary 2.6.5 *Let L be a finite-dimensional Lie algebra of characteristic 0. Then L is solvable if and only if $\mathrm{Tr}(\mathrm{ad}x \cdot \mathrm{ad}y) = 0$ for all $x \in L$ and $y \in [L, L]$*

Proof. This immediately follows from Proposition 2.6.4 together with Corollary 2.5.6. □

Let L be a Lie algebra of characteristic 0. Here we use Proposition 2.6.4 to give a simple algorithm for calculating the solvable radical of L.

Let $\{x_1, \ldots, x_n\}$ be a basis of L and let $\{y_1, \ldots, y_s\}$ be a basis of $[L, L]$. Then by Proposition 2.6.4, $x = \sum_i \alpha_i x_i$ is an element of $SR(L)$ if and only if

$$\sum_{i=1}^{n} \mathrm{Tr}(\mathrm{ad}x_i \cdot \mathrm{ad}y_j)\alpha_i = 0 \quad \text{for } 1 \leq j \leq s. \tag{2.1}$$

So we have an algorithm SolvableRadical. The input of this algorithm is a finite-dimensional Lie algebra L of characteristic 0. Using the algorithm ProductSpace we calculate a basis of $[L, L]$. Subsequently we calculate the equations (2.1) and solve them by a Gaussian elimination.

In the case where L is defined over a field of characteristic $p > 0$ the situation is much more difficult. To tackle this case we define a series of ideals $R_k \subset L$ by

$$R_1 = NR(L), \quad R_{k+1}/R_k = NR(L/R_k). \tag{2.2}$$

We claim that R_k is solvable for $k \geq 1$. This is certainly true for $k = 1$. Suppose that R_k is solvable. We have that R_{k+1}/R_k is nilpotent and hence solvable. Therefore, by Proposition 2.3.1, R_{k+1} is solvable. So our claim follows by induction on k. Let u be the integer such that $R_u = R_{u+1}$. Then $\mathrm{NR}(L/R_u) = 0$. This implies that $\mathrm{SR}(L/R_u) = 0$. Indeed, set $I = \mathrm{SR}(L/R_u)$. If $I \neq 0$, then there is an $m > 0$ such that $I^{(m)} \neq 0$ and $I^{(m+1)} = 0$ (where $I^{(m)}$ denotes the m-th term of the derived series of I). Hence $I^{(m)}$ is a commutative ideal of $\mathrm{SR}(L/R_u)$. Therefore it is contained in $\mathrm{NR}(L/R_u)$, implying that $I^{(m)} = 0$; and we derived a contradiction. Consequently, $R_u = \mathrm{SR}(L)$.

So by using an algorithm for calculating the nilradical we find an algorithm SolvableRadical for calculating the solvable radical of a Lie algebra, that also works over the fields of characteristic $p > 0$ for which there are algorithms for calculating the nilradical.

2.7   Finding a non-nilpotent element

Let L be a Lie algebra. An element $x \in L$ is said to be *nilpotent* if the endomorphism $\mathrm{ad}x$ is nilpotent. By Engel's theorem (Theorem 2.1.5), L contains non-nilpotent elements if and only if L is not a nilpotent Lie algebra. In this section we give an algorithm for finding a non-nilpotent element in L, if L is not nilpotent. It is based on the following two propositions.

Proposition 2.7.1 *Let K be a proper subalgebra of L and suppose that $\mathrm{ad}_L y$ is nilpotent for all $y \in K$. Let x be an element of $N_L(K) \setminus K$ and let $\overline{K}$ be the subalgebra spanned by K together with x. Then either $\mathrm{ad}_L x$ is not nilpotent or $\mathrm{ad}_L u$ is nilpotent for all $u \in \overline{K}$.*

Proof. First we note that $N_L(K)$ is strictly larger than K because $\mathrm{ad}_L y$ is a nilpotent linear transformation for $y \in K$. Indeed, since K is a subalgebra of L, we have that the adjoint representation induces a representation of K on L/K. By Proposition 2.1.2, we see that the elements of K have a common eigenvector (with eigenvalue 0) in L/K. Any pre-image of this vector lies in $N_L(K)$ (but not in K).

Suppose that $\mathrm{ad}_L x$ is nilpotent. Let $y \in K$ and set $u = x + y$. We prove that $\mathrm{ad}_L u$ is nilpotent. First we note that $\mathrm{ad}_L : \overline{K} \to \mathfrak{gl}(L)$ is a representation of $\overline{K}$. Let $0 = L_0 \subset L_1 \subset \cdots \subset L_{s+1} = L$ be a composition series of L with respect to the action of $\overline{K}$. Since $x \in N_L(K)$, we have that K is an ideal in $\overline{K}$. Hence, by Proposition 2.1.4, $\mathrm{ad}_L y(L_{i+1}) \subset L_i$ for $0 \leq i \leq s$. So $(\mathrm{ad}_L u)^m L_{i+1} \equiv (\mathrm{ad}_L x)^m L_{i+1} \bmod L_i$. Since $\mathrm{ad}_L x$ is nilpotent it follows that there is an $m > 0$ such that $(\mathrm{ad}_L x)^m = 0$. Hence $(\mathrm{ad}_L u)^m (L_{i+1}) \subset L_i$ for $0 \leq i \leq s$. The conclusion is that $\mathrm{ad}_L u$ is nilpotent. $\qquad\square$

If we start with $K = 0$ and repeatedly apply Proposition 2.7.1 then we either find a non-nilpotent element, or after $\dim L$ steps we have that $K = L$, implying that L is nilpotent. An element $x \in N_L(K) \setminus K$ can be found by calculating $N_L(K)$. Alternatively, we can construct a sequence of elements in the following way. First we fix a basis of K. Let x be an element of L not lying in K. If for some basis element y of K we have that $[x, y] \notin K$, then replace x by $[x, y]$. Since K acts nilpotently on L we need no more than $\dim L - 1$ such replacement operations to obtain an element lying in $N_L(K) \setminus K$.

Proposition 2.7.1 yields an algorithm for finding a non-nilpotent element in L. However, if L is defined over a field of characteristic 0, then there is a much simpler method available.

Proposition 2.7.2 *Let L be a non-nilpotent Lie algebra over a field of characteristic 0 with basis $\{x_1, \dots, x_n\}$, then the set*

$$\{x_1, \dots, x_n\} \cup \{x_i + x_j \mid 1 \leq i < j \leq n\}$$

contains a non-nilpotent element.

Proof. If L is solvable but not nilpotent then by Proposition 2.3.6 we see that the nilradical of L is the set of all nilpotent elements of L. Hence there must be a basis element x_i such that x_i is not nilpotent. On the other hand, if L is not solvable, then there exist basis elements x_i and x_j for such that $\mathrm{Tr}(\mathrm{ad}x_i \cdot \mathrm{ad}x_j) \neq 0$. (Otherwise $\mathrm{Tr}(\mathrm{ad}x \cdot \mathrm{ad}y) = 0$ for all $x, y \in L$, implying that L is solvable (Corollary 2.5.6)). From

$$\mathrm{Tr}((\mathrm{ad}x_i + \mathrm{ad}x_j)^2) - \mathrm{Tr}((\mathrm{ad}x_i)^2) - \mathrm{Tr}((\mathrm{ad}x_j)^2) =$$
$$= \mathrm{Tr}(\mathrm{ad}x_i \cdot \mathrm{ad}x_j) + \mathrm{Tr}(\mathrm{ad}x_j \cdot \mathrm{ad}x_i) = 2\mathrm{Tr}(\mathrm{ad}x_i \cdot \mathrm{ad}x_j) \neq 0$$

we infer that the elements x_i, x_j and $x_i + x_j$ cannot be all nilpotent. $\square$

Propositions 2.7.1 and 2.7.2 lead to the following algorithm.
Algorithm NonNilpotentElement
Input: a finite-dimensional Lie algebra L.
Output: a non-nilpotent element of L, or 0 if L is nilpotent.

Step 1 If L is of characteristic 0, then go to Step 2, else go to Step 4.

Step 2 Let $\{x_1, \dots, x_n\}$ be a basis of L and set $A = \{x_1, \dots, x_n\} \cup \{x_i + x_j \mid 1 \leq i < j \leq n\}$.

Step 3 For elements $x \in A$ test whether $\text{ad}_L x$ is nilpotent, until a non-nilpotent element is found, or A is exhausted. If A contains a non-nilpotent element then return this element, otherwise return 0.

Step 4 Set $K = 0$.

Step 5 Find an element $x \in N_L(K) \setminus K$. If $\text{ad}_L x$ is not nilpotent, then return x.

Step 6 Replace K by the subalgebra spanned by K and x. If $\dim K = \dim L$, then return 0. Otherwise go to Step 5.

2.8   Notes

The algorithm for calculating the solvable radical of a Lie algebra of characteristic 0 is described in [7]. In [7], Proposition 2.2.2 together with Lie's theorem (Theorem 2.4.5) are used to derive an algorithm for calculating the nilradical of a solvable Lie algebra L of characteristic 0, such that $\text{ad} L$ is split. It is based on finding a common eigenvector of the elements of $\text{ad} L$. By using the algorithm for the solvable radical (in characteristic 0), this yields an algorithm for calculating the nilradical of a finite dimensional Lie algebra of characteristic 0 having a split solvable radical.

The algorithm for calculating the nilradical given in Section 2.2 and the algorithm for calculating the solvable radical of a Lie algebra defined over a field of characteristic $p > 0$ are taken from [46], [73].

Our description of the algorithm for finding a non-nilpotent element follows [37]. In [7] a different (more complicated) method is described.

Chapter 3

Cartan subalgebras

The adjoint representation of a Lie algebra L (see Example 1.12.2), encodes its multiplicative structure (modulo the centre). This allows us to investigate the structure of L by investigating its adjoint representation. In this way the tools of linear algebra (matrices, eigenvalues et cetera) become available to us. We have used this successfully in Chapter 2, where most results on the structure of solvable and nilpotent Lie algebras where obtained by looking at the adjoint representation. As an example we mention Lie's theorem (Theorem 2.4.5).

In this chapter we restrict the adjoint representation of a Lie algebra L to particular subalgebras of it. Furthermore, if K is a subalgebra of L, then we decompose L as a direct sum of K-submodules. Of particular interest are those subalgebras K that yield a so-called primary decomposition. In Section 3.1 we first study linear Lie algebras acting on a vector space V. We show that a nilpotent linear Lie algebra yields a primary decomposition of V. Then we consider the restriction of the adjoint representation of a Lie algebra L to a nilpotent subalgebra K. This gives us a primary decomposition of L relative to K. We also describe a second decomposition of a vector space relative to the action of a nilpotent linear Lie algebra, namely the Fitting decomposition.

In Section 3.2 we introduce Cartan subalgebras; these are nilpotent subalgebras that yield a particularly interesting primary decomposition. We show that Cartan subalgebras exist and we give algorithms for calculating a Cartan subalgebra.

The subject of Section 3.3 is the primary decomposition of L relative to a "split" Cartan subalgebra. In this case the primary decomposition is called the root space decomposition. We show how the root space decomposition encodes part of the multiplicative structure of L.

Cartan subalgebras are in general not unique, i.e., a Lie algebra usually has more than one Cartan subalgebra. If L is defined over an algebraically closed field of characteristic 0, then this non-uniqueness does not bother us too much since in this case all Cartan subalgebras of L are conjugate under the automorphism group of L. The proof of this forms the subject of Sections 3.4, 3.5. In Section 3.6 we show that for the case where the Lie algebra is solvable we may drop the assumption that the ground field is algebraically closed.

Finally in Section 3.7 we apply the theory of Cartan subalgebras and Fitting decompositions to obtain two algorithms for calculating the nilradical.

3.1 Primary decompositions

In this section V will be a finite-dimensional vector space over a field F, and $K \subset \mathfrak{gl}(V)$ a linear Lie algebra.

Definition 3.1.1 *A decomposition*

$$V = V_1 \oplus \cdots \oplus V_s$$

of V into K-submodules V_i is said to be primary *if the minimum polynomial of the restriction of x to V_i is a power of an irreducible polynomial for all $x \in K$ and $1 \leq i \leq s$. The subspaces V_i are called* primary components.

In general V will not have a primary decomposition relative to K. Using Lemma A.2.2 we derive a sufficient condition on K to yield a primary decomposition of V. From Appendix A we recall that for a polynomial $p \in F[X]$ and $x \in K$ the space $V_0(p(x))$ is defined by

$$V_0(p(x)) = \{ v \in V \mid p(x)^r(v) = 0 \text{ for some } r > 0 \}.$$

Proposition 3.1.2 *Suppose that for all $x \in K$ and any polynomial $p \in F[X]$ the space $V_0(p(x))$ is invariant under K. Then V has a primary decomposition with respect to K.*

Proof. The proof is by induction on the dimension of V. If every element of K has a minimum polynomial that is a power of an irreducible polynomial, then there is nothing to be proved. Otherwise there is an $x \in K$ such that the minimum polynomial of x has at least two distinct factors. We apply Lemma A.2.2 to V and the linear transformation x. It is seen that V decomposes as

$$V = V_0(p_1(x)) \oplus \cdots \oplus V_0(p_s(x))$$

where the p_i are the irreducible factors of the minimum polynomial of x. By assumption the subspaces $V_0(p_i(x))$ are invariant under K. Furthermore, the dimension of these subspaces is strictly less than $\dim V$. Hence by induction they all admit a primary decomposition relative to K, and by summing these we get a primary decomposition of V with respect to K. $\square$

Corollary 3.1.3 *Suppose that K is Abelian. Then V admits a primary decomposition relative to K.*

Proof. Let $x, y \in K$ and let $p \in F[X]$ be a polynomial. Let $v \in V_0(p(x))$, then there is an $m > 0$ such that $p(x)^m v = 0$. Since $x \cdot y = y \cdot x$ we have $p(x)^m yv = yp(x)^m v = 0$. As a consequence the subspace $V_0(p(x))$ is invariant under y and the result follows by Proposition 3.1.2. $\square$

It turns out that we can generalize Corollary 3.1.3 to the case where K is nilpotent. To see this we need some technical lemmas. For $a \in \mathrm{End}(V)$ we define a linear map $d_a : \mathrm{End}(V) \to \mathrm{End}(V)$ by $d_a(b) = ab - ba$.

Lemma 3.1.4 *Let $a \in \mathrm{End}(V)$. Let $f = \sum_{k=0}^n \alpha_k X^k$ be a univariate polynomial. Then for all $b \in \mathrm{End}(V)$,*

$$f(a)b = \sum_{i=0}^n d_a^i(b)\left(\sum_{k=i}^n \alpha_k \binom{k}{i} a^{k-i} \right).$$

Proof. For $k \geq 0$ we prove the identity

$$a^k b = \sum_{i=0}^k \binom{k}{i} d_a^i(b) a^{k-i}, \tag{3.1}$$

by induction on k. For $k = 0, 1$, (3.1) is easily verified. Let $k \geq 1$. Assuming that (3.1) holds for k and for all $b \in \mathrm{End}(V)$, we calculate

$$a^{k+1} b = a^k a b = a^k (ba + d_a(b))$$

$$= \left(\sum_{i=0}^k \binom{k}{i} d_a^i(b) a^{k-i} \right) a + \sum_{i=0}^k \binom{k}{i} d_a^{i+1}(b) a^{k-i}$$

$$= \sum_{i=0}^{k+1} \binom{k+1}{i} d_a^i(b) a^{k+1-i}.$$

The result now follows by a second calculation:

$$f(a)b = \sum_{k=0}^{n} \alpha_k a^k b = \sum_{k=0}^{n} \sum_{i=0}^{k} \binom{k}{i} \alpha_k d_a^i(b) a^{k-i}$$

$$= \sum_{i=0}^{n} d_a^i(b) \left(\sum_{k=i}^{n} \alpha_k \binom{k}{i} a^{k-i} \right).$$

□

For $r \geq 0$ let δ_r be the linear mapping on the polynomial ring $F[X]$ defined by $\delta_r(X^m) = \binom{m}{r} X^{m-r}$ (where we set $\binom{m}{r} = 0$ if $r > m$). If $f = \sum_{k=0}^{n} \alpha_k X^k$ is a polynomial, then

$$\delta_r(f) = \sum_{k=r}^{n} \alpha_k \binom{k}{r} X^{k-r}.$$

So the conclusion of Lemma 3.1.4 reads $f(a)b = \sum_{i=0}^{n} d_a^i(b) \delta_i(f)(a)$.

Lemma 3.1.5 *Let f, g be univariate polynomials, then we have $\delta_r(fg) = \sum_{i=0}^{r} \delta_i(f) \delta_{r-i}(g)$.*

Proof. We first prove the statement for the case where $f = X^m$ and $g = X^n$:

$$\sum_{i=0}^{r} \delta_i(f) \delta_{r-i}(g) = \sum_{i=0}^{r} \binom{m}{i} \binom{n}{r-i} X^{m-i} X^{n-r+i}$$

$$= X^{m+n-r} \sum_{i=0}^{r} \binom{m}{i} \binom{n}{r-i} = \binom{n+m}{r} X^{m+n-r} = \delta_r(X^{m+n}).$$

The next to last equality follows from

$$\sum_{i=0}^{r} \binom{m}{i} \binom{n}{r-i} = \binom{n+m}{r},$$

which is proved by equating the coefficients of x^r in the expression $(x+1)^m (x+1)^n = (x+1)^{m+n}$. Now the general result follows from the linearity of the maps δ_i.   □

Lemma 3.1.6 *Let h, p be univariate polynomials such that h^{m+1} divides p for an $m \geq 0$. Then h divides $\delta_r(p)$ for $0 \leq r \leq m$.*

Proof. The proof is by induction on m. The statement is clearly true for $m = 0$ since $\delta_0(p) = p$. For the induction step write $p = qh$ where h^m divides q. Then we may assume that h divides $\delta_i(q)$ for $0 \le i \le m - 1$ and h divides $\delta_j(p)$ for $0 \le j \le m - 1$. Then by Lemma 3.1.5,

$$\delta_m(p) = \delta_m(qh) = \sum_{i=0}^{m} \delta_i(q)\delta_{m-i}(h)$$

and hence h also divides $\delta_m(p)$. □

Proposition 3.1.7 *Let $a, b \in End(V)$ and suppose that $d_a^{n+1}(b) = 0$. Let p be a polynomial, then $V_0(p(a))$ is invariant under b.*

Proof. Let $v \in V_0(p(a))$, then there is an $r > 0$ such that $p^r(a)v = 0$. Now set $q = (p^r)^{n+1}$. Then by Lemma 3.1.4,

$$q(a)bv = \sum_{i=0}^{d} d_a^i(b)\delta_i(q)(a)v = \sum_{i=0}^{n} d_a^i(b)\delta_i(q)(a)v,$$

where d is the degree of q. By Lemma 3.1.6 we have that p^r divides $\delta_i(q)$ for $0 \le i \le n$. Hence $q(a)bv = 0$ and $bv \in V_0(p(a))$. □

Corollary 3.1.8 *Suppose that K is nilpotent. Then V has a primary decomposition with respect to K.*

Proof. Let c be the length of the lower central series of K (i.e., $K^{c+1} = 0$). Let $x, y \in K$, then $[x, [x, [x, \cdots [x, y] \cdots]]] = 0$ (c factors x). This is the same as saying that $d_x^c(y) = 0$, and Proposition 3.1.7 applies to x and y. The statement now follows by Proposition 3.1.2. □

It is easily seen that a primary decomposition is in general not unique. Let $K \subset \mathfrak{gl}(V)$ be a 1-dimensional Lie algebra spanned by the element x. Suppose that in a primary decomposition of V relative to K there occurs a component V_i such that the restriction of x relative to V_i is the identity. Suppose further that $\dim V_i > 1$. Let $V_i = W_1 \oplus W_2$ be any decomposition of V_i into a direct sum of subspaces. Then by replacing V_i by the two components W_1 and W_2 we obtain a different primary decomposition of V relative to K. To avoid situations like this we "collect" the components in a primary decomposition. The resulting primary decomposition turns out to be unique.

Definition 3.1.9 *A primary decomposition of V relative to K is called collected if for any two primary components V_i and V_j ($i \neq j$), there is an $x \in K$ such that the minimum polynomials of the restrictions of x to V_i and V_j are powers of different irreducible polynomials.*

Theorem 3.1.10 *Let K be nilpotent. Then V has a unique collected primary decomposition relative to K.*

Proof. It is easy to see that V has a collected primary decomposition with respect to K. Indeed, let

$$V = V_1 \oplus \cdots \oplus V_s \qquad (3.2)$$

be a primary decomposition of V with respect to K. If there are two components V_i and V_j such that for all $x \in K$, the minimum polynomials of $x|_{V_i}$ and $x|_{V_j}$ are powers of the same irreducible polynomial, then replace V_i and V_j in the decomposition by their direct sum and obtain a primary decomposition with one component less. Continuing this process we obtain a collected primary decomposition.

Now suppose that (3.2) is collected. For $x \in K$ and $1 \leq i \leq s$ define $p_{x,i}$ to be the irreducible polynomial such that the minimum polynomial of x restricted to V_i is a power of $p_{x,i}$. We claim that

$$V_i = \{v \in V \mid \text{for all } x \in K \text{ there is an } m > 0 \text{ such that } p_{x,i}(x)^m(v) = 0\}. \qquad (3.3)$$

First of all, V_i is certainly contained in the right-hand side. To see the other inclusion, let $v \in V$ be an element of the right-hand side of (3.3). Write $v = v_1 + \cdots + v_s$ where $v_j \in V_j$. Fix a $j \neq i$ between 1 and s and choose $x \in K$ such that $p_{x,i} \neq p_{x,j}$ (such x exist because (3.2) is collected). Then there is an $m > 0$ such that

$$0 = p_{x,i}(x)^m v = p_{x,i}(x)^m v_1 + \cdots + p_{x,i}(x)^m v_s.$$

It follows that $p_{x,i}(x)^m v_j = 0$. But also $p_{x,j}(x)^n v_j = 0$, for some $n > 0$. Now because $p_{x,i}$ and $p_{x,j}$ are relatively prime we have $v_j = 0$. So $v = v_i \in V_i$.

Now suppose that there is a second collected primary decomposition,

$$V = W_1 \oplus \cdots \oplus W_t.$$

Let $1 \leq i \leq t$ be such that W_i does not occur among the components V_j. For $x \in K$, define q_x to be the irreducible polynomial such that the minimum polynomial of the restriction of x to W_i is a power of q_x. Let $v \in W_i$ and write $v = v_1 + \cdots + v_s$, where $v_j \in V_j$. Fix a j between 1 and s, and let $x \in K$ be such that $q_x \neq p_{x,j}$ (such an x exists because otherwise $q_x = p_{x,j}$ for all $x \in K$ and by (3.3), $W_i = V_j$). Then there is an $m > 0$ such that

$$0 = q_x(x)^m(v) = q_x(x)^m(v_1) + \cdots + q_x(x)^m(v_s)$$

and because the V_k are invariant under x and the sum (3.2) is direct we infer that $q_x(x)^m v_j = 0$. But also $p_{x,j}(x)^n v_j = 0$. Now because q_x and $p_{x,j}$ are relatively prime it follows that $v_j = 0$. If we vary j between 1 and s, then it is seen that $v = 0$ so that $W_i = 0$. $\qquad\square$

Let $V = V_1 \oplus \cdots \oplus V_s$ be the collected primary decomposition of V with respect to K. Let X be an indeterminate. Let $x \in K$. If the minimum polynomial of the restriction of x to V_i is a power of X, then x acts nilpotently on V_i. On the other hand, if this minimum polynomial is a power of any other irreducible polynomial, then x is non-singular on V_i, i.e., $x \cdot V_i = V_i$. As the primary decomposition is collected there is at most one component V_i such that all elements $x \in K$ act nilpotently on V_i. We denote this component by $V_0(K)$. Furthermore, if we let $V_1(K)$ be the sum of the remaining primary components, then $V = V_0(K) \oplus V_1(K)$ and by (3.3),

$$V_0(K) = \{v \in V \mid \text{ for all } x \in K \text{ there is an } m > 0 \text{ such that } x^m \cdot v = 0\}.$$

And also $K \cdot V_1(K) = V_1(K)$. The components $V_0(K)$ and $V_1(K)$ are called the Fitting-null and -one component respectively. The decomposition $V = V_0(K) \oplus V_1(K)$ is called the *Fitting decomposition* of V with respect to K.

Now we change the setting a little bit. Let L be a finite-dimensional Lie algebra over the field F and let $K \subset L$ be a nilpotent subalgebra of L. By restricting the adjoint representation of L to K we get a representation of K:

$$\mathrm{ad}_L : K \longrightarrow \mathfrak{gl}(L).$$

By Theorem 3.1.10 L has a unique collected primary decomposition relative to K. Also L has a Fitting decomposition relative to K, which reads $L = L_0(K) \oplus L_1(K)$, where

$$L_0(K) = \{y \in L \mid \text{ for all } x \in K \text{ there is a } t > 0 \text{ such that } (\mathrm{ad}x)^t(y) = 0\},$$

and $[K, L_1(K)] = L_1(K)$. These two decompositions of L relative to K will be important tools for investigating the structure of L.

We end this section by giving an algorithm for calculating the Fitting-one component of a Lie algebra L relative to a nilpotent subalgebra K.

For a subalgebra K of L we write

$$[K^m, L] = [K, [K, \cdots [K, L] \cdots]] \ (m \text{ factors } K).$$

Then $[K^{m+1}, L] \subset [K^m, L]$ so that the subspaces $[K^m, L]$ form a decreasing series.

Lemma 3.1.11 *Let L be a finite-dimensional Lie algebra and let $K \subset L$ be a nilpotent subalgebra. Let $L = L_0(K) \oplus L_1(K)$ be the Fitting decomposition of L with respect to K. Let $m \geq 1$ be such that $[K^m, L] = [K^{m+1}, L]$, then $L_1(K) = [K^m, L]$.*

Proof. Since $[K, L_1(K)] = L_1(K)$, we have

$$[K^r, L] = [K^r, L_0(K)] + [K^r, L_1(K)] = [K^r, L_0(K)] + L_1(K).$$

From the fact that $\mathrm{ad}x$ acts nilpotently on $L_0(K)$ for all $x \in K$ we have that the series of subspaces $[K^r, L_0(K)]$ form a decreasing series, ending in 0 (Proposition 2.1.4). Hence $[K^m, L] = [K^{m+1}, L]$ implies $[K^m, L] = L_1(K)$. $\square$

Lemma 3.1.11 yields an easy algorithm FittingOneComponent for calculating the Fitting-one component of a Lie algebra L with respect to a nilpotent subalgebra K. It calculates the decreasing series of subspaces $[K^m, L]$ (by calls to ProductSpace). When the point is reached where $[K^m, L] = [K^{m+1}, L]$, the space $[K^m, L]$ is returned.

3.2 Cartan subalgebras

Throughout this section L will be a finite-dimensional Lie algebra over a field F. Let K be a nilpotent subalgebra of L. Suppose that $\mathrm{ad}_L x$ is a nilpotent linear transformation for all $x \in K$. Then we have that $L_0(K) = L$ and the primary decomposition is not very revealing. We want to avoid this situation as much as possible. Therefore we try to find subalgebras K such that $L_0(K)$ is a small as possible. Such subalgebras are called *Cartan subalgebras*.

Definition 3.2.1 *A nilpotent subalgebra H of L is called a* Cartan subalgebra *if $L_0(H) = H$.*

The next lemma provides a convenient tool for proving that a certain nilpotent subalgebra is a Cartan subalgebra.

Lemma 3.2.2 *Let H be a nilpotent subalgebra of L. Then H is a Cartan subalgebra if and only if $N_L(H) = H$.*

Proof. Suppose that H is a Cartan subalgebra. Let $x \in N_L(H)$. Let $L = H \oplus L_1(H)$ be the Fitting decomposition of L with respect to H. Write $x = h + y$ where $h \in H$ and $y \in L_1(H)$. As $x \in N_L(H)$ we have $[x, h'] \in H$ for all $h' \in H$. But $[x, h'] = [h, h'] + [y, h']$, so that $[y, h'] \in H \cap L_1(H)$, which means that $[y, h'] = 0$. Since this holds for all $h' \in H$ we must have $y = 0$. It follows that $N_L(H) = H$.

Now suppose that $N_L(H) = H$. We consider the (adjoint) action of H on the space $L_0(H)$. Since H is nilpotent, $L_0(H)$ contains H. Because H is stable under H we have that H acts on the quotient space $L_0(H)/H$. Also for all $h \in H$ the restriction of $\operatorname{ad} h$ to $L_0(H)$ is a nilpotent transformation. Hence also the induced action of $h \in H$ on $L_0(H)/H$ is nilpotent. Suppose that $L_0(H)$ is strictly larger than H. Then by Proposition 2.1.2, it follows that $L_0(H)/H$ contains a non-zero element $\bar{x}$ mapped to 0 by all elements of H. Let x be a pre-image of $\bar{x}$. Then $[x, H] \subset H$ and hence $x \in H$. But that means that $\bar{x} = 0$ and therefore we have $L_0(H) = H$. $\square$

Proposition 3.2.3 *Let L be a Lie algebra defined over the field F. Let $\tilde{F}$ be an extension field of F. If $H \subset L$ is a Cartan subalgebra of L, then $\tilde{H} = H \otimes_F \tilde{F}$ is a Cartan subalgebra of $\tilde{L} = L \otimes_F \tilde{F}$.*

Proof. As seen in Section 1.14, $\tilde{H}$ is nilpotent. Let $\{x_1, \ldots, x_n\}$ be a basis of L and $\{h_1, \ldots, h_l\}$ a basis of H. Then $\{x_1 \otimes 1, \ldots, x_n \otimes 1\}$ and $\{h_1 \otimes 1, \ldots, h_l \otimes 1\}$ are bases of $\tilde{L}$ and $\tilde{H}$ respectively. Now if we provide these bases as input to the algorithm **Normalizer**, then exactly the same equation system will be solved as when we input the bases $\{x_1, \ldots, x_n\}$ and $\{h_1, \ldots, h_l\}$ of L and H. Hence the solution space is defined over F and we have that $N_{\tilde{L}}(\tilde{H}) = N_L(H) \otimes_F \tilde{F} = \tilde{H}$. Now by Lemma 3.2.2, $\tilde{H}$ is a Cartan subalgebra of $\tilde{L}$. $\square$

From Definition 3.2.1 it is not clear whether an arbitrary Lie algebra has a Cartan subalgebra. Here we prove that Lie algebras defined over a big field and restricted Lie algebras over a field of characteristic $p > 0$ possess Cartan subalgebras. The proofs yield algorithms for calculating a Cartan subalgebra.

Lemma 3.2.4 *Let K be a nilpotent subalgebra of L. Let $x \in L_0(K)$, then the primary components in the collected primary decomposition of L relative to K are invariant under $\operatorname{ad} x$.*

Proof. Let $h \in K$. Then $[h, [h, \cdots [h, x] \cdots]] = 0$ and hence by Proposition 3.1.7, we have that the primary components of L relative to $\mathrm{ad}h$ are invariant under $\mathrm{ad}x$. By (3.3) the primary components of L relative to K are intersections of the primary components of L relative to $\mathrm{ad}h$ for $h \in K$. It follows that the primary components of L relative to K are invariant under $\mathrm{ad}x$. $\qquad\qquad\square$

Proposition 3.2.5 *Let K be a nilpotent subalgebra of L. Then*

1. $L_0(K)$ *is a subalgebra of L,*

2. $[L_0(K), L_1(K)] \subset L_1(K)$,

3. $N_L(L_0(K)) = L_0(K)$.

Proof. We have that $L_0(K)$ is a primary component and $L_1(K)$ is a sum of primary components. The first two statements now follow from Lemma 3.2.4. Let $x \in L$ lie in $N_L(L_0(K))$. Write $x = x_0 + x_1$, where $x_0 \in L_0(K)$ and $x_1 \in L_1(K)$. Then for $y \in L_0(K)$ we have $[x, y] = [x_0, y] + [x_1, y]$ so that $[x_1, y] \in L_0(K)$. However, by statement 2., $[x_1, y] \in L_1(K)$ and hence $[x_1, y] = 0$. So $[x_1, L_0(K)] = 0$ and in particular $[x_1, K] = 0$ implying that $x_1 \in L_0(K)$. It follows that $x_1 = 0$ and $x \in L_0(K)$. $\qquad\qquad\square$

Proposition 3.2.6 *Let Ω be a subset of the field F of size at least $\dim L + 1$. Let $x \in L$ and set $A = L_0(\mathrm{ad}x)$. Suppose that there is a $y \in A$ such that $\mathrm{ad}_A y$ is not a nilpotent linear transformation. Then there is a $c_0 \in \Omega$ such that $L_0(\mathrm{ad}(x + c_0(y - x)))$ is properly contained in A.*

Proof. Let $L = A \oplus L_1(\mathrm{ad}x)$ be the Fitting decomposition of L relative to the subalgebra spanned by x. Then by Proposition 3.2.5, the transformations $\mathrm{ad}x$ and $\mathrm{ad}y$ both stabilize A and $L_1(\mathrm{ad}x)$, hence so does $\mathrm{ad}(x + c(y - x))$ for all $c \in F$. Let T be an indeterminate and let $f(T)$ be the characteristic polynomial of $\mathrm{ad}(x + c(y - x))$. Then $f(T) = g(T)h(T)$ where g is the characteristic polynomial of the restriction of $\mathrm{ad}(x + c(y - x))$ to A and h the characteristic polynomial of the restriction of $\mathrm{ad}(x + c(y - x))$ to $L_1(\mathrm{ad}x)$. Furthermore

$$g(T) = T^d + g_1(c)T^{d-1} + \cdots + g_d(c)$$

and

$$h(T) = T^e + h_1(c)T^{e-1} + \cdots + h_e(c),$$

where g_i and h_i are polynomials in c. Also, if $g_i \neq 0$ then $\deg g_i = i$ and likewise for h_i. Now because $\operatorname{ad}_A y$ is not nilpotent, there is an i such that $g_i(1) \neq 0$. And since $A = L_0(\operatorname{ad} x)$, we have that $h_e(0) \neq 0$. In particular g_i and h_e are not the zero polynomial. Since $\deg g_i h_e = i + e \leq d + e = \dim L$, there is a $c_0 \in \Omega$ such that $g_i h_e(c_0) \neq 0$. From $h_e(c_0) \neq 0$ it follows that $L_0(\operatorname{ad}(x + c_0(y - x)))$ is contained in A. From $g_i(c_0) \neq 0$ it follows that this containment is proper. □

Definition 3.2.7 *An element $x \in L$ is called* regular *if the dimension of $L_0(\operatorname{ad} x)$ is minimal. If $x \in L$ is regular, then* $\dim L_0(\operatorname{ad} x)$ *is called the* rank *of L.*

Corollary 3.2.8 *Suppose that L is defined over a field of size at least $\dim L + 1$. Let $x \in L$ be a regular element, then $L_0(\operatorname{ad} x)$ is a Cartan subalgebra.*

Proof. Set $H = L_0(\operatorname{ad} x)$. Then by Proposition 3.2.5, H is a subalgebra of L and $N_L(H) = H$. Also if H is not nilpotent, then by Engel's theorem (Theorem 2.1.5), there is an element $h \in H$ such that $\operatorname{ad}_H h$ is not nilpotent. Hence, by Proposition 3.2.6, there is a $c_0 \in F$ such that $\dim L_0(\operatorname{ad}(x + c_0(h - x))) < \dim L_0(\operatorname{ad} x)$. But since x is regular, this is not possible. So H is nilpotent, and by Lemma 3.2.2 it is a Cartan subalgebra. □

Using Proposition 3.2.6 we formulate an algorithm for calculating a Cartan subalgebra of a Lie algebra defined over a big field (i.e., a field of size at least $\dim L + 1$).

Algorithm CartanSubalgebraBigField
Input: a finite-dimensional Lie algebra L, and a subset $\Omega \subset F$ of size at least $\dim L + 1$.
Output: a Cartan subalgebra of L.

Step 1 If L is nilpotent, then return L.

Step 2 $x := $ NonNilpotentElement(L);

Step 3 If $L_0(\operatorname{ad} x)$ is nilpotent then return $L_0(\operatorname{ad} x)$.

Step 4 Let $y := $ NonNilpotentElement$(L_0(\operatorname{ad} x))$. Find a $c_0 \in \Omega$ such that the dimension of $L_0(\operatorname{ad}(x + c_0(y - x)))$ is strictly less than $\dim L_0(\operatorname{ad} x)$. Set $x := x + c_0(y - x)$ and go to Step 3.

Comments: This algorithm terminates because dim $L_0(\mathrm{ad}x)$ decreases every round of the iteration. Also by 3. of Proposition 3.2.5 and Lemma 3.2.2, the subalgebra that is returned is a Cartan subalgebra. We note that every step is computable. This is clear for Steps 1 and 2. Calculating (a basis of) a subalgebra of the form $L_0(\mathrm{ad}u)$ requires solving a system of linear equations. In Step 4, after at most dim $L + 1$ computations of the dimension of a subalgebra of the form $L_0(\mathrm{ad}(x+c_0(y-x)))$, we find a $c_0 \in \Omega$ such that dim $L_0(\mathrm{ad}(x+c_0(y-x)))$ is $<$ dim $L_0(\mathrm{ad}x)$ (cf. Proposition 3.2.6).

Proposition 3.2.9 *Let L be a restricted Lie algebra over the field F of characteristic $p > 0$. Let $x \mapsto x^p$ be a fixed p-th power mapping. Let K be a nilpotent subalgebra of L. Then $L_0(K)$ is closed under the p-th power mapping. Furthermore, every Cartan subalgebra of $L_0(K)$ is a Cartan subalgebra of L.*

Proof. Set $A = L_0(K)$ and let $x, y \in A$. Then $[x^p, y] = (\mathrm{ad}x)^p(y)$ which lies in A by 1. of Proposition 3.2.5. Hence $x^p \in N_L(A) = A$ by 3. of Proposition 3.2.5. It follows that A is closed under the p-th power mapping of L.

Let H be a Cartan subalgebra of A. Then H is nilpotent and $A_0(H) = H$. We have to prove that $L_0(H) = H$. Let $x \in K$ and choose an integer m such that $(\mathrm{ad}x)^{p^m} A = 0$ (such an m exists because x acts nilpotently on A). Set $y = x^{p^m}$ then, as seen above, $y \in A$. Furthermore, $[y, A] = 0$ and in particular $[y, H] = 0$. So because H is a Cartan subalgebra of A, we have $y \in H$. Hence $L_0(H) \subset L_0(\mathrm{ad}y) \subset L_0(\mathrm{ad}x)$ and as a consequence

$$L_0(H) \subset \bigcap_{x \in K} L_0(\mathrm{ad}x) = L_0(K) = A.$$

And it follows that $L_0(H) = A_0(H) = H$, which is what we wished to prove. $\square$

The previous result yields an algorithm for computing a Cartan subalgebra of a restricted Lie algebra of characteristic $p > 0$.

Algorithm CartanSubalgebraRestricted
Input: a restricted finite-dimensional Lie algebra L of characteristic $p > 0$.
Output: a Cartan subalgebra of L.

Step 1 Set $K := L$.

Step 2 If K is nilpotent then return K.

Step 3 $x := \mathsf{NonNilpotentElement}(K)$. Replace K by $K_0(\mathrm{ad}_K x)$ and return
   to Step 2.

Comments: This algorithm terminates because the quantity $\dim K$
decreases every round of the iteration. Furthermore, the replacement of Step
3 is justified by Proposition 3.2.9. We note that the p-th power mapping is
not needed in the algorithm. It is only needed in the proof of Proposition
3.2.9.

Corollary 3.2.10 *If L is defined over a field of size at least $\dim L + 1$, or
L is a restricted Lie algebra of characteristic $p > 0$, then L has a Cartan
subalgebra.*

Remark. We remark that a Lie algebra L in general has more than one
Cartan subalgebra. Indeed, let L be the 3-dimensional Lie algebra $\mathfrak{sl}_2(\mathbb{Q})$
with basis x, y, h and multiplication table

$$[h, x] = 2x, \ [h, y] = -2y, \ [x, y] = h,$$

(see Example 1.3.6). Then the subalgebras of L spanned by h and $x + y$
respectively are both Cartan subalgebras of L.

3.3 The root space decomposition

Let L be a Lie algebra over the field F. A Cartan subalgebra H of L is
said to be *split* if F contains the eigenvalues of the $\mathrm{ad}_L h$ for all $h \in H$ (see
also Definition 2.4.1). In this section we suppose that L has a split Cartan
subalgebra and we let H be such a split Cartan subalgebra.
   Let

$$L = H \oplus L_1 \oplus \cdots \oplus L_s$$

be the (collected) primary decomposition of L with respect to H. Let $h \in
H$, then the minimum polynomial of the restriction of $\mathrm{ad}h$ to a primary
component L_i is a power of an irreducible polynomial. Since H is split,
this irreducible polynomial is of the form $X - \alpha_i(h)$, where $\alpha_i(h)$ is a scalar
depending on i and h. By fixing the primary component L_i, we get a
function $\alpha_i : H \to F$. This function is called a *root* (because the $\alpha_i(h)$ are
roots of the characteristic polynomial of $\mathrm{ad}h$). The corresponding primary
component L_i is called a *root space*. In the sequel it will be convenient to

index a root space by the corresponding root, i.e.,

$$L_{\alpha_i} = L_i =$$
$$\{x \in L \mid \text{for all } h \in H \text{ there is a } k > 0 \text{ such that } (\mathrm{ad}h - \alpha_i(h))^k(x) = 0\}.$$

The primary decomposition

$$L = H \oplus L_{\alpha_1} \oplus \cdots \oplus L_{\alpha_s}$$

is called the *root space decomposition* of L. We note that $H = L_0$, the primary component corresponding to the function $\alpha_0 : H \to F$ given by $\alpha_0(h) = 0$ for all $h \in H$. However, usually α_0 is not called a root. We let $\Phi = \{\alpha_1, \ldots, \alpha_s\}$ be the set of non-zero roots. The next result shows how part of the multiplicative structure of L is encoded in relations satisfied by the roots.

Proposition 3.3.1 *Let* $\alpha, \beta \in \Phi$ *and let* $x \in L_\alpha$ *and* $y \in L_\beta$. *Then* $[x, y] \in L_{\alpha+\beta}$ *if* $\alpha + \beta \in \Phi$ *and* $[x, y] \in H$ *if* $\alpha + \beta = 0$. *In all other cases* $[x, y] = 0$.

Proof. Let $h \in H$, then $\mathrm{ad}h$ is a derivation of L. A straightforward induction on m (cf. (4.4) in Chapter 4) establishes

$$(\mathrm{ad}h - (\alpha(h) + \beta(h)))^m([x, y]) =$$
$$\sum_{i=0}^{m} \binom{m}{i} [(\mathrm{ad}h - \alpha(h))^{m-i}(x), (\mathrm{ad}h - \beta(h))^i(y)].$$

And for m big enough this is 0. So if $\alpha + \beta \in \Phi$, then $[x, y] \in L_{\alpha+\beta}$. Also if $\alpha + \beta = 0$, then $[x, y] \in L_0 = H$. In all other cases $\mathrm{ad}h - (\alpha(h) + \beta(h))$ is nonsingular and it follows that $[x, y] = 0$. $\qquad\square$

Example 3.3.2 Let L be the 8-dimensional Lie algebra over $\mathbb{Q}$ with basis $\{x_1, \ldots, x_8\}$ and multiplication table as shown in Table 3.1.

We compute a Cartan subalgebra H of L and the corresponding root space decomposition. First, $x = x_1$ is a non-nilpotent element and $L_0(\mathrm{ad}x)$ is spanned by $\{x_1, x_4, x_6, x_7\}$. In this subalgebra $y = x_4$ is not nilpotent, so according to Step 4 of the algorithm CartanSubalgebraBigField we have to find a $c_0 \in \mathbb{Q}$ such that the dimension of $L_0(\mathrm{ad}(x + c_0(y - x)))$ is smaller than $\dim L_0(\mathrm{ad}x)$. It is easily seen that $c_0 = 1$ does not work, so we try $c_0 = 2$ which does the job. The subalgebra $L_0(\mathrm{ad}(2y - x))$ is spanned by $\{x_1, x_4\}$ and is nilpotent (even Abelian). So we have found a Cartan subalgebra.

	x_1	x_2	x_3	x_4	x_5	x_6	x_7	x_8
x_1	0	$2x_2$	$-2x_3$	0	$-x_5$	0	0	x_8
x_2	$-2x_2$	0	x_1	0	x_8	0	0	0
x_3	$2x_3$	$-x_1$	0	0	0	0	0	x_5
x_4	0	0	0	0	$-x_5$	$-2x_6$	$-x_6 - 2x_7$	$-x_8$
x_5	x_5	$-x_8$	0	x_5	0	0	0	$-x_6$
x_6	0	0	0	$2x_6$	0	0	0	0
x_7	0	0	0	$x_6 + 2x_7$	0	0	0	0
x_8	$-x_8$	0	$-x_5$	x_8	x_6	0	0	0

Table 3.1: Multiplication table of a 8-dimensional Lie algebra.

Set $h_1 = x_1$ and $h_2 = x_4$ and let H be the Cartan subalgebra spanned by h_1, h_2. The matrices of $\mathrm{ad}h_1$ and $\mathrm{ad}h_2$ are easily computed and it is seen that the primary decomposition of L is

$$L = H \oplus L_{\alpha_1} = \langle x_2 \rangle \oplus L_{\alpha_2} = \langle x_3 \rangle \oplus L_{\alpha_3} = \langle x_5 \rangle \oplus L_{\alpha_4} = \langle x_6, x_7 \rangle \oplus L_{\alpha_5} = \langle x_8 \rangle,$$

(where $\langle v_1, v_2, \dots, v_k \rangle$ denotes the space spanned by $v_1, \dots, v_k$). Now we turn our attention towards the roots α_i. Set $h = \lambda_1 h_1 + \lambda_2 h_2$. Then the matrix of the restriction of $\mathrm{ad}h$ to L_{α_4} is

$$\begin{pmatrix} -2\lambda_2 & -\lambda_2 \\ 0 & -2\lambda_2 \end{pmatrix}.$$

The minimum polynomial of this matrix is $(X + 2\lambda_2)^2$. Hence $\alpha_4(h) = -2\lambda_2$. In similar fashion we determine the other roots; we have

$$\alpha_1(h) = 2\lambda_1, \ \alpha_2(h) = -2\lambda_1, \ \alpha_3(h) = -\lambda_1 - \lambda_2, \ \alpha_4(h) = -2\lambda_2,$$
$$\alpha_5(h) = \lambda_1 - \lambda_2.$$

We see that $\alpha_3 + \alpha_5 = \alpha_4$. This corresponds to the commutation relation $[x_5, x_8] = -x_6 \in L_{\alpha_4}$.

We have seen that part of the multiplication table of a Lie algebra L is encoded in relations satisfied by the roots. Moreover, as will be seen in Chapters 4 and 5, in the special case of semisimple Lie algebras, the set of roots determines the entire Lie algebra structure. However, in general L has many Cartan subalgebras and the root space decomposition of L relative to one Cartan subalgebra differs from the root space decomposition of L relative to another Cartan subalgebra. So the question presents itself as to which Cartan subalgebras of L are best for our purpose of investigating the structure of L. Fortunately, in the case where L is defined over an algebraically closed field of characteristic 0 all Cartan subalgebras are conjugate under the automorphism group of L. This means that the root space decomposition of L relative to a Cartan subalgebra H_1 is mapped onto the root space decomposition of L relative to a different Cartan subalgebra H_2 by an automorphism. So in this case it does not matter which Cartan subalgebra we take. The proof of this result is the subject of the next sections.

3.4   Polynomial functions

Let V be a finite-dimensional vector space over the field F with basis $\{v_1, \dots, v_n\}$. Let $p : V \to V$ be a map defined by

$$p\Big(\sum_{i=1}^{n} \lambda_i v_i\Big) = \sum_{i=1}^{n} p_i(\lambda_1, \dots, \lambda_n) v_i \qquad (3.4)$$

where $p_i \in F[X_1, \dots, X_n]$ are polynomials. A map of this form is called a *polynomial map*.

Let $p : V \to V$ be a polynomial map defined by (3.4) and let $v = \lambda_1 v_1 + \cdots + \lambda_n v_n \in V$. Then we define the *differential* of p at v to be the linear map $d_v p : V \to V$ given by

$$d_v p\Big(\sum_{i=1}^{n} \mu_i v_i\Big) = \sum_{i=1}^{n} \Big(\sum_{j=1}^{n} \Big(\frac{\partial p_i}{\partial X_j}\Big)\Big|_{(X_k = \lambda_k)} \mu_j\Big) v_i.$$

Example 3.4.1 Let V be a two-dimensional vector space over $\mathbb{Q}$ with basis $\{v_1, v_2\}$. Then the map $p : V \to V$ defined by

$$p(\lambda_1 v_1 + \lambda_2 v_2) = (\lambda_1 \lambda_2 + \lambda_2^2) v_1 + (\lambda_1^2 + \lambda_2^2) v_2$$

is a polynomial map. Let $v = \lambda_1 v_1 + \lambda_2 v_2$ be an element of V. Then for $w = \mu_1 v_1 + \mu_2 v_2$ we have

$$d_v p(w) = (\lambda_2 \mu_1 + (\lambda_1 + 2\lambda_2)\mu_2) v_1 + (2\lambda_1 \mu_1 + 2\lambda_2 \mu_2) v_2.$$

Lemma 3.4.2 *Let $p : V \to V$ be a polynomial map. Let t be an indeterminate. Then $p(v + tw) = p(v) + t d_v p(w) \pmod{t^2}$.*

Proof. Let w be given by $w = \sum_{j=1}^{n} \mu_j v_j$. Let $f : V \to F$ be a polynomial function from V to F (i.e., $f(\sum \lambda_i v_i) = p_f(\lambda_1, \dots, \lambda_n)$ for a $p_f \in F[X_1, \dots, X_n]$). Then by Taylor's theorem in n variables (see, e.g., [30]), we have

$$f(v + tw) = f(v) + t \sum_{j=1}^{n} \frac{\partial f}{\partial X_j}(v) \mu_j \pmod{t^2}.$$

The lemma is a straightforward consequence of this.   □

A subset $S \subset V$ is said to be *open* if there is a polynomial f in the ring $F[X_1, \dots, X_n]$ such that $S = \{v \in V \mid f(v) \neq 0\}$.

Theorem 3.4.3 *Let V be a vector space defined over an algebraically closed field F. Let $p : V \to V$ be a polynomial map. Suppose that there is a $v \in V$ such that $d_v p$ is a surjective linear map. Let $S \subset V$ be an open subset of V. Then the image $p(S)$ contains an open subset of V.*

The proof of this theorem belongs to algebraic geometry; it is beyond the scope of this book. For a proof we refer the reader to, e.g., [48], or [14].

3.5 Conjugacy of Cartan subalgebras

Let L be a Lie algebra. From Section 1.11 we recall that $\exp \operatorname{ad} x$ is an automorphism of L if $x \in L$ is such that $\operatorname{ad} x$ is nilpotent. The group generated by these automorphisms is called the inner automorphism group of L and denoted by $\operatorname{Int}(L)$.

In the rest of this section we assume that the ground field F is algebraically closed and of characteristic 0. Under this hypothesis we prove that all Cartan subalgebras are conjugate under $\operatorname{Int}(L)$.

Let H be a Cartan subalgebra of L and let

$$L = H \oplus L_{\alpha_1} \oplus \cdots \oplus L_{\alpha_s}$$

be the root space decomposition of L with respect to H. (Note that H is split as F is algebraically closed.) Set $L_0 = H$. Let $x \in L_{\alpha_i}$ and $y \in L_{\alpha_j}$ then by Proposition 3.3.1 we see that $(\operatorname{ad} x)^k(y) \in L_{k\alpha_i + \alpha_j}$ if $k\alpha_i + \alpha_j$ is a root or if it is 0, and otherwise $(\operatorname{ad} x)^k(y) = 0$. So since there are only a finite number of roots we see that $(\operatorname{ad} x)^k(y) = 0$ for some $k > 0$. Also by a similar argument we have that $(\operatorname{ad} x)^m(h) = 0$ for $h \in H$ and an $m > 0$. It follows that $\operatorname{ad} x$ is nilpotent and hence $\exp \operatorname{ad} x \in \operatorname{Int}(L)$. We let $\operatorname{E}(H)$ denote the subgroup of $\operatorname{Int}(L)$ generated by $\exp \operatorname{ad} x$ for $x \in L_{\alpha_i}$, where i ranges from 1 to s.

Now let $\{h_1, \ldots, h_l, x_{l+1}, \cdots, x_n\}$ be a basis of L such that $h_1, \ldots, h_l$ form a basis of H and $x_{l+1}, \ldots, x_n$ span the root spaces. Let $x = \sum_{i=1}^{l} \mu_i h_i + \sum_{i=l+1}^{n} \lambda_i x_i$ be an element of L. Then we define a function $p_H : L \to L$ by

$$p_H(x) = \exp(\lambda_{l+1}\mathrm{ad}x_{l+1}) \cdots \exp(\lambda_n\mathrm{ad}x_n)\left(\sum_{i=1}^{l} \mu_i h_i\right).$$

(Note that for fixed $\lambda_{l+1}, \ldots, \lambda_n$ this is an element of $\mathrm{E}(H)$ acting on an element of H.) It is clear that p_H is a polynomial function. We calculate its differential at the point $h_0 \in H$. For this let $h = \sum_{i=1}^{l} \mu_i h_i$ and $u = \sum_{i=l+1}^{n} \lambda_i x_i$. Let t be an indeterminate. Then

$$
\begin{aligned}
p_H(h_0 + t(h + u)) &= \exp(\lambda_{l+1}t\mathrm{ad}x_{l+1}) \cdots \exp(\lambda_n t\mathrm{ad}x_n)(h_0 + th) \\
&= (1 + t\lambda_{l+1}\mathrm{ad}x_{l+1}) \cdots (1 + t\lambda_n\mathrm{ad}x_n)(h_0 + th) \quad (\mathrm{mod}\ t^2) \\
&= h_0 + th + t \sum_{i=l+1}^{n} \lambda_i\mathrm{ad}x_i(h_0) \quad (\mathrm{mod}\ t^2) \\
&= h_0 + t(h + \mathrm{ad}u(h_0)) \quad (\mathrm{mod}\ t^2).
\end{aligned}
$$

So by Lemma 3.4.2 we see that the differential $d_{h_0}p_H$ is the linear map $h + u \mapsto h + [u, h_0]$. (Note that $P_H(h_0) = h_0$.)

A root $\alpha_i : H \to F$ is a non-zero polynomial function. Hence also the product $f = \alpha_1 \cdots \alpha_s$ is a non-zero polynomial function. So there are $h \in H$ such that $f(h) \neq 0$. By H_{reg} we denote the set of all $h \in H$ such that $f(h) \neq 0$. Let $L_1(H) = L_{\alpha_1} \oplus \cdots \oplus L_{\alpha_s}$ be the Fitting-one component of L relative to H. We define a polynomial function $\tilde{f} : L \to F$ by $\tilde{f}(h+u) = f(h)$ for $h \in H$ and $u \in L_1(H)$. Then

$$O_H = \{h + u \mid h \in H_{\mathrm{reg}} \text{ and } u \in L_1(H)\}$$

is the open set in L corresponding to $\tilde{f}$ (where "open" is defined as in Section 3.4). If $h_0 \in H_{\mathrm{reg}}$ then $\alpha_i(h_0) \neq 0$ for $1 \leq i \leq s$ so that $H = L_0(\mathrm{ad}h_0)$. Therefore the restriction of $\mathrm{ad}h_0$ to $L_1(H)$ is non-singular. So by the above calculation we see that the differential $d_{h_0}p_H : L \mapsto L$ is surjective. Consequently, by Theorem 3.4.3, the image set $p_H(O_H)$ contains an open set. Now let H' be a second Cartan subalgebra. Then we define $p_{H'}$ using the root space decomposition of L relative to H'. And in the same way we see that $p_{H'}(O_{H'})$ contains an open set. But two open sets always have a non-empty intersection. This means that there are $g \in \mathrm{E}(H)$, $h \in H_{\mathrm{reg}}$, $g' \in \mathrm{E}(H')$ and $h' \in H'_{\mathrm{reg}}$ such that $g(h) = g'(h')$. But then

$$g(H) = g(L_0(\mathrm{ad}h)) = L_0(\mathrm{ad}g(h)) = L_0(\mathrm{ad}g'(h')) = g'(L_0(\mathrm{ad}h')) = g'(H').$$

Hence the element $g'^{-1}g \in \mathrm{Int}(L)$ maps H to H'. This means that we have proved the following theorem.

Theorem 3.5.1 *Let L be a finite-dimensional Lie algebra over an algebraically closed field of characteristic 0. Let H, H' be two Cartan subalgebras of L. Then there is an element $g \in \mathrm{Int}(L)$ such that $g(H) = H'$.*

Proposition 3.5.2 *Let L be a finite-dimensional Lie algebra over a field of characteristic 0 (not necessarily algebraically closed). Then any Cartan subalgebra H of L is of the form $L_0(\mathrm{ad}x)$ where $x \in L$ is a regular element.*

Proof. Let $\{x_1, \ldots, x_n\}$ be a basis of L. Let $F(Y_1, \ldots, Y_n)$ be the field of rational functions in n indeterminates. Put $x = \sum_{i=1}^{n} Y_i x_i$ which is an element of $L \otimes_F F(Y_1, \ldots, Y_n)$. Then the characteristic polynomial of $\mathrm{ad}x$ is of the form

$$\det(T - \mathrm{ad}x) = T^n + f_1(Y_1, \ldots, Y_n)T^{n-1} + \cdots + f_{n-l}(Y_1, \ldots, Y_n)T^l,$$

where the f_i are polynomials in the $Y_1, \ldots, Y_n$. We see that an element $\sum \mu_i x_i \in L$ is regular if and only if $f_{n-l}(\mu_1, \ldots, \mu_n) \neq 0$. So since F is infinite we have that L contains regular elements.

Suppose first that F is algebraically closed. Let $y \in L$ be regular. By Corollary 3.2.8 we see that $L_0(\mathrm{ad}y)$ is a Cartan subalgebra of L. By Theorem 3.5.1 there is an element $g \in \mathrm{Int}(L)$ such that $g(L_0(\mathrm{ad}y))$ equals the given Cartan subalgebra H. Hence $H = L_0(\mathrm{ad}g(y))$ and $g(y)$ is a regular element.

Now we drop the assumption that F is algebraically closed, and we show that H contains regular elements. Let $\{h_1, \ldots, h_l\}$ be a basis of H. Let $Y_1, \ldots, Y_l$ be l indeterminates over F. We consider the element $h = \sum_{i=1}^{l} Y_i h_i$ which lies in $L \otimes_F F(Y_1, \ldots, Y_l)$. Then the characteristic polynomial of $\mathrm{ad}h$ is

$$\det(T - \mathrm{ad}h) = T^n + g_1(Y_1, \ldots, Y_l)T^{n-1} + \cdots + g_{n-l}(Y_1, \ldots, Y_l)T^l.$$

It follows that an element $\sum_{i=1}^{l} \alpha_i h_i \in H$ is a regular element if and only if $g_{n-l}(\alpha_1, \ldots, \alpha_l) \neq 0$. By the first part, the Cartan subalgebra H contains regular elements over the algebraic closure of F. Hence g_{n-l} is not the zero polynomial. So, since F is infinite, there are $\alpha_1, \ldots, \alpha_l \in F$ such that $g_{n-l}(\alpha_1, \ldots, \alpha_l) \neq 0$ and H contains regular elements. $\qquad \square$

Example 3.5.3 Let L be the Lie algebra $\mathfrak{sl}_2(\mathbb{Q})$ of Example 1.3.6. This Lie algebra has basis x, y, h and multiplication table

$$[h, x] = 2x, \ [h, y] = -2y, \ [x, y] = h.$$

A Cartan subalgebra of L is spanned by h. Denote this Cartan subalgebra by H. Then there are two root spaces relative to H, spanned by x and y respectively. We have $(\mathrm{ad}x)^3 = (\mathrm{ad}y)^3 = 0$ and

$$p_H(\alpha x + \beta y + \gamma h) = (\exp \alpha \mathrm{ad}x)(\exp \beta \mathrm{ad}y)(\gamma h)$$

$$= \gamma(1 + \alpha \mathrm{ad}x + \frac{1}{2}\alpha^2(\mathrm{ad}x)^2)(1 + \beta \mathrm{ad}y + \frac{1}{2}\beta^2(\mathrm{ad}y)^2)(h)$$

$$= -\gamma(2\alpha^2\beta + 2\alpha)x + 2\gamma\beta y + \gamma(1 + 2\alpha\beta)h.$$

Now a second Cartan subalgebra H' is spanned by $x + y$. The two root spaces relative to H' are spanned by $x' = x - y - h$ and $y' = x - y + h$. If we set $h' = x + y$ then

$$[h', x'] = 2x', \ [h', y'] = -2y', \ [x', y'] = -4h'.$$

Also

$$p_{H'}(\alpha'x' + \beta'y' + \gamma'h') =$$

$$\gamma'(1 + \alpha'\mathrm{ad}x' + \frac{1}{2}\alpha'^2(\mathrm{ad}x')^2)(1 + \beta'\mathrm{ad}y' + \frac{1}{2}\beta'^2(\mathrm{ad}y')^2)(h') =$$

$$\gamma'(8\alpha'^2\beta' - 2\alpha')x' + 2\gamma'\beta'y' + \gamma'(1 - 8\alpha'\beta')h'.$$

To find an automorphism mapping H onto H' we need to find an element in the intersection of the images of p_H and $p_{H'}$. For example we can set $\alpha' = \beta' = 0$ and $\gamma' = 1$, then $p_{H'}(\alpha'x' + \beta'y' + \gamma'h') = h'$. Now we have to find α, β, γ such that $p_H(\alpha x + \beta y + \gamma h) = h'$. This is equivalent to

$$-\gamma(2\alpha^2\beta + 2\alpha) = 2\gamma\beta = 1 \text{ and } \gamma(1 + 2\alpha\beta) = 0.$$

If we set $\gamma = 1$, then we see that $\beta = \frac{1}{2}$, and $\alpha = -1$. It follows that

$$\exp(-\mathrm{ad}x)\exp(\frac{1}{2}\mathrm{ad}y)(H) = H'.$$

3.6 Conjugacy of Cartan subalgebras of solvable Lie algebras

As seen in the last section, we need the ground field to be algebraically closed and of characteristic 0 in order to have that all Cartan subalgebras are conjugate. However, in an important special case, where the Lie algebra is solvable, we can drop the assumption that the ground field be algebraically closed. First we need some lemmas that are of general nature.

Lemma 3.6.1 *Let L be a finite-dimensional Lie algebra. Let K be a subalgebra of L containing a subalgebra of the form $L_0(\operatorname{ad} x)$ for an $x \in L$. Then $N_L(K) = K$.*

Proof. We consider the vector space $V = N_L(K)/K$. Since $x \in K$ we have that $\operatorname{ad}_L x$ stabilizes K and $N_L(K)$; hence $\operatorname{ad}_L x$ induces a linear map $a_x : V \to V$. The characteristic polynomial of $\operatorname{ad}_{N_L(K)} x$ is the product of the characteristic polynomials of $\operatorname{ad}_K x$ and a_x. So because $L_0(\operatorname{ad} x) \subset K$ we see that a_x has no eigenvalue 0 on V. Also $x \in K$ implies that a_x maps $N_L(K)$ into K, i.e., $a_x(V) = 0$. It follows that $V = 0$ and we are done. □

Lemma 3.6.2 *Let L_1, L_2 be finite-dimensional Lie algebras over a field of characteristic 0. Let $\phi : L_1 \to L_2$ be a surjective homomorphism. If H is a Cartan subalgebra of L_1, then $\phi(H)$ is a Cartan subalgebra of L_2.*

Proof. Let H^k be the k-th term of the lower central series of H, and similarly for $\phi(H)^k$. Then $\phi(H)^k = \phi(H^k) = 0$ for k large enough. Hence $\phi(H)$ is nilpotent. Now suppose $[\phi(x), \phi(H)] \subset \phi(H)$, for an $x \in L_1$; then $[x, H] \subset H + \ker \phi$. Hence $[x, H + \ker \phi] \subset H + \ker \phi$, i.e., $x \in N_{L_1}(H + \ker \phi)$. Now since the ground field is of characteristic 0, the Cartan subalgebra H is of the form $L_0(\operatorname{ad} u)$ for some $u \in L_1$ (Proposition 3.5.2). So by Lemma 3.6.1 we see that $x \in H + \ker \phi$. And consequently $\phi(x) \in \phi(H)$. The conclusion is that $N_{L_2}(\phi(H)) = \phi(H)$ and by Lemma 3.2.2 $\phi(H)$ is a Cartan subalgebra of L_2. □

Lemma 3.6.3 *Let L_1, L_2 be finite-dimensional Lie algebras over a field of characteristic 0. Let $\phi : L_1 \to L_2$ be a surjective homomorphism. Let H_2 be a Cartan subalgebra of L_2 and set $K = \phi^{-1}(H_2)$. Then every Cartan subalgebra of K is also a Cartan subalgebra of L_1.*

Proof. Let H_1 be a Cartan subalgebra of K. Then H_1 is a nilpotent Lie algebra. By Lemma 3.6.2, $\phi(H_1)$ is a Cartan subalgebra of $\phi(K) = H_2$. Hence $\phi(H_1) = H_2$. Suppose that $[x, H_1] \subset H_1$ for an $x \in L_1$. Then $[\phi(x), H_2] \subset H_2$ and $\phi(x) \in H_2$. It follows that $x \in K$. But since H_1 is a Cartan subalgebra of K we must have $x \in H_1$; so $N_{L_1}(H_1) = H_1$. By

Lemma 3.2.2, H_1 is a Cartan subalgebra of L_1. $\square$

Now let L be a solvable Lie algebra over a field of characteristic 0. Let $x \in [L, L]$, then by Lemma 2.6.2 we see that $\mathrm{ad}_L x$ is nilpotent. Hence $\exp \mathrm{ad} x$ is an element of the inner automorphism group $\mathrm{Int}(L)$. By $\mathrm{D}(L)$ we denote the subgroup of $\mathrm{Int}(L)$ generated by all elements $\exp \mathrm{ad} x$ for $x \in [L, L]$.

Theorem 3.6.4 *Let L be a solvable Lie algebra of characteristic 0. Let H, H' be two Cartan subalgebras of L. Then there is a $g \in \mathrm{D}(L)$ such that $g(H') = H$.*

Proof. The proof is by induction on $\dim L$. The case $\dim L = 0$ is trivial. Also we may assume that L is not nilpotent. Then L has non-trivial proper commutative ideals (e.g., the last non-zero term of the derived series). Choose a non-trivial commutative ideal I of L of minimal dimension. Then by Lemma 3.6.2, $(H + I)/I$ and $(H' + I)/I$ are Cartan subalgebras of L/I. Hence by induction there is a $\bar{g} \in \mathrm{D}(L/I)$ such that $\bar{g}((H' + I)/I) = (H + I)/I$. Let $\pi : L \to L/I$ be the projection map. Then π induces a group homomorphism

$$\phi : \mathrm{D}(L) \longrightarrow \mathrm{D}(L/I)$$

defined by $\phi(\exp \mathrm{ad} x) = \exp(\mathrm{ad} \pi(x))$. If $x \in [L, L]$, then $\pi(x) \in [L/I, L/I]$ so ϕ is well defined. Also because π is surjective we have the same for ϕ. It follows that there is a $g_1 \in \mathrm{D}(L)$ such that $\phi(g_1) = \bar{g}$. Set $H_1 = g_1(H')$. We prove that H_1 and H are conjugate under $\mathrm{D}(L)$.

First we note that H_1 is a Cartan subalgebra of L and $H_1 + I = H + I$. So H_1 and H are Cartan subalgebras of $H + I$. If $H + I$ is properly contained in L, then by induction there is a $g_2 \in \mathrm{D}(H + I)$ such that $g_2(H_1) = H$. Since $H + I$ is a subalgebra of L the group $\mathrm{D}(H + I)$ can be viewed as a subgroup of $\mathrm{D}(L)$ (this follows from the fact that $[H + I, H + I]$ is a subalgebra of $[L, L]$). Hence g_2 lies in $\mathrm{D}(L)$ and we are done in this case.

Now suppose that $L = H + I$. Let $u \in H$ be a regular element of L (such u exists by Proposition 3.5.2). Let

$$L = H \oplus L_1(H)$$

be the Fitting decomposition of L with respect to H. Then since u is regular, $\mathrm{ad} u : L_1(H) \to L_1(H)$ is non-singular. Let $x \in L_1(H)$, then because $L = H + I$ we can write $x = h + a$ where $h \in H$ and $a \in I$. Let $k > 0$ be an integer such that $(\mathrm{ad} u)^k(H) = 0$; then $(\mathrm{ad} u)^k(x) = (\mathrm{ad} u)^k(a)$

which lies in I since I is an ideal. So $(\operatorname{ad}u)^k(L_1(H)) \subset I$ and because also $(\operatorname{ad}u)^k$ is non-singular on $L_1(H)$ we see that $L_1(H) \subset I$. In particular $[L_1(H), L_1(H)] = 0$. Therefore

$$[L, L_1(H)] = [H + L_1(H), L_1(H)] \subset L_1(H).$$

It follows that $L_1(H)$ is a commutative ideal of L contained in I. Because I is minimal we must have $L_1(H) = 0$ or $L_1(H) = I$. In the first case $L = H$ and L is nilpotent, which was excluded.

On the other hand, if $L_1(H) = I$, then $L = H_1 + I = H_1 + L_1(H)$. So we can write $u = h' + y$ where $h' \in H_1$ and $y \in L_1(H)$. Also we can write $y = [z, u]$ for some $z \in L_1(H)$. Since $[u, L_1(H)] = L_1(H)$ we have that $L_1(H) \subset [L, L]$ and $g_3 = \exp \operatorname{ad}z \in D(L)$. Furthermore, since $(\operatorname{ad}z)^2 = 0$ we have that $g_3 = 1 + \operatorname{ad}z$. From the fact that $L_1(H)$ is commutative we get that $[z, y] = 0$. Hence

$$g_3(h') = g_3(u - y) = u - y + [z, u - y] = u.$$

The conclusion is that $g_3(H_1)$ is a Cartan subalgebra of L having a regular element in common with H. Hence $g_3(H_1) = H$. □

3.7   Calculating the nilradical

In Section 2.2 we described an algorithm for calculating the nilradical of Lie algebras defined over a field of characteristic 0 or over a finite field. However this algorithm calculates a basis of the associative algebra $(\operatorname{ad}L)^*$. The dimension of that algebra may be substantially bigger than the dimension of L. Therefore it is desirable to have an algorithm that works inside the Lie algebra L. In this section we describe an algorithm of this kind. We also describe a second algorithm that mostly works inside the Lie algebra (it computes the radical of an associative algebra that is much smaller than $(\operatorname{ad}L)^*$). However, they both only work over fields of characteristic 0.

Lemma 3.7.1 *Let L be a solvable Lie algebra over the field F of characteristic 0. Let d be a derivation of L. Then $d(\operatorname{NR}(L)) \subset \operatorname{NR}(L)$.*

Proof. Let Fx denote the 1-dimensional vector space over F with basis x. We consider the direct sum $K = Fx \oplus L$, which is made into a Lie algebra by setting $[\alpha x + y_1, \beta x + y_2] = \alpha d(y_2) - \beta d(y_1) + [y_1, y_2]$ for $y_1, y_2 \in L$. This is a special case of the construction of a semidirect product in Section 1.10. Then L is an ideal in K and K/L is commutative.

Hence K is solvable. Therefore by Corollary 2.6.3, $[K, K]$ is a nilpotent ideal of K. And since $[K, K] \subset L$ it is also a nilpotent ideal of L, i.e., $[K, K] \subset \mathrm{NR}(L)$. But $[K, K] = [L, L] + d(L)$. Hence $d(L) \subset \mathrm{NR}(L)$, and in particular $d(\mathrm{NR}(L)) \subset \mathrm{NR}(L)$. $\qquad\square$

Proposition 3.7.2 *Let L be a solvable Lie algebra of characteristic 0. Let I be an ideal of L and let M be the ideal of L (containing I) such that $\mathrm{NR}(L/I) = M/I$. Then $\mathrm{NR}(L) = \mathrm{NR}(M)$.*

Proof. First we have that $\mathrm{NR}(M)$ is a nilpotent Lie algebra. Furthermore, since M is an ideal of L we have that $\mathrm{ad}x \in \mathrm{Der}(M)$ for all $x \in L$. Hence by Lemma 3.7.1, $[x, \mathrm{NR}(M)] \subset \mathrm{NR}(M)$ for all $x \in L$. It follows that $\mathrm{NR}(M)$ is an ideal of L and hence $\mathrm{NR}(M) \subset \mathrm{NR}(L)$.

For the other inclusion we note that $\mathrm{NR}(L) \subset M$ and hence $\mathrm{NR}(L) \subset \mathrm{NR}(M)$. $\qquad\square$

Proposition 3.7.3 *Let L be a solvable Lie algebra of characteristic 0. Set $I = [[L, L], [L, L]]$. Then $\mathrm{NR}(L/I) = \mathrm{NR}(L)/I$.*

Proof. First we note that by Corollary 2.6.3 we have that $I \subset [L, L] \subset \mathrm{NR}(L)$. So $\mathrm{NR}(L)/I$ is a nilpotent ideal of L/I hence $\mathrm{NR}(L)/I \subset \mathrm{NR}(L/I)$.

Now suppose that $\bar{x} \in \mathrm{NR}(L/I)$; then $\mathrm{ad}_{L/I}\bar{x}$ is nilpotent. Let $x \in L$ be a pre-image in L of $\bar{x}$. We claim that $\mathrm{ad}_L x$ is nilpotent. Because $\mathrm{ad}_{L/I}\bar{x}$ is nilpotent we have $(\mathrm{ad}_L x)^m(L) \subset I$ for some $m > 0$. Now we set $J_1 = [L, L]$, and for $k \geq 2$, $J_k = [J_1, J_{k-1}]$. So $J_2 = I$ and $(\mathrm{ad}_L x)^m(L) \subset J_2$. We show by induction on k that for $k \geq 2$ there is an integer s_k such that $(\mathrm{ad}_L x)^{s_k}(L) \subset J_k$. For $k = 2$ we take $s_2 = m$. So let $k \geq 2$, then by induction we have $(\mathrm{ad}x)^{s_k}(L) \subset J_k$. Now let $a \in J_1$ and $b \in J_{k-1}$, then $[a, b] \in J_k$ and by Leibniz' formula (1.11),

$$(\mathrm{ad}x)^t([a, b]) = \sum_{i=0}^{t} \binom{t}{i} \left[(\mathrm{ad}x)^i(a), (\mathrm{ad}x)^{t-i}(b)\right].$$

We take $t = 2s_k$. This means that if $t - i \geq s_k$, then $(\mathrm{ad}x)^{t-i}(b) \in J_k$ and $(\mathrm{ad}x)^i(a) \in J_1$. On the other hand if $t - i < s_k$, then $i > s_k$ so that $(\mathrm{ad}x)^i(a) \in J_k$ and $(\mathrm{ad}x)^{t-i}(b) \in J_1$. In both cases we see that

$$\left[(\mathrm{ad}x)^i(a), (\mathrm{ad}x)^{t-i}(b)\right] \in J_{k+1}.$$

As a consequence we have that $(\mathrm{ad}x)^t(J_k) \subset J_{k+1}$. Hence $(\mathrm{ad}x)^{t+s_k}(L) \subset J_{k+1}$. Now because J_1 is a nilpotent Lie algebra, there is a $k > 0$ such that $J_k = 0$. Hence $\mathrm{ad}_L x$ is nilpotent and $x \in \mathrm{NR}(L)$ by Proposition 2.3.6. Therefore $\bar{x} \in \mathrm{NR}(L)/I$ and we are done. □

Proposition 3.7.4 *Let L be a solvable Lie algebra of characteristic* 0. *Let $C_m(L)$ be the hypercentre of L. Then* $\mathrm{NR}(L/C_m(L)) = \mathrm{NR}(L)/C_m(L)$.

Proof. Let

$$C(L) = C_1(L) \subset C_2(L) \subset \cdots \subset C_m(L)$$

be the upper central series of L. Just as in the proof of Proposition 3.7.3 it can be shown that $\mathrm{NR}(L)/C_m(L) \subset \mathrm{NR}(L/C_m(L))$. Now let $\bar{x}$ be an element of $\mathrm{NR}(L/C_m(L))$ and let $x \in L$ be a pre-image of $\bar{x}$. Then $(\mathrm{ad}_L x)^n(L) \subset C_m(L)$ for some $n > 0$. And since

$$C_m(L)/C_{m-1}(L) = C(L/C_{m-1}(L))$$

(by the definition of upper central series) we see that $\mathrm{ad} x(C_m(L)) \subset 0 + C_{m-1}(L)$. Now, by going down the upper central series we arrive at

$$(\mathrm{ad} x)^{m+n}(L) = 0.$$

So $x \in \mathrm{NR}(L)$ by Proposition 2.3.6. □

Proposition 3.7.5 *Let L be a Lie algebra such that* $[[L, L], [L, L]] = 0$ *and $C(L) = 0$. Then* $C_L([L, L]) = [L, L]$.

Proof. Set $K = C_L([L, L])$, then we define a representation $\phi : L \to \mathfrak{gl}(K)$ of L by $\phi(x)(v) = [x, v]$ for $x \in L$ and $v \in K$. (Because $[[L, L], [L, L]] = 0$ we have that $[x, v] \in C_L([L, L])$.) The kernel of ϕ contains $[L, L]$ so $\phi(L) \cong L/\ker \phi$ is commutative. Let

$$K = K_0(\phi(L)) \oplus K_1(\phi(L))$$

be the Fitting decomposition of K with respect to $\phi(L)$. Then since $\phi(L) \cdot K_1(\phi(L)) = K_1(\phi(L))$ we have $K_1(\phi(L)) \subset [L, L]$. Furthermore, every element of $\phi(L)$ acts nilpotently on $K_0(\phi(L))$. Suppose that $K_0(\phi(L)) \neq 0$. Then by Proposition 2.1.2 there is a non-zero $v \in K_0(\phi(L))$ such that $\phi(L)v = 0$. But then $v \in C(L)$ and we have reached a contradiction. So $K_0(\phi(L)) = 0$ and $K \subset [L, L]$. It follows that $K = [L, L]$. □

Proposition 3.7.6 *Let L be a Lie algebra such that $[[L, L], [L, L]] = 0$ and $C(L) = 0$. Then $[L, [L, L]] = [L, L]$.*

Proof. Set $K = [L, L]$ and let ϕ be defined as in the proof of Proposition 3.7.5. In the same way as seen in that proof, $K = K_1(\phi(L))$. So $\phi(L) \cdot K = K$, but that is the same as $[L, [L, L]] = [L, L]$. □

Now the algorithm reads as follows:

Algorithm NilRadical
Input: a finite-dimensional Lie algebra L of characteristic 0.
Output: NR(L).

Step 1 Compute the solvable radical $R = \mathrm{SR}(L)$. If $R = 0$ then return R, otherwise continue with R in place of L.

Step 2 Compute the ideal $I = [[L, L], [L, L]]$. If $I \neq 0$ then compute (by a recursive call) the nilradical $\overline{N}$ of L/I and return the inverse image of N in L. Otherwise proceed to Step 3.

Step 3 Compute the hypercentre $C_m(L)$ of L. If $C_m(L) \neq 0$ then proceed as in Step 2 where $C_m(L)$ plays the role of I. Otherwise go to Step 4.

Step 4 Compute a basis of L of the form $\{x_1, \dots, x_s, y_1, \dots, y_t\}$ where $[L, L]$ is spanned by $\{x_1, \dots, x_s\}$. Set $i := 1$;

Step 5 Let a be the matrix of the adjoint action of y_i on $[L, L]$. If the rank of a is strictly smaller than s, then compute the ideal $J = \mathrm{ad} y_i([L, L])$. Compute recursively the nilradical of L/J and let M be the ideal of L containing J such that $\mathrm{NR}(L/J) = M/J$. Compute (by a recursive call) $\mathrm{NR}(M)$ and return this ideal. If the rank of a is equal to s, then go to Step 6.

Step 6 Let f be the minimum polynomial of a. If f is not square-free then set $g = f / \gcd(f, f')$. Compute the ideal $I = g(a) \cdot [L, L]$ and proceed as in Step 5. If f is square-free proceed to Step 7.

Step 7 If $i < t$ then set i equal to $i + 1$ and go to Step 5. Otherwise, $\mathrm{NR}(L) = [L, L]$.

Comments:

Step 1 Since $NR(R) = NR(L)$, (cf. Proposition 2.3.6) we may replace L by R.

Step 2 This step is justified by Proposition 3.7.3.

Step 3 This step is justified by Proposition 3.7.4.

Step 5 The rank of a is not 0 by Proposition 3.7.5 (the conditions of this proposition are fulfilled by Steps 2 and 3). If it is less than s, then $J = \mathrm{ad}y_i([L, L])$ will be an ideal of L properly contained in $[L, L]$. Hence Proposition 3.7.6 (ensuring that L/J is not nilpotent, and hence $M \neq L$) and Proposition 3.7.2 justify the recursive calls.

Step 6 For $z, u, v \in L$ we have

$$[z, (\mathrm{ad}y_i)^m([u, v])] = (\mathrm{ad}y_i)^m([z, [u, v]])$$

which is proved by induction on m. From this it follows that the space $h(\mathrm{ad}y_i)([L, L])$ is an ideal of L for every polynomial h. In particular $g(\mathrm{ad}y_i)([L, L])$ is an ideal of L and it is properly contained in $[L, L]$ because $g(\mathrm{ad}y_i)$ is nilpotent.

Step 7 If $i = t$ then all elements y_k act by a semisimple matrix on $[L, L]$. Furthermore these matrices commute. So any nilpotent element of L is contained in the span of $x_1, \ldots, x_s$. It follows that $NR(L)$ is contained in $[L, L]$. By Step 1 we have that L is solvable so that $NR(L) = [L, L]$.

Example 3.7.7 Let L be the 5-dimensional Lie algebra over $\mathbb{Q}$ with basis $\{x_1, x_2, x_3, x_4, x_5\}$ and multiplication table

$$[x_1, x_5] = x_1, \ [x_2, x_5] = x_1 + x_2, \ [x_3, x_4] = x_1, \ [x_3, x_5] = x_3.$$

Then L is solvable. Furthermore, both $[[L, L], [L, L]]$ and $C(L)$ are zero. A basis as in Step 4 of the algorithm is given by $\{x_1, x_2, x_3, y_1, y_2\}$, where $y_1 = x_4$ and $y_2 = x_5$. The matrix of the restriction of $\mathrm{ad}y_1$ to $[L, L]$ is

$$a = \begin{pmatrix} 0 & 0 & -1 \\ 0 & 0 & 0 \\ 0 & 0 & 0 \end{pmatrix}.$$

The rank of a is 1, so in Step 5 we calculate the ideal J spanned by $a \cdot [L, L] = \langle x_1 \rangle$. Let z_i be the image of x_i in L/J for $i = 2, 3, 4, 5$. Then a multiplication table of $K_1 = L/J$ is

$$[z_2, z_5] = z_2, \ [z_3, z_5] = z_3.$$

We calculate the nilradical of K_1. The centre $C(K_1)$ is spanned by z_4. Set $K_2 = K_1/C(K_1)$; then the centre of K_2 is zero. So the hypercentre of K_1 is equal to $C(K_1)$. Let u_i be the image of z_i in K_2 for $i = 2, 3, 5$. Then the u's satisfy the same commutation relations as the z's. Hence $C(K_2) = [[K_2, K_2], [K_2, K_2]] = 0$, so for the calculation of the nilradical of K_2 we proceed to Step 4. We have that $[K_2, K_2]$ is spanned by u_2, u_3. The matrix of the action of u_5 on this space is

$$\begin{pmatrix} -1 & 0 \\ 0 & -1 \end{pmatrix}.$$

The minimum polynomial of this matrix is $X + 1$, which is square-free. By Step 7 we now see that $\mathrm{NR}(K_2) = [K_2, K_2]$. By Step 3, $\mathrm{NR}(K_1)$ is spanned by z_2, z_3, z_4.

We now continue calculating $\mathrm{NR}(L)$. We left this computation in Step 5. We have calculated the nilradical of K_1. Now M is spanned by x_1, x_2, x_3, x_4. According to Step 5, we have $\mathrm{NR}(L) = \mathrm{NR}(M)$. But M is nilpotent. So $\mathrm{NR}(L) = M$.

Lemma 3.7.8 *Let $L \subset \mathfrak{gl}(V)$ be a solvable linear Lie algebra of characteristic 0. Let A be the associative algebra generated by L together with the identity on V. Then $A/\mathrm{Rad}(A)$ is a commutative associative algebra.*

Proof. Let $0 = V_0 \subset V_1 \subset \cdots \subset V_{s+1} = V$ be a composition series of V with respect to the action of L. Then all elements of A stabilize the modules V_i. Let $I \subset A$ be the set of all $a \in A$ such that $aV_{i+1} \subset V_i$ for $0 \leq i \leq s$. Then I is a nilpotent ideal of A and hence $I \subset \mathrm{Rad}(A)$. Because V_{i+1}/V_i is an irreducible L-module, by Proposition 2.3.3 we have that $[x, y] \cdot V_{i+1} \subset V_i$ for all $x, y \in L$. Hence $[x, y] \in \mathrm{Rad}(A)$. So the generators of A commute modulo $\mathrm{Rad}(A)$. Hence all elements of A commute modulo $\mathrm{Rad}(A)$. $\square$

Theorem 3.7.9 *Let L be a finite-dimensional Lie algebra of characteristic 0. Set $R = \mathrm{SR}(L)$. Let H be a Cartan subalgebra of R. Let $R_1 = R_1(H)$ be the Fitting-one component of R relative to H. Let A denote the associative algebra generated by $\mathrm{ad}_{R_1} h$ for $h \in H$ along with the identity on R_1. Then we have*

$$\mathrm{NR}(L) = R_1 \oplus \{h \in H \mid \mathrm{ad}_{R_1} h \in \mathrm{Rad}(A)\}.$$

Proof. First we have that $R_1 = [H, R_1] \subset [R, R] \subset \mathrm{NR}(L)$ (the second inclusion follows from Corollary 2.6.3). Furthermore, if h is an element of

H such that $\operatorname{ad}_{R_1} h \in \operatorname{Rad}(A)$, then $\operatorname{ad}_{R_1} h$ is nilpotent. Since H is nilpotent, there is a $k > 0$ such that $(\operatorname{ad}_R h)^k(H) = 0$. Hence for m big enough,

$$(\operatorname{ad}_R h)^m(R) = (\operatorname{ad}_R h)^m(H) + (\operatorname{ad}_R h)^m(R_1) = 0.$$

So $\operatorname{ad}_R h$ is nilpotent, forcing $h \in \operatorname{NR}(L)$ (Proposition 2.3.6). So everything on the right hand side is contained in $\operatorname{NR}(L)$.

Now let $x \in \operatorname{NR}(L)$. Then also $x \in R$, and consequently there are unique $h \in H$ and $y \in R_1$ such that $x = h + y$. We must show that $\operatorname{ad}_{R_1} h \in \operatorname{Rad}(A)$. By the Wedderburn-Malcev principal theorem (Theorem A.1.5) there is a semisimple subalgebra S of A such that $A = S \oplus \operatorname{Rad}(A)$. By Lemma 3.7.8 we have that S is commutative. Therefore any nilpotent element of S will generate a nilpotent ideal of S. It follows that S does not contain nilpotent elements. This implies that all elements of S are semisimple linear transformations. Indeed, let $a \in S$ and let $a = s + n$ be its Jordan decomposition. Then both $s, n \in S$ so that $n = 0$. Now write $\operatorname{ad}_{R_1} h = s + r$ where $s \in S$ and $r \in \operatorname{Rad}(A)$. Then since $\operatorname{ad}_{R_1} h$ is nilpotent, we must have that $s = 0$, and we are done. $\qquad\square$

On the basis of this theorem we formulate the following algorithm for calculating the nilradical of a Lie algebra of characteristic 0.

Algorithm NilRadical
Input: a finite-dimensional Lie algebra L of characteristic 0.
Output: a basis of $\operatorname{NR}(L)$.

Step 1 Calculate the solvable radical $R = \operatorname{SR}(L)$ of L.

Step 2 Calculate a Cartan subalgebra H of R.

Step 3 Set $R_1 :=$ FittingOneComponent(R, H) and calculate a basis of the associative algebra A generated by $\operatorname{ad}_{R_1} h$ for $h \in H$.

Step 4 Calculate $\operatorname{Rad}(A)$ and calculate the space N of all $h \in H$ such that $\operatorname{ad}_{R_1} h \in \operatorname{Rad}(A)$.

Step 5 Return $N \oplus R_1$.

Example 3.7.10 Let L be the 5-dimensional Lie algebra of Example 3.7.7. A non-nilpotent element if L is x_5. We have that $L_0(\operatorname{ad} x_5)$ is spanned by x_4, x_5. This subalgebra is nilpotent (even commutative) so $H = L_0(\operatorname{ad} x_5)$ is a Cartan subalgebra of L. The Fitting-one component $L_1(H)$ is spanned by x_1, x_2, x_3. The matrices of the restrictions of $\mathrm{ad}x_4$ and $\mathrm{ad}x_5$ to $L_1(H)$ are

$$\mathrm{ad}_{L_1(H)}x_4 = \begin{pmatrix} 0 & 0 & -1 \\ 0 & 0 & 0 \\ 0 & 0 & 0 \end{pmatrix} \text{ and } \mathrm{ad}_{L_1(H)}x_5 = \begin{pmatrix} -1 & -1 & 0 \\ 0 & -1 & 0 \\ 0 & 0 & -1 \end{pmatrix}.$$

Let A be the associative algebra (with one) generated by these matrices. Then A is 3-dimensional. The radical of A is spanned by $\mathrm{ad}_{L_1(H)}x_4$ along with

$$\begin{pmatrix} 0 & 1 & 0 \\ 0 & 0 & 0 \\ 0 & 0 & 0 \end{pmatrix}.$$

So $\mathrm{NR}(L)$ is spanned by x_1, x_2, x_3, x_4.

Some practical experiences with the algorithms for calculating the nilradical are reported in [35]. The algorithms of this section turn out to be markedly more efficient than the algorithm described in Section 2.2. Furthermore, the second algorithm of this section (using Cartan subalgebras) runs faster than the first algorithm on the examples considered in [35].

3.8   Notes

The algorithms for calculating a Cartan subalgebra are taken from [37]. In that paper it is shown that these algorithms run in polynomial time. A different algorithm is described in [7]. There the computation of a Cartan subalgebra is split into two separate cases. In the first case the Lie algebra is semisimple. In this case a characterization of Cartan subalgebras as *maximal tori* is used (see also Section 4.9). The resulting algorithm is somewhat similar in nature to the algorithms described in Section 3.2. The construction of a maximal torus starts by constructing a semisimple element s_1. A maximal torus containing s_1 is contained in the centralizer of s_1. So the centralizer of s_1 is constructed. If this centralizer is not nilpotent, then it contains a semisimple element s_2 independent from s_1. Now the centralizer of the subalgebra generated by s_1, s_2 is constructed. This process repeats until a Cartan subalgebra is found. In the second case the Lie algebra is solvable. For this case a rather complicated procedure is used. Furthermore, it is shown that we can compute a Cartan subalgebra of any Lie algebra using the algorithms for these two cases.

Also [91] contains an algorithm for calculating a Cartan subalgebra. The strategy used there consists of trying to find a nilpotent subalgebra K of L

such that $L_0(K)$ is a proper subalgebra of L. When such a subalgebra is found, recursion is applied to find a Cartan subalgebra H of $L_0(K)$ and by Proposition 3.2.9 (we refer to [88] for an extension of this result to characteristic 0), H is also a Cartan subalgebra of L. The algorithm starts with an arbitrary nilpotent subalgebra K. If $L_0(K)$ happens to be equal to L, then two strategies for replacing K are possible. First of all, if the centralizer of K in L is bigger than K we can add an element of the complement to K in the centralizer and produce a bigger nilpotent subalgebra. We can do the same with the normalizer. However in this case, in order to get a nilpotent subalgebra, we must make sure that the element x of the complement acts nilpotently on K. If this happens not to be the case then x is a non-nilpotent element and hence the nilpotent subalgebra K spanned by x will have the property that $L_0(K) \neq L$.

The Lie algebra of Table 3.1 is taken from [84] (it is the Lie algebra with name $L_{8,11}$ in that paper). The proof of Theorem 3.6.4 follows C. Chevalley ([19]). The first algorithm for calculating the nilradical in Section 3.7 is based on [71]. The second algorithm is taken from [35].

Chapter 4

Lie algebras with non-degenerate Killing form

The Killing form of a Lie algebra is the trace form corresponding to the adjoint representation. The requirement that the Killing form of a Lie algebra L be non-degenerate poses strong restrictions on the structure of L. For instance, we will show that the solvable radical of L is zero (Section 4.2). Furthermore, in Section 4.3 we show that L is a direct sum of ideals that are simple (i.e., they have no nontrivial proper ideals). In characteristic 0 all L-modules satisfy an analogous property, namely they are direct sums of irreducible submodules. This is known as Weyl's theorem (Section 4.4). Also all derivations of L must be of the form $\mathrm{ad}x$ for some $x \in L$ (Section 4.5). This fact will be used to show that the elements of L have a decomposition analogous to the Jordan decomposition of matrices (Section 4.6). In characteristic 0 this Jordan decomposition is compatible with the usual Jordan decomposition of linear transformations in the following sense: if $x = x_s + x_n$ is the Jordan decomposition of $x \in L$ and $\rho : L \to \mathfrak{gl}(V)$ is a finite-dimensional representation of L, then $\rho(x) = \rho(x_s) + \rho(x_n)$ is the Jordan decomposition of $\rho(x)$.

The subject of Section 4.7 is the so-called Levi decomposition: we prove that a finite-dimensional Lie algebra of characteristic 0 is the direct sum of its solvable radical and a subalgebra that has a non-degenerate Killing form. A semisimple subalgebra provided by this theorem is called a Levi subalgebra.

In Section 4.9 we carry out a first investigation into the root space decomposition of a Lie algebra L with a non-degenerate Killing form, relative to a split Cartan subalgebra. Among other things we prove that any Cartan subalgebra of L is commutative and consists of semisimple elements.

Furthermore we show that the root spaces are all 1-dimensional.

In Section 4.11 we use splitting and decomposing elements of a Cartan subalgebra H to calculate the collected primary decomposition of L with respect to H. We use this in Section 4.12 to formulate an algorithm for calculating the direct sum decomposition of L. In the next section we describe algorithms to calculate a Levi subalgebra of a Lie algebra L. Then in Section 4.14 we show how a Cartan subalgebra of a Lie algebra L carries information about its Levi decomposition. In the last section we use this to formulate a different algorithm for calculating a Levi subalgebra.

4.1   Trace forms and the Killing form

Let L be a finite-dimensional Lie algebra over the field F and let $\rho : L \to \mathfrak{gl}(V)$ be a finite-dimensional representation of L. Then we define a bilinear form $f_\rho : L \times L \to F$ by

$$f_\rho(x, y) = \mathrm{Tr}(\rho(x)\rho(y)) \text{ for } x, y \in L.$$

The form f_ρ is called the *trace form* corresponding to the representation ρ. Let $B = \{x_1, \dots, x_n\}$ be a basis of L, then $(f_\rho(x_i, x_j))_{i,j=1}^n$ is the *matrix* of f_ρ relative to the basis B. The *radical* S_ρ of f_ρ is the subspace

$$S_\rho = \{x \in L \mid f_\rho(x, y) = 0 \text{ for all } y \in L\}.$$

The form f_ρ is said to be *non-degenerate* if $S_\rho = 0$. It is straightforward to see that f_ρ is non-degenerate if and only if the matrix of f_ρ (relative to any basis) is nonsingular.

Lemma 4.1.1 *Let f_ρ be a trace form of the Lie algebra L. Then we have $f_\rho([x, y], z) = f_\rho(x, [y, z])$ for all $x, y, z \in L$.*

Proof. This is a consequence Lemma 2.5.4 and ρ being a morphism of Lie algebras. $\square$

Proposition 4.1.2 *Let f_ρ be a trace form of the Lie algebra L. Then the radical S_ρ of f_ρ is an ideal of L. Furthermore, if L is of characteristic 0 and ρ is faithful, then S_ρ is solvable.*

Proof. The fact that S_ρ is an ideal of L follows immediately from Lemma 4.1.1. Now suppose that L is of characteristic 0 and ρ is faithful. Let $x \in [S_\rho, S_\rho]$ and $y \in S_\rho$, then $x \in S_\rho$ and hence $\mathrm{Tr}(\rho(x)\rho(y)) = 0$. So by

Cartan's criterion (Theorem 2.5.5), $\rho(S_\rho)$ is solvable. Since ρ is faithful, this implies that S_ρ is solvable. □

The trace form corresponding to the adjoint representation of L is called the *Killing form*. It is denoted by κ_L, i.e., $\kappa_L(x, y) = \mathrm{Tr}(\mathrm{ad}_L x \cdot \mathrm{ad}_L y)$. If it is clear from the context which Lie algebra we mean, we also write κ instead of κ_L. In this chapter we will investigate the structure of Lie algebras that have a non-degenerate Killing form. A first result is the following.

Lemma 4.1.3 *Let L be a Lie algebra with a non-degenerate Killing form. Then the centre of L is 0.*

Proof. Let $x \in C(L)$. Then $\mathrm{ad}x = 0$, and hence $\kappa(x, y) = 0$ for all $y \in L$. It follows that $x = 0$. □

As a consequence the adjoint representation $\mathrm{ad} : L \to \mathfrak{gl}(L)$ is faithful if the Killing form of L is non-degenerate.

4.2 Semisimple Lie algebras

This section contains the so-called Cartan's criterion for semisimplicity, which yields a straightforward way of checking whether a given Lie algebra is semisimple.

Definition 4.2.1 *A Lie algebra L said to be* semisimple *if the solvable radical* $\mathrm{SR}(L)$ *is zero.*

Proposition 4.2.2 (Cartan's criterion for semisimplicity) *Let L be a Lie algebra. If the Killing form κ_L is non-degenerate then L is semisimple. Conversely, if L is semisimple and of characteristic 0, then κ_L is non-degenerate.*

Proof. Set $I = \mathrm{SR}(L)$ and suppose that $I \neq 0$. Then there is an integer $m \geq 1$ such that $I^{(m+1)} = 0$ while $I^{(m)} \neq 0$ (where $I^{(m)}$ denotes the m-th term of the derived series of I). Now let $x \in I^{(m)}$ and let $y \in L$. Then $\mathrm{ad}x\, \mathrm{ad}y(L) \subset I^{(m)}$. And since $\mathrm{ad}x(I^{(m)}) \subset I^{(m+1)} = 0$ we have that $(\mathrm{ad}x\, \mathrm{ad}y)^2 = 0$. It follows that $\kappa_L(x, y) = 0$ and consequently $I^{(m)}$ is contained in the radical of κ_L. But this means that $I^{(m)} = 0$ and from the assumption that $I \neq 0$ we have reached a contradiction. The conclusion is that $I = 0$ and L is semisimple.

Suppose that L is semisimple and of characteristic 0. Let S_κ be the radical of κ. Since $\mathrm{SR}(L) = 0$ also the centre of L must be 0. Hence the adjoint representation of L is faithful. So by Proposition 4.1.2, S_κ is a solvable ideal of L. Consequently $S_\kappa = 0$ and κ is non-degenerate.   □

Example 4.2.3 Let $L = \mathfrak{sl}_2(F)$. As seen in Example 1.3.6, this Lie algebra has basis $\{h, x, y\}$ with $[h, x] = 2x$, $[h, y] = -2y$ and $[x, y] = h$. The matrices of $\mathrm{ad}x$, $\mathrm{ad}y$ and $\mathrm{ad}h$ with respect to this basis are easily computed:

$$\mathrm{ad}x = \begin{pmatrix} 0 & 0 & 1 \\ -2 & 0 & 0 \\ 0 & 0 & 0 \end{pmatrix}, \mathrm{ad}y = \begin{pmatrix} 0 & -1 & 0 \\ 0 & 0 & 0 \\ 2 & 0 & 0 \end{pmatrix}, \mathrm{ad}h = \begin{pmatrix} 0 & 0 & 0 \\ 0 & 2 & 0 \\ 0 & 0 & -2 \end{pmatrix}.$$

Then the matrix of κ equals

$$\begin{pmatrix} 8 & 0 & 0 \\ 0 & 0 & 4 \\ 0 & 4 & 0 \end{pmatrix}.$$

The determinant of this matrix is -128, hence by Proposition 4.2.2, L is semisimple if the characteristic of F is not 2.

4.3 Direct sum decomposition

Definition 4.3.1 *A finite-dimensional Lie algebra L is said to be* simple *if* $\dim L > 1$ *and L has no ideals except 0 and L.*

If L is simple, then $\mathrm{SR}(L) = L$ or $\mathrm{SR}(L) = 0$. In the first case we have that L is solvable and $[L, L]$ is an ideal of L not equal to L. Hence $[L, L] = 0$ and L is commutative. It follows that every 1-dimensional subspace is an ideal of L, which is not possible because L is simple (note that here we use the condition $\dim L > 1$). Therefore $\mathrm{SR}(L) = 0$. So a simple Lie algebra is also semisimple.

The converse does not hold. Let K be a Lie algebra with a non-degenerate Killing form (so that K is certainly semisimple by Proposition 4.2.2) and set $L = K \oplus K$ (direct sum of Lie algebras, see Section 1.10). Then the Killing form of L is also non-degenerate (this will follow from Lemma 4.3.4); but L is not simple. In this section we prove that a Lie algebra with non-degenerate Killing form is always of this form: a direct sum of simple ideals.

Lemma 4.3.2 *Suppose that L is a direct sum of ideals, $L = I \oplus J$. Then κ_I is the restriction of κ_L to I (and likewise for κ_J).*

Proof. Let $x, y \in L$. Then $\mathrm{ad}x$ and $\mathrm{ad}y$ leave the subspaces I and J invariant; so this also holds for $\mathrm{ad}x \cdot \mathrm{ad}y$. Hence

$$\kappa_L(x,y) = \mathrm{Tr}_L(\mathrm{ad}x\mathrm{ad}y) = \mathrm{Tr}_I(\mathrm{ad}_I x \mathrm{ad}_I y) + \mathrm{Tr}_J(\mathrm{ad}_J x \mathrm{ad}_J y).$$

But if $x \in I$, then $\mathrm{ad}_J x = 0$, whence the lemma. $\qquad\qquad\square$

Lemma 4.3.3 *Let $L = I \oplus J$, a direct sum of ideals. Then $\kappa_L(I, J) = 0$.*

Proof. Let $x \in I$ and $y \in J$, then $\mathrm{ad}x \cdot \mathrm{ad}y(L) \subset I \cap J = 0$. Hence $\kappa_L(x,y) = 0$. $\qquad\qquad\square$

Lemma 4.3.4 *Let $L = I \oplus J$ be the direct sum of two ideals I and J. Then κ_L is non-degenerate if and only if κ_I and κ_J are non-degenerate.*

Proof. Suppose κ_L is non-degenerate and κ_I is degenerate. Then there is an $x_0 \in I$ such that $\kappa_I(x_0, y) = 0$ for all $y \in I$. For all $z \in J$ we have $\kappa_L(x_0, z) = 0$ (Lemma 4.3.3). Let $u \in L$. We write $u = u_1 + u_2$ where $u_1 \in I$ and $u_2 \in J$. Then $\kappa_L(x_0, u) = \kappa_L(x_0, u_1) + \kappa_L(x_0, u_2) = 0$. So κ_L is degenerate, a contradiction. An analogous argument proves that κ_J is non-degenerate.

Now suppose that κ_I and κ_J are non-degenerate. Let $x \in L$ be such that $\kappa_L(x, y) = 0$ for all $y \in L$. Write $x = x_1 + x_2$ where $x_1 \in I$, $x_2 \in J$. Then using Lemmas 4.3.2 and 4.3.3 we see that for $y \in I$,

$$0 = \kappa_L(x, y) = \kappa_L(x_1 + x_2, y) = \kappa_L(x_1, y) = \kappa_I(x_1, y).$$

Hence $x_1 = 0$. Similarly it can be seen that $x_2 = 0$; i.e., $x = 0$ and κ_L is non-degenerate. $\qquad\qquad\square$

If $V \subset L$ is a subspace of L, then the set

$$V^{\perp} = \{x \in L \mid \kappa_L(x, y) = 0 \text{ for all } y \in V\}$$

is called the orthogonal complement of V in L with respect to κ_L.

Lemma 4.3.5 *Let L be a finite-dimensional Lie algebra, and suppose that the Killing form of L is non-degenerate. Let I be an ideal of L, then the orthogonal complement $I^{\perp}$ of I with respect to κ_L is an ideal of L and $L = I \oplus I^{\perp}$.*

Proof. By Lemma 4.1.1 we have that $I^\perp$ is an ideal in L. Set $J = I \cap I^\perp$, and let $x, y \in J$. Then for all $z \in L$ we have by Lemma 4.1.1,

$$\kappa_L([x, y], z) = \kappa_L(x, [y, z]) = 0$$

(since $x \in I$ and $[y, z] \in I^\perp$). And because κ_L is non-degenerate, $[x, y] = 0$. So J is a solvable ideal of L and since L is semisimple (Proposition 4.2.2) we must have $J = 0$. The non-degeneracy of κ_L implies that $\dim I + \dim I^\perp = \dim L$. Hence $L = I \oplus I^\perp$. $\qquad\square$

Proposition 4.3.6 *Let L be a finite-dimensional Lie algebra with a non-degenerate Killing form. Then*

$$L = J_1 \oplus \cdots \oplus J_s$$

where the J_k are ideals of L that are simple (for $1 \leq k \leq s$).

Proof. If L is simple, then we can take $s = 1$ and $J_1 = L$. On the other hand, if L is not simple, then L has an ideal I not equal to L, and not zero. Let $I^\perp$ be the orthogonal complement to I with respect to κ_L. Then by Lemma 4.3.5, $I^\perp$ is an ideal of L and $L = I \oplus I^\perp$. By Lemma 4.3.4, κ_I and $\kappa_{I^\perp}$ are non-degenerate. Hence, by induction on the dimension, I and $I^\perp$ are direct sums of simple ideals. So also L is a direct sum of simple ideals. $\qquad\square$

The following proposition says that the decomposition of Proposition 4.3.6 is unique.

Proposition 4.3.7 *Let L be a Lie algebra with non-degenerate Killing form. Suppose that*

$$L = J_1 \oplus \cdots \oplus J_m \quad and \quad L = K_1 \oplus \cdots \oplus K_n$$

are two decompositions of L into simple ideals. Then $m = n$ and every ideal J_i is equal to an ideal K_j.

Proof. Fix an index $i \in \{1, \ldots, m\}$. We prove that J_i is equal to an ideal K_j. Set $I_j = J_i \cap K_j$ for $1 \leq j \leq n$. Then I_j is an ideal of L and it is contained in both J_i and K_j. Suppose that $I_j \neq 0$; then since J_i and K_j are simple, we must have $J_i = I_j = K_j$. On the other hand, if $I_j = 0$, then $[J_i, K_j] \subset I_j = 0$. So if $I_j = 0$ for all j, then $[J_i, L] = [J_i, K_1] + \cdots + [J_i, K_n] = 0$. But this is impossible in view of

Lemma 4.1.3. Hence $I_j \neq 0$ for exactly one j.   □

Given a semisimple Lie algebra L we may calculate its decomposition into a direct sum of simple ideals using the algorithm of Section 1.15. However, later in this chapter (Section 4.12) we will give an algorithm especially designed for this case.

4.4   Complete reducibility of representations

In this section L is a semisimple Lie algebra over a field F of characteristic 0. We recall that a representation $\rho : L \to \mathfrak{gl}(V)$ of L is completely reducible if V is a direct sum of irreducible submodules (see Section 1.12). The purpose of this section is to show that every representation of L is completely reducible.

The plan of the proof is to show that if $W \subset V$ is a submodule of V, then there is a submodule $W' \subset V$ such that $V = W \oplus W'$ (and then to proceed by induction on the dimension). This will be proved by constructing projections of V onto W and W'. These are elements of $\mathrm{Hom}_F(V, V)$.

In the sequel $\rho : L \to \mathfrak{gl}(V)$ will be a finite-dimensional representation of L. For convenience we sometimes also refer to V as an L-module. From Section 1.12 we recall that if V, W are L-modules, then $\mathrm{Hom}_F(V, W)$ (the set of all linear maps from V into W) becomes an L-module by setting

$$(x \cdot \phi)(v) = x \cdot (\phi(v)) - \phi(x \cdot v) \text{ for all } x \in L, \ \phi \in \mathrm{Hom}_F(V, W) \text{ and } v \in V.$$

Also by $\mathrm{Hom}_L(V, W)$ we denote the set of all $\phi \in \mathrm{Hom}_F(V, W)$ such that $x \cdot \phi = 0$ for all $x \in L$.

Lemma 4.4.1 *We have*

$$\mathrm{Hom}_L(V, V) = \{a \in \mathrm{Hom}_F(V, V) \mid [a, \rho(L)] = 0\}.$$

Furthermore, $\mathrm{Hom}_L(V, V)$ is an associative subalgebra of $\mathrm{Hom}_F(V, V)$ containing the identity.

Proof. Let $a \in \mathrm{Hom}_L(V, V)$, then $x \cdot a = 0$ for all $x \in L$. But this is equivalent to saying that for all $v \in V$:

$$0 = (x \cdot a)(v) = x \cdot (av) - a(x \cdot v) = \rho(x)a(v) - a\rho(x)(v).$$

Which is equivalent to $[a, \rho(x)] = 0$ for all $x \in L$.

Let $a, b \in \mathrm{Hom}_L(V, V)$ then it is straightforward to see that $[ab, \rho(L)] = 0$. Hence $\mathrm{Hom}_L(V, V)$ is a subalgebra of the associative algebra $\mathrm{Hom}_F(V, V)$. It evidently contains the identity. $\qquad\square$

The existence of a complementary submodule W' to W is closely related to the existence of idempotents in the subalgebra $\mathrm{Hom}_L(V, V)$.

Lemma 4.4.2 *Let W be a submodule of V. Then there is a submodule $W' \subset V$ such that $V = W \oplus W'$ if and only if there is an idempotent $e \in \mathrm{Hom}_L(V, V)$ such that $W' = \ker e$ and $W = \mathrm{im}\, e$.*

Proof. First we remark that since $e \in \mathrm{Hom}_L(V, V)$ we have that the kernel and image of e are submodules of V.

Suppose that $V = W \oplus W'$. An element $v \in V$ can uniquely be written as $v = w + w'$ where $w \in W$ and $w' \in W'$. Let e be the linear map defined by $e(v) = e(w + w') = w$. Then $e^2 = e$ and $W = \mathrm{im}\, e$ and $W' = \ker e$. Also for $x \in L$ we calculate

$$(x \cdot e)(w + w') = x \cdot (e(w + w')) - e(x \cdot (w + w'))$$
$$= x \cdot w - e(x \cdot w + x \cdot w')$$
$$= x \cdot w - x \cdot w = 0.$$

Hence $e \in \mathrm{Hom}_L(V, V)$.

On the other hand suppose that $e \in \mathrm{Hom}_L(V, V)$ is an idempotent with the listed properties. Then for $v \in V$ we have $ev \in W$ and $v - ev \in W'$, and consequently $V = W + W'$. Let $v \in W \cap W'$ then $v = eu$ for some $u \in V$ and also $ev = 0$. Hence $v = eu = e^2 u = ev = 0$. It follows that $W \cap W' = 0$ and V is the direct sum of these submodules. $\qquad\square$

In the sequel we will construct idempotents in $\mathrm{Hom}_L(V, V)$. First we will do this by constructing a so-called *Casimir operator* corresponding to the representation ρ.

Since the representation ρ is a morphism of Lie algebras, $\ker \rho$ is an ideal of L. Hence, by Lemma 4.3.5 we see that L is a direct sum of ideals $L = \ker \rho \oplus L_1$. And by Lemma 4.3.4 and Proposition 4.2.2, L_1 is semisimple. Furthermore, the restriction of ρ to L_1 is faithful. Let $f_\rho : L \times L \to F$ be the trace form of ρ. Then by Proposition 4.1.2, the restriction of f_ρ to L_1 is non-degenerate.

Now let $x_1, \ldots, x_m$ be a basis of L_1. Then, since the restriction of f_ρ to L_1 is non-degenerate, there is a basis $y_1, \ldots, y_m$ of L_1 such that

$f_\rho(x_i, y_j) = \delta_{ij}$. Now the operator

$$c_\rho = \sum_{i=1}^{m} \rho(x_i)\rho(y_i)$$

is called a *Casimir operator* corresponding to the representation ρ. We note
that $\mathrm{Tr}(c_\rho) = \sum_i \mathrm{Tr}(\rho(x_i)\rho(y_i)) = \sum_i f_\rho(x_i, y_i) = \dim L_1$.

Proposition 4.4.3 c_ρ is an element of $\mathrm{Hom}_L(V, V)$, i.e., $[c_\rho, \rho(L)] = 0$.

Proof. Let $x \in L_1$ and define $\alpha_{ik} \in F$ and $\beta_{ik} \in F$ by

$$[x, x_i] = \sum_{k=1}^{m} \alpha_{ik}x_k \quad \text{and} \quad [x, y_i] = \sum_{k=1}^{m} \beta_{ik}y_k.$$

Then using Lemma 4.1.1 we have

$$\alpha_{ik} = f_\rho([x, x_i], y_k) = -f_\rho([x_i, x], y_k) = -f_\rho(x_i, [x, y_k]) = -\beta_{ki}. \qquad (4.1)$$

We calculate

$$[\rho(x), c_\rho] = \sum_{i=1}^{m} [\rho(x), \rho(x_i)\rho(y_i)]$$

$$= \sum_{i=1}^{m} [\rho(x), \rho(x_i)]\rho(y_i) + \rho(x_i)[\rho(x), \rho(y_i)]$$

$$= \sum_{i,k=1}^{m} \alpha_{ik}\rho(x_k)\rho(y_i) + \beta_{ik}\rho(x_i)\rho(y_k)$$

$$= \sum_{i,k=1}^{m} \alpha_{ik}\rho(x_k)\rho(y_i) - \sum_{i,k=1}^{m} \alpha_{ki}\rho(x_i)\rho(y_k) \quad \text{(by (4.1))}$$

$$= 0.$$

□

We now find a complement W' of a submodule W by constructing a
suitable idempotent in $\mathrm{Hom}_L(V, V)$. First we have a lemma dealing with a
particular case, that can be solved by using a Casimir operator. Then there
is a proposition that proves the statement in general.

Lemma 4.4.4 Let W be a submodule of V such that $\rho(L)V \subset W$. Then
there is a submodule $W' \subset V$ such that $V = W \oplus W'$.

Proof. First of all, if $\rho(L) = 0$, then we can take any complementary subspace W'. The rest of the proof deals with the case where $\rho(L) \neq 0$.

First we suppose that W is irreducible. Let c_ρ be a Casimir operator corresponding to ρ. Let L_1 be the ideal complementary to $\ker \rho$. Then $\mathrm{Tr}(c_\rho) = \dim L_1$ which is non-zero because $\rho(L) \neq 0$. Also there is no non-zero $w \in W$ such that $c_\rho w = 0$; otherwise the set of all such w is a non-zero submodule (this follows from Proposition 4.4.3) which contradicts the assumption that W is irreducible. Hence the characteristic polynomial of the restriction of c_ρ to W has a non-zero constant term. It follows that there is a polynomial $p \in F[X]$ without constant term such that $e = p(c_\rho)$ is the identity on W. Since p does not have a constant term, we have $eV = W$. Hence $e^2 V = e(eV) = eV$ so that e is an idempotent. Furthermore, since $c_\rho \in \mathrm{Hom}_L(V, V)$ also $e \in \mathrm{Hom}_L(V, V)$. Set $W' = \ker e$, then by Lemma 4.4.2, W' is a submodule of V and $V = W \oplus W'$.

Now if W is not irreducible, then we prove the statement by induction on $\dim W$. Let $U \subset W$ be an irreducible submodule of W. We consider the L-module V/U and denote the induced representation also by ρ. Now $\rho(L)(V/U) \subset W/U$ so by induction there is a complement X/U to W/U. Furthermore, $U \subset X$ is irreducible, so by the first part of the proof there is a submodule $W' \subset X$ such that $X = W' \oplus U$. Finally, $V = W \oplus W'$. $\qquad\square$

Proposition 4.4.5 *Let L be a semisimple Lie algebra of characteristic 0. Let $\rho : L \to \mathfrak{gl}(V)$ be a finite-dimensional representation of L. If W is a submodule of V, then there is a submodule $W' \subset V$ such that $V = W \oplus W'$.*

Proof. Let $r : \mathrm{Hom}_F(V, W) \to \mathrm{Hom}_F(W, W)$ be the restriction map. It is straightforward to see that r is a morphism of L-modules. Also $\mathrm{Hom}_L(W, W)$ is contained in $\mathrm{Hom}_F(W, W)$ and furthermore L acts trivially on $\mathrm{Hom}_L(W, W)$. Set

$$U = \{\phi \in \mathrm{Hom}_F(V, W) \mid r(\phi) \in \mathrm{Hom}_L(W, W)\}.$$

Then $\ker r \subset U$ and $\ker r$ is an L-submodule because r is a morphism of L-modules. If $x \in L$ and $\phi \in U$, then $r(x \cdot \phi) = x \cdot r(\phi) = 0$. This shows that U is an L-module and $L \cdot U \subset \ker r$. So by Lemma 4.4.4, there is a complementary submodule $X \subset U$ such that $U = X \oplus \ker r$. And the restriction of r to X is a bijection onto $\mathrm{Hom}_L(W, W)$. Let e be the element of X such that $r(e)$ is the identity mapping on W. Then because $e \in \mathrm{Hom}_F(V, W)$, $e(V) \subset W$. So since $e(w) = w$ for $w \in W$ we have $e^2 = e$. Let $x \in L$, then because X is an L-module, also $x \cdot e \in X$. Furthermore, $r(x \cdot e) = x \cdot r(e) = 0$, so that $x \cdot e = 0$ and $e \in \operatorname{Hom}_L(V, V)$. Lemma 4.4.2 now finishes the proof. $\square$

Theorem 4.4.6 (Weyl) *Let L be a semisimple Lie algebra of characteristic 0. Then every finite-dimensional representation of L is completely reducible.*

Proof. Let V be a finite-dimensional L-module. If V is irreducible, then there is nothing to prove. On the other hand, if V contains a proper submodule W, then by Proposition 4.4.5, there is a submodule $W' \subset V$ such that $V = W \oplus W'$. Now by induction on the dimension W and W' both split as a direct sum of irreducible submodules. Hence the same holds for V. $\square$

Example 4.4.7 We remark that the decomposition of the L-module V into a direct sum of irreducible submodules in general is *not* unique. Let L be the Lie algebra $\mathfrak{sl}_2(F)$, with basis $\{x, y, h\}$ (see Example 1.3.6). By Example 4.2.3, this Lie algebra is semisimple. Let V be a 4-dimensional module for L with basis $\{v_1, v_2, w_1, w_2\}$. We let the basis elements of L act as follows:

$$h \cdot v_1 = v_1, \ h \cdot v_2 = -v_2, \ x \cdot v_2 = v_1, \ y \cdot v_1 = v_2, \ x \cdot v_1 = y \cdot v_2 = 0,$$

and

$$h \cdot w_1 = w_1, \ h \cdot w_2 = -w_2, \ x \cdot w_2 = w_1, \ y \cdot w_1 = w_2, \ x \cdot w_1 = y \cdot w_2 = 0.$$

For $\alpha \neq 0$ let U_α be the space spanned by $v_1 + \alpha w_1$ and $v_2 + \alpha w_2$ and let W_α be the space spanned by $v_1 - \alpha w_1$ and $v_2 - \alpha w_2$. Then $V = U_\alpha \oplus W_\alpha$ is a decomposition of V into irreducible submodules. Furthermore, if $\alpha, \beta > 0$ and $\alpha \neq \beta$, then U_α is not equal to either U_β or W_β (and similarly for W_α). So we have found an infinite series of decompositions of V into submodules.

4.5 All derivations are inner

Let L be a finite-dimensional Lie algebra. Recall that a derivation $d \in \operatorname{Der}(L)$ is called inner if $d = \operatorname{ad} x$ for an element $x \in L$. Here we show that for Lie algebras with non-degenerate Killing form all derivations are inner.

Lemma 4.5.1 *Let L be a finite-dimensional Lie algebra over the field F. And let $\phi : L \times L \to F$ be a non-degenerate bilinear form on L. Let $f : L \to F$ be a linear function. Then there is a unique $x_f \in L$ such that $f(y) = \phi(x_f, y)$ for all $y \in L$.*

Proof. Let $\{x_1, \ldots, x_n\}$ be a basis of L. Let $x_f = \sum_{k=1}^{n} \alpha_k x_k$ be an element of L, where the α_k are unknowns to be determined. Then $f(y) = \phi(x_f, y)$ for all $y \in L$ if and only if $f(x_i) = \phi(x_f, x_i)$ for $1 \leq i \leq n$. But this is the same as

$$\sum_{k=1}^{n} \phi(x_k, x_i)\alpha_k = f(x_i) \text{ for } 1 \leq i \leq n.$$

It follows that we have n linear equations for n unknowns α_k. Since ϕ is non-degenerate, the matrix of this equation system is non-singular. Hence these equations have a unique solution. $\qquad\qquad\qquad\qquad\square$

Proposition 4.5.2 *Let L be a finite-dimensional Lie algebra with non-degenerate Killing form. Then all derivations of L are inner.*

Proof. Let $d \in \mathrm{Der}(L)$ be a derivation of L, and let $x \in L$. Then

$$[d, \mathrm{ad}x](y) = d(\mathrm{ad}x(y)) - \mathrm{ad}x(d(y)) = [d(x), y] + [x, d(y)] - [x, d(y)] = [d(x), y],$$

i.e.,

$$[d, \mathrm{ad}x] = \mathrm{ad}d(x). \qquad\qquad\qquad\qquad (4.2)$$

Let F be the ground field of L. Consider the linear function $f : L \to F$ defined by $f(y) = \mathrm{Tr}(d \cdot \mathrm{ad}y)$. Then by Lemma 4.5.1, there is an element $x_f \in L$ such that $f(y) = \kappa_L(x_f, y)$ for all $y \in L$. Set $e = d - \mathrm{ad}x_f$, then

$$\mathrm{Tr}(e \cdot \mathrm{ad}y) = \mathrm{Tr}(d \cdot \mathrm{ad}y) - \mathrm{Tr}(\mathrm{ad}x_f \cdot \mathrm{ad}y) = 0 \text{ for all } y \in L. \qquad (4.3)$$

For $x, y \in L$ we calculate

$$\begin{aligned}
\kappa_L(e(x), y) &= \mathrm{Tr}(\mathrm{ad}e(x) \cdot \mathrm{ad}y) \\
&= \mathrm{Tr}([e, \mathrm{ad}x] \cdot \mathrm{ad}y) \text{ (by (4.2))} \\
&= \mathrm{Tr}(e \cdot \mathrm{ad}x \cdot \mathrm{ad}y) - \mathrm{Tr}(\mathrm{ad}x \cdot e \cdot \mathrm{ad}y) \\
&= \mathrm{Tr}(e \cdot \mathrm{ad}x \cdot \mathrm{ad}y) - \mathrm{Tr}(e \cdot \mathrm{ad}y \cdot \mathrm{ad}x) \\
&= \mathrm{Tr}(e \cdot \mathrm{ad}[x, y]) = 0 \text{ (by (4.3))}.
\end{aligned}$$

Therefore, since κ_L is non-degenerate, $e(x) = 0$ for all $x \in L$, i.e., $e = 0$. It follows that $d = \mathrm{ad}x_f$. $\qquad\qquad\qquad\qquad\square$

4.6 The Jordan decomposition

Let V be a finite-dimensional vector space defined over a perfect field. As seen in Proposition A.2.6, for an element $x \in \text{End}(V)$ there are unique $s, n \in \text{End}(V)$ such that $x = s + n$, s is semisimple, n is nilpotent and $[s, n] = 0$. In this section we prove that a similar statement holds for Lie algebras over a perfect field with a non-degenerate Killing form.

Let A be a finite-dimensional algebra over an arbitrary field F and $d \in \text{Der}(A)$ a derivation of A. We claim that

$$(d - (\lambda + \mu))^n (a \cdot b) =$$
$$\sum_{i=0}^{n} \binom{n}{i} (d - \lambda)^{n-i}(a) \cdot (d - \mu)^i(b) \quad \text{for } a, b \in A \text{ and } \lambda, \mu \in F. \quad (4.4)$$

It is straightforward to check this formula by induction:

$$(d - (\lambda + \mu))^{n+1}(a \cdot b) = (d - (\lambda + \mu)) \sum_{i=0}^{n} \binom{n}{i} (d - \lambda)^{n-i}(a) \cdot (d - \mu)^i(b)$$

$$= \sum_{i=0}^{n} \binom{n}{i} \left((d - \lambda)^{n-i+1}(a) \cdot (d - \mu)^i(b) + (d - \lambda)^{n-i}(a) \cdot (d - \mu)^{i+1}(b) \right)$$

$$= \sum_{i=0}^{n+1} \binom{n+1}{i} (d - \lambda)^{n+1-i}(a) \cdot (d - \mu)^i(b).$$

In the rest of this section F will be a perfect field.

Lemma 4.6.1 *Let A be a finite-dimensional algebra over F. Let $d \in \text{Der}(A)$ and let $d = d_s + d_n$ be the Jordan decomposition of d (where d_s is semisimple and d_n is nilpotent). Then $d_s, d_n \in \text{Der}(A)$.*

Proof. We prove that $d_s \in \text{Der}(A)$ (then necessarily also $d_n \in \text{Der}(A)$). For this we may assume that the ground field is algebraically closed. Indeed, let $\overline{F}$ be the algebraic closure of F. Then d can also be viewed as a derivation of $\overline{A} = A \otimes \overline{F}$. Let $\{a_1, \ldots, a_n\}$ be a basis of A. Now if we can prove that $d_s \in \text{Der}(\overline{A})$ then also $d_s \in \text{Der}(A)$ because the matrix of d_s relative to the basis $\{a_1 \otimes 1, \ldots, a_n \otimes 1\}$ is defined over F.

Let $\lambda \in F$ be an eigenvalue of d. We set

$$A_\lambda = \{a \in A \mid (d - \lambda)^k(a) = 0 \text{ for some } k > 0\}.$$

Then d_s acts as multiplication by λ on A_λ (see Proposition A.2.6). Now let $\lambda, \mu \in F$ be two eigenvalues of d_s and let a_λ and a_μ be corresponding eigenvectors. From (4.4) it follows that $A_\lambda \cdot A_\mu \subset A_{\lambda+\mu}$. Hence $d_s(a_\lambda \cdot a_\mu) = (\lambda+\mu)(a_\lambda \cdot a_\mu)$. But also $d_s(a_\lambda) \cdot a_\mu + a_\lambda \cdot d_s(a_\mu) = (\lambda+\mu)(a_\lambda \cdot a_\mu)$. Since A is the direct sum of subspaces of the form A_λ (Lemma A.2.2), we conclude that d_s is a derivation of A and we are done. $\square$

Proposition 4.6.2 *Let L be a Lie algebra over F with non-degenerate Killing form. Let $x \in L$, then there are unique $x_s, x_n \in L$ such that $x = x_s + x_n$, $\mathrm{ad}_L x_s$ is semisimple, $\mathrm{ad}_L x_n$ is nilpotent and $[x_s, x_n] = 0$.*

Proof. We consider the derivation $d = \mathrm{ad}x$ of L. Let $d = d_s + d_n$ be the Jordan decomposition of d. Then by Lemma 4.6.1, d_s and d_n are derivations of L. Because the Killing form of L is non-degenerate, all derivations of L are inner (Proposition 4.5.2), and hence there are $x_s, x_n \in L$ such that $\mathrm{ad}x_s = d_s$ and $\mathrm{ad}x_n = d_n$. As the centre of L is zero (Lemma 4.1.3) we see that the map $\mathrm{ad} : L \to \mathfrak{gl}(L)$ is injective. So x_s and x_n are the only elements of L such that $\mathrm{ad}x_s = d_s$ and $\mathrm{ad}x_n = d_n$. Also from $\mathrm{ad}x = \mathrm{ad}(x_s + x_n)$ it follows that $x = x_s + x_n$. Finally, $[x_s, x_n] = 0$ also follows directly from the injectivity of ad. $\square$

Theorem 4.6.3 *Let F be a field of characteristic 0. Let V be a finite-dimensional vector space over F and $L \subset \mathfrak{gl}(V)$ a semisimple linear Lie algebra. Let $x \in L$ and let $x = s + n$ be its Jordan decomposition, where $s \in \mathrm{End}(V)$ is semisimple and $n \in \mathrm{End}(V)$ is nilpotent. Then $s, n \in L$.*

Proof. Instead of proving the statement directly we take a slight detour: we prove that s and n lie in a Lie algebra that is maybe bigger than L. Then we prove that this Lie algebra is in fact equal to L.

Set $N = N_{\mathfrak{gl}(V)}(L)$, the normalizer of L in $\mathfrak{gl}(V)$. By Lemma 2.5.2, $\mathrm{ad}x = \mathrm{ad}s + \mathrm{ad}n$ is the Jordan decomposition of $\mathrm{ad}x$ in $\mathrm{End}(\mathfrak{gl}(V))$. Now, $\mathrm{ad}x(L) \subset L$ and since $\mathrm{ad}s$ and $\mathrm{ad}n$ are polynomials in $\mathrm{ad}x$ (Proposition A.2.6) we see that $\mathrm{ad}s(L) \subset L$ and $\mathrm{ad}n(L) \subset L$, i.e., s and n lie in N.

Let $\overline{F}$ be the algebraic closure of F. Set $\overline{L} = L \otimes_F \overline{F}$ and $\overline{V} = V \otimes_F \overline{F}$. Then $\overline{V}$ is an $\overline{L}$-module. For an $\overline{L}$-submodule W of $\overline{V}$ we set

$$L_W = \{y \in \mathfrak{gl}(V) \mid y(W) \subset W \quad \text{and} \quad \mathrm{Tr}(y|_W) = 0\}.$$

And we put

$$L^* = N \cap \Big(\bigcap_W L_W\Big),$$

where the intersection is taken over all $\overline{L}$-submodules W of $\overline{V}$. Note that the spaces L_W and L^* are subspaces of $\mathfrak{gl}(V)$. Now L^* is the Lie algebra we want. We prove that it contains L (even as an ideal) and that it contains s and n. Then we show that L^* is equal to L.

It is clear that $L \subset N$. Secondly, L is a direct sum of simple ideals (Proposition 4.3.6). And since $[K, K] = K$ if K is a simple Lie algebra we have $L = [L, L]$. Now because the trace of a commutator is always 0 we see that $L \subset L_W$ for every $\overline{L}$-submodule W of $\overline{V}$. It follows that L is a subalgebra of L^*. Also, because L^* is contained in N, we have $[L, L^*] \subset L$, i.e., L is an ideal in L^*.

We show that s and n are elements of L^*. If x maps a subspace W of $\overline{V}$ into itself then, since s and n are polynomials in x, also s and n map W into itself. Furthermore, $\operatorname{Tr}(s|_W) = \operatorname{Tr}(x|_W) = 0$. So s and n are elements of all L_W. Combined with the fact that $s, n \in N$, this leads to the desired conclusion.

Finally we show that $L = L^*$. Since L^* is a finite-dimensional L-module, by Proposition 4.4.5, we may write $L^* = L \oplus M$ (direct sum of L-modules). Because L is an ideal in L^*, we have $[L, M] \subset L \cap M = 0$. So also $[\overline{L}, M] = 0$. Now let $y \in M$ and let W be an irreducible $\overline{L}$-submodule of $\overline{V}$. Let λ be an eigenvalue of y on W (this exists because W is defined over $\overline{F}$). Then since $[\overline{L}, y] = 0$ it is easily seen that the eigenspace of λ is an $\overline{L}$-submodule of W. Hence, since W is irreducible, this eigenspace equals W. But $y \in L^*$ so that $\operatorname{Tr}(y|_W) = 0$ and consequently $\lambda = 0$. It follows that y acts nilpotently on W. In particular $y \cdot W$ is properly contained in W. But $y \cdot W$ is an $\overline{L}$-submodule of W; hence $y \cdot W = 0$. Now, thanks to Weyl's theorem, $\overline{V}$ is the direct sum of irreducible $\overline{L}$-submodules and hence $y \cdot \overline{V} = 0$ and $y = 0$. The conclusion is that $M = 0$ and $L = L^*$. $\square$

Corollary 4.6.4 *Let L be a finite-dimensional semisimple Lie algebra over a field of characteristic 0 and $\rho : L \to \mathfrak{gl}(V)$ a finite-dimensional representation of L. Let $x \in L$ and write $x = x_s + x_n$, where $\operatorname{ad} x_s$ is semisimple, $\operatorname{ad} x_n$ is nilpotent and $[x_s, x_n] = 0$ (cf. Proposition 4.6.2). Then $\rho(x) = \rho(x_s) + \rho(x_n)$ is the Jordan decomposition of $\rho(x) \in \mathfrak{gl}(V)$.*

Proof. As in the proof of Lemma 4.6.1 we may assume that the ground field is algebraically closed. Set $K = \rho(L)$ which is a semisimple subalgebra of $\mathfrak{gl}(V)$. If $y \in L$ is an eigenvector of $\operatorname{ad}_L x_s$, then $\rho(y)$ is an eigenvector of $\operatorname{ad}_K \rho(x_s)$. Now L is spanned by the eigenvectors of $\operatorname{ad}_L x_s$, so K is spanned by the eigenvectors of $\operatorname{ad}_K \rho(x_s)$. Hence $\operatorname{ad}_K \rho(x_s)$ is a semisimple linear transformation. Secondly, $(\operatorname{ad}_K \rho(x_n))^k(\rho(y)) = \rho((\operatorname{ad}_L x_n)^k(y))$

which is zero for k large enough. Hence $\mathrm{ad}_K \rho(x_n)$ is nilpotent. Finally, $[\rho(x_s), \rho(x_n)] = \rho([x_s, x_n]) = 0$ so $\rho(x_s)$ and $\rho(x_n)$ commute.

Let $\rho(x) = s + n$ be the Jordan decomposition of $\rho(x)$ in $\mathrm{End}(V)$. Then by Theorem 4.6.3, $s, n \in K$. Lemma 2.5.2 implies that $\mathrm{ad}_K s$ is semisimple and $\mathrm{ad}_K n$ is nilpotent. Hence by Proposition 4.6.2, $s = x_s$ and $n = x_n$. $\quad\square$

There is a straightforward algorithm for computing the decomposition $x = x_s + x_n$ of Proposition 4.6.2. First we compute the Jordan decomposition $\mathrm{ad}_L x = s + n$ (see Section A.2). Then by solving a system of linear equations we compute elements $x_s, x_n \in L$ such that $\mathrm{ad} x_s = s$ and $\mathrm{ad} x_n = n$ (these equations have a solution by Proposition 4.6.2).

4.7   Levi's theorem

Let L be a Lie algebra over a field of characteristic 0 and suppose that L is not solvable. Then $L/\mathrm{SR}(L)$ is a semisimple Lie algebra. Now Levi's theorem states that this semisimple Lie algebra occurs as a subalgebra of L. It is the analogue for Lie algebras of the theorem of Malcev-Wedderburn for associative algebras (Theorem A.1.5).

Theorem 4.7.1 (Levi) *Let L be a finite-dimensional Lie algebra of characteristic 0. If $\mathrm{SR}(L) \neq L$, then L contains a semisimple subalgebra K such that $L = K \oplus \mathrm{SR}(L)$ (direct sum of subspaces).*

Proof. We distinguish three cases. In the first case we assume that L has no ideals I such that $I \neq 0$ and I is properly contained in $\mathrm{SR}(L)$. We also assume that $[L, \mathrm{SR}(L)] \neq 0$. From the first assumption it follows that $[\mathrm{SR}(L), \mathrm{SR}(L)] = 0$, because otherwise $[\mathrm{SR}(L), \mathrm{SR}(L)]$ would be an ideal of L properly contained in $\mathrm{SR}(L)$.

Now set $V = \mathrm{Hom}_F(L, L)$ (i.e., all linear maps from L into L). Since L is an L-module by the adjoint representation, also V can be made into an L-module by setting (see Section 1.12),

$$(x \cdot \phi)(y) = [x, \phi(y)] - \phi([x, y]) \text{ for } x, y \in L \text{ and } \phi \in V.$$

We consider the following L-submodules of V:

$A = \{\mathrm{ad}_L x \mid x \in \mathrm{SR}(L)\},$

$B = \{\phi \in V \mid \phi(L) \subset \mathrm{SR}(L) \text{ and } \phi(\mathrm{SR}(L)) = 0\},$

$C = \{\phi \in V \mid \phi(L) \subset \mathrm{SR}(L) \text{ and } \phi|_{\mathrm{SR}(L)} \text{ is multiplication by a scalar}\}.$

We have $C \supset B$, and because $[\mathrm{SR}(L), \mathrm{SR}(L)] = 0$ also $B \supset A$.

Let ϕ_λ be an element of C acting on $\mathrm{SR}(L)$ as multiplication by the scalar λ. Then for $x \in L$ and $y \in \mathrm{SR}(L)$ we have

$$(x \cdot \phi_\lambda)(y) = [x, \phi_\lambda(y)] - \phi_\lambda([x,y]) = [x, \lambda y] - \lambda[x,y] = 0. \qquad (4.5)$$

Hence $x \cdot \phi_\lambda \in B$ and we conclude that $L \cdot C \subset B$.

If $x \in \mathrm{SR}(L)$, then because $[\mathrm{SR}(L), \mathrm{SR}(L)] = 0$ we have for all $y \in L$,

$$(x \cdot \phi_\lambda)(y) = [x, \phi_\lambda(y)] - \phi_\lambda([x,y]) = [-\lambda x, y] \qquad (4.6)$$

Consequently, $x \cdot \phi_\lambda = \mathrm{ad}(-\lambda x)$.

So we have derived the following:

$$L \cdot C \subset B \quad \text{and} \quad \mathrm{SR}(L) \cdot C \subset A. \qquad (4.7)$$

Now $\mathrm{SR}(L) \cdot C \subset A$ implies that C/A is a $L/\mathrm{SR}(L)$-module. It contains B/A as a proper submodule. Since $L/\mathrm{SR}(L)$ is semisimple, by Proposition 4.4.5 there is a complementary submodule D/A. Here $D \subset C$ is the full inverse image so $B + D = C$ and $B \cap D = A$. Since $L \cdot C \subset B$ we have that $L/\mathrm{SR}(L)$ must map D/A into B/A. But D/A is a complementary submodule to B/A and hence $L/\mathrm{SR}(L)$ maps D/A to 0. By (4.7), $\mathrm{SR}(L)$ maps D into A, and consequently $L \cdot D \subset A$. Now choose a non-zero ψ in $D \setminus A$. It is not an element of B, so after modifying it by a scalar, we may assume that $\psi|_{\mathrm{SR}(L)}$ is the identity on $\mathrm{SR}(L)$. Furthermore $L \cdot \psi \subset A$ because $L \cdot D \subset A$.

Set

$$K = \{x \in L \mid x \cdot \psi = 0\}.$$

Let x be an element of L. Then since $x \cdot \psi \in A$, we have $x \cdot \psi = \mathrm{ad}y$ for some $y \in \mathrm{SR}(L)$. By (4.6), $y \cdot \psi = -\mathrm{ad}y$, so that $(x + y) \cdot \psi = 0$. It follows that $x + y$ is an element of K and therefore $x = (x + y) - y$ is an element of $K + \mathrm{SR}(L)$. Hence $L = K + \mathrm{SR}(L)$.

We now show that $K \cap \mathrm{SR}(L) = 0$. Let x be a non-zero element of the intersection. Then by (4.6) we have $x \cdot \psi = \mathrm{ad}(-x)$ and hence $\mathrm{ad}(-x) = 0$. So x spans an ideal I of L lying inside $\mathrm{SR}(L)$, and consequently $I = \mathrm{SR}(L)$ or $I = 0$. If $I = \mathrm{SR}(L)$, then since $\mathrm{ad}x = 0$ we have $[L, \mathrm{SR}(L)] = 0$ which was excluded. So $I = 0$ and $x = 0$. The conclusion is that $L = K \oplus \mathrm{SR}(L)$ and we are done in the first case.

In the second case we suppose that $[L, \mathrm{SR}(L)] = 0$. Then $\mathrm{SR}(L) = C(L)$ and $\mathrm{ad}L \cong L/\mathrm{SR}(L)$ (see Example 1.8.3). Hence the adjoint representation of L induces a representation

$$\rho : L/\mathrm{SR}(L) \longrightarrow \mathfrak{gl}(L)$$

of the semisimple Lie algebra $L/\mathrm{SR}(L)$. The radical $\mathrm{SR}(L)$ is a nontrivial submodule, and hence, Proposition 4.4.5, there is a complementary submodule which is the required K.

In the final case we suppose that L has an ideal I properly contained in $\mathrm{SR}(L)$. In this case, by induction, we may assume that the theorem holds for all Lie algebras with a solvable radical of dimension less than $\dim \mathrm{SR}(L)$. Let

$$\pi : L \longrightarrow L/I$$

be the projection map. The solvable radical of L/I is $\mathrm{SR}(L)/I$, so by the induction hypothesis, L/I has a subalgebra K'' complementary to $\mathrm{SR}(L)/I$. Set $K' = \pi^{-1}(K'')$. It is a subalgebra of L containing I with the property that $K'/I = K''$ is semisimple. Therefore the radical of K' is I and hence, by induction,

$$K' = I + K$$

where K is a semisimple subalgebra. Now, by applying π^{-1} to $L/I = \mathrm{SR}(L)/I + K''$ we conclude that

$$L = \mathrm{SR}(L) + K' = \mathrm{SR}(L) + I + K = \mathrm{SR}(L) + K.$$

Furthermore, $\mathrm{SR}(L) \cap K = 0$ because K is semisimple. The conclusion is that $L = K \oplus \mathrm{SR}(L)$.   □

Definition 4.7.2 *Let L be a finite-dimensional Lie algebra of characteristic 0. Let K be a subalgebra provided by Theorem 4.7.1. Then K is called a Levi subalgebra of L.*

Example 4.7.3 Let L be the Lie algebra with basis $\{x_1, \ldots, x_5\}$ and multiplication table as shown in Table 4.1.

	x_1	x_2	x_3	x_4	x_5
x_1	0	$2x_1 - 3x_4$	$-x_2$	x_5	0
x_2	$-2x_1 + 3x_4$	0	$2x_3$	x_4	$-x_5$
x_3	x_2	$-2x_3$	0	0	x_4
x_4	$-x_5$	$-x_4$	0	0	0
x_5	0	x_5	$-x_4$	0	0

Table 4.1: Multiplication table of a 5-dimensional Lie algebra.

It is straightforward to see that the subspace I spanned by x_4 and x_5 is a commutative ideal of L. The quotient algebra L/I is spanned by $\bar{x}_1$,

$\bar{x}_2$ and $\bar{x}_3$ (where $\bar{x}_i$ is the image of x_i in L/I). These elements multiply as follows:

$$[\bar{x}_1, \bar{x}_2] = 2\bar{x}_1, \quad [\bar{x}_1, \bar{x}_3] = -\bar{x}_2, \quad [\bar{x}_2, \bar{x}_3] = 2\bar{x}_3.$$

By comparing with Example 1.3.6 we see $L/I \cong \mathfrak{sl}_2(F)$, which is semisimple (Example 4.2.3). Hence $\mathrm{SR}(L) = I$. We try to find a Levi subalgebra of L. A complement to $\mathrm{SR}(L)$ is spanned by x_1, x_2, x_3. Hence a Levi subalgebra is spanned by y_1, y_2, y_3, where $y_i = x_i + r_i$ for certain $r_i \in \mathrm{SR}(L)$, i.e.,

$$y_1 = x_1 + \alpha x_4 + \beta x_5$$
$$y_2 = x_2 + \gamma x_4 + \delta x_5$$
$$y_3 = x_3 + \epsilon x_4 + \eta x_5.$$

We want to determine the unknowns $\alpha, \dots, \eta$ such that y_1, y_2, y_3 span a subalgebra of L isomorphic to $L/\mathrm{SR}(L)$. This is satisfied if

$$[y_1, y_2] = 2y_1, \quad [y_1, y_3] = -y_2, \quad [y_2, y_3] = 2y_3.$$

The first of these requirements is equivalent to

$$2x_1 - (3 + \alpha)x_4 + (\gamma + \beta)x_5 = 2x_1 + 2\alpha x_4 + 2\beta x_5,$$

which boils down to the equations $\alpha = -1$ and $\beta = \gamma$. The second requirement is equivalent to

$$-x_2 + \epsilon x_5 - \beta x_4 = -x_2 - \gamma x_4 - \delta x_5,$$

i.e., $\delta = -\epsilon$ and $\gamma = \beta$. In the same way it can be seen that the last requirement is equivalent to the equations $\eta = 0$ and $\delta = -\epsilon$. These equations have a 2-parameter family of solutions given by $\alpha = -1$, $\eta = 0$ and $\gamma = \beta$ and $\epsilon = -\delta$.

In particular we see that a Levi subalgebra in general is not unique.

From Example 4.7.3 we can easily distill an algorithm for computing a Levi subalgebra in the case where $\mathrm{SR}(L)$ is Abelian. Later, in Section 4.13, we will generalize such a procedure to an algorithm for the general case.

4.8   Existence of a Cartan subalgebra

Let L be a Lie algebra with a non-degenerate Killing form. In Section 4.9 we will use the primary decomposition of L with respect to a Cartan subalgebra H of L to obtain information about the structure of L. However, if L is defined over a small finite field (of size less than $\dim L$), then it is not clear whether or not L has a Cartan subalgebra. In this section we prove that L is restricted, thereby ensuring that L has a Cartan subalgebra (see Corollary 3.2.10). From Section 1.13 we recall that a Lie algebra of characteristic $p > 0$ is said to be restricted if $(\mathrm{ad}x)^p$ is an inner derivation for all $x \in L$.

Proposition 4.8.1 *Let L be a Lie algebra over a field of characteristic $p > 0$. Suppose that the Killing form of L is non-degenerate. Then L is restricted.*

Proof. Let $x \in L$; then $(\mathrm{ad}x)^p$ is a derivation of L (see Section 1.13). As all derivations of L are inner (Proposition 4.5.2), we have that $(\mathrm{ad}x)^p = \mathrm{ad}y$ for some $y \in L$. But this means that L is restricted. □

Corollary 4.8.2 *Let L be a Lie algebra with a non-degenerate Killing form. Then L has a Cartan subalgebra.*

Proof. L is restricted (Proposition 4.8.1), hence L has a Cartan subalgebra (Corollary 3.2.10). □

4.9   Facts on roots

If L is a Lie algebra with a non-degenerate Killing form then by Corollary 4.8.2, L has a Cartan subalgebra. In this section L will be a Lie algebra over the field F with a non-degenerate Killing form. Throughout we assume that F is perfect. Furthermore we assume that L has a *split* Cartan subalgebra H.

From Section 3.3 we recall that there are functions $\alpha_i : H \to F$ for $1 \leq i \leq s$ such that
$$L = H \oplus L_{\alpha_1} \oplus \cdots \oplus L_{\alpha_s},$$
where $L_{\alpha_i} = \{x \in L \mid$ for all $h \in H$ there is a $k > 0$ such that $(\mathrm{ad}h - \alpha_i(h))^k(x) = 0\}$. This decomposition is called the root space decomposition; the functions $\alpha_i : H \to F$ are roots and the subspaces L_{α_i} are root spaces.

In this section we collect a number of facts on roots and root spaces. These will be used in Sections 4.11 and 4.12, where we describe certain algorithms operating on Lie algebras with a non-degenerate Killing form. Moreover, the results of this section will be of vital importance for the classification of semisimple Lie algebras of characteristic 0, undertaken in Chapter 5. To give the flavour of the results we start with an example.

Example 4.9.1 We consider the Lie algebra $L = \mathfrak{o}_5(\mathbb{Q})$ (see Example 1.2.4). This Lie algebra has non-degenerate Killing form (this will follow from Proposition 5.12.1). Now L is spanned by the elements A_{11}, A_{22}, A_{12}, A_{21}, B_{12}, C_{12}, p_1, p_2, q_1, q_2 (notation as in Example 1.2.4). The multiplication table of L relative to this basis is shown in Table 4.2.

	A_{11}	A_{22}	A_{12}	A_{21}	B_{12}	C_{12}	p_1	p_2	q_1	q_2
A_{11}	0	0	A_{12}	$-A_{21}$	B_{12}	$-C_{12}$	$-p_1$	0	q_1	0
A_{22}	·	0	$-A_{12}$	A_{21}	B_{12}	$-C_{12}$	0	$-p_2$	0	q_2
A_{12}	·	·	0	$A_{11} - A_{22}$	0	0	$-p_2$	0	0	q_1
A_{21}	·	·	·	0	0	0	0	$-p_1$	q_2	0
B_{12}	·	·	·	·	0	$-A_{11} - A_{22}$	$-q_2$	q_1	0	0
C_{12}	·	·	·	·	·	0	0	0	$-p_2$	p_1
p_1	·	·	·	·	·	·	0	$-C_{12}$	A_{11}	A_{21}
p_2	·	·	·	·	·	·	·	0	A_{12}	A_{22}
q_1	·	·	·	·	·	·	·	·	0	$-B_{12}$
q_2	·	·	·	·	·	·	·	·	·	0

Table 4.2: Multiplication table of $L = \mathfrak{o}_5(\mathbb{Q})$.

A Cartan subalgebra is found by applying the algorithm CartanSubalgebraBigField. First $x = A_{11}$ is a non-nilpotent element of L. The subalgebra $L_0(\mathrm{ad}x)$ is spanned by A_{11}, A_{22}, p_2, q_2. This subalgebra is not nilpotent, and $y = A_{22}$ is a non-nilpotent element of $L_0(\mathrm{ad}x)$. If we set $z = x + 2(y - x) = -x + 2y$, then we see that $L_0(\mathrm{ad}z)$ is spanned by A_{11}, A_{22}. From Table 4.2 we infer that $L_0(\mathrm{ad}z)$ is nilpotent and hence it is a Cartan subalgebra.

Let $H = L_0(\mathrm{ad}z)$ be the Cartan subalgebra found above. Then H is Abelian and the two basis elements act diagonally on L. It follows that the root spaces are the common eigenspaces of $\mathrm{ad}A_{11}$ and $\mathrm{ad}A_{22}$. It is seen that every basis element other than A_{11} and A_{22} spans a root space and hence the root spaces are 1-dimensional. Let $h \in H$, and let α be a root. Then $\alpha(h)$ is the eigenvalue of $\mathrm{ad}h$ on the root space corresponding to α. Hence $\alpha : H \to \mathbb{Q}$ is a linear function. So the roots are elements of the dual space $H^* = \{\lambda : H \to \mathbb{Q} \mid \lambda \text{ is linear}\}$. If we represent a $\lambda \in H^*$ by a vector $(\lambda(A_{11}), \lambda(A_{22}))$, then the roots of L relative to H are seen to be

$$(1, -1), \ (-1, 1), \ (1, 1), \ (-1, -1), \ (-1, 0), \ (0, -1), \ (1, 0), \ (0, 1).$$

In particular we see that the set of roots spans H^* and if α is a root, then so is $-\alpha$.

Let $\Phi = \{\alpha_1, \ldots, \alpha_s\}$ be the set of roots. The zero function is not said to be a root, but the space $L_0(H) = H$ occurs in the root space decomposition, and therefore it is often convenient to include the zero function in the arguments. For this reason we set $\Phi^0 = \Phi \cup \{0\}$ (so $L_0 = H$). The first fact is an echo of Proposition 3.3.1; we repeat it here because it is of fundamental importance.

Root fact 1 *Let $\alpha, \beta \in \Phi^0$ and let $x \in L_\alpha$ and $y \in L_\beta$. Then $[x, y] \in L_{\alpha+\beta}$ if $\alpha + \beta \in \Phi^0$, otherwise $[x, y] = 0$.*

Root fact 2 *Let $\alpha, \beta \in \Phi^0$ be such that $\beta \neq -\alpha$. Then $\kappa_L(L_\alpha, L_\beta) = 0$.*

Proof. First suppose that $\beta = 0$. Then we have to prove that $\kappa_L(L_\alpha, H) = 0$. For this choose arbitrary $x_\alpha \in L_\alpha$ and $h \in H$. Since $\alpha \neq 0$ there is a $g \in H$ such that $\alpha(g) \neq 0$. Since $\alpha(g)$ is the single eigenvalue of the restriction of $\mathrm{ad}g$ to L_α, we have that this restriction is a nonsingular linear map. Hence the same holds for $(\mathrm{ad}g)^k$ for $k > 0$. It follows that for any $k \geq 1$, there is a $y_k \in L_\alpha$ such that $x_\alpha = (\mathrm{ad}g)^k y_k$. Now we apply Lemma 4.1.1 k times to find

$$\kappa_L(x_\alpha, h) = \kappa_L((\mathrm{ad}g)^k y_k, h) = (-1)^k \kappa_L(y_k, (\mathrm{ad}g)^k h).$$

But since H is nilpotent, this last element is 0 for k large enough.

Now we assume that $\beta \neq 0$ and choose $x_\beta \in L_\beta$. By the first part of the proof we may assume that $\alpha \neq 0$. Let $g \in H$ be such that $\alpha(g) \neq 0$. As seen above there is a $y_1 \in L_\alpha$ such that $x_\alpha = [g, y_1]$. Hence, by Lemma 4.1.1,

$$\kappa_L(x_\alpha, x_\beta) = \kappa_L([g, y_1], x_\beta) = \kappa_L(g, [y_1, x_\beta]).$$

We have $\alpha + \beta \neq 0$ and if $\alpha + \beta \in \Phi^0$ then $[y_1, x_\beta] \in L_{\alpha+\beta}$, so that by the first part of the proof $\kappa_L(g, [y_1, x_\beta]) = 0$. On the other hand, if $\alpha + \beta \notin \Phi^0$, then $[y_1, x_\beta] = 0$ by Root fact 1, and we reach the same conclusion.   □

Root fact 3 *Let $\alpha \in \Phi$ and let x_α be a non-zero element of L_α. Then there is a $x_{-\alpha} \in L_{-\alpha}$ such that $\kappa_L(x_\alpha, x_{-\alpha}) \neq 0$.*

Proof. Suppose that $\kappa_L(x_\alpha, y) = 0$ for all $y \in L_{-\alpha}$. Then by Root fact 2, $\kappa_L(x_\alpha, L) = 0$ contradicting the non-degeneracy of κ_L.   □

Root fact 4 *The restriction of κ_L to H is non-degenerate.*

Proof. Let $h \in H$ and suppose that $\kappa_L(h, g) = 0$ for all $g \in H$. By Root fact 2 we have that $\kappa_L(h, L_\alpha) = 0$ for all $\alpha \in \Phi$. From this it follows that $\kappa_L(h, L) = 0$. And because κ_L is non-degenerate we must have $h = 0$. $\square$

Root fact 5 *If $\alpha \in \Phi$, then also $-\alpha \in \Phi$.*

Proof. Suppose $-\alpha \notin \Phi$. Then by Root fact 2 we infer that $\kappa_L(L_\alpha, L_\beta) = 0$ for all $\beta \in \Phi^0$. Hence $\kappa_L(L_\alpha, L) = 0$ contradicting the non-degeneracy of κ_L. $\square$

Lemma 4.9.2 *Let V be a finite-dimensional vector space and let $K \subset \mathfrak{gl}(V)$ be a commutative linear Lie algebra. Suppose that K is split. Then there is a basis of V relative to which the matrices of all elements of K are all in upper triangular form.*

Proof. The proof is completely analogous to the proof of Lie's theorem (Theorem 2.4.4). Here we don't need the ground field to be of characteristic 0 in order to prove a statement analogous to Proposition 2.4.2. $\square$

Root fact 6 *H is commutative (i.e., $[H, H] = 0$).*

Proof. Since H is nilpotent, there is an integer $c > 0$ such that $H^{c+1} = 0$ while $H^c \neq 0$. Suppose $c \geq 2$ and choose a non-zero $h \in H^c$. Let $g \in H$, then $[h, g] \in H^{c+1} = 0$. Hence $[\mathrm{ad}h, \mathrm{ad}g] = 0$ and $\mathrm{ad}h$, $\mathrm{ad}g$ span a commutative and split linear Lie algebra. Let $\alpha \in \Phi^0$ and set $V = L_\alpha$. Then by Lemma 4.9.2, there is a basis of V relative to which $\mathrm{ad}_V h$ and $\mathrm{ad}_V g$ both have upper triangular matrices (with $\alpha(h)$, $\alpha(g)$ on the respective diagonals). So $\mathrm{Tr}(\mathrm{ad}_V h \cdot \mathrm{ad}_V g) = (\dim V)\alpha(h)\alpha(g)$. But $h \in H^c$ implies $h \in [H, H]$ so that $\mathrm{Tr}(\mathrm{ad}_V h) = 0$. But this means that $(\dim V)\alpha(h) = 0$, and hence $\mathrm{Tr}(\mathrm{ad}_V h \cdot \mathrm{ad}_V g) = 0$. And because L is the direct sum of the root spaces L_α for $\alpha \in \Phi^0$ we infer that $\mathrm{Tr}(\mathrm{ad}h \cdot \mathrm{ad}g) = 0$. Since this holds for any $g \in H$, we see that the restriction of κ_L to H is degenerate, contradicting Root fact 4. $\square$

Now let $\alpha \in \Phi$ and let L_α be the corresponding root space. Then by Lemma 4.9.2 together with Root fact 6, there is a basis of L_α such that for

$h \in H$ the matrix of $\mathrm{ad}_{L_\alpha} h$ takes the form

$$
\mathrm{ad}_{L_\alpha} h = \begin{pmatrix} \alpha(h) & & * \\ & \ddots & \\ 0 & & \alpha(h) \end{pmatrix}. \tag{4.8}
$$

Let $h, g \in H$, then by (4.8) we have

$$
\kappa_L(h, g) = \mathrm{Tr}(\mathrm{ad}_L h \cdot \mathrm{ad}_L g) = \sum_{\alpha \in \Phi} (\dim L_\alpha) \alpha(h) \alpha(g). \tag{4.9}
$$

Root fact 7 *If $h \in H$ is such that $\alpha(h) = 0$ for all $\alpha \in \Phi$, then $h = 0$.*

Proof. If $h \in H$ satisfies the hypothesis, then $\kappa_L(h, g) = 0$ for all $g \in H$, by (4.9). Hence by Root fact 4, $h = 0$. $\quad\square$

Root fact 8 *Let $h \in H$, then $\mathrm{ad}_L h$ is a semisimple linear transformation.*

Proof. By Proposition 4.6.2 there are unique $h_s, h_n \in L$ such that $h = h_s + h_n$ and $\mathrm{ad}_L h_s$ is semisimple and $\mathrm{ad}_L h_n$ is nilpotent. (Here we use the assumption that the ground field is perfect.) Also, since $\mathrm{ad} h = \mathrm{ad} h_s + \mathrm{ad} h_n$ is the Jordan decomposition of $\mathrm{ad} h$ we have that $\mathrm{ad} h_s$ and $\mathrm{ad} h_n$ are polynomials in $\mathrm{ad} h$ (Proposition A.2.6). Hence $\mathrm{ad} h_s(H) \subset H$ so that $h_s \in N_L(H) = H$ (the last equality is Lemma 3.2.2), and similarly $h_n \in H$. But because $\mathrm{ad} h_n$ is nilpotent, $\alpha(h_n) = 0$ for all $\alpha \in \Phi$. Now Root fact 7 implies $h_n = 0$. $\quad\square$

Root fact 9 *Let $\alpha \in \Phi^0$ and $x_\alpha \in L_\alpha$, then $[h, x_\alpha] = \alpha(h) x_\alpha$ for all $h \in H$.*

Proof. Let $h \in H$, and let X be an indeterminate. Since $\mathrm{ad}_L h$ is semisimple (Root fact 8) we have that the minimum polynomial of $\mathrm{ad}_{L_\alpha} h$ is $X - \alpha(h)$. So on L_α the endomorphism $\mathrm{ad} h$ acts as $\alpha(h)$ times the identity. $\quad\square$

By H^* we denote the dual space of H, i.e.,

$$
H^* = \{\phi : H \to F \mid \phi \text{ is linear}\}.
$$

The space H^* has the same dimension as H. Furthermore, by Root fact 9 we see that $\alpha(h_1 + h_2) = \alpha(h_1) + \alpha(h_2)$ and $\alpha(\lambda h) = \lambda \alpha(h)$ for $\alpha \in \Phi$ and $h_1, h_2, h \in H$ and $\lambda \in F$. It follows that the roots are elements of H^*. By the next fact they even span the space H^*.

Root fact 10 *There are* dim H *linearly independent roots.*

Proof. Suppose that the space spanned by the roots is of strictly smaller dimension than H. This implies that there is a non-zero element $h \in H$ such that $\alpha(h) = 0$ for all $\alpha \in \Phi$. Indeed, let $\{h_1, \dots, h_l\}$ be a basis of H, and let $\{\alpha_1, \dots, \alpha_r\}$ be a basis of the subspace of H^* spanned by the roots. Set $h = \sum_i \lambda_i h_i$, where the λ_i are unknown scalars that are to be determined. Then $\alpha_j(h) = 0$ is equivalent to $\sum_i \alpha_j(h_i)\lambda_i = 0$. Collecting these equations for $1 \leq j \leq r$ together we get an equation system for the λ_i, of rank $r < \dim H$. Hence a non-zero solution can be found. This however contradicts Root fact 7. $\qquad\square$

The non-degeneracy of κ_L allows us to identify the spaces H^* and H. Let $\sigma \in H^*$ then we define a corresponding element $h_\sigma \in H$ by the equation

$$\kappa_L(h, h_\sigma) = \sigma(h) \quad \text{for all } h \in H. \tag{4.10}$$

The fact that κ_L is non-degenerate ensures the existence and uniqueness of h_σ (see Lemma 4.5.1). So the map $\sigma \mapsto h_\sigma$ is a bijective linear map.

Example 4.9.3 Let L be the Lie algebra of Example 4.9.1. Set $h_1 = A_{11}$ and $h_2 = A_{22}$. Then h_1, h_2 span a Cartan subalgebra H of L. Let α be a root of L relative to H. We want to calculate h_α. For this set $h_\alpha = ah_1 + bh_2$, where $a, b \in \mathbb{Q}$ are scalars to be determined. Now (4.10) is equivalent to $\kappa_L(h_i, h_\alpha) = \alpha(h_i)$ for $1 \leq i \leq 2$. Hence a, b must satisfy the linear equations

$$\kappa_L(h_1, h_1)a + \kappa_L(h_1, h_2)b = \alpha(h_1)$$
$$\kappa_L(h_2, h_1)a + \kappa_L(h_2, h_2)b = \alpha(h_2).$$

Since h_1 and h_2 act diagonally, $\kappa_L(h_i, h_j)$ is easily read off from the multiplication table (Table 4.2). We have $\kappa_L(h_1, h_1) = \kappa_L(h_2, h_2) = 6$ and $\kappa_L(h_1, h_2) = \kappa_L(h_2, h_1) = 0$. Therefore, $a = \alpha(h_1)/6$ and $b = \alpha(h_2)/6$.

Root fact 11 *Let* $x_\alpha \in L_\alpha$ *and* $x_{-\alpha} \in L_{-\alpha}$. *Then we have* $[x_\alpha, x_{-\alpha}] = \kappa_L(x_\alpha, x_{-\alpha})h_\alpha$.

Proof. Let $h \in H$; using Lemma 4.1.1 and Root fact 9 we calculate

$$\kappa_L([x_\alpha, x_{-\alpha}], h) = \kappa_L(x_\alpha, [x_{-\alpha}, h]) = \alpha(h)\kappa_L(x_\alpha, x_{-\alpha}),$$

and using (4.10),

$$\kappa_L(\kappa_L(x_\alpha, x_{-\alpha})h_\alpha, h) = \kappa_L(x_\alpha, x_{-\alpha})\kappa_L(h_\alpha, h) = \alpha(h)\kappa_L(x_\alpha, x_{-\alpha}).$$

Now the non-degeneracy of κ_L on H (Root fact 4) gives the desired result. $\qquad\square$

Example 4.9.4 Again we let L be the Lie algebra from Example 4.9.1. Let α be the root corresponding to the vector $(1, -1)$ (i.e., $\alpha(A_{11}) = 1$ and $\alpha(A_{22}) = -1$). Then α is a root of L and A_{12} spans the corresponding root space. Furthermore A_{21} spans the root space corresponding to $-\alpha$. By Example 4.9.3 we see that

$$h_\alpha = \frac{1}{6}A_{11} - \frac{1}{6}A_{22}.$$

Also $[A_{12}, A_{21}] = A_{11} - A_{22}$. Hence by Root fact 11, $\kappa_L(A_{12}, A_{21}) = 6$, a fact which may be verified by direct calculation.

The map $\sigma \mapsto h_\sigma$ gives rise to a bilinear form $(\ ,\ )$ in H^* defined by $(\sigma, \rho) = \kappa_L(h_\sigma, h_\rho)$. We have that $(\ ,\ )$ is a non-degenerate symmetric bilinear form in H^* (this follows immediately from the corresponding properties of κ_L). We note that $(\sigma, \sigma) = \kappa_L(h_\sigma, h_\sigma) = \sigma(h_\sigma)$; this fact will be used frequently.

Example 4.9.5 We consider the Lie algebra L from Example 4.9.1. As in Example 4.9.3 we set $h_1 = A_{11}$ and $h_2 = A_{22}$. Using Example 4.9.3 it is easy to calculate (α, β) for roots α, β of L. For instance let $\alpha = (1, 1)$ and $\beta = (-1, -1)$, then

$$(\alpha, \beta) = \kappa_L(h_\alpha, h_\beta) = \kappa_L\left(\frac{1}{6}h_1 + \frac{1}{6}h_2, -\frac{1}{6}h_1 - \frac{1}{6}h_2\right) = -\frac{1}{3}.$$

Root fact 12 *Let $\alpha \in \Phi$ be such that $(\alpha, \alpha) \neq 0$. Set*

$$h = \frac{2h_\alpha}{(\alpha, \alpha)},$$

and choose a non-zero $x \in L_\alpha$. Then there is a $y \in L_{-\alpha}$ such that

$$[h, x] = 2x, \quad [h, y] = -2y, \quad [x, y] = h,$$

(i.e., $\{x, y, h\}$ spans a subalgebra of L isomorphic to $\mathfrak{sl}_2(F)$).

Proof. By Root fact 3 there is a $y \in L_{-\alpha}$ such that $\kappa_L(x, y) \neq 0$. After modifying y by a scalar we may assume that

$$\kappa_L(x, y) = \frac{2}{(\alpha, \alpha)}.$$

Then by Root fact 11 it follows that $[x, y] = h$. The other two product relations follow from $\alpha(h) = 2$.   □

Now we prove two facts for the case where the field of definition is of characteristic 0. The statements also hold over fields of positive characteristic (not equal to $2, 3$); but the proofs are considerably longer. Therefore we postpone the proofs for the other characteristics until Section 4.10.

Root fact 13 *Assume that the ground field F is of characteristic 0. Then $(\alpha, \alpha) \neq 0$ for $\alpha \in \Phi$.*

Proof. Let x_α be a non-zero element of L_α. By Root fact 3 there is an $x_{-\alpha} \in L_{-\alpha}$ such that $\kappa_L(x_\alpha, x_{-\alpha}) \neq 0$. After modifying $x_{-\alpha}$ by a scalar we may assume that $\kappa_L(x_\alpha, x_{-\alpha}) = 1$.

Now suppose that $(\alpha, \alpha) = \alpha(h_\alpha) = 0$, then $[h_\alpha, x_\alpha] = [h_\alpha, x_{-\alpha}] = 0$. Furthermore, by Root fact 11, $[x_\alpha, x_{-\alpha}] = h_\alpha$ so that $\{h_\alpha, x_\alpha, x_{-\alpha}\}$ spans a solvable subalgebra K of L. We consider the representation $\mathrm{ad}_L : K \to \mathfrak{gl}(L)$ of K. By Lemma 2.6.2 the elements of $\mathrm{ad}_L[K, K]$ are nilpotent, in particular $\mathrm{ad}_L h_\alpha$ is nilpotent. By Root fact 8 we see that $\mathrm{ad}_L h_\alpha$ is semisimple. The conclusion is that $h_\alpha = 0$, which implies $\alpha = 0$. So from the assumption $(\alpha, \alpha) = 0$ we have reached a contradiction. □

Root fact 14 *Suppose that the ground field F is of characteristic 0. Let $\alpha \in \Phi$, then $\dim L_\alpha = 1$ and the only integral multiples of α which are roots are α and $-\alpha$.*

Proof. For $\sigma \in H^*$ we set $L_\sigma = 0$ if $\sigma \notin \Phi^0$. Let N be the subspace of L spanned by all root spaces of the form $L_{-k\alpha}$ for $k \geq 1$, i.e.,

$$N = \bigoplus_{k \geq 1} L_{-k\alpha}.$$

Then by Root fact 1 we see that N is closed under multiplication, so that N is a subalgebra of L. Let x_α be a non-zero element of L_α and let $K \subset L$ be the subspace spanned by x_α, h_α together with N. By Root fact 9 it is seen that $\mathrm{ad} h_\alpha(x_\alpha) = \alpha(h_\alpha)x_\alpha$ and if $-k\alpha$ is a root, $\mathrm{ad} h_\alpha(x_{-k\alpha}) = -k\alpha(h_\alpha)x_{-k\alpha}$ for $x_{-k\alpha} \in L_{-k\alpha}$. Hence $\mathrm{ad} h_\alpha$ maps K into itself and if we set $n_\beta = \dim L_\beta$, then

$$\mathrm{Tr}(\mathrm{ad}_K h_\alpha) = \alpha(h_\alpha)(1 - n_{-\alpha} - 2n_{-2\alpha} - \cdots). \tag{4.11}$$

Choose $x_{-\alpha} \in L_{-\alpha}$ such that $\kappa_L(x_\alpha, x_{-\alpha}) \neq 0$ (such $x_{-\alpha}$ exists by Root fact 3); and after modifying $x_{-\alpha}$ by a scalar we may assume that $\kappa_L(x_\alpha, x_{-\alpha}) = 1$. By Root fact 1 we have that K is a subalgebra. In particular K is invariant under $\mathrm{ad}_K x_\alpha$ and $\mathrm{ad}_K x_{-\alpha}$. Also by Root fact 11 we have that $[x_\alpha, x_{-\alpha}] = h_\alpha$. So $\mathrm{ad}_K h_\alpha = [\mathrm{ad}_K x_\alpha, \mathrm{ad}_K x_{-\alpha}]$ and it is seen that $\mathrm{Tr}(\mathrm{ad}_K h_\alpha) = 0$. By Root fact 13 we have that $\alpha(h_\alpha) = (\alpha, \alpha) \neq 0$. So from (4.11) we infer that $1 - n_{-\alpha} - 2n_{-2\alpha} - \cdots = 0$. But this can only happen if $n_{-\alpha} = 1$ and $n_{-2\alpha} = n_{-3\alpha} = \ldots = 0$. So $\dim L_{-\alpha} = 1$ and $-2\alpha, -3\alpha, \cdots$ are not roots. We can replace α by $-\alpha$ in the above argument and obtain $\dim L_\alpha = 1$ and $2\alpha, 3\alpha, \cdots$ are not roots.   □

4.10   Some proofs for modular fields

In this section we prove that the statements of Root facts 13 and 14 also hold over fields of positive characteristic $p \neq 2, 3$.

Throughout this section we assume that F is a perfect field of characteristic $p > 0$ and $p \neq 2, 3$. Furthermore, L is a Lie algebra over F with a non-degenerate Killing form and H is a split Cartan subalgebra of L. We recall that Φ is the set of roots of L (with respect to H).

Let $\alpha \in \Phi$ be a non-zero root such that $(\alpha, \alpha) \neq 0$. Then by Root fact 12, there are $x \in L_\alpha$, $y \in L_{-\alpha}$ and $h \in H$ such that $[x, y] = h$, $[h, x] = 2x$, $[h, y] = -2y$. Let K be the subalgebra of L spanned by x, y, h. Then the adjoint representation of L induces a representation $\mathrm{ad}_L : K \to \mathfrak{gl}(L)$ of K. Representations of these kind will be one of our main tools in this section.

Proposition 4.10.1 *Let $\rho : K \to \mathfrak{gl}(V)$ be a finite-dimensional representation of K. Suppose that there is a basis of V relative to which $\rho(h)$ has a diagonal matrix. Suppose further that there is an eigenvector v_0 of $\rho(h)$ such that $\rho(x)v_0 = 0$. Set $v_i = \rho(y)^i v_0$ for $i \geq 0$ and let $\lambda \in F$ be the eigenvalue corresponding to v_0. Then $\rho(h)v_i = (\lambda - 2i)v_i$ and $\rho(x)v_i = i(\lambda - i + 1)v_{i-1}$ for $i \geq 0$.*

Proof. Let v_μ be an eigenvector of $\rho(h)$ with eigenvalue μ. Then we calculate $\rho(h)\rho(y)v_\mu = \rho(y)\rho(h)v_\mu + \rho([h, y])v_\mu = (\mu - 2)\rho(y)v_\mu$. Hence $\rho(v_\mu)$ is an eigenvector of $\rho(h)$ with eigenvalue $\mu - 2$. The first equality follows from this. For the second equality we use induction on i. Setting $v_{-1} = 0$, it certainly holds for $i = 0$; so suppose $i \geq 0$ and $\rho(x)v_i = i(\lambda - i + 1)v_{i-1}$. Then

$$
\begin{aligned}
\rho(x)v_{i+1} = \rho(x)\rho(y)v_i &= \rho(y)\rho(x)v_i + \rho([x, y])v_i \\
&= i(\lambda - i + 1)\rho(y)v_{i-1} + \rho(h)v_i \\
&= \big(i(\lambda - i + 1) + \lambda - 2i\big)v_i \\
&= (i + 1)(\lambda - i)v_i.
\end{aligned}
$$

So by induction we have the second equality.   □

Proposition 4.10.2 *Let* $\alpha \in \Phi$ *be a non-zero root. If* $\alpha(h_\alpha) \neq 0$, *then not all integral multiples of* α *are roots.*

Proof. Set $h = \frac{2}{\alpha(h_\alpha)}h_\alpha$ and choose an arbitrary non-zero $x \in L_\alpha$. Then by Root fact 12, there is an $y \in L_{-\alpha}$ such that

$$[h,x] = 2x, \ [h,y] = -2y, \ [x,y] = h.$$

Hence the subspace K spanned by h, x, y is a subalgebra of L isomorphic to $\mathfrak{sl}_2(F)$.

Now assume that $\alpha, 2\alpha, \dots, (p-1)\alpha$ are all roots. This is equivalent to assuming that, $-\alpha, -2\alpha, \dots, -(p-1)\alpha$ are all roots. Let $2 \leq k \leq p-2$ and suppose that there is a non-zero $e \in L_{-k\alpha}$ such that $[x,e] = 0$. Then since $y \in L_{-k\alpha}$ we have $(\mathrm{ad}y)^{p-k-1}e \in L_\alpha$ and by Root fact 11 we see that $(\mathrm{ad}y)^{p-k}e = \mu h_\alpha$ for a $\mu \in F$. Now using Lemma 4.1.1,

$$\mu\alpha(h_\alpha) = \mu\kappa_L(h_\alpha, h_\alpha) = \kappa_L((\mathrm{ad}y)^{p-k}e, h_\alpha) = \pm\kappa_L(e, (\mathrm{ad}y)^{p-k}h_\alpha) = 0,$$

(the last equality follows from the fact that $(\mathrm{ad}y)^2 h_\alpha = 0$ and $p - k \geq 2$). Since $\alpha(h_\alpha) \neq 0$ by hypothesis, we have that $(\mathrm{ad}y)^{p-k}e = 0$.

Let $u \in \{0, 1, \cdots, p-k-1\}$ be the smallest integer such that $(\mathrm{ad}y)^{u+1}e = 0$. Set $v_0 = e$ and for $i \geq 0$, $v_i = (\mathrm{ad}y)^i v_0$. Then $v_{u+1} = 0$ and $v_i \neq 0$ for $0 \leq i \leq u$. Furthermore, since $\mathrm{ad}x(v_0) = 0$ we are in the situation of Proposition 4.10.1 with $\lambda = -2k$. By Proposition 4.10.1 we now have

$$0 = \mathrm{ad}x(v_{u+1}) = (u+1)(-2k - (u+1) + 1)v_u = (u+1)(-2k-u)v_u.$$

And since $u + 1 \neq 0$ we see that $2k \equiv -u \pmod{p}$.

The conclusion is that if there are $0 \neq x \in L_\alpha$ and $0 \neq e \in L_{-k\alpha}$ for a $k \in \{2, \dots, p-2\}$, such that $[x,e] = 0$, then it follows that $2k \equiv -u \pmod{p}$ for some $u \in \{0, \dots, p-k-1\}$.

Now set $\beta = 2\alpha$, then also $\beta(h_\beta) \neq 0$ and by assumption all integral multiples of β are roots. Therefore the above conclusion is also valid for β. Let $m = \frac{p+1}{2}$, then $m\beta = \alpha$. So we can turn things around: our x above lies in $L_{m\beta}$ and if in the above we take $k = 2$, then the element e such that $[x,e] = 0$ lies in $L_{-\beta}$. So the conclusion in this case (with m instead of k) reads $p + 1 \equiv -t \pmod{p}$ with $t \in \{0, 1, \dots, \frac{p-1}{2}\}$. This is clearly not possible and from the assumption that there is an $e \in L_{-2\alpha}$ such that $[x,e] = 0$ we have reached a contradiction. As a consequence $\mathrm{ad}x : L_{-2\alpha} \to L_{-\alpha}$ is injective.

From the injectivity of ad$x$ on $L_{-2\alpha}$ it follows that $\dim L_{-2\alpha} \leq \dim L_{-\alpha}$ and applying this repeatedly we get $\dim L_{-2^j\alpha} \leq \dim L_{-2^{j-1}\alpha} \leq \cdots \leq \dim L_{-2\alpha} \leq \dim L_{-\alpha}$. Now take $j = p - 1$, then we see that $\dim L_{-2\alpha} = \dim L_{-\alpha}$ and ad$x : L_{-2\alpha} \to L_{-\alpha}$ is a bijection. So $y = [x, e]$ for an $e \in L_{-2\alpha}$. Now finally

$$0 \neq \kappa_L(x, y) = \kappa_L(x, [x, e]) = \kappa_L([x, x], e) = 0.$$

We have reached a contradiction and $\alpha, 2\alpha, \ldots, (p-1)\alpha$ are not all roots. $\square$

Lemma 4.10.3 *Let $\alpha \in \Phi$ be such that $\alpha(h_\alpha) \neq 0$. Then 2α is not a root.*

Proof. Suppose that $2\alpha \in \Phi$. Let $r \geq 2$ be an integer such that $\alpha, 2\alpha, \ldots, r\alpha$ are roots, but $(r + 1)\alpha$ is not a root (such r exists by Proposition 4.10.2). Then since $(p - 1)\alpha = -\alpha$ is a root, we must have $r \leq p - 3$. Let K be the subalgebra spanned by h, x, y where h, x, y are as in the proof of Proposition 4.10.2. Now also $-\alpha, -2\alpha, \ldots, -r\alpha$ are roots, but $-(r + 1)\alpha$ is not. Set

$$V = K \oplus L_{-\alpha} \oplus L_{-2\alpha} \oplus \cdots L_{-r\alpha}$$

then by Root facts 1 and 11 we have that V is stable under ad$K$. Let $0 \neq e \in L_{-2\alpha}$ and suppose that $[x, e] = 0$. Set $w_0 = e$ and $w_i = (\text{ad}y)^i e$. We are now in the situation of Proposition 4.10.1 where $\lambda = -4$. There is an $s > 0$ such that $w_s \neq 0$ and $w_{s+1} = 0$. By Proposition 4.10.1 we calculate $0 = \text{ad}x(w_{s+1}) = (s + 1)(-4 - s)w_s$. Since $s \leq r - 2$ we have that $s + 1 \neq 0$. Hence $s = p - 4$ and $w_s \in L_{-(p-2)\alpha}$. But this last space does not lie in V and we have a contradiction. So $[x, e] \neq 0$.

Now set $v_0 = [x, e]$. Then $v_0 \in L_{-\alpha}$ and $\kappa_L(x, v_0) = \kappa_L(x, [x, e]) = \kappa_L([x, x], e) = 0$. So by Root fact 11 we see that $[x, v_0] = 0$. Set $v_i = (\text{ad}y)^i v_0$ and again we are in the situation of Proposition 4.10.1 where this time $\lambda = -2$. There is an integer $s \leq r$ such that $v_s \neq 0$ and $v_{s+1} = 0$. By Proposition 4.10.1 we infer that $0 = \text{ad}x(v_{s+1}) = (s + 1)(-2 - s)v_s$. It follows that $s = p - 2$ which is not possible in view of $r \leq p - 3$. So again we have a contradiction and 2α is not a root. $\square$

Root fact 15 *For $\alpha \in \Phi$ we have that $(\alpha, \alpha) = \alpha(h_\alpha) \neq 0$.*

Proof. Suppose that $\alpha(h_\alpha) = 0$. Since $h_\alpha \neq 0$, by Root fact 7 we see that there is a $\beta \in \Phi$ such that $\beta(h_\alpha) \neq 0$. We claim that $\beta + k\alpha$ are roots for $k \geq 0$.

It is enough to prove that $\beta + \alpha$ is a root, for then we can continue with $\beta + \alpha$ in place of β (since also $(\beta+\alpha)(h_\alpha) \neq 0$). Suppose that $\beta + \alpha$ is not a root. Choose non-zero $x_\alpha \in L_\alpha$ and $x_{-\alpha} \in L_{-\alpha}$ such that $\kappa_L(x_\alpha, x_{-\alpha}) = 1$ (so that $[x_\alpha, x_{-\alpha}] = h_\alpha$ by Root fact 11). Furthermore choose a non-zero $x_\beta \in L_\beta$. Then by Root fact 1, $[x_\alpha, x_\beta] = 0$. Set $v_i = (\mathrm{ad}x_{-\alpha})^i x_\beta$ for $i \geq 0$. Then $\mathrm{ad}x_\alpha(v_i) = i\beta(h_\alpha)v_{i-1}$ as is easily proved by induction. Then because $v_{p-1} \in L_{\beta+\alpha} = 0$ there is an integer $k \in \{0, \dots, p-2\}$ such that $v_k \neq 0$ while $v_{k+1} = 0$. Hence

$$0 = \mathrm{ad}x_\alpha(v_{k+1}) = (k+1)\beta(h_\alpha)v_k.$$

From which it follows that $k + 1 = 0$, which is a contradiction. It follows that $\alpha + \beta$ is a root.

Now we prove that $\beta(h_\beta) \neq 0$. Suppose on the contrary that $\beta(h_\beta) = 0$. Then

$$(\beta + \alpha)(h_{\beta+\alpha}) = (\beta + \alpha)(h_\beta + h_\alpha) = 2\kappa_L(h_\alpha, h_\beta) = 2\beta(h_\alpha) \neq 0,$$

and by Lemma 4.10.3, $2(\alpha + \beta)$ is not a root. Now by the above, $\beta + 2\alpha$ is a root. And $(\beta + 2\alpha)(h_\beta) = 2\kappa_L(h_\alpha, h_\beta) \neq 0$; so by the first part of the proof $(\beta + 2\alpha) + \beta$ is a root and we have obtained a contradiction.

Let $\gamma = \alpha + 2\beta$, then γ is not a root. Indeed, suppose that γ is a root. We note that $\gamma(h_\alpha) \neq 0$ and hence by the first part of the proof $\gamma - \alpha = 2\beta$ is a root. But this contradicts Lemma 4.10.3, since $\beta(h_\beta) \neq 0$. In the same way it can be seen that $\alpha - 2\beta$ is not a root.

Let $x \in L_\beta$, $y \in L_{-\beta}$ and $h = ah_\beta$ be chosen as in Root fact 12 and let K be the subalgebra spanned by these elements. Set

$$V = L_{\alpha-\beta} \oplus L_\alpha \oplus L_{\alpha+\beta}$$

then by the above V is stable under $\mathrm{ad}_L K$. Let $0 \neq v_0 \in L_{\alpha+\beta}$. Then v_0 is an eigenvector of $\mathrm{ad}h$ with eigenvalue $(\alpha + \beta)(h) = \alpha(h) + 2$. Set $v_1 = \mathrm{ad}y(v_0)$ and $v_2 = (\mathrm{ad}y)^2(v_0)$. Then v_1, v_2 are eigenvectors of $\mathrm{ad}h$ with eigenvalues $\alpha(h)$ and $\alpha(h) - 2$ respectively. Now the assumptions $v_1 = 0$ and $v_2 = 0$ both lead to contradictions. (The first assumption implies $\alpha(h) = -2$ and in the second case we have $\alpha(h) = -1$. In both cases we consider a non-zero vector $w_0 \in L_{\alpha-\beta}$ and the K-module generated by it; and in both cases we obtain a contradiction.) Now $v_3 = (\mathrm{ad}y)^3 v_0 = 0$ and Proposition 4.10.1 implies that $0 = \mathrm{ad}x(v_3) = 3\alpha(h)v_2$. Since by assumption $3 \neq 0$ we must have $\alpha(h) = 0$, but $\alpha(h) = a\alpha(h_\beta) = a\kappa_L(h_\alpha, h_\beta) = a\beta(h_\alpha) \neq 0$. It follows that $\alpha(h_\alpha) \neq 0$. $\qquad\square$

Root fact 16 *Let $\alpha \in \Phi$. Then* $\dim L_\alpha = 1$ *and the only integral multiples of α which are roots are $\pm\alpha$.*

Proof. Since $\alpha(h_\alpha) \neq 0$ (Root fact 15), we can choose h, x, y as in Root fact 12. Suppose that $\dim L_\alpha > 1$. This implies that there is a non-zero $e \in L_\alpha$ such that $\kappa_L(e, y) = 0$; so that $[e, y] = 0$ (Root fact 11). Also, since 2α is not a root by Lemma 4.10.3 we have that $[x, e] = 0$. Now by the Jacobi identity,

$$[e, h] = [e, [x, y]] = -[x, [y, e]] - [y, [e, x]] = 0.$$

But this is a contradiction since $[h, e] = \alpha(h)e = 2e$.

Suppose that $k\alpha$ is a root, where $k \in \{2, 3, \cdots, p - 1\}$. Also, since $k\alpha$ is a root if and only if $-k\alpha$ is a root (Root fact 5), we may assume that $k \in \{2, 3, \cdots, \frac{p-1}{2}\}$. Furthermore, by Lemma 4.10.3 we have that 2α is not a root. So $k \neq 2$. Set $\beta = \frac{p-1}{2}\alpha$; then if β is a root, $-\alpha = 2\beta$ is not (Lemma 4.10.3). But $-\alpha$ is a root so that β is not a root and $k \neq \frac{p-1}{2}$. Now let k be the smallest integer > 1 such that $k\alpha$ is a root. Then

$$2 < k < \frac{p-1}{2}.$$

Let $r \geq 0$ be an integer such that $k\alpha, (k+1)\alpha, \ldots, (k+r)\alpha$ are roots, but $(k + r + 1)\alpha$ is not a root. Then

$$k + r < \frac{p-1}{2}.$$

Set $V = L_{k\alpha} \oplus L_{(k+1)\alpha} \oplus \cdots \oplus L_{(k+r)\alpha}$. Let K be the subalgebra of L spanned by h, x, y; then V is stable under $\mathrm{ad}K$. Let $0 \neq v_0 \in L_{(k+r)\alpha}$ and set $v_i = (\mathrm{ad}y)^i v_0$. Then we are once more in the situation of Proposition 4.10.1 with $\lambda = 2(k + r)$. Let $s \geq 0$ be the smallest integer such that $v_s \neq 0$ and $v_{s+1} = 0$. Then by Proposition 4.10.1 we calculate $0 = \mathrm{ad}x(v_{s+1}) = (s+1)(2k+2r-s)v_s$. Since $s \leq r$, we see that $s+1 \neq 0$ and hence $2(k + r) \equiv s \pmod{p}$. But $0 < 2(k + r) < p - 1$ and $s \leq r < (p - 1)$ so that $2(k + r) = s$ which is absurd in view of $s \leq r$. $\qquad\square$

4.11   Splitting and decomposing elements

Let L be a Lie algebra with a non-degenerate Killing form. In this section we consider the problem of computing the (collected) primary decomposition of L relative to a Cartan subalgebra H. Throughout we assume that L is defined over a perfect field F of characteristic $p \neq 2, 3$. We do *not* assume that H is split.

The first algorithm for computing the primary decomposition that comes into mind is to pick an element $h_1 \in H$ and compute the primary decomposition of L relative to $\mathrm{ad}h_1$. We can do this by factorizing the characteristic polynomial of $\mathrm{ad}h_1$. Then for every irreducible factor f_i we calculate the space $L_0(f_i(\mathrm{ad}h_1))$. Now the spaces $L_0(f_i(\mathrm{ad}h_1))$ are the primary components of L relative to $\mathrm{ad}h_1$ (Lemma A.2.2). Then by Proposition 3.1.7 all components that we have obtained are invariant under $\mathrm{ad}H$. So we can continue and pick a second element $h_2 \in H$ and decompose each primary component $L_0(f_i(\mathrm{ad}h_1))$ relative to $\mathrm{ad}h_2$. Continuing like this we will find the primary decomposition of L relative to H. The problem however is to find a good stopping criterion: how can we ascertain that the restrictions of *all* elements of H to a certain component have a minimum polynomial that is a power of an irreducible polynomial? Here we show that this can be achieved by looking at particular elements, namely splitting elements (if the ground field is big) and decomposing elements (if the ground field is small). If these elements have a minimum polynomial that is a power of an irreducible polynomial, then this will hold for all elements of H.

Let $h \in H$. Then all eigenvalues of the restriction of $\mathrm{ad}h$ to H are 0 (Root fact 6). Furthermore, since all root spaces are 1-dimensional (Root facts 14, 16), we have that $\mathrm{ad}h$ has no more than $\dim L - \dim H + 1$ different eigenvalues. The element h is called a *splitting element* if it has exactly that number of eigenvalues. Since $\mathrm{ad}h$ is a semisimple linear transformation (Root fact 8) we have that $\mathrm{ad}h$ has $\dim L - \dim H + 1$ different eigenvalues if and only if the minimum polynomial of $\mathrm{ad}h$ has degree $\dim L - \dim H + 1$ (Proposition A.2.4). Hence we have a good criterion for deciding whether or not an element $h \in H$ is a splitting element, without having to calculate the eigenvalues of $\mathrm{ad}h$ (which might involve factorizing the minimum polynomial of $\mathrm{ad}h$ over a big extension field of the ground field F).

Let $\tilde{F}$ be an extension field of F such that H splits over $\tilde{F}$. Set $\tilde{L} = L \otimes_F \tilde{F}$. We note that H can be viewed as a subset of $\tilde{L}$ and $\tilde{H} = H \otimes_F \tilde{F}$ is a Cartan subalgebra of $\tilde{L}$ (Proposition 3.2.3). The Lie algebra $\tilde{L}$ has a root space decomposition relative to $\tilde{H}$. Let Φ be the set of roots of $\tilde{L}$ relative to $\tilde{H}$ and set $\Phi^0 = \Phi \cup \{0\}$. Let $\alpha \in \Phi$, then since H is a subset of $\tilde{H}$ we can restrict α to H and we get an F-linear function $\alpha : H \to \tilde{F}$.

The next lemma follows immediately from the definition of splitting element.

Lemma 4.11.1 *We have that $h \in H$ is a splitting element if and only if* *all elements $\alpha(h)$ are different for $\alpha \in \Phi^0$.*

Proposition 4.11.2 *Set $N = \dim L - \dim H$ and $m = N(N+1)/2$. Let $\{h_1, \ldots, h_l\}$ be a basis of H. Let $0 < \epsilon < 1$ and let Ω be a subset of F of size at least m/ϵ. Let $\lambda_1, \ldots, \lambda_l$ be random elements chosen uniformly and independently from Ω. Then the probability that $h = \sum \lambda_i h_i$ is a splitting element is at least $1 - \epsilon$.*

Proof. Denote the elements of Φ by $\alpha_1, \ldots, \alpha_N$. By α_0 we denote the element $0 \in \Phi^0$. Let $X_1, \ldots, X_l$ be l indeterminates and for $0 \leq i, j \leq N$ put

$$f_{ij}(X_1, \ldots, X_l) = \sum_{k=1}^{l} (\alpha_i - \alpha_j)(h_k) X_k,$$

and set

$$g(X_1, \ldots, X_l) = \prod_{0 \leq i < j \leq N} f_{ij}(X_1, \ldots, X_n).$$

Then $g \in \tilde{F}[X_1, \ldots, X_n]$ is a polynomial of degree m. We claim that $\sum \lambda_i h_i$ is not a splitting element if and only if $g(\lambda_1, \ldots, \lambda_l) = 0$. Indeed, this last condition is the same as saying that there are $0 \leq i < j \leq N$ such that $f_{ij}(\lambda_1, \ldots, \lambda_l) = 0$. But this is equivalent to $\alpha_i(h) = \alpha_j(h)$, which by Lemma 4.11.1 is equivalent to h not being a splitting element. Now since $\deg g = m$, Corollary 1.5.2 implies that the probability that $g(\lambda_1, \ldots, \lambda_l) = 0$ is less than $m/|\Omega| \leq \epsilon$.   □

Corollary 4.11.3 *Let m be as in Proposition 4.11.2. If the size of F is strictly bigger than m, then H contains splitting elements.*

Proof. Choose $0 < \epsilon < 1$ such that the size of F is at least m/ϵ. Then by Proposition 4.11.2 the probability that a randomly chosen element of H is a splitting element is not 0. Hence H contains splitting elements.   □

Proposition 4.11.2 yields a powerful randomized (Las Vegas type) algorithm for finding a splitting element in H.

Algorithm SplittingElementRandom
Input: a Lie algebra L defined over a perfect field F with a non-degenerate Killing form, a basis $\{h_1, \ldots, h_l\}$ of a Cartan subalgebra H of L and a subset Ω of F of size at least $N(N+1)$, where $N = \dim L - \dim H$.
Output: a splitting element of H.

Step 1 Select randomly and uniformly l elements $\lambda_1, \dots, \lambda_l$ from Ω.

Step 2 Compute the minimum polynomial of $\mathrm{ad}_L h$, where $h = \sum \lambda_i h_i$; if it is of degree $\dim L - \dim H + 1$ then return h, otherwise return to Step 1.

Comments: by Proposition 4.11.2 we see that the probability that h is a splitting element is at least $\frac{1}{2}$. Hence we expect to find a splitting element in no more than two steps.

We can also construct splitting elements by a deterministic method:

Algorithm SplittingElementDeterministic
Input: a Lie algebra L with non-degenerate Killing form, a Cartan subalgebra H of L and a subset Ω of F of size at least $m(\dim H - 1) + 1$, where m is as in Proposition 4.11.2.
Output: a splitting element of H.

Step 1 Let $\{h_1, \dots, h_l\}$ be a basis of H. Denote the elements of Ω by $\omega_1, \dots, \omega_t$. Set $k := 1$.

Step 2 Set $\lambda_i := \omega_k^{i-1}$ for $1 \leq i \leq l$.

Step 3 Compute the minimum polynomial of $h = \sum \lambda_i h_i$; if it is of degree $\dim L - \dim H + 1$ then return h, otherwise set $k := k + 1$ and return to Step 2.

Comments: Let $g \in \tilde{F}[X_1, \dots, X_l]$ be the polynomial in the proof of Proposition 4.11.2. Let Y be another indeterminate and substitute Y^{i-1} for X_i in g. This yields a polynomial in $\tilde{F}[Y]$ of degree at most $m(\dim H - 1)$. Hence by trying at most $m(\dim H - 1) + 1$ values for Y, we obtain a number ξ such that $g(1, \xi, \dots, \xi^{l-1}) \neq 0$.

Now we turn our attention towards calculating the primary decomposition of L relative to H.

Proposition 4.11.4 *Let $h_0 \in H$ be a splitting element. Let X be an indeterminate. Let*

$$L = L_0 \oplus L_1 \oplus \cdots \oplus L_s \qquad (4.12)$$

be the primary decomposition of L relative to $\mathrm{ad} h_0$. Let the primary component corresponding to the polynomial X be L_0. Then $H = L_0$ and the decomposition (4.12) is the (collected) primary decomposition of L with respect to H.

Proof. We note that $H \subset L_0$ since H is nilpotent. Furthermore, Lemma 4.11.1 implies that $\dim L_0 = \dim H$, and hence $L_0 = H$. For $h \in H$ we have that $\mathrm{ad}h$ is a semisimple linear transformation (Root fact 8). Hence saying that $\mathrm{ad}h$ has a minimum polynomial that is a power of an irreducible polynomial is the same as saying that the minimum polynomial of $\mathrm{ad}h$ is irreducible. Suppose that there is an $h \in H$ such that the restriction of $\mathrm{ad}h$ to L_j has a reducible minimum polynomial $f = f_1 f_2$. Then $L_j = V_1 \oplus V_2$ where $V_1 = (L_j)_0(f_1(\mathrm{ad}h))$ and $V_2 = (L_j)_0(f_2(\mathrm{ad}h))$. Now by Proposition 3.1.7, $\mathrm{ad}h_0$ stabilizes both V_1 and V_2. Hence, since the minimum polynomial of the restriction of $\mathrm{ad}h_0$ to L_j is irreducible, $\mathrm{ad}_{V_1}h_0$ and $\mathrm{ad}_{V_2}h_0$ have the same minimum polynomial. So $\mathrm{ad}h_0$ has an eigenvalue of multiplicity at least 2. But this contradicts Lemma 4.11.1. It is also clear that (4.12) is collected because for primary components $L_i \neq L_j$ we have that $\mathrm{ad}_{L_i}h_0$ and $\mathrm{ad}_{L_j}h_0$ have different minimum polynomials. $\qquad\square$

The conclusion is that in the case where F is a big field (of size at least $N(N+1)$, where $N = \dim L - \dim H$) we have an algorithm for calculating the (collected) primary decomposition of L relative to H.

Algorithm PrimaryDecompositionBigField
Input: a Lie algebra L with non-degenerate Killing form over a perfect field F of size at least $N(N+1)$ and a Cartan subalgebra H of L.
Output: the (collected) primary decomposition of L relative to H.

Step 1 Let Ω be a subset of F of size at least $N(N+1)$. Calculate a splitting element $h \in H$ using the algorithm SplittingElementRandom.

Step 2 Compute the irreducible factors $f_0, \ldots, f_s$ of the minimum polynomial of $\mathrm{ad}h$.

Step 3 For $0 \leq i \leq s$ compute the space $L_i = \{x \in L \mid f_i(\mathrm{ad}h)x = 0\}$. Return $\{L_0, \ldots, L_s\}$.

Comment: In Step 1 it is of course also possible to use SplittingElementDeterministic, but then we must take a set Ω of a different size.

Now we suppose that L is defined over a small finite field (i.e., so small that a splitting element is not guaranteed to exist). In this case we work with elements satisfying a weaker condition; they are called decomposing elements.

Definition 4.11.5 *Let V be a subspace of L stable under $\mathrm{ad}H$. Let T_V be the associative algebra generated by 1 and $\mathrm{ad}h|_V$ for $h \in H$. Let $x \in T_V$*

and let f be the minimum polynomial of x. Then x is called decomposing
(on V) if f is reducible. And x is called good *(with respect to V) if f is*
irreducible and deg $f = \dim T_V$.

We note that if an element $x \in T_V$ is good, then it generates the whole
algebra T_V. Also, since its minimum polynomial f is irreducible we see
that T_V is isomorphic to $F(\xi)$, where ξ is a root of f. Hence T_V is a field.
Therefore every element of T_V has an irreducible minimum polynomial.
Consequently, V is a primary component. On the basis of this we formulate
an algorithm for finding the primary decomposition of L with respect to H.

Algorithm PrimaryDecompositionSmallField
  Input: a Lie algebra L with non-degenerate Killing form, defined over a
field F with q elements and a Cartan subalgebra H of L.
Output: the (collected) primary decomposition of L with respect to H.

Step 1 Set to-do:= {FittingOneComponent(L, H)} and primary-components:=
    {H}.

Step 2 Let V be an element from to-do. Let T_V be the associative algebra
    over F generated by 1 and $\mathrm{ad}_V H$.

Step 3 Let x be a random element from T_V and f the minimum polynomial
    of x. Let $f_1, \cdots, f_m$ be the irreducible factors of f. Now there are
    three cases:

    Step 3a In this case f is irreducible (i.e., $m = 1$), and $\deg(f) =$
        $\dim T_V$. Then add V to primary-components and delete V from to-
        do. If at this point to-do is empty, then return primary-components.
        Otherwise (i.e., to-do is not empty) return to Step 2.

    Step 3b Here f is irreducible, but $\deg(f) < \dim T_V$. Then return to
        Step 3.

    Step 3c Here f is reducible (i.e., $m > 1$). Then set $V_i = V_0(f_i(x))$ for
        $1 \leq i \leq m$. Erase V from to-do; and add all V_i for $1 \leq i \leq m$ to
        to-do. Return to Step 2.

Proposition 4.11.6 *The algorithm* PrimaryDecompositionSmallField *termi-*
nates in a finite number of steps, and outputs the collected primary decom-
position of L with respect to H.

Proof. First we prove the correctness of the algorithm. At termination we
have that for every element V of the set primary-components there exists an

$x_V \in T_V$ such that x_V is good with respect to V. This implies that T_V is a field and every element of T_V has an irreducible minimum polynomial. Hence the minimum polynomial of $\mathrm{ad}h|_V$ is irreducible for every $h \in H$. So the decomposition returned by the algorithm is primary. We show that it is collected as well. Let V_1, V_2 be two primary components from the output. Suppose that for all elements $h \in H$ the minimum polynomials of $\mathrm{ad}_{V_1}h$ and $\mathrm{ad}_{V_2}h$ are equal. As seen above, T_{V_1} and T_{V_2} are fields. Now let $h \in H$ be such that $\mathrm{ad}_{V_1}h$ is not contained in any proper subfield of T_{V_1}. Then $\mathrm{ad}_{V_1}h$ generates T_{V_1}. As the minimum polynomial of $\mathrm{ad}_{V_2}h$ is equal to the minimum polynomial of $\mathrm{ad}_{V_1}h$, we see that $\mathrm{ad}_{V_2}h$ generates a subfield of T_{V_2} isomorphic to T_{V_1}. By an analogous argument we have that T_{V_1} contains a subfield isomorphic to T_{V_2}. It follows that T_{V_1} and T_{V_2} are isomorphic fields; furthermore, they are both isomorphic to T_V, where $V = V_1 \oplus V_2$. So for every $x \in T_V$ the minimum polynomials of $x|_{V_1}$ and $x|_{V_2}$ are equal. But this is not possible in view of Step 3c. It follows that there is an $h \in H$ such that the minimum polynomials of $\mathrm{ad}_{V_1}h$ and $\mathrm{ad}_{V_2}h$ are different.

To prove termination we must show that the random element x chosen in the algorithm is either decomposing or good with sufficiently high probability. Let V be a subspace of L that is stable under H. Let T_V be the associative algebra generated by 1 and $\mathrm{ad}h|_V$ for $h \in H$. By Root facts 6 and 8 we have that T_V is a semisimple commutative associative algebra. So by Wedderburn's structure theorem (Theorem A.1.4), we have that T_V is the direct sum of ideals that are full matrix algebras over division algebras over F. But the commutativity of T_V implies that the matrix algebras consist of 1×1-matrices and the division algebras are field extensions of F. So we have

$$T_V = F_1 \oplus \cdots \oplus F_m$$

where the F_i are finite extensions of the ground field F.

If the minimum polynomial f of the randomly chosen element x is irreducible, then $m = 1$ and $T_V = \mathbb{F}_{q^n}$. We estimate the probability that x is good. First of all, if $n = 1$ then all elements of T_V are good. Now suppose that $n > 1$. Let E be the subset of T_V consisting of all elements x of T_V that do not lie in a proper subfield of T_V. To every monic irreducible polynomial g of degree n over $\mathbb{F}_q$ corresponds a subset of n elements of E (namely the set of the roots of g). Also the sets corresponding to different monic irreducible polynomials do not intersect (because an element $\xi \in E$ has a unique minimum polynomial over $\mathbb{F}_q$). Let $N_q(n)$ be the number of monic irreducible polynomials of degree n over $\mathbb{F}_q$. A well-known formula reads

$$N_q(n) = \frac{1}{n} \sum_{d|n} \mu(d) q^{n/d}$$

where $\mu : \mathbb{N} \to \{0, \pm 1\}$ is the Möbius function (see for example [59], Theorem 3.25). Write $n = ab$ where b is the largest proper divisor of n. Then we use the fact that $\mu(r) \geq -1$ for all natural numbers r to estimate

$$
\begin{aligned}
|E| = nN_q(n) &= \sum_{d|n} \mu(d) q^{n/d} \\
&\geq q^n - q^b - q^{b-1} - \cdots - q \\
&= q^n - q\frac{q^b - 1}{q - 1}.
\end{aligned}
$$

An element $x \in T_V$ is good if and only if $x \in E$. And the probability that a randomly chosen element $x \in T_V$ lies in E is

$$\frac{|E|}{q^n} \geq (q^n - q\frac{q^b - 1}{q - 1})/q^n = 1 - \frac{1}{q^{n-b-1}}\frac{1 - 1/q^b}{q - 1} \geq 1 - \frac{1}{q^{n-b-1}(q - 1)}.$$

And this is $\geq \frac{1}{2}$ unless $(n, q) = (2, 2)$, but in that case it can be checked directly that $\frac{|E|}{q^n} \geq \frac{1}{2}$.

Now let $m > 1$. We estimate the probability that x is decomposing. First we have that $x = x_1 + \cdots + x_m$ where $x_i \in F_i$ are randomly and independently chosen elements. The minimum polynomial of x is the least common multiple of the minimum polynomials of the x_i. So if x is *not* decomposing then all x_i have the same minimum polynomial. It follows that the subfields of the F_i generated by the x_i are all isomorphic. Let this subfield be $\mathbb{F}_{q^n}$. We may suppose that all F_i are equal to $\mathbb{F}_{q^n}$; otherwise the probability that $x_i \in \mathbb{F}_{q^n}$ is less. Now we assume that we have chosen an element $x_1 \in F_1$ with an irreducible minimum polynomial f of degree n. Because there are exactly n elements in F_2 with minimum polynomial equal to f, we have that the probability that a randomly chosen element $x_2 \in F_2$ also has minimum polynomial f is equal to

$$\frac{n}{q^n} \leq \frac{1}{2}.$$

It follows that the probability that x is decomposing is $\geq 1/2$. The conclusion is that the probability that a randomly chosen element is either good or decomposing is $\geq 1/2$. Hence we expect to find such an element in at most two steps. $\square$

4.12   Direct sum decomposition

Let L be a Lie algebra with a non-degenerate Killing form and a Cartan subalgebra H. By Proposition 4.3.6, L is a direct sum of simple ideals, $L = J_1 \oplus \cdots \oplus J_r$. In this section we use the primary decomposition of L with respect to H to find this direct sum decomposition.

The next theorem states that the primary decomposition of L with respect to H is compatible with the direct sum decomposition of L.

Theorem 4.12.1 *Let L be a Lie algebra defined over a perfect field with a non-degenerate Killing form and Cartan subalgebra H and let*

$$L = H \oplus L_1 \oplus \cdots \oplus L_s$$

be the (collected) primary decomposition of L with respect to H. Suppose that L decomposes as a direct sum of ideals, $L = I_1 \oplus I_2$. Then $H = H_1 \oplus H_2$, where H_k is a Cartan subalgebra of I_k for $k = 1, 2$. Furthermore, every L_i is contained in either I_1 or I_2.

Proof. By Proposition 1.15.1 there are two non-trivial orthogonal idempotents e_1, e_2 commuting with ad$L$ such that $e_1 + e_2$ is the identity on L and $I_k = e_k L$ for $k = 1, 2$. Hence

$$H = (e_1 + e_2)H \subset e_1 H \oplus e_2 H.$$

Let $g \in H$, then as H is commutative (Root fact 6), $(\mathrm{ad}h)e_1(g) = e_1 \mathrm{ad}h(g) = 0$ for all $h \in H$ so $e_1 g \in H = L_0(H)$. Therefore we have that $e_1 H \subset H$ and similarly $e_2 H \subset H$. So also $e_1 H \oplus e_2 H \subset H$ and hence $H = e_1 H \oplus e_2 H$. Set $H_k = e_k H$ for $k = 1, 2$; then $H = H_1 \oplus H_2$ and H_k is a Cartan subalgebra of I_k for $k = 1, 2$.

Let L_i be a primary component, and let X be an indeterminate. We claim that there is an element $h \in H_1 \cup H_2$ such that the restriction of ad$h$ to L_i is nonsingular. Indeed, otherwise the minimum polynomial of the restriction of every element of a basis of H to L_i would be X. This implies that $[H, L_i] = 0$ and by Lemma 3.2.2 we have $L_i \subset H$, a contradiction. First suppose that $h \in H_1$. Then also $h \in I_1$ so that $\mathrm{ad}h(L) \subset I_1$ and in particular $\mathrm{ad}h(L_i) \subset I_1$. Now the fact that ad$h$ is nonsingular on L_i implies that $L_i = [h, L_i] \subset I_1$. In the same way $h \in H_2$ implies that L_i is contained in I_2. $\square$

This theorem implies that the following algorithm is correct.

Algorithm DirectSumDecomposition
 Input: a Lie algebra L with a non-degenerate Killing form defined over a perfect field F of characteristic not $2, 3$.
 Output: a list of the direct summands of L.

Step 1 Compute the primary decomposition $L = H \oplus L_1 \oplus \cdots \oplus L_s$ (Section 4.11).

Step 2 For $1 \leq i \leq s$ determine a basis of the ideal of L generated by L_i. Delete multiple instances from the list, and return it.

Example 4.12.2 Let L be a Lie algebra with basis $\{h_1, x_1, y_1, h_2, x_2, y_2\}$ and multiplication table

$$
\begin{array}{llll lll}
[h_1, x_1] &=& 2x_1 & [h_2, x_1] &=& 2x_1 \\
[h_1, y_1] &=& -2y_1 & [h_2, y_1] &=& -2y_1 \\
[h_1, x_2] &=& 2x_2 & [h_2, x_2] &=& -2x_2 \\
[h_1, y_2] &=& -2y_2 & [h_2, y_2] &=& 2y_2 \\
[x_1, y_1] &=& \frac{1}{2}h_1 + \frac{1}{2}h_2 & [x_2, y_2] &=& \frac{1}{2}h_1 - \frac{1}{2}h_2.
\end{array}
$$

Brackets of pairs of basis elements that are not present are assumed to be 0. The determinant of the matrix of the Killing form is 2^{16}, hence the Killing form of L is non-degenerate if the characteristic of the ground field F is not 2. As is easily verified, $H = \langle h_1, h_2 \rangle$ is a Cartan subalgebra.

First we take the ground field to be equal to $\mathbb{Q}$. Then the minimum polynomial of $\mathrm{ad}(h_1 + 2h_2)$ is $X(X + 6)(X - 6)(X + 2)(X - 2)$ so that $h_1 + 2h_2$ is a splitting element. The primary decomposition of L relative to $\mathrm{ad}(h_1 + 2h_2)$ is

$$L = \langle h_1, h_2 \rangle \oplus \langle x_1 \rangle \oplus \langle x_2 \rangle \oplus \langle y_1 \rangle \oplus \langle y_2 \rangle. \tag{4.13}$$

Now the ideal generated by x_1 is spanned by $\{x_1, y_1, (h_1 + h_2)/2\}$. Similarly, the ideal generated by x_2 is spanned by $\{x_2, y_2, (h_1 - h_2)/2\}$. It follows that we have found the decomposition of L into simple ideals.

The structure constants of L can also be viewed as elements of $\mathbb{F}_5$. So now we take $\mathbb{F}_5$ as the ground field. Then the Killing form is non-degenerate so that we can apply the algorithm PrimaryDecompositionSmallField. The Fitting-one component $L_1(H)$ is spanned by $\{x_1, y_1, x_2, y_2\}$. The minimum polynomial of the restriction of $\mathrm{ad}(h_1 + h_2)$ to $L_1(H)$ is $X(X - 1)(X + 1)$. So $h_1 + h_2$ is a decomposing element and the corresponding decomposition of $L_1(H)$ is

$$L_1(H) = \langle x_1 \rangle \oplus \langle y_1 \rangle \oplus \langle x_2, y_2 \rangle.$$

Now we turn our attention to the space $V = \langle x_2, y_2 \rangle$ (the other two spaces are 1-dimensional and hence irreducible). The minimum polynomial of the restriction of $\mathrm{ad}(h_1 + 2h_2)$ to V is $(X - 2)(X - 3)$ so that $h_1 + 2h_2$ is a decomposing element. We again find the primary decomposition (4.13).

Example 4.12.3 Let L be a Lie algebra over $\mathbb{Q}$ with basis $\{x_1, \ldots, x_6\}$ and multiplication table as shown in Table 4.3. This Lie algebra is a Levi subalgebra of the so-called Poincaré algebra (see, e.g., [68]).

	x_1	x_2	x_3	x_4	x_5	x_6
x_1	0	0	$2x_4$	$-2x_3$	$-2x_6$	$2x_5$
x_2	0	0	$2x_3$	$2x_4$	$-2x_5$	$-2x_6$
x_3	$-2x_4$	$-2x_3$	0	0	x_2	x_1
x_4	$2x_3$	$-2x_4$	0	0	x_1	$-x_2$
x_5	$2x_6$	$2x_5$	$-x_2$	$-x_1$	0	0
x_6	$-2x_5$	$2x_6$	$-x_1$	x_2	0	0

Table 4.3: Multiplication table of a 6-dimensional Lie algebra.

The determinant of the matrix of the Killing form is -2^{20} so that L is semisimple. A Cartan subalgebra of L is spanned by $\{x_1, x_2\}$. The minimum polynomial of $\mathrm{ad}(x_1 + x_2)$ is $X(X^2 - 4X + 8)(x^2 + 4X + 8)$. Hence $x_1 + x_2$ is a splitting element. The corresponding primary decomposition is

$$L = L_{1,2} \oplus L_{3,4} \oplus L_{5,6},$$

where $L_{i,j}$ is the subspace spanned by $\{x_i, x_j\}$. From the multiplication table it follows that the ideals generated by $L_{3,4}$ and $L_{5,6}$ are both equal to L. Hence, by Theorem 4.12.1 we have that L is a simple Lie algebra.

Now let K be the Lie algebra with basis $\{x_1, \ldots, x_6\}$ and the same multiplication table as L, but defined over $\mathbb{Q}(i)$, where $i^2 = -1$. Then the minimum polynomial of $\mathrm{ad}(x_1 + x_2)$ has irreducible factors X, $X \pm (2 - 2i)$, $X \pm (2 + 2i)$. The corresponding primary decomposition is

$$K = \langle x_1, x_2 \rangle \oplus \langle x_3 - ix_4 \rangle \oplus \langle x_3 + ix_4 \rangle \oplus \langle x_5 - ix_6 \rangle \oplus \langle x_5 + ix_6 \rangle.$$

Now $K = I_1 \oplus I_2$ where I_1 is spanned by $x_2 - ix_1, x_3 - ix_4, x_5 - ix_6$ and I_2 by $x_2 + ix_1, x_3 + ix_4, x_5 + ix_6$.

4.13   Computing a Levi subalgebra

In this section we consider the problem of computing a Levi subalgebra of a Lie algebra of characteristic 0. We give two related algorithms for doing this. In subsequent sections we show that we can make good use of a Cartan subalgebra to find a Levi subalgebra.

In this section L will be a Lie algebra over a field F of characteristic 0 and R its solvable radical. First we remark that in the case where R is commutative there is an easy algorithm along the lines of Example 4.7.3. It starts with computing a basis $\{x_1, \dots, x_m\}$ of a complement to R in L. Then we set $y_i = x_i + r_i$, where r_i are unknown elements of R. For $1 \leq i \leq m$ we let $\bar{x}_i$ be the image of x_i in L/R. Then $[\bar{x}_i, \bar{x}_j] = \sum_k \gamma_{ij}^k \bar{x}_k$. We require that the elements y_i satisfy exactly this commutation relations which, because R is commutative, is equivalent to

$$[x_i, r_j] + [r_i, x_j] - \sum_{k=1}^{m} \gamma_{ij}^k r_k = -[x_i, x_j] + \sum_{k=1}^{m} \gamma_{ij}^k x_k. \qquad (4.14)$$

But this amounts to linear equations for the r_i, which we can solve. We note that the equations have a solution due to Levi's theorem.

The algorithm for the general case is a generalization of the algorithm for the commutative case. Let

$$R = R_1 \supset R_2 \supset \cdots \supset R_d \supset R_{d+1} = 0$$

be a descending series of ideals of R such that $[R_i, R_i] \subset R_{i+1}$ (the derived series is an example). As above, let $\{x_1, \dots, x_m\}$ be a basis of a complement in L to R. Then there are $\gamma_{ij}^k \in F$ such that

$$[x_i, x_j] = \sum_{k=1}^{m} \gamma_{ij}^k x_i \bmod R.$$

We phrase this by saying that the x_i span a Levi subalgebra modulo R. For $t = 1, 2, \dots$ we successively we construct $u_i^t \in R$ such that the elements $y_i^t = x_i + u_i^t$ span a Levi subalgebra modulo R_t. When we arrive at $t = d+1$ we have that the y_i^t span a Levi subalgebra modulo $R_{d+1} = 0$, i.e., they span a Levi subalgebra.

Initially we set $t = 1$ and $y_i^1 = x_i$. Write $R_t = V_t \oplus R_{t+1}$, where V_t is a complementary vector space. Then iteratively set $y_i^{t+1} = y_i^t + r_i^t$, where $r_i^t \in V_t$. We require that the y_i^{t+1} span a Levi subalgebra modulo R_{t+1}, i.e., $[y_i^{t+1}, y_i^{t+1}] = \sum_k \gamma_{ij}^k y_k^{t+1} \bmod R_{t+1}$. Since $[R_t, R_t] \subset R_{t+1}$ this is equivalent to

$$[y_i^t, r_j^t] + [r_i^t, y_j^t] - \sum_{k=1}^{m} \gamma_{ij}^k r_k^t = -[y_i^t, y_j^t] + \sum_{k=1}^{m} \gamma_{ij}^k y_k^t \bmod R_{t+1}. \qquad (4.15)$$

These are linear equations for the r_i^t. We note that since the equations are modulo R_{t+1} the left-hand side as well as the right-hand side can be viewed as elements of V_t. So when solving the equations we can work inside this space. Finally we remark that by Levi's theorem applied to the Lie algebra L/R_{t+1} the equations (4.15) have a solution.

This discussion leads to the following algorithm:

Algorithm LeviSubalgebra

Input: a Lie algebra L of characteristic 0, the solvable radical R and a series $R = R_1 \supset \cdots \supset R_{d+1} = 0$ satisfying $[R_t, R_t] \subset R_{t+1}$ for $1 \le t \le d$.

Output: a Levi subalgebra of L.

Step 1 Compute a basis $\{x_1, \ldots, x_m\}$ of a complement in L to R. For $1 \le i \le m$ set $y_i^1 = x_i$.

Step 2 For $1 \le t \le d$ do the following

    Step2.1 Compute a complement V_t in R_{t+1} to R_t.

    Step2.2 Set $y_i^{t+1} = y_i^t + r_i^t$, where r_i^t are unknown elements of V_t.

    Step2.3 Compute the equations (4.15) for the r_i^t and solve them.

Step 2 Return the subalgebra spanned by the y_i^{d+1}.

When we input the derived series of R, then this is a straightforward algorithm to calculate a Levi subalgebra.

Of course one could input a different a descending series. If the radical R happens to be nilpotent, then we can take the lower central series. In this case the ideals R_t satisfy the stronger property $[R_t, R] \subset R_{t+1}$. In the algorithm we always have that $y_i^t = x_i + u_i$, where $u_i \in R$. Hence $[y_i^t, r_j^t] = [x_i, r_j^t] \bmod R_{t+1}$. So the equations (4.15) transform into

$$[x_i, r_j^t] + [r_i^t, x_j] - \sum_{k=1}^{m} \gamma_{ij}^k r_k^t = -[y_i^t, y_j^t] + \sum_{k=1}^{m} \gamma_{ij}^k y_k^t \bmod R_{t+1}. \qquad (4.16)$$

Now by the following lemma this leads to an algorithm for the general case. The proof is exactly the same as the proof for the third case in the proof of Levi's theorem. We leave the details to the reader.

Lemma 4.13.1 *Let K_1 be the inverse image in L of a Levi subalgebra of $L/[R, R]$. Let K be a Levi subalgebra of K_1, then K is a Levi subalgebra of L.*

The radical of K_1 (which is $[R, R]$) and the radical of $L/[R, R]$ (which is $R/[R, R]$ and hence Abelian) are nilpotent. So we can reduce the general case to two calls to LeviSubalgebra, where we input the lower central series of the Lie algebras and take the equations (4.16) instead of (4.15). We call this algorithm LeviSubalgebraByLCSeries.

Example 4.13.2 Let L be a 6-dimensional Lie algebra over the field F of characteristic 0 with basis $\{x_1, x_2, x_3, x_4, x_5, x_6\}$. The multiplication table of L is displayed in Table 4.4.

	x_1	x_2	x_3	x_4	x_5	x_6
x_1	0	0	x_1	x_5	$-\frac{1}{2}x_6$	0
x_2	0	0	$2x_2$	$x_3 - \frac{1}{2}x_6$	x_1	0
x_3	$-x_1$	$-2x_2$	0	$2x_4$	x_5	0
x_4	$-x_5$	$-x_3 + \frac{1}{2}x_6$	$-2x_4$	0	0	0
x_5	$\frac{1}{2}x_6$	$-x_1$	$-x_5$	0	0	0
x_6	0	0	0	0	0	0

Table 4.4: Multiplication table of a 6-dimensional Lie algebra.

It is easily seen that x_1, x_5, x_6 span a nilpotent ideal R of L. Furthermore, if we let $\bar{x}_2, \bar{x}_3, \bar{x}_4$ be the images of x_2, x_3, x_4 in L/R then

$$[\bar{x}_2, \bar{x}_3] = 2\bar{x}_2, \quad [\bar{x}_2, \bar{x}_4] = \bar{x}_3, \quad [\bar{x}_3, \bar{x}_4] = 2\bar{x}_4.$$

Hence L/R is isomorphic to $\mathfrak{sl}_2(F)$, which is semisimple. So $\mathrm{SR}(L) = R$. Now $[R, R]$ is spanned by x_6 so that $[[R, R], [R, R]] = 0$. We perform the algorithm LeviSubalgebra with the derived series.

Firstly, x_1, x_5 span a complement to $[R, R]$ in R. So we set $y_i^1 = x_i$ for $i = 2, 3, 4$, and

$$y_2^2 = x_2 + \alpha x_1 + \beta x_5$$
$$y_3^2 = x_3 + \gamma x_1 + \delta x_5$$
$$y_4^2 = x_4 + \epsilon x_1 + \eta x_5.$$

Then the requirement $[y_2^2, y_3^2] = 2y_2^2 \bmod x_6$ is equivalent to

$$2x_2 + (\delta + \alpha)x_1 - \beta x_5 + \frac{1}{2}(\beta\gamma - \alpha\delta)x_6 = 2x_2 + 2\alpha x_1 + 2\beta x_5 \bmod x_6.$$

Which is equivalent to the equations $\delta = \alpha$ and $\beta = 0$. Note that by calculating modulo x_6 we get rid of the non-linearity. Continuing, $[y_2^2, y_4^2] =$

$y_3^2 \bmod x_6$ is equivalent to

$$x_3 + \eta x_1 + \alpha x_5 + \frac{1}{2}(\beta\epsilon - \alpha\eta - 1)x_6 = x_3 + \gamma x_1 + \delta x_5 \bmod x_6.$$

This leads to the equations $\delta = \alpha$ and $\eta = \gamma$. Furthermore, from $[y_3^2, y_4^2] = 2y_4^2 \bmod x_6$ we get $\epsilon = 0$ and $\gamma = \eta$. We see that we can choose values $\alpha = \alpha_0$ and $\gamma = \gamma_0$ freely; all other variables are then determined. The solution is

$$y_2^2 = x_2 + \alpha_0 x_1, \ \ y_3^2 = x_3 + \gamma_0 x_1 + \alpha_0 x_5, \ \ y_4^2 = x_4 + \gamma_0 x_5.$$

Now for the next step we set

$$y_2^3 = y_2^2 + ax_6, \ \ y_3^3 = y_3^2 + bx_6, \ \ y_4^3 = y_4^2 + cx_6,$$

and again write down the linear equations. First $[y_2^3, y_3^3] = 2y_2^3$ is the same as

$$2x_2 + 2\alpha_0 x_1 - \frac{1}{2}\alpha_0^2 x_6 = 2y_2^2 + 2ax_6,$$

i.e., $a = -\frac{1}{4}\alpha_0^2$. Going on like this we find $b = -\frac{1}{2}(1 + \alpha_0\gamma_0)$ and $c = -\frac{1}{4}\gamma_0^2$. Hence the full solution reads

$$y_2^3 = x_2 + \alpha_0 x_1 - \frac{1}{4}\alpha_0^2 x_6$$

$$y_3^3 = x_3 + \gamma_0 x_1 + \alpha_0 x_5 - \frac{1}{2}(1 + \alpha_0\gamma_0)x_6$$

$$y_4^3 = x_4 + \gamma_0 x_5 - \frac{1}{4}\gamma_0^2 x_6.$$

4.14   A structure theorem of Cartan subalgebras

In this section we show that a Cartan subalgebra carries information on a Levi decomposition of L. We start with two lemmas.

Lemma 4.14.1 *Let L be a Lie algebra of characteristic 0. Let K be a Levi subalgebra of L, and let H be a Cartan subalgebra of K. Then $\mathrm{ad}_L h$ is a semisimple linear transformation for $h \in H$.*

Proof.   Write $L = K \oplus R$, where R is the solvable radical of L. Let $h \in H$. Then by Root fact 8 we have that $\mathrm{ad}_K h$ is semisimple. Since R is an ideal we have that the adjoint representation of L induces a representation $\mathrm{ad}_R : K \to \mathfrak{gl}(R)$ of K on R. Hence by Corollary 4.6.4 we infer that $\mathrm{ad}_R h$ is semisimple. Now the minimum polynomial of $\mathrm{ad}_L h$ is the least common multiple of the minimum polynomials of $\mathrm{ad}_K h$ and $\mathrm{ad}_R h$. Hence the minimum polynomial of $\mathrm{ad}_L h$ is square-free. It follows that $\mathrm{ad}_L h$ is a semisimple transformation. $\square$

Lemma 4.14.2 *Let L be a Lie algebra of characteristic 0. Let S be a commutative subalgebra of L such that $\mathrm{ad}_L s$ is a semisimple linear transformation for $s \in S$. Then S is contained in a Cartan subalgebra of L.*

Proof. First we note that since all elements of $\mathrm{ad}_L S$ are simultaneously diagonalizable (over the algebraic closure of the ground field), we have that the Fitting-null component $L_0(S)$ is equal to $C_L(S)$, the centralizer of S in L. Hence the Fitting decomposition of L with respect to S reads

$$L = C_L(S) \oplus L_1(S).$$

For $s \in S$, let $\rho(s)$ be the restriction of $\mathrm{ad}_L s$ to $L_1(S)$. Then there are no non-zero $x \in L_1(S)$ such that $\rho(s)x = 0$ for all $s \in S$. Hence there is an $s_0 \in S$ such that $\rho(s_0)$ is nonsingular (this can be seen by diagonalizing all $\rho(s)$ for s in a basis of S; most linear combinations of basis elements have no eigenvalues that are zero). Then $L_0(\mathrm{ad} s_0) = C_L(S)$; and by repeatedly applying Proposition 3.2.6 we see that $L_0(\mathrm{ad} s_0)$ contains a subalgebra $H = L_0(\mathrm{ad} x)$ for a certain $x \in C_L(S)$, such that H is nilpotent. But then $N_L(H) = H$ (Proposition 3.2.5) and hence H is a Cartan subalgebra of L (Lemma 3.2.2). But since $H \subset C_L(S)$ we have $[H, S] = 0$ and hence $S \subset H$ (cf. Lemma 3.2.2). $\square$

Theorem 4.14.3 *Let L be a Lie algebra of characteristic 0. Let K be a Levi subalgebra of L and H_1 a Cartan subalgebra of K. Then there exists a Cartan subalgebra H of L containing H_1. Conversely, if H is any Cartan subalgebra of L, then there is a Levi subalgebra K of L having a Cartan subalgebra contained in H.*

Proof. By Lemma 4.14.1 together with Root fact 6 we have that H_1 is a commutative subalgebra of L such that $\mathrm{ad}_L h$ is a semisimple linear transformation for $h \in H_1$. Hence by Lemma 4.14.2 there is a Cartan subalgebra H of L such that $H_1 \subset H$.

Now let H be any Cartan subalgebra of L and let R be the solvable radical of L. Let $\pi : L \to L/R$ be the projection map. Then by Lemma 3.6.2, $\pi(H)$ is a Cartan subalgebra of L/R. Now let K_1 be any Levi subalgebra of L. Then the restriction of π to K_1 is an isomorphism of K_1 onto L/R.

Hence K_1 has a Cartan subalgebra H_1 such that $\pi(H_1) = \pi(H)$. By the first part of the proof there is a Cartan subalgebra H' of L such that $H_1 \subset H'$.

We construct an automorphism of L mapping H' onto H. First of all, we have that $\pi(H) = \pi(H_1) \subseteq \pi(H')$. But by Lemma 3.6.2, $\pi(H')$ is a Cartan subalgebra of L/R. So $\pi(H') = \pi(H)$ and hence $H + R = H' + R$. Set $B = H + R$, then since R is an ideal, B is a subalgebra of L. Furthermore, B/R is commutative so that $[B, B] \subset R$. Consequently B is a solvable Lie algebra. Now from Section 3.6 we recall that $D(B)$ is the subgroup of $\mathrm{Int}(B)$ generated by all $\exp \mathrm{ad}_B x$ for $x \in [B, B]$. Also, if $x \in [B, B]$ then $x \in [L, L] \cap R$ and by Lemma 2.6.2 we see that $\mathrm{ad}_L x$ is nilpotent. So $\exp \mathrm{ad}_L x$ is an element of $\mathrm{Int}(L)$ whose restriction to B equals $\exp \mathrm{ad}_B x$. The conclusion is that every element of $D(B)$ naturally extends to an element of $\mathrm{Int}(L)$.

Now by Theorem 3.6.4 there is a $g' \in D(B)$ such that $g'(H') = H$. Let $g \in \mathrm{Int}(L)$ be the extension of g' to an automorphism of L. Set $K = g(K_1)$; then K is a Levi subalgebra having a Cartan subalgebra $g(H_1)$ contained in H.   □

4.15   Using Cartan subalgebras to compute Levi subalgebras

In this section we put Theorem 4.14.3 to use. By this theorem any Cartan subalgebra H of a Lie algebra L of characteristic 0 contains a Cartan subalgebra H_1 of a certain Levi subalgebra of L. By Lemma 4.14.1, $\mathrm{ad}_L H_1$ consists of semisimple linear transformations. Also if we suppose the solvable radical R to be nilpotent (which by Lemma 4.13.1 we can do without loss of generality), then $\mathrm{ad}_L x$ is nilpotent for $x \in R$. Hence every semisimple element of $\mathrm{ad} H$ "comes from" a Levi subalgebra of L. Here we show how we can go after these semisimple elements directly.

In the sequel we let L be a Lie algebra over the field F of characteristic 0, not equal to its solvable radical R. Throughout we assume that R is nilpotent.

Definition 4.15.1 *A commutative subalgebra T of L is said to be* toral *if $\dim T = \dim \mathrm{ad}_L T$ and the associative algebra $(\mathrm{ad}_L T)^*$, generated by 1 together with $\mathrm{ad}_L T$, is commutative and semisimple.*

The first condition is included to avoid calling the centre of a Lie algebra toral.

Proposition 4.15.2 *Let K be a Levi subalgebra of L. Let H_1 be a Cartan subalgebra of K. Then H_1 is a toral subalgebra of L. Put $H = C_L(H_1)$, the centralizer of H_1 in L. Then H is a Cartan subalgebra of L, and $H = H_1 \oplus C_R(H_1)$.*

Proof. Lemma 4.14.1 together with Root fact 6 imply that H_1 is a commutative subalgebra of L such that $\mathrm{ad}_L h$ is a semisimple for $h \in H_1$. Also since K does not contain elements of the centre of L we have that the adjoint representation of L is faithful on H_1, hence H_1 is a toral subalgebra of L.

Let $x \in C_L(H_1)$, then $x = y + r$, where $y \in K$ and $r \in R$. So for $h \in H_1$ we have $0 = [h, x] = [h, y] + [h, r]$. But the first element is in K and the second in R. Therefore $[h, y] = [h, r] = 0$ for all $h \in H_1$. Since H_1 is a Cartan subalgebra of K we have $y \in H_1$. It follows that $C_L(H_1) = H_1 \oplus C_R(H_1)$.

As in the proof of Lemma 4.14.2 we see that $H = C_L(H_1) = L_0(H_1)$. Hence by Proposition 3.2.5 we have $N_L(H) = H$. Furthermore, let H^k denote the k-th term of the lower central series of H. Then

$$H^k = [H_1 \oplus C_R(H_1), [H_1 \oplus C_R(H_1), \cdots, [H_1 \oplus C_R(H_1), H_1 \oplus C_R(H_1)] \cdots]]$$
$$= [C_R(H_1), [C_R(H_1), \cdots [C_R(H_1), C_R(H_1)] \cdots]] = C_R(H_1)^k \subset R^k,$$

and hence H is nilpotent. Now by Lemma 3.2.2, H is a Cartan subalgebra of L. $\qquad\square$

Corollary 4.15.3 *Let H be a Cartan subalgebra of L, then H contains a Cartan subalgebra H_1 of a Levi subalgebra of L. Furthermore, for any such H_1 we have that $H = H_1 \oplus C_R(H_1)$.*

Proof. The first assertion follows from Theorem 4.14.3. Set $H' = H_1 \oplus C_R(H_1)$. By Proposition 4.15.2, H' is a Cartan subalgebra of L. We have to show that $H' = H$. As in the proof of Lemma 4.14.2 we have that $H' = C_L(H_1) = L_0(\mathrm{ad}\, h_1)$ for a certain $h_1 \in H_1$. Hence h_1 is a regular element. So H and H' have a regular element in common and hence they are equal. $\qquad\square$

Now we consider the problem of calculating a toral subalgebra inside a given Cartan subalgebra H of L. The next proposition yields a way of doing this. Furthermore it states that a maximal toral subalgebra of H is "almost" (possibly modulo elements of the centre of L), equal to a Cartan subalgebra of a Levi subalgebra of L.

Proposition 4.15.4 *Let H be a Cartan subalgebra of L. Let T be maximal (with respect to inclusion) among all toral subalgebras of L contained in H. Then there is a Levi subalgebra K of L and a Cartan subalgebra H_1 of K such that $\mathrm{ad}_L H_1 = \mathrm{ad}_L T$ and $H_1 \subset H$. Let $x \in H$ and let $\mathrm{ad}_L x = s + n$ be the Jordan decomposition of $\mathrm{ad}_L x$. Then there is an $h \in H$ such that $\mathrm{ad}_L h = s$.*

Proof. By Corollary 4.15.3 there is a Levi subalgebra K of L having a Cartan subalgebra H_1 such that $H_1 \subset H$. Furthermore, $H = H_1 \oplus C_R(H_1)$. Let $t \in T$ and write $t = h + r$ where $h \in H_1$ and $r \in C_R(H_1)$. Then $[h, r] = 0$ and $\mathrm{ad} t = \mathrm{ad} h + \mathrm{ad} r$. Now $\mathrm{ad} h$ is semisimple and $\mathrm{ad} r$ is nilpotent (because the radical R is nilpotent). So $\mathrm{ad} t = \mathrm{ad} h + \mathrm{ad} r$ is the Jordan decomposition of $\mathrm{ad} t$. But $\mathrm{ad} t$ is semisimple and hence $\mathrm{ad} r = 0$. So T consists of elements $h + r$ for $h \in H_1$ and r lies in the centre of L. Now since T is maximal and commutes with H_1 it follows that for all $h \in H_1$ there is an r in the centre of L such that $h + r \in T$. We conclude that $\mathrm{ad} T = \mathrm{ad} H_1$.

For the last statement write $x = h + r$, where $h \in H_1$ and $r \in R$. Again we have that $\mathrm{ad} x = \mathrm{ad} h + \mathrm{ad} r$ is the Jordan decomposition of $\mathrm{ad} x$. So by the uniqueness of the Jordan decomposition we infer that $s = \mathrm{ad} h$. $\square$

On the basis of Proposition 4.15.4 we formulate an algorithm:

Algorithm ToralSubalgebra
Input: a Lie algebra L of characteristic 0 such that its solvable radical is nilpotent.
Output: a toral subalgebra T of L such that $\mathrm{ad} T = \mathrm{ad} H_1$ for a Cartan subalgebra H_1 of a Levi subalgebra K.

Step 1 Set R :=SolvableRadical(L), H :=CartanSubalgebraBigField(L), and C :=Centre(H).

Step 2 Let $\{h_1, \dots, h_m\}$ be a basis of C. For $1 \le i \le m$ compute the Jordan decomposition $\mathrm{ad}_L h_i = \mathrm{ad}_L s_i + \mathrm{ad}_L n_i$ of $\mathrm{ad}_L h_i$; where $s_i, n_i \in H$.

Step 3 Let T be the span of all s_i for $1 \le i \le m$. Return T.

Comments: Let T be maximal among all toral subalgebras of L contained in H. Let $H_1 \subset H$ be a Cartan subalgebra of a Levi subalgebra of L such that $\mathrm{ad}_L H_1 = \mathrm{ad}_L T$ (cf. Proposition 4.15.4). Then since $H = H_1 \oplus C_R(H_1)$ (Corollary 4.15.3) we have that H_1 is contained in the centre of H. The same holds for T because $\mathrm{ad}_L T = \mathrm{ad}_L H_1$. By Proposition 4.15.4 we see that the elements s_i, n_i exist. We use the algorithm Jordan-Decomposition (see Section A.2) to compute the Jordan decomposition of

$\mathrm{ad}_L h_i$. Then by solving a system of linear equations, we find s_i. (Note that s_i is not necessarily unique.) Now since $\mathrm{ad}_L s_i$ and $\mathrm{ad}_L n_i$ are polynomials in $\mathrm{ad}_L h_i$ without constant term we see that s_i, n_i commute with everything that commutes with h_i. In particular s_i, n_i lie in the centre of H.

We show that the span of the s_i is a maximal toral subalgebra contained in H. Let $s \in H$ be such that $\mathrm{ad}_L s$ is semisimple. Write $s = h + r$ where $h \in H_1$ and $r \in C_R(H_1)$. Then in the same way as in the proof of Proposition 4.15.4 we see that $\mathrm{ad}_L r = 0$. Write $h = \sum_{i=1}^m \alpha_i h_i$ for some $\alpha_i \in F$. Then $\mathrm{ad}_L s = \sum_{i=1}^m \alpha_i \mathrm{ad}_L h_i = \sum_i \alpha_i \mathrm{ad}_L s_i + \sum_i \alpha_i \mathrm{ad}_L n_i$. The first summand is a sum of commuting semisimple transformations and hence semisimple itself. Similarly the second summand is nilpotent. Furthermore they commute with each other. Hence this is the Jordan decomposition of $\mathrm{ad}_L s$ and it follows that $\mathrm{ad}_L(\sum_i \alpha_i n_i) = 0$. So $\mathrm{ad}_L s = \mathrm{ad}_L s'$ for some $s' \in T$. It follows that $s \notin T$ implies that the subalgebra generated by T together with s contains elements of the centre and is therefore not toral. The conclusion is that T is a maximal toral subalgebra of H.

Proposition 4.15.5 *Let H be a Cartan subalgebra of L. Let T and K be as in Proposition 4.15.4. Let*

$$L = L_1 \oplus \cdots \oplus L_n$$

be the (collected) primary decomposition of L relative to T. Write $L_i = V_i \oplus R_i$, where $R_i = R \cap L_i$ and V_i is a complementary subspace. Then there is a basis of K consisting of elements of the form $v_i + r_i$, where $v_i \in V_i$ and $r_i \in R_i$.

Proof. Let $K = K_1 \oplus \cdots \oplus K_m$ be the primary decomposition of K with respect to T. Also $R = R_1 \oplus \cdots \oplus R_n$ is the primary decomposition of R with respect to T. Adding these decompositions, and taking the subspaces K_i and R_j such that the minimum polynomials of the restrictions of $\mathrm{ad}\, t$ to K_i and R_j are equal for all $t \in T$, together, we obtain the (collected) primary decomposition of L relative to T. Hence, if L_i is not contained in R, then $L_i = K_{k_i} \oplus R_i$ and it is seen that V_i has a basis consisting of elements of the form $w_i + r_i$, where $w_i \in K_{k_i}$ and $r_i \in R_i$, and the result follows. $\square$

Let T, L_i, V_i, R_i be as in Proposition 4.15.5. Let H_1 be a Cartan subalgebra of a Levi subalgebra K of L such that $\mathrm{ad}_L H_1 = \mathrm{ad}_L T$. We note the following:

- The centralizer $C_L(T)$ occurs among the primary components of L. It contains H_1. The root spaces of the Levi subalgebra K relative to H_1

are contained in the other primary components. Furthermore, these root spaces generate K (this follows from Root fact 11). Suppose that $L_1 = C_L(T)$. If for $2 \leq i \leq n$ we have that $R_i = 0$ whenever $V_i \neq 0$, then it follows that the subalgebra generated by the V_i is a Levi subalgebra.

- If we are not so fortunate, then in the equation systems (4.16), we can reduce the number of variables. We start with a basis $x_1, \ldots, x_m$ of a complement in L to R, consisting of elements of the spaces V_i. Then in the iteration of the algorithm LeviSubalgebra we add elements r_i^t to x_i. But by Proposition 4.15.5, we can take these elements from R_k where k is such that $x_i \in V_k$.

Now we formulate an algorithm for calculating a Levi subalgebra of L, in the case where the radical R is nilpotent. By Lemma 4.13.1, this also gives an algorithm for the general case.

Algorithm LeviSubalgebra
Input: a finite-dimensional Lie algebra L of characteristic 0 such that the solvable radical $\mathrm{SR}(L)$ is nilpotent.
Output: a basis of a Levi subalgebra of L.

Step 1 Set $T := $ ToralSubalgebra(L).

Step 2 Calculate the primary decomposition $L_1 \oplus \cdots \oplus L_s$ of L with respect to $\mathrm{ad}T$.

Step 3 Calculate spaces V_i and $R_i \subset R$ such that $L_i = V_i \oplus R_i$. If $R_i = 0$ for all i such that $V_i \neq 0$, then return the subalgebra generated by the V_i.

Step 4 Take bases of the V_i together to obtain a basis $x_1, \ldots, x_m$ of a complement in L to R. Iteratively calculate the equation systems (4.16), where the r_i^t are taken from R_k, where k is such that $x_i \in V_k$. Solve the systems and return the resulting subalgebra.

Comment: for the calculation of the primary decomposition of L with respect to T we can proceed exactly as in the algorithm PrimaryDecompositionSmallField; the randomly chosen element x in Step 3 is either decomposing or good with high probability. We leave the details to the reader.

Example 4.15.6 Let L be the 6-dimensional Lie algebra of Example 4.13.2. Then x_3 is a non-nilpotent element and $H = L_0(\mathrm{ad}x_3)$ is a commutative subalgebra, and hence it is a Cartan subalgebra (cf. the algorithm Cartan-SubalgebraBigField). Now H is spanned by x_3, x_6; and x_3 spans a maximal torus T. The primary decomposition of L with respect to T reads

$$L = H \oplus L_1 \oplus L_2 \oplus L_4 \oplus L_5,$$

where L_i is spanned by x_i for $i = 1, 2, 4, 5$. The primary components not contained in the radical are L_2 and L_4. We have that $L_2 \cap R = L_4 \cap R = 0$ and hence it immediately follows that they are contained in a Levi subalgebra of L. So the subalgebra K they generate is a Levi subalgebra. It is spanned by $x_2, x_4, x_3 - \frac{1}{2} x_6$.

4.16   Notes

The Lie algebra of Table 4.4 is a symmetry algebra admitted by the heat equation (see [10], §4.2.4).

The proofs in Section 4.10 (for Lie algebras over modular fields) follow [76]. Theorem 4.14.3 is a result of Chevalley; for the proof we followed [19]. For a different approach see [34].

The notion of splitting element appears in [27] (see also Appendix A). There they are elements of a semisimple associative algebra with a maximal number of eigenvalues. Since splitting elements of Cartan subalgebras in semisimple Lie algebras satisfy the analogous property, we have adopted the same terminology. Also elements of an associative algebra with a reducible minimum polynomial are called *decomposable* in [28]. In Section 4.11 we use the term *decomposing* to denote the same property.

The algorithm in Section 4.13 for calculating a Levi subalgebra using the derived series of the solvable radical is from [71]. The algorithm LeviSubalgebraByLCSeries is taken from [36]. In this paper it is also proved that this algorithm runs in polynomial time (since the left hand side of (4.16) does not depend on the output of the previous round, it can be proved that the coefficients of the solutions do not blow up). The algorithm for calculating a Levi subalgebra using the information carried by a Cartan subalgebra is taken from [34].

Chapter 5

The classification of the simple Lie algebras

This chapter is entirely devoted to the classification of the (isomorphism classes of) Lie algebras with a non-degenerate Killing form defined over a field of characteristic 0. In this case the non-degeneracy of the Killing form is equivalent to L being semisimple (Proposition 4.2.2). Therefore in this chapter we will speak of semisimple Lie algebras, rather than of Lie algebras having a non-degenerate Killing form.

When classifying Lie algebras we need a tool for deciding whether two Lie algebras are isomorphic or not. To this end we use *structural invariants*. A structural invariant of Lie algebras is a function

$$f : \{ \text{ Lie algebras } \} \longrightarrow \{ \text{ objects } \}$$

such that for two Lie algebras L_1, L_2 we have that $L_1 \cong L_2$ implies $f(L_1) \sim f(L_2)$, where $\sim$ is some equivalence relation on the image of f. Examples are the dimension of a Lie algebra (where $\sim$ is just equality of integers), the solvable radical (where $\sim$ is isomorphism of Lie algebras) and so on. We say that a structural invariant f is *complete* if $f(L_1) \sim f(L_2)$ implies $L_1 \cong L_2$.

We prove that the set of roots of a semisimple Lie algebra L forms a root system. It will be shown that the root system is a complete structural invariant of semisimple Lie algebras. It follows that we can classify semisimple Lie algebras by classifying their root systems. We do this by attaching an integral matrix to a root system, called the Cartan matrix. We will show that the Cartan matrix is a complete invariant of root systems. We then replace the Cartan matrix by a graph, the Dynkin diagram. In Section 5.9 we show by some elementary considerations that a number of Dynkin diagrams cannot exist. In the next section we then exhibit a root system for each remaining Dynkin diagram (thereby showing that each such Dynkin diagram in fact does exist). Subsequently we construct for each root system Φ a semisimple Lie algebra having Φ as a root system (Sections 5.12, 5.13, 5.15). This results in an algorithm for constructing these Lie algebras, given the Cartan matrix of the root system. Other algorithms that we describe in this chapter include an algorithm for constructing a root system (given the Cartan matrix) (Section 5.6), an algorithm for calculating a set of canonical generators (Section 5.11) and an algorithm for constructing an isomorphism of (isomorphic) semisimple Lie algebras with split Cartan subalgebras (Section 5.11). In the final section we discuss the algorithmic problem of determining the isomorphism class of a semisimple Lie algebra (with maybe a non-split Cartan subalgebra).

5.1 Representations of $\mathfrak{sl}_2(F)$

Let L be a semisimple Lie algebra, and let Φ be the set of roots of L with respect to a split Cartan subalgebra H. Let $\alpha \in \Phi$ then by Root facts 12 and 13 we see that there are $x \in L_\alpha$, $y \in L_{-\alpha}$, $h \in H$ that span a subalgebra K of L isomorphic to $\mathfrak{sl}_2(F)$. The adjoint representation of L induces a representation $\mathrm{ad}_L : K \to \mathfrak{gl}(L)$. In Section 4.10 we made good use of these representations to investigate the structure of Lie algebras of characteristic $p > 0$ having a non-degenerate Killing form. In characteristic 0 however, this tool is even more powerful because the representations of K have a well determined structure. It is the objective of this section to explore that structure.

Let F be a field of characteristic 0; and let K be the Lie algebra over F with basis $\{h, x, y\}$ and multiplication table

$$[h, x] = 2x, \quad [h, y] = -2y, \quad [x, y] = h,$$

(see Examples 1.3.6, 4.2.3). Let $\rho : K \to \mathfrak{gl}(V)$ be a finite-dimensional representation of K. Then by Corollary 4.6.4 together with Root fact 8 (note that K is semisimple by Example 4.2.3, and that h spans a Cartan subalgebra) we have that $\rho(h)$ is a semisimple linear transformation. In this section we assume that $\rho(h)$ is split (i.e., there is a basis of V relative to which the matrix of $\rho(h)$ is diagonal). Let $\lambda_1, \ldots, \lambda_r \in F$ be the distinct eigenvalues of $\rho(h)$. Then V decomposes as

$$V = \bigoplus_{i=1}^{r} V_{\lambda_i} \quad \text{where } V_{\lambda_i} = \{v \in V \mid \rho(h)v = \lambda_i v\}.$$

Now let $v \in V_{\lambda_i}$; then

$$\rho(h)\rho(x)v = \rho(x)\rho(h)v + \rho([h, x])v = (\lambda_i + 2)\rho(x)v,$$

and hence $\rho(x)v \in V_{\lambda_i + 2}$. A similar calculation shows that $\rho(y)v \in V_{\lambda_i - 2}$.

Definition 5.1.1 *Let λ be an eigenvalue of $\rho(h)$. Then the space V_λ is called a* weight space *and the eigenvalue λ is called a* weight. *The elements of V_λ are called* weight vectors *of weight λ. Furthermore, an eigenvalue λ such that $V_\lambda \neq 0$ and $V_{\lambda+2} = 0$ is called a* highest weight *of V. Furthermore, if λ is a highest weight, then a non-zero vector $v \in V_\lambda$ is called a* highest-weight vector.

Since there are only a finite number of distinct eigenvalues (V being finite-dimensional), there is an eigenvalue λ of $\rho(h)$ such that $\lambda + 2$ is not an eigenvalue. Hence V has at least one highest weight λ. Let $v_0 \in V_\lambda$, then $\rho(x)v_0 = 0$. Furthermore we set $v_i = \rho(y)^i v_0$ for $i \geq 0$.

Lemma 5.1.2 *We have $\rho(h)v_i = (\lambda - 2i)v_i$ and $\rho(x)v_i = i(\lambda - i + 1)v_{i-1}$.*

Proof. This is the same as Proposition 4.10.1. $\qquad\qquad\qquad\square$

Lemma 5.1.3 *Let $\rho : K \to \mathfrak{gl}(W)$ be a finite-dimensional irreducible representation of K. Let λ be a highest weight of W and v_0 a corresponding highest-weight vector. Set $v_i = \rho(y)^i v_0$ for $i \geq 0$. Then W is spanned by the v_i for $i \geq 0$; and if we set $n = \dim W - 1$, then $\lambda = n$ and $\rho(h)$ has exactly $n + 1$ distinct eigenvalues which are $n, n - 2, n - 4, \ldots, -n + 2, -n$.*

Proof. By Lemma 5.1.2, the space spanned by the v_i is a submodule of W; hence it is equal to W because W is irreducible. As W is finite-dimensional, there must be an $n \geq 0$ such that $v_{n+1} = 0$. Then W is spanned by $v_0, v_1, \ldots, v_n$ so that $n = \dim W - 1$. (Note that the v_i are linearly independent as they are eigenvectors with different eigenvalues). Furthermore, using Lemma 5.1.2 we calculate

$$0 = \rho(x)v_{n+1} = (n + 1)(\lambda - n)v_n.$$

Hence $\lambda = n$ and the statement of the lemma follows by Lemma 5.1.2. $\quad\square$

Theorem 5.1.4 *Let $\rho : K \to \mathfrak{gl}(V)$ be a finite-dimensional representation of K. Then V decomposes as a direct sum of irreducible submodules. Let*

*W be an irreducible submodule occurring in this decomposition. Then the
eigenvalues of $\rho(h)$ restricted to W are $n, n-2, \ldots, -n$, where $\dim W = n +
1$. Furthermore, the number of summands in the direct sum decomposition
of V is equal to $\dim V_0 + \dim V_1$.*

Proof. By Weyl's theorem (Theorem 4.4.6) V decomposes as a direct
sum of irreducible submodules. Let W be an irreducible submodule of V
and let $w_0 \in W$ be a highest-weight vector of W. Set $n = \dim W - 1$.
Then by Lemma 5.1.3, the eigenvalues of the restriction of $\rho(h)$ to W
are $n, n - 2, \ldots, -n$. It is clear that each such summand has either a
1-dimensional eigenspace with eigenvalue 1, or a 1-dimensional eigenspace
with eigenvalue 0 (but not both). Hence the last statement follows.   □

Remark. Let $\rho : K \to \mathfrak{gl}(V)$ be a finite-dimensional representation of K.
Then by Theorem 5.1.4, the eigenvalues of $\rho(h)$ are integers. It follows that
our assumption that $\rho(h)$ be split is no real restriction because $\rho(h)$ is always split.

5.2   Some more root facts

In this section we continue where we left off in Section 4.9. As before Φ
is the set of roots of a semisimple Lie algebra L with respect to a split
Cartan subalgebra H. Here we prove some root facts that pave the way
for the (abstract) root systems. We define a vector space over the rational
numbers that contains all roots. This will allow us to attach a root system
(in a Euclidean space over the reals) to a semisimple Lie algebra.

We recall that for $\alpha \in H^*$ the element $h_\alpha \in H$ is defined by (4.10).
Furthermore, we have a symmetric non-degenerate bilinear form $(\ ,\ )$ on
H^*, defined by $(\alpha, \beta) = \kappa_L(h_\alpha, h_\beta)$.

Root fact 17 *Suppose α and β are elements of Φ with $\beta \neq \pm\alpha$. Let r and
q be (respectively) the largest integers for which $\beta - r\alpha$ and $\beta + q\alpha$ are roots.
Then $\beta + i\alpha$ is a root for $-r \leq i \leq q$ and*

$$\frac{2(\beta, \alpha)}{(\alpha, \alpha)} = r - q.$$

Proof. Set

$$h = \frac{2h_\alpha}{(\alpha, \alpha)},$$

then, according to Root fact 12 (together with Root fact 13), there are $x \in L_\alpha$, $y \in L_{-\alpha}$ such that x, y, h span a subalgebra K_α of L isomorphic to $\mathfrak{sl}_2(F)$. Set

$$V = L_{\beta - r\alpha} \oplus L_{\beta - (r-1)\alpha} \oplus \cdots \oplus L_{\beta + q\alpha}.$$

Then by Root fact 1 we have that V is a K_α-module. For $v \in L_{\beta + i\alpha}$ we calculate:

$$[h, v] = (\beta + i\alpha)(h)v = (\beta(h) + 2i)v.$$

Hence $L_{\beta + i\alpha}$ is a weight space with weight $\beta(h) + 2i$. These weights are all distinct, and it is not possible that both 0 and 1 occur as weights of this form, so by Theorem 5.1.4 we conclude that V is irreducible.

Let the highest weight of h on V be m, then by Theorem 5.1.4 the set of weights of h is equal to $\{m, m - 2, \dots, -(m - 2), -m\}$. Therefore we have

$$\beta(h) + 2q = m$$

and

$$\beta(h) - 2r = -m$$

from which it follows that $\beta(h) = r - q$. Now since $\beta(h_\alpha) = \kappa_L(h_\beta, h_\alpha)$ (cf. (4.10)), we have that

$$r - q = \beta(h) = \frac{2(\beta, \alpha)}{(\alpha, \alpha)}.$$

Furthermore, all weights $m - 2j$ for $0 \le j \le m$ must occur, so all $\beta + i\alpha$ are roots for $-r \le i \le q$. □

Definition 5.2.1 Let $\alpha, \beta \in \Phi$ and let q, r be the largest integers such that $\beta - r\alpha$ and $\beta + q\alpha$ are roots. The string of roots $\beta - r\alpha, \dots, \beta + q\alpha$ is called the α-string containing β.

Example 5.2.2 Let L be the 10-dimensional Lie algebra of Example 4.9.1. Let $\alpha = (1, 0)$ and $\beta = (-1, 1)$. Then the α-string containing β is seen to be $\beta, \beta + \alpha, \beta + 2\alpha$. So in this case $r = 0$ and $q = 2$. Furthermore from Example 4.9.3 we see that

$$h_\alpha = \frac{1}{6} A_{11}, \quad h_\beta = -\frac{1}{6} A_{11} + \frac{1}{6} A_{22}.$$

And hence

$$(\beta, \alpha) = \kappa_L(-\frac{1}{6} A_{11} + \frac{1}{6} A_{22}, \frac{1}{6} A_{11}) = -\frac{1}{6}.$$

Similarly $(\alpha, \alpha) = 1/6$. It follows that

$$2\frac{(\beta, \alpha)}{(\alpha, \alpha)} = -2 = r - q.$$

Root fact 18 *Let $\alpha \in \Phi$ be a root. Then no scalar multiple of α is a root except $\pm \alpha$.*

Proof. Suppose that $\beta = \lambda \alpha$ is a root, where $\lambda \neq \pm 1$. By Root fact 14 we may assume that λ is not an integer. However, by Root fact 17 we have that

$$\frac{2(\beta, \alpha)}{(\alpha, \alpha)} = 2\lambda$$

is an integer. So $\lambda = \frac{1}{2}k$, where k is an odd integer. Furthermore, since also $-\beta$ is a root (Root fact 5), we may assume that $k > 0$. Now let $\beta - r\alpha, \ldots, \beta + q\alpha$ be the α-string containing β. Then by Root fact 17 we see that $r - q = k$, and in particular $\beta - i\alpha$ is a root for $0 \leq i \leq k + q$. Hence also

$$\frac{1}{2}\alpha = \beta - \frac{k-1}{2}\alpha$$

is a root. But by Root fact 14 this implies that $\alpha = 2\frac{1}{2}\alpha$ cannot be a root, which is a contradiction. $\qquad \square$

By Root fact 10 we have that Φ spans H^*, so we can choose a basis $\{\alpha_1, \ldots, \alpha_l\}$ of H^* consisting of roots. Now set

$$V_{\mathbb{Q}} = \mathbb{Q}\,\alpha_1 \oplus \cdots \oplus \mathbb{Q}\,\alpha_l$$

which is the vector space over $\mathbb{Q}$ spanned by $\alpha_1, \ldots, \alpha_l$.

Root fact 19 *For $\alpha, \beta \in H^*$ we have*

$$(\alpha, \beta) = \sum_{\gamma \in \Phi} (\alpha, \gamma)(\beta, \gamma).$$

Proof. First we recall that $(\alpha, \beta) = \kappa_L(h_\alpha, h_\beta)$. By (4.9) and Root fact 14 we see that

$$\kappa_L(h_\alpha, h_\beta) = \sum_{\gamma \in \Phi} \gamma(h_\alpha)\gamma(h_\beta) = \sum_{\gamma \in \Phi} (\alpha, \gamma)(\beta, \gamma).$$

$\qquad \square$

Root fact 20 *For $\alpha, \beta \in \Phi$ we have that $(\alpha, \beta) \in \mathbb{Q}$.*

Proof. By Root fact 17 we see that

$$\frac{2(\rho, \sigma)}{(\sigma, \sigma)}$$

is an integer for all $\rho, \sigma \in \Phi$. So also

$$\frac{4}{(\alpha, \alpha)} = \left(\frac{2\alpha}{(\alpha, \alpha)}, \frac{2\alpha}{(\alpha, \alpha)} \right)$$
$$= \sum_{\gamma \in \Phi} \frac{2(\gamma, \alpha)}{(\alpha, \alpha)} \cdot \frac{2(\gamma, \alpha)}{(\alpha, \alpha)} \quad \text{(Root fact 19)}$$

is an integer. It follows that

$$(\alpha, \beta) = \frac{2(\alpha, \beta)}{(\alpha, \alpha)} \cdot \frac{(\alpha, \alpha)}{2}$$

is a rational number. $\square$

Root fact 21 *The vector space $V_{\mathbb{Q}}$ contains Φ. Furthermore, the bilinear form (,) is positive definite on $V_{\mathbb{Q}}$.*

Proof. If β is an element of Φ, then there are unique $\lambda_1, \ldots, \lambda_l$ in F such that

$$\beta = \sum_{i=1}^{l} \lambda_i \alpha_i.$$

We must prove that $\lambda_i \in \mathbb{Q}$ for $1 \leq i \leq l$. First we have

$$(\beta, \alpha_j) = \sum_{i=1}^{l} (\alpha_i, \alpha_j) \lambda_i \quad \text{for } j = 1, \ldots, l.$$

This is a system of l equations for the l scalars λ_i. Its matrix $((\alpha_i, \alpha_j))$ is non-singular because $\alpha_1, \ldots, \alpha_l$ is a basis of H^* and the form (,) is non-degenerate. Root fact 20 tells us that the entries of this matrix are all rational, so that the equations have a unique solution over $\mathbb{Q}$. Therefore, β is an element of $V_{\mathbb{Q}}$. For $\sigma \in V_{\mathbb{Q}}$ we have by Root fact 19,

$$(\sigma, \sigma) = \sum_{\gamma \in \Phi} (\sigma, \gamma)^2$$

which is nonnegative by Root fact 20. Furthermore, $(\sigma, \sigma) = 0$ is equivalent to $(\sigma, \gamma) = 0$ for all $\gamma \in \Phi$. By the definition of $(\ ,\ )$ this is equivalent to $\gamma(h_\sigma) = 0$ for all $\gamma \in \Phi$. By Root fact 7 this implies that $h_\sigma = 0$, i.e., $\sigma = 0$. □

5.3 Root systems

In this section we introduce the abstract notion of a root system in a Euclidean space. Subsequently we show that to a semisimple Lie algebra we can attach a root system and that this root system is a structural invariant of semisimple Lie algebras.

Let V be a Euclidean space, i.e., a finite-dimensional vector space over $\mathbb{R}$ together with a bilinear form

$$(\ ,\ ) : V \times V \longrightarrow \mathbb{R}$$

that is positive definite and symmetric.

Definition 5.3.1 *A* reflection *r in V is a linear map $r : V \longrightarrow V$ that leaves some hyperplane P pointwise fixed, and sends each vector perpendicular to P to its negative. The hyperplane P is called the* reflecting hyperplane *of r.*

Let v be a non-zero element of V and let the map

$$r_v : V \longrightarrow V$$

be defined as

$$r_v(w) = w - \frac{2(w, v)}{(v, v)} v.$$

Then r_v is the identity on $P_v = \{ w \in V \mid (w, v) = 0 \}$, i.e., the hyperplane perpendicular to v. We also have that r_v maps every vector perpendicular to P_v to its negative. Hence r_v is a reflection of V. Furthermore, it is clear that all reflections of V are obtained in this way.

For $u, v, w \in V$ we have

$$(r_u(v), r_u(w)) = \left(v - \frac{2(v, u)}{(u, u)} u, w - \frac{2(w, u)}{(u, u)} u\right) = (v, w). \tag{5.1}$$

Which means that the map r_u leaves the bilinear form $(\ ,\ )$ invariant.

Definition 5.3.2 *A subset* Φ *of* V *is called a* root system *if the following conditions are satisfied:*

(R_1) Φ *is finite, spans* V *and does not contain* 0.

(R_2) *Let* $\alpha \in \Phi$ *and* $\lambda \in \mathbb{R}$*; then* $\lambda\alpha \in \Phi$ *if and only if* $\lambda = \pm 1$.

(R_3) *For every* $\alpha \in \Phi$*, the reflection* r_α *leaves* Φ *invariant.*

(R_4) *For all* $\alpha, \beta \in \Phi$ *the number*

$$\frac{2(\beta, \alpha)}{(\alpha, \alpha)}$$

is an integer.

Definition 5.3.3 *Let* Φ *be a root system in the Euclidean space* V. *Then* $\dim V$ *is called the* rank *of* Φ.

Lemma 5.3.4 *Let* Φ *be a root system and* $\alpha \in \Phi$*, then* r_α *is a bijection from* Φ *onto* Φ.

Proof. This follows from the fact that r_α^2 is the identity on V.   □

Let L be a semisimple Lie algebra with a split Cartan subalgebra H. As in Section 5.2 we let Φ be the set of roots of L with respect to H and we let $V_\mathbb{Q}$ be the vector space over $\mathbb{Q}$ spanned by Φ (cf. Root fact 21). The restriction of the form $(\ ,\ )$ (which is defined on H^*) to $V_\mathbb{Q}$ has values in $\mathbb{Q}$ by Root fact 20. Moreover, by Root fact 21 the form $(\ ,\ )$ is positive definite on $V_\mathbb{Q}$. Set $V = V_\mathbb{Q} \otimes \mathbb{R}$, and extend $(\ ,\ )$ to V by setting $(v \otimes \lambda, w \otimes \mu) = (v, w) \otimes \lambda\mu$ for $v, w \in V_\mathbb{Q}$ and $\lambda, \mu \in \mathbb{R}$. Then $(\ ,\ )$ is also positive definite on V. Hence V is a real Euclidean space.

Theorem 5.3.5 *Let* V, Φ *be as above; then* Φ *is a root system in* V.

Proof. (R_1) is immediate. (R_2) follows from Root fact 18. (R_3) and (R_4) are contained in Root fact 17.   □

Example 5.3.6 Let $V = \mathbb{R}^2$ and put

$$\Phi = \{(1,0), (0,1), (-1,0), (0,-1), (1,1), (-1,1), (-1,-1), (1,-1)\}.$$

Then Φ is the root system corresponding to the Lie algebra L of Example 4.9.1. Set $\alpha = (1,0)$ and $\beta = (0,1)$. Then in the same way as in Example 4.9.5 we calculate

$$(\alpha, \alpha) = (\beta, \beta) = \frac{1}{6} \quad \text{and} \quad (\alpha, \beta) = (\beta, \alpha) = 0.$$

Note that the bilinear form $(\ ,\ )$ is determined by this. Using this we can check that Φ is a root system in V.

Because the number

$$\frac{2(\beta, \alpha)}{(\alpha, \alpha)}$$

occurs frequently, we abbreviate it by $\langle \beta, \alpha \rangle$.

Now let V_1 and V_2 be Euclidean spaces with root systems Φ_1 and Φ_2 respectively. Then these root systems are called *isomorphic* if there is a bijective linear map

$$f : V_1 \longrightarrow V_2$$

such that $f(\Phi_1) = \Phi_2$ and $\langle \alpha, \beta \rangle = \langle f(\alpha), f(\beta) \rangle$ for all $\alpha, \beta \in \Phi_1$. If Φ_1, Φ_2 are isomorphic, then we write $\Phi_1 \cong \Phi_2$. The next proposition implies that the root system is a structural invariant of semisimple Lie algebras.

Proposition 5.3.7 *Let L_1 and L_2 be semisimple Lie algebras over F with root systems Φ_1 and Φ_2 respectively, relative to split Cartan subalgebras H_1, H_2. If L_1 and L_2 are isomorphic then Φ_1 and Φ_2 are isomorphic.*

Proof. For this we may assume that the ground field is algebraically closed. Indeed, let $\overline{F}$ be the algebraic closure of F, then $H_i \otimes_F \overline{F}$ is a Cartan subalgebra of $L_i \otimes_F \overline{F}$ (Proposition 3.2.3) yielding the same root systems as H_i in L_i. Let $f : L_1 \longrightarrow L_2$ be an isomorphism of Lie algebras. We recall that all Cartan subalgebras of L_2 are conjugate under the automorphism group of L_2 (Theorem 3.5.1). So after (maybe) composing f with an automorphism of L_2, we may assume that $f(H_1) = H_2$. We can extend f to a map from H_1^* into H_2^* by setting $f(\alpha)(h_2) = \alpha(f^{-1}(h_2))$ for $\alpha \in H_1^*$ and $h_2 \in H_2$. Let $\alpha \in \Phi_1$ and let $x \in (L_1)_\alpha$ be a root vector. Then $[h_1, x] = \alpha(h_1)x$ for all $h_1 \in H_1$ (Root fact 9), so

$$[f(h_1), f(x)] = f([h_1, x]) = \alpha(h_1)f(x) = f(\alpha)(f(h_1))f(x),$$

i.e., $f(x)$ is a root vector for $f(\alpha)$. In particular $f(\alpha)$ is a root. From this it also follows that f maps distinct roots to distinct roots so that f is a bijection from Φ_1 onto Φ_2.

Let V_1, V_2 be the real Euclidean spaces associated with Φ_1 and Φ_2. Because Φ_i spans V_i we can extend f to a linear map from V_1 into V_2. Then f maps root strings to root strings, and by Root fact 17 it follows that $\langle \alpha, \beta \rangle = \langle f(\alpha), f(\beta) \rangle$ for $\alpha, \beta \in \Phi_1$. □

It turns out that the root system is a complete invariant of semisimple Lie algebras. This will be shown in Section 5.11.

Let Φ be a root system and suppose that $\Phi = \Phi_1 \cup \Phi_2$ where $\Phi_1, \Phi_2 \subset \Phi$ are such that $\langle \alpha_1, \alpha_2 \rangle = 0$ for all $\alpha_1 \in \Phi_1$ and $\alpha_2 \in \Phi_2$. Then we say that Φ is the *direct sum* of Φ_1 and Φ_2 and we write

$$\Phi = \Phi_1 \oplus \Phi_2.$$

A root system Φ that cannot be decomposed as a direct sum is called *irreducible*.

Proposition 5.3.8 *Let L be a semisimple Lie algebra and let Φ be the root system corresponding to L. Then L is a direct sum of non-zero ideals, $L = J_1 \oplus J_2$ if and only if $\Phi = \Phi_1 \oplus \Phi_2$, where Φ_1 and Φ_2 are the root systems of J_1 and J_2 respectively.*

Proof. Let H be a fixed split Cartan subalgebra of L.

First suppose that $L = J_1 \oplus J_2$. Then by Theorem 4.12.1, $H = H_1 \oplus H_2$, where H_k is a Cartan subalgebra of J_k. Now let $\Phi_1 = \{\alpha_1, \dots, \alpha_s\}$ be the set of roots of J_1 relative to H_1 and $\Phi_2 = \{\beta_1, \dots, \beta_t\}$ the set of roots of J_2 relative to H_2. We extend the functions α_i and β_j to functions on H by setting $\alpha_i(h_1 + h_2) = \alpha_i(h_1)$ and $\beta_j(h_1 + h_2) = \beta_j(h_2)$ for $h_1 \in H_1$ and $h_2 \in H_2$. Then

$$L = H_1 \oplus H_2 \oplus L_{\alpha_1} \oplus \cdots \oplus L_{\alpha_s} \oplus L_{\beta_1} \oplus \cdots \oplus L_{\beta_t}$$

is the root space decomposition of L relative to $H = H_1 \oplus H_2$. Let $\alpha_i \in \Phi_1$ and $\beta_j \in \Phi_2$, then we claim that $\alpha_i + \beta_j$ cannot be a root. Suppose it is a root, then it equals some α_k or some β_k. Suppose that $\alpha_i + \beta_j = \alpha_k$ for an $\alpha_k \in \Phi_1$. Choose an $h_2 \in H_2$ such that $\beta_j(h_2) \neq 0$, then

$$0 = \alpha_k(h_2) = \alpha_i(h_2) + \beta_j(h_2) \neq 0,$$

a contradiction. In the same way $\alpha_i + \beta_j$ cannot be equal to a β_k. By an analogous argument it can be shown that $\alpha_i - \beta_j$ cannot be a root. Now by Root fact 17 we see that $\langle \alpha_i, \beta_j \rangle = 0$. Hence $\Phi = \Phi_1 \oplus \Phi_2$.

On the other hand, suppose that $\Phi = \Phi_1 \oplus \Phi_2$ where $\Phi_1 = \{\alpha_1, \dots, \alpha_s\}$ and $\Phi_2 = \{\beta_1, \dots, \beta_t\}$. Furthermore $\langle \alpha_i, \beta_j \rangle = 0$ for $\alpha_i \in \Phi_1$ and $\beta_j \in \Phi_2$.

Also $\langle \alpha_i + \beta_j, \alpha_i \rangle \neq 0$ so that $\alpha_i + \beta_j \notin \Phi_2$ and $\langle \alpha_i + \beta_j, \beta_j \rangle \neq 0$ from which $\alpha_i + \beta_j \notin \Phi_1$. It follows that $\alpha_i + \beta_j$ is not a root for $\alpha_i \in \Phi_1$ and $\beta_j \in \Phi_2$.

Let J_1 be the subalgebra generated by the L_{α_i} for $\alpha_i \in \Phi_1$. Let H_1 be the subspace of H spanned by the h_{α_i} (which are defined by (4.10)). Then by Root facts 1, 6, 9, 11 we have that

$$J_1 = H_1 \oplus L_{\alpha_1} \oplus \cdots \oplus L_{\alpha_s}.$$

Furthermore, because $\alpha_i + \beta_j$ is not a root we have that $[L_{\beta_j}, J_1] = 0$ and J_1 is an ideal of L. Let J_2 be the subalgebra generated by L_{β_k} for $\beta_k \in \Phi_2$. Then

$$J_2 = H_2 \oplus L_{\beta_1} \oplus \cdots \oplus L_{\beta_t},$$

where H_2 is spanned by the h_{β_j}. Also J_2 is an ideal of L and $L = J_1 + J_2$. Furthermore, $[J_1, J_2] = 0$ (note that $[h_{\alpha_i}, x_{\beta_j}] = \beta_j(h_{\alpha_i})x_{\alpha_i} = (\beta_j, \alpha_i)x_{\alpha_i} = 0$). Therefore $J_1 \cap J_2$ is contained in the centre of L which means that it is zero. The conclusion is that $L = J_1 \oplus J_2$. $\qquad\square$

5.4   Root systems of rank two

In the previous section we associated a root system to a semisimple Lie algebra. We showed that the root system is a structural invariant of semisimple Lie algebras. In this and subsequent sections we prepare the way for a classification of all root systems (which will be performed in Section 5.9). For the moment we abandon all thought of Lie algebras and investigate abstract root systems in their own right.

Let Φ be a root system in the Euclidean space V over $\mathbb{R}$ with positive definite bilinear form (,). For $v \in V$ we define the norm $\|v\|$ of v by setting $\|v\| = \sqrt{(v, v)}$. As V is a Euclidean space we have the inequality of Cauchy-Schwarz:

$$|(v, w)| \leq \|v\|\|w\| \quad \text{for all } v, w \in V. \tag{5.2}$$

Let $v, w \in V$. Then the angle between v and w is defined to be the real number $\theta \in [0, \pi]$ such that

$$\cos\theta = \frac{(v, w)}{\|v\|\|w\|}.$$

(Note that by (5.2) this number lies between -1 and 1.) Now Let $\alpha, \beta \in \Phi$ and let θ be the angle between them. Then

$$\langle \beta, \alpha \rangle = \frac{2(\beta, \alpha)}{(\alpha, \alpha)} = 2\frac{\|\beta\|}{\|\alpha\|}\cos\theta.$$

So $\langle \alpha, \beta \rangle \langle \beta, \alpha \rangle = 4 \cos^2 \theta$ which must be an integer because $\langle \alpha, \beta \rangle$ and $\langle \beta, \alpha \rangle$ are integers. Now, because $0 \leq \cos^2 \theta \leq 1$, we have that $\langle \alpha, \beta \rangle$ and $\langle \beta, \alpha \rangle$ have the same sign and are equal to 0, ± 1, ± 2 or ± 3. So Table 5.1 gives the only possibilities for the angle between α and β when $\alpha \neq \pm \beta$ (this rules out $\theta = 0, \pi$) and $\|\beta\| \geq \|\alpha\|$.

$\langle \alpha, \beta \rangle$	$\langle \beta, \alpha \rangle$	θ	$\|\beta\|^2 / \|\alpha\|^2$
0	0	$\frac{\pi}{2}$	undetermined
1	1	$\frac{\pi}{3}$	1
-1	-1	$\frac{2\pi}{3}$	1
1	2	$\frac{\pi}{4}$	2
-1	-2	$\frac{3\pi}{4}$	2
1	3	$\frac{\pi}{6}$	3
-1	-3	$\frac{5\pi}{6}$	3

Table 5.1: Possible angles θ between roots α and β.

Proposition 5.4.1 *Let α and β be two roots with $\alpha \neq \pm \beta$. If $(\alpha, \beta) > 0$ then $\alpha - \beta$ is a root, and if $(\alpha, \beta) < 0$ then $\alpha + \beta$ is a root.*

Proof. If $(\alpha, \beta) > 0$ then also $\langle \alpha, \beta \rangle > 0$. Now Table 5.1 implies that $\langle \alpha, \beta \rangle = 1$ or $\langle \beta, \alpha \rangle = 1$. In the first case we have $r_\beta(\alpha) = \alpha - \beta$ lies in Φ. On the other hand, if $\langle \beta, \alpha \rangle = 1$ then by the same argument we have $\beta - \alpha \in \Phi$. Hence also $\alpha - \beta = -(\beta - \alpha)$ lies in Φ. Finally, if $(\alpha, \beta) < 0$ then we apply the above argument to $-\beta$. $\square$

Let $\alpha, \beta \in \Phi$. We recall that the string of roots $\beta - r\alpha, \beta - (r - 1)\alpha, \dots, \beta + q\alpha$ is called the α-string containing β. The first two parts of the next proposition have been proved for root systems arising from semisimple Lie algebras. But since we are dealing with abstract root systems that may not be connected to any Lie algebra, we have to prove them again.

Proposition 5.4.2 *Let $\alpha, \beta \in \Phi$ be two roots such that $\alpha \neq \pm \beta$. Let r, q be the largest integers such that $\beta - r\alpha$ and $\beta + q\alpha$ are roots. Then*

1. *$\beta + i\alpha$ are roots for $-r \leq i \leq q$,*

2. *$r - q = \langle \beta, \alpha \rangle$,*

3. *the length of the α-string containing β is at most 4, i.e., $r + q + 1 \leq 4$.*

Proof.

1. If $r = q = 0$, then there is nothing to prove. So suppose that not both r and q are 0. Suppose further that there is an integer j with $-r < j < q$ such that $\beta + j\alpha$ is not a root. Set $\sigma = \beta - r\alpha$. Then there are integers $0 \leq s < t$ such that $\sigma + s\alpha$ and $\sigma + t\alpha$ are roots while $\sigma + (s + 1)\alpha$ and $\sigma + (t - 1)\alpha$ are not roots. Proposition 5.4.1 implies that $(\sigma + s\alpha, \alpha) \geq 0$ and $(\sigma + t\alpha, \alpha) \leq 0$. This means that

$$(\beta, \alpha) + (s - r)(\alpha, \alpha) \geq 0 \quad \text{while} \quad (\beta, \alpha) + (t - r)(\alpha, \alpha) \leq 0.$$

Multiplying the first inequality by -1 and adding we get $(t-s)(\alpha, \alpha) \leq 0$. But this is a contradiction since $\alpha \neq 0$.

2. The reflection r_α adds an integer multiple of α to a root, hence the α-string containing β is invariant under r_α. Furthermore, since $r_\alpha : \Phi \to \Phi$ is a bijection (Lemma 5.3.4), it also maps the α-string containing β bijectively onto itself. The reflection r_α acts as follows:

$$r_\alpha(\beta + i\alpha) = r_\alpha(\beta) - i\alpha.$$

From this it follows that the pre-image of $\beta - r\alpha$ must be $\beta + q\alpha$. Hence

$$\beta - r\alpha = r_\alpha(\beta + q\alpha) = (\beta + q\alpha) - \langle \beta + q\alpha, \alpha \rangle \alpha.$$

And this is equivalent to $r - q = \langle \beta, \alpha \rangle$.

3. By 2. we have that $\langle \beta + q\alpha, \alpha \rangle = \langle \beta, \alpha \rangle + 2q = q + r$. So by Table 5.1 we see that $q + r = 0, 1, 2, 3$. Hence $r + q + 1 \leq 4$.

$\square$

Example 5.4.3 We can use Table 5.1 to classify all possible root systems Φ of rank 2. For that choose two linearly independent roots α and β from Φ in such a way that $\langle \beta, \alpha \rangle$ is minimal. Then $\langle \beta, \alpha \rangle$ can only be 0, -1, -2 or -3, and for each of these values we have a unique root system (up to isomorphism). They are shown in Figure 5.1. We observe that the root system of type B_2 is the one connected to the Lie algebra of Example 4.9.1.

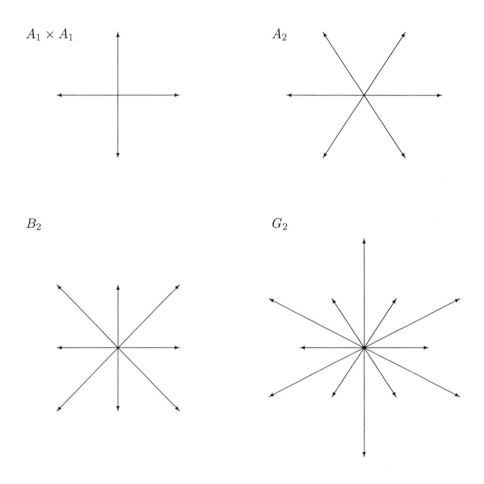

Figure 5.1: All possible root systems of rank 2.

5.5   Simple systems

In this section we will make a first step towards classifying root systems. We define a special type of basis consisting of roots, called simple system. Then in the next section we put the integers $\langle \alpha, \beta \rangle$ for α, β in a simple system into a matrix, called a Cartan matrix. We will show that a Cartan matrix is a complete structural invariant of root systems. This will allow us to classify root systems by classifying Cartan matrices.

Let Φ be a root system in the Euclidean space V. We say that a partial order $<$ on V is a *root order* if

1. every root $\alpha \in \Phi$ is comparable to zero (i.e., either $\alpha > 0$ or $\alpha < 0$),

2. if $v \in V$ is such that $v > 0$, then for $\lambda \in \mathbb{R}$, $\lambda v > 0$ if $\lambda > 0$ and $\lambda v < 0$ if $\lambda < 0$,

3. if for $u, v \in V$ we have $u < v$, then $u + w < v + w$ for all $w \in V$.

Let $<$ be a root order. We note that if $v < 0$, then by adding $-v$ and applying 3., we get $-v > 0$.

To give an example of a root order we first choose a basis $\{v_1, \ldots, v_l\}$ of V. Let $v \in V$, then v can be written as a linear combination of the basis elements:

$$v = \sum_{i=1}^{l} \lambda_i v_i \quad \text{where } \lambda_1, \ldots, \lambda_l \in \mathbb{R}.$$

Then $v > 0$ if the first non-zero λ_i is positive. Also for $v, w \in V$ we set $v > w$ if $v - w > 0$. We have that $>$ is a total order on V and it is clearly a root order. We call $<$ the *lexicographical order* relative to $v_1, \ldots, v_l$.

Now let $<$ be a root order. A root $\alpha \in \Phi$ is said to be *positive* if $\alpha > 0$, and *negative* if $\alpha < 0$. By Φ^+ we denote the set of positive roots and by Φ^- the set of negative roots. Since roots are either positive or negative, $\Phi = \Phi^+ \cup \Phi^-$. Note that $\Phi^- = -\Phi^+$.

Definition 5.5.1 *A root α is said to be* simple *if α is positive and α cannot be written as a sum $\alpha = \beta + \gamma$ where $\beta, \gamma \in \Phi$ are both positive.*

Example 5.5.2 Let $V = \mathbb{R}^2$ and let

$$\Phi = \{(1,0), (0,1), (-1,0), (0,-1), (1,1), (-1,1), (-1,-1), (1,-1)\}$$

be the root system of Example 5.3.6. Set $\alpha = (1,0)$ and $\beta = (0,1)$ and let $<$ be the lexicographical order relative to α, β. Then the positive roots are seen to be $(1,0)$, $(0,1)$, $(1,1)$, and $(1,-1)$. Furthermore, $(1,0) = (1,-1) + (0,1)$ and $(1,1) = (1,0) + (0,1)$. The other two positive roots cannot be written as sums of positive roots; so the simple roots are $(0,1)$ and $(1,-1)$.

Lemma 5.5.3 *Let α and β be simple roots. If $\alpha \neq \beta$ then $\alpha - \beta$ is not a root and $(\alpha, \beta) \leq 0$.*

Proof. Suppose that $\alpha - \beta$ is a root. If it is positive then $\alpha = (\alpha - \beta) + \beta$ and α is not simple. On the other hand, if it is negative, then $\beta = -(\alpha - \beta) + \alpha$

is not simple. And we have obtained a contradiction. The last assertion
now follows from Proposition 5.4.1.   □

Lemma 5.5.4 *Let* $\alpha_1, \dots, \alpha_t \in V$ *be such that* $\alpha_i > 0$ *and* $(\alpha_i, \alpha_j) \leq 0$ *for*
$1 \leq i \neq j \leq n$. *Then these vectors are linearly independent.*

Proof. Suppose that there exists a relation $\sum_i \lambda_i \alpha_i = 0$. Without loss of
generality we may assume that $\lambda_1 \neq 0$. After dividing by λ_1 we obtain a
relation

$$\alpha_1 = \sum_{i=2}^{t} \mu_i \alpha_i.$$

Now let Δ_1 be the set of the α_i such that $\mu_i > 0$, and Δ_2 the set of those
α_i such that $\mu_i \leq 0$. Set

$$\sigma = \sum_{\alpha_i \in \Delta_1} \mu_i \alpha_i \quad \text{and} \quad \rho = \sum_{\alpha_i \in \Delta_2} \mu_i \alpha_i.$$

Then $\alpha_1 = \sigma + \rho$. Furthermore, $\sigma \neq 0$ because $\alpha_1 > 0$. Now $(\sigma, \rho) = \sum \nu_{ij}(\alpha_i, \alpha_j)$ where $\nu_{ij} \leq 0$, and because $(\alpha_i, \alpha_j) \leq 0$ we have $(\sigma, \rho) \geq 0$. So

$$(\alpha_1, \sigma) = (\sigma, \sigma) + (\rho, \sigma) > 0,$$

but also $(\alpha_1, \sigma) = \sum_{\alpha_i \in \Delta_1} \mu_i(\alpha_1, \alpha_i) \leq 0$. And from the assumption that
the α_i are linearly dependent we have derived a contradiction.   □

Proposition 5.5.5 *Let* $\Delta = \{\alpha_1, \dots, \alpha_l\}$ *be the set of all simple roots.*
Then Δ *is a basis of* V. *Moreover, if* α *is a positive root, then*

$$\alpha = \sum_{i=1}^{l} k_i \alpha_i$$

where the k_i *are non-negative integers.*

Proof. The fact that Δ is linearly independent follows from Lemmas 5.5.3
and 5.5.4.

   We prove that each positive root is a non-negative integral combination
of the simple roots. Let $\alpha > 0$ and suppose we already know the result for
all roots $< \alpha$. If α is simple, then there is nothing to prove. Otherwise
$\alpha = \beta + \gamma$ where β and γ are positive roots. Then $\alpha - \beta > 0$ so that $\alpha > \beta$

and likewise $\alpha > \gamma$. By induction, β and γ can be written as linear combinations of the simple roots with non-negative integral coefficients. Hence the same holds for α. $\qquad \square$

Corollary 5.5.6 *Let β be a positive root that is not simple. Then there is a simple root α such that $\beta - \alpha$ is a root.*

Proof. Let Δ be the set of all simple roots, and suppose that $(\beta, \alpha) \leq 0$ for all $\alpha \in \Delta$. Then by Lemma 5.5.4 $\Delta \cup \{\beta\}$ is linearly independent which is impossible because by Proposition 5.5.5, Δ is a basis of V. Hence there is an $\alpha \in \Delta$ such that $(\beta, \alpha) > 0$. Then by Proposition 5.4.1, $\beta - \alpha$ is a root. $\quad \square$

Corollary 5.5.7 *Let $\Delta = \{\alpha_1, \dots, \alpha_l\}$ be the set of simple roots. Let β be a root and write $\beta = \sum_i k_i \alpha_i$. Then k_i is an integer for $1 \leq i \leq l$ and either all $k_i \geq 0$ or all $k_i \leq 0$.*

Proof. This follows at once from Proposition 5.5.5. $\qquad \square$

Let Δ be the set of all simple roots relative to some root order $<$. Then Δ is called the *simple system* relative to $<$, of the root system Φ. By Proposition 5.5.5 it forms a basis of V. We remark that Φ has no unique simple system of roots. Indeed, Δ depends on the order $<$. In general a different order yields a different simple system.

By the next result we have that the statement of Corollary 5.5.7 characterizes simple systems.

Proposition 5.5.8 *Let $\alpha_1, \dots, \alpha_l \in \Phi$, where l is the rank of Φ, be such that every element of Φ can be written as a $\mathbb{Z}$-linear combination of the α_i such that all coefficients are either non-negative or non-positive. Then there exists a root order $<$ such that $\{\alpha_1, \dots, \alpha_l\}$ is the simple system relative to $<$ of Φ.*

Proof. Since every root is a linear combination of the elements of Δ, we have that Δ is a basis of V. Now let $<$ be the lexicographical order of V defined by this basis (i.e., $v = \sum_i \lambda_i \alpha_i > 0$ if the left-most non-zero λ_i is positive). Then a root $\beta = \sum_i k_i \alpha_i$ is positive if $k_i \geq 0$ for all i. So Δ consists of positive roots, and clearly no $\alpha_i \in \Delta$ can be written as a sum of positive roots. Hence Δ is the simple system of Φ relative to $<$. $\quad \square$

We end this section with a definition that will be useful later on.

Definition 5.5.9 *Let β be a positive root and write β as an integral linear combination of simple roots,*

$$\beta = \sum_{i=1}^{l} k_i \alpha_i$$

where all $k_i \geq 0$. Then the number $\mathrm{ht}(\beta) = \sum_{i=1}^{l} k_i$ *is called the* height *of β.*

5.6   Cartan matrices

An $l \times l$-matrix with integer coefficients is called a *Cartan matrix* if there are vectors $v_1, \ldots, v_l$ in a Euclidean space V such that

- $C(i, j) = \langle v_i, v_j \rangle$ for $1 \leq i, j \leq l$ and

- $(v_i, v_j) \leq 0$ for $1 \leq i \neq j \leq l$.

We note that the vectors $v_1, \ldots, v_l$ are necessarily independent by Lemma 5.5.4. The diagonal entries of a Cartan matrix are $\langle v_i, v_i \rangle = 2$. Furthermore, since $(v_i, v_j) \leq 0$ for $i \neq j$ we have that the off-diagonal entries are non-positive. Because $\langle v_i, v_j \rangle$ is an integer, by the same argument as used in Section 5.4 we see that $\langle v_i, v_j \rangle = 0, -1, -2, -3$ for $i \neq j$.

If we permute the vectors $v_1, \ldots, v_l$, then also the corresponding Cartan matrix will change. Let C' be a second Cartan matrix, then C' is said to be *equivalent* to C (we write $C \sim C'$) if a permutation of the vectors $v_1, \ldots, v_l$ carries C to C'.

Let Φ be a root system in the Euclidean space V and let $\Delta = \{\alpha_1, \ldots, \alpha_l\}$ be a simple system of Φ (relative to some root order). Then Lemma 5.5.3 implies that the matrix

$$C = \left(\langle \alpha_i, \alpha_j \rangle \right)_{i,j=1}^{l}$$

is a Cartan matrix; it is called a Cartan matrix of the root system Φ relative to Δ. The objective of this and the next section is to show that the root system Φ and the Cartan matrix C of Φ determine each other upto equivalence.

Example 5.6.1 Let Φ be the root system of Example 5.5.2. Set $\alpha_1 = (0, 1)$ and $\alpha_2 = (1, -1)$; then as seen in Example 5.5.2, $\Delta_1 = \{\alpha_1, \alpha_2\}$ forms a simple system for Φ. By computing the various root strings involving α_1

and α_2 and using Proposition 5.4.2, we see that the Cartan matrix relative to Δ_1 is

$$C_1 = \begin{pmatrix} 2 & -1 \\ -2 & 2 \end{pmatrix}.$$

However, if we set $\beta_1 = \alpha_2$ and $\beta_2 = \alpha_1$ and $\Delta_2 = \{\beta_1, \beta_2\}$, then the Cartan matrix relative to Δ_2 is

$$C_2 = \begin{pmatrix} 2 & -2 \\ -1 & 2 \end{pmatrix}.$$

And these are equivalent Cartan matrices: $C_1 \sim C_2$.

A Cartan matrix C together with the corresponding simple system $\Delta = \{\alpha_1, \ldots, \alpha_l\}$ determines the root system Φ completely. In order to show this we give an algorithm for determining Φ from C and Δ. We recall that the height of a root is defined in Definition 5.5.9.

Algorithm CartanMatrixToRootSystem
Input: a simple system $\Delta = \{\alpha_1, \ldots, \alpha_l\}$ of the root system Φ and an $l \times l$-matrix C that is the Cartan matrix of Φ relative to Δ.
Output: the set of vectors Φ.

Step 1 Set $\Phi^+ := \Delta$ and $n := 1$.

Step 2 For all $\gamma \in \Phi^+$ of height n and for all $\alpha_j \in \Delta$ we do the following:

    1. Write

$$\gamma = \sum_{i=1}^{l} k_i \alpha_i. \tag{5.3}$$

    2. Determine the largest integer $r \geq 0$ such that $\gamma - r\alpha_j \in \Phi^+$.

    3. Set

$$q := r - \sum_{i=1}^{l} k_i C(i, j).$$

    4. If $q > 0$ then set $\Phi^+ := \Phi^+ \cup \{\gamma + \alpha_j\}$.

Step 3 If in Step 2 Φ^+ has been enlarged, then set $n := n + 1$ and return to the beginning of Step 2. Otherwise return $\Phi^+ \cup -\Phi^+$.

Proposition 5.6.2 *Let Φ be a root system and let Δ be a simple system of Φ with Cartan matrix C. Then* CartanMatrixToRootSystem(Δ, C) *returns the set Φ.*

Proof. Let β be a positive root. By induction on $\operatorname{ht}(\beta)$ we prove that after $\operatorname{ht}(\beta) - 1$ rounds of the iteration, the set Φ^+ will contain β. For $\operatorname{ht}(\beta) = 1$ this is clear since the roots of height 1 are the simple roots in Δ.

Now suppose that $\operatorname{ht}(\beta) = n + 1$ and Φ^+ already contains the positive roots of height $\leq n$. Then by Corollary 5.5.6, we see that $\beta = \gamma + \alpha_j$ where $\gamma \in \Phi$ is of height n and $\alpha_j \in \Delta$. So at some stage in Step 2, γ and α_j are considered. Since $\gamma + \alpha_j$ is a root we have that $\gamma \neq \alpha_j$. We let $\gamma - r\alpha_j, \dots, \gamma + q\alpha_j$ be the α_j-string containing γ. As $\gamma \neq \alpha_j$, in (5.3) there is some $k_i > 0$ with $i \neq j$. Hence all elements of the α_j-string containing γ are positive roots by Corollary 5.5.7. So by induction, the first part of the string, $\gamma - r\alpha_j, \dots, \gamma$ is contained in Φ^+. Hence in Step 2.2 the maximal integer r is determined such that $\gamma - r\alpha_j$ is a root. Then by Proposition 5.4.2 we have

$$q = r - \langle \gamma, \alpha_j \rangle = r - \sum_{i=1}^{l} k_i \langle \alpha_i, \alpha_j \rangle = r - \sum_{i=1}^{l} k_i C(i, j).$$

Now since $\beta \in \Phi$, we have $q > 0$ and β will be added to Φ^+.

Also it is clear that only roots are added to Φ^+. Therefore after a finite number of steps, Φ^+ will contain all positive roots. Hence $\Phi = \Phi^+ \cup -\Phi^+$. $\square$

Corollary 5.6.3 *Let Φ_1 and Φ_2 be root systems in the Euclidean spaces V_1 and V_2 respectively. Let Δ_1 and Δ_2 be simple systems of respectively Φ_1 and Φ_2 and let C_1 and C_2 be their respective Cartan matrices. Then $C_1 \sim C_2$ implies $\Phi_1 \cong \Phi_2$.*

Proof. Let $\Delta_1 = \{\alpha_1, \dots, \alpha_l\}$ and $\Delta_2 = \{\beta_1, \dots, \beta_l\}$. Because $C_1 \sim C_2$ we may suppose that Δ_2 has been permuted in such a way that $\langle \alpha_i, \alpha_j \rangle = \langle \beta_i, \beta_j \rangle$ for $1 \leq i, j \leq l$. Let $f : V_1 \to V_2$ be the linear map defined by $f(\alpha_i) = \beta_i$. By the algorithm CartanMatrixToRootSystem we see that

$$\sum_{i=1}^{l} k_i \alpha_i \in \Phi_1 \iff \sum_{i=1}^{l} k_i \beta_i \in \Phi_2.$$

In particular, the root strings in Φ_1 and Φ_2 match. So by Proposition 5.4.2, we have that f is an isomorphism of root systems. $\square$

5.7   Simple systems and the Weyl group

The objective of this section is to show the converse of Corollary 5.6.3: namely that two isomorphic root systems have equivalent Cartan matrices. First we show that two simple systems Δ_1 and Δ_2 of the same root system Φ can be mapped onto each other by an automorphism of Φ. The desired result will be an easy consequence of this. In the proof we make use of a group of automorphisms of Φ, called the Weyl group. This group is of paramount importance in the theory of semisimple Lie algebras. At the end of the section we exhibit a convenient set of generators of it.

First we introduce one more example of a root order. For that let $v_0 \in V$ be such that $(v_0, \alpha) \neq 0$ for all $\alpha \in \Phi$. As such a v_0 must lie outside a finite number of hyperplanes, such v_0 evidently exist. Now for $u, v \in V$ such that $(u, v_0) \neq (v, v_0)$ we put $u < v$ if $(u, v_0) < (v, v_0)$. It is immediate that this is a root order. We say that $<$ is the root order defined by v_0. As seen in Section 5.5, this order defines a simple system. Conversely, let Δ be a simple system of Φ (relative to some root order). Choose $v_0 \in V$ such that $(\alpha, v_0) > 0$ for all $\alpha \in \Delta$ (from the non-degeneracy of $(,)$ it follows that we can choose v_0 such that, for example, $(\alpha, v_0) = 1$ for all $\alpha \in \Delta$). Then by Corollary 5.5.7 it is clear that $(v_0, \alpha) \neq 0$ for all $\alpha \in \Phi$. Hence v_0 defines a root order. It is straightforward to see that Δ is the set of simple roots relative to this order.

Let $\alpha \in \Phi$ and consider the reflection r_α. By Lemma 5.3.4 together with (5.1) we see that r_α is an automorphism of the root system Φ. Also products of reflections are automorphisms of Φ. So it is natural to consider the group generated by all reflections r_α for $\alpha \in \Phi$. This group is called the *Weyl group*; it is denoted by $W(\Phi)$.

Since an element of $W(\Phi)$ is entirely determined by the way in which it acts on Φ, the Weyl group $W(\Phi)$ can be viewed as a subgroup of the symmetric group on Φ. In particular $W(\Phi)$ is finite. Furthermore, because the generators of $W(\Phi)$ leave the inner product invariant by (5.1), the same holds for all elements of $W(\Phi)$.

Now we fix a root order $<$ and let Δ be the simple system of Φ, relative to $<$.

Lemma 5.7.1 *Let $\alpha \in \Phi$ be a simple root then r_α permutes the set $\Phi^+ \backslash \{\alpha\}$.*

Proof. Let β be a positive root distinct from α. We must prove that $r_\alpha(\beta)$ is again an element of $\Phi^+ \setminus \{\alpha\}$. By Corollary 5.5.7 we have

$$\beta = \sum_{\gamma \in \Delta} k_\gamma \gamma$$

where all k_γ are non-negative integers. Because β cannot be a scalar multiple of α, some k_{γ_0} for a $\gamma_0 \neq \alpha$ must be positive. Now $r_\alpha(\beta) = \beta - \langle \beta, \alpha \rangle \alpha$; hence the coefficient of γ_0 in $r_\alpha(\beta)$ still is k_{γ_0}. So $r_\alpha(\beta)$ has at least one positive coefficient, and therefore it is a positive root. Also $r_\alpha(\beta) \neq \alpha$ because $k_{\gamma_0} > 0$.   □

Set

$$\rho = \frac{1}{2} \sum_{\beta \in \Phi^+} \beta.$$

The vector $\rho \in V$ is called the *Weyl vector*.

Corollary 5.7.2 *Let α be a simple root. Then $r_\alpha(\rho) = \rho - \alpha$.*

Proof. First we have

$$\rho = \frac{1}{2}\alpha + \frac{1}{2} \sum_{\beta \in \Phi^+ \setminus \alpha} \beta.$$

Hence the corollary follows from the fact that r_α maps α to $-\alpha$ and permutes the other positive roots (Lemma 5.7.1).   □

The next theorem implies that the Cartan matrix does not depend on the simple system chosen.

Theorem 5.7.3 *Let Δ_1 and Δ_2 be two simple systems of Φ. Then there is a $g \in W(\Phi)$ such that $g(\Delta_1) = \Delta_2$.*

Proof. Let $v_0 \in V$ be such that $(\alpha, v_0) > 0$ for all $\alpha \in \Delta_1$. Then as seen at the beginning of this section, Δ_1 is the simple system of Φ relative to the order defined by v_0. Let ρ be the Weyl vector corresponding to Δ_2 (i.e., half the sum of the roots that are positive relative to the order that defines Δ_2). Now choose a $g \in W(\Phi)$ such that $(g(v_0), \rho)$ is maximal. (Because $W(\Phi)$ is finite such a g exists.) Then for $\alpha \in \Delta_2$ we have

$$\begin{aligned}
(g(v_0), \rho) &\geq (r_\alpha g(v_0), \rho) \\
&= (g(v_0), r_\alpha(\rho)) \quad \text{(by (5.1) and } r_\alpha^2 = 1) \\
&= (g(v_0), \rho - \alpha) \quad \text{(by Corollary 5.7.2)} \\
&= (g(v_0), \rho) - (g(v_0), \alpha).
\end{aligned}$$

Hence $(g(v_0), \alpha) \geq 0$ for all $\alpha \in \Delta_2$. Furthermore, $(g(v_0), \alpha) = 0$ for some $\alpha \in \Delta_2$ implies that $(v_0, g^{-1}(\alpha)) = 0$ which is impossible because $(v_0, \beta) \neq 0$ for all $\beta \in \Phi$ by the choice of v_0. So the order defined by $g(v_0)$ yields the simple system Δ_2. But for $\beta \in \Delta_1$ we have $(g(\beta), g(v_0)) = (\beta, v_0) > 0$. Hence $g(\Delta_1) = \Delta_2$. $\square$

Corollary 5.7.4 *Let Φ_1 and Φ_2 be root systems in the Euclidean spaces V_1 and V_2 respectively. Let Δ_1 and Δ_2 be simple systems of Φ_1 and Φ_2 respectively and let C_1 and C_2 be their respective Cartan matrices. Then $C_1 \sim C_2$ if and only if $\Phi_1 \cong \Phi_2$.*

Proof. One direction was performed in Corollary 5.6.3. For the other direction, suppose that $\Phi_1 \cong \Phi_2$. Let $f : V_1 \to V_2$ be an isomorphism of root systems (i.e., a bijective linear map such that $\langle f(\alpha), f(\beta) \rangle = \langle \alpha, \beta \rangle$ for $\alpha, \beta \in \Phi$). Then by Theorem 5.7.3, after composing f with a suitable element of $W(\Phi_2)$ we may suppose that $f(\Delta_1) = \Delta_2$. Hence $C_1 \sim C_2$. $\square$

Let Δ be a fixed simple system of Φ and let $\alpha \in \Delta$. Then the reflection r_α is called a *simple reflection*. We show that the Weyl group $W(\Phi)$ is generated by the simple reflections. For that let $W_0(\Phi)$ be the subgroup of $W(\Phi)$ generated by all simple reflections.

Lemma 5.7.5 *Let $\beta \in \Phi$ be a root; then there exist $\alpha \in \Delta$ and $g \in W_0(\Phi)$ such that $g(\alpha) = \beta$.*

Proof. First we suppose that β is a positive root. We use induction on the number $\mathrm{ht}(\beta)$. If $\mathrm{ht}(\beta) = 1$, then $\beta \in \Delta$ and we can take $g = 1$. Suppose that $\mathrm{ht}(\beta) > 1$. Then there is an $\alpha \in \Delta$ such that $(\alpha, \beta) > 0$ (otherwise the set $\Delta \cup \{\beta\}$ is linearly independent by Lemma 5.5.4; and this is absurd). Then also $\langle \beta, \alpha \rangle > 0$. Set $\gamma = r_\alpha(\beta) = \beta - \langle \beta, \alpha \rangle \alpha$; then γ is a positive root and $\mathrm{ht}(\gamma) < \mathrm{ht}(\beta)$. So by induction there is a $h \in W_0(\Phi)$ and an $\alpha' \in \Delta$ such that $\gamma = h(\alpha')$. Hence $\beta = r_\alpha(\gamma) = r_\alpha h(\alpha')$ and we take $g = r_\alpha h$.

Now if β is negative, then $-\beta$ is positive and hence $-\beta = g(\alpha)$ for some $\alpha \in \Delta$ and $g \in W_0(\Phi)$. So $\beta = g(-\alpha) = g r_\alpha(\alpha)$ and the result follows. $\square$

Theorem 5.7.6 *The Weyl group $W(\Phi)$ is generated by the simple reflections.*

Proof. Let $\beta \in \Phi$, then by Lemma 5.7.5 there is a $g \in W_0(\Phi)$ such that $g(\beta) = \alpha$ for some simple root $\alpha \in \Delta$. So $r_\alpha = r_{g(\beta)} = g r_\beta g^{-1}$ (the last equality can be established by a straightforward calculation, using the fact that g leaves the inner product $(\ ,\ )$ invariant). Hence $r_\beta = g^{-1} r_\alpha g \in W_0(\Phi)$ and the result follows. $\square$

5.8 Dynkin diagrams

In Section 5.7 we showed that a Cartan matrix is a complete invariant of root systems (Corollary 5.7.4). In particular, a classification of Cartan matrices will yield a classification of root systems. In this section we replace the Cartan matrix by a graph, called a *Dynkin diagram*. We show that a Cartan matrix and the corresponding Dynkin diagram determine each other. Finally in the next section we will classify Dynkin diagrams.

Let C be a Cartan matrix and let $\alpha_1, \ldots, \alpha_l$ be elements of a Euclidean space V such that $C(i,j) = \langle \alpha_i, \alpha_j \rangle$ and $(\alpha_i, \alpha_j) \leq 0$ for $i \neq j$. We define the Dynkin diagram of C to be the graph on l points with labels $\alpha_1, \ldots, \alpha_l$. Two points α_i and α_j will be connected by $\langle \alpha_i, \alpha_j \rangle \langle \alpha_j, \alpha_i \rangle = 0, 1, 2, 3$ lines. If the number of such lines is greater than one, then the two elements α_i and α_j have unequal length. In this case we put an arrow pointing from the *longer* to the *shorter* vector.

Example 5.8.1 Suppose that the Cartan matrix corresponding to the vectors $\alpha_1, \alpha_2, \alpha_3, \alpha_4$ is

$$C = \begin{pmatrix} 2 & -1 & 0 & 0 \\ -1 & 2 & -2 & 0 \\ 0 & -1 & 2 & -1 \\ 0 & 0 & -1 & 2 \end{pmatrix}.$$

We see that

$$\langle \alpha_2, \alpha_3 \rangle = \frac{2(\alpha_2, \alpha_3)}{(\alpha_3, \alpha_3)} = -2,$$

$$\langle \alpha_3, \alpha_2 \rangle = \frac{2(\alpha_3, \alpha_2)}{(\alpha_2, \alpha_2)} = -1.$$

Hence α_2 is longer than α_3. From this it follows that the Dynkin diagram of C is

We show that from the Dynkin diagram we can recover the Cartan matrix. Let C be a Cartan matrix corresponding to the vectors $\alpha_1, \ldots, \alpha_l$. Let $i \neq j \in \{1, \ldots, l\}$. Then we know that $\langle \alpha_i, \alpha_j \rangle \leq 0$. Let k be the number of lines connecting α_i and α_j in the Dynkin diagram. If $k = 0$, then $\langle \alpha_i, \alpha_j \rangle = 0$, and if $k = 1$, then $\langle \alpha_i, \alpha_j \rangle = -1$. If $k \geq 2$, then α_i and

α_j have unequal length. From the diagram it can be determined which of the two is the shortest. Suppose $k = 2$; then $\langle \alpha_i, \alpha_j \rangle$ is -1 or -2. If α_i is shorter than α_j, then $\langle \alpha_i, \alpha_j \rangle = -1$. On the other hand, if α_i is longer than α_j, then $\langle \alpha_i, \alpha_j \rangle = -2$. By a similar reasoning we see that we can determine $\langle \alpha_i, \alpha_j \rangle$ in the case where $k = 3$. So from the Dynkin diagram we can determine the off-diagonal elements of the Cartan matrix. Furthermore the diagonal elements are all equal to 2. The conclusion is that the Dynkin diagram determines the Cartan matrix.

Now let Φ be a root system with Cartan matrix C. Let D be the Dynkin diagram of C; then we also say that D is the Dynkin diagram of Φ. Since the Cartan matrix is a complete invariant of root systems and the Cartan matrix and the Dynkin diagram determine each other we see that a root system has a uniquely determined Dynkin diagram. By Proposition 5.3.8, a direct sum decomposition of a semisimple Lie algebra corresponds to a direct sum decomposition of its root system. By the next result, this in turn corresponds to a decomposition of the Dynkin diagram into connected components.

Lemma 5.8.2 *Let Φ be a root system with Dynkin diagram D. Then $\Phi = \Phi_1 \oplus \Phi_2$ if and only if D can be decomposed as the union of two components D_1 and D_2 that are not connected and such that D_i is the Dynkin diagram of Φ_i for $i = 1, 2$.*

Proof. First suppose that $\Phi = \Phi_1 \oplus \Phi_2$. Let $\Delta_1 = \{\alpha_1, \ldots, \alpha_s\}$ and $\Delta_2 = \{\beta_1, \ldots, \beta_t\}$ be simple systems of Φ_1 and Φ_2 respectively. Then $\Delta = \Delta_1 \cup \Delta_2$ is a simple system of Φ. Let D be the Dynkin diagram of Φ. Then the vertices of D are labelled $\alpha_1, \ldots, \alpha_s, \beta_1, \ldots, \beta_t$. Furthermore none of the α_i can be connected to a β_j. Hence D is the union of two components D_1 and D_2 that are the Dynkin diagrams of Φ_1 and Φ_2 respectively and D_1 and D_2 are not connected.

Now suppose that D is the union of two components D_1, D_2 that are not connected. Let Δ be a simple system of Φ. Then $\Delta = \Delta_1 \cup \Delta_2$, where Δ_1, Δ_2 are non-empty and $\langle \alpha, \beta \rangle = 0$ for $\alpha \in \Delta_1$ and $\beta \in \Delta_2$. Let C be the Cartan matrix corresponding to D. We apply the algorithm CartanMatrix-ToRootSystem on Δ and C. Then the positive roots that are constructed are all of the form $\sum_\gamma k_\gamma \gamma$, where all $\gamma \in \Delta_1$ or all $\gamma \in \Delta_2$. Hence $\Phi = \Phi_1 \oplus \Phi_2$ where Φ_i is the root system with simple system Δ_i for $i = 1, 2$. $\square$

5.9 Classifying Dynkin diagrams

In this section we classify all possible Dynkin diagrams, thereby classifying all possible root systems. By Lemma 5.8.2 we may restrict our attention to connected Dynkin diagrams. We then obtain a classification of all root systems that are not direct sums. All other root systems are direct sums of these.

For simplicity we first drop the arrows in the Dynkin diagram. The resulting graph is called a *Coxeter diagram*. We will determine all possible Coxeter diagrams and than add the arrows on again.

Let C be a Cartan matrix corresponding to the vectors $\alpha_1, \ldots, \alpha_l$ in the Euclidean space V. Then for $1 \leq i \neq j \leq l$ we have

$$\frac{4(\alpha_i, \alpha_j)^2}{(\alpha_i, \alpha_i)(\alpha_j, \alpha_j)} = 0, 1, 2, 3, \quad \text{and} \quad (\alpha_i, \alpha_j) \leq 0.$$

Now we replace α_i by $v_i = \lambda_i \alpha_i$ where $\lambda_i \in \mathbb{R}$ is a positive number such that v_i has length 1. Then we have the simpler conditions

$$(v_i, v_i) = 1, \ 4(v_i, v_j)^2 = 0, 1, 2, 3, \quad \text{and} \quad (v_i, v_j) \leq 0, \tag{5.4}$$

for $1 \leq i \neq j \leq l$. A set of vectors $A = \{v_1, \ldots, v_l\}$ in V is called an *admissible configuration* if the conditions (5.4) are satisfied. The Coxeter diagram of an admissible configuration $A = \{v_1, \ldots, v_l\}$ consists of l points labelled $v_1, \ldots, v_l$, where the points v_i and v_j are connected by $4(v_i, v_j)^2$ edges. It is clear that the Coxeter diagram of the original Cartan matrix C is the same as the Coxeter diagram of A.

We now prove a series of lemmas that will lead to the classification of the Coxeter diagrams of admissible configurations. The first of these lemmas is clear without proof.

Lemma 5.9.1 *Let Γ be the Coxeter diagram of the admissible configuration A. Let A' be a subset of A and let Γ' be the diagram obtained from Γ by deleting all points corresponding to the elements that are not in A' and all lines incident with them. Then A' is an admissible configuration and Γ' is its Coxeter diagram.*

Lemma 5.9.2 *Let $A = \{v_1, \ldots, v_l\}$ be an admissible system. Then the number of pairs v_i, v_j such that $i < j$ and $(v_i, v_j) \neq 0$ is less than l.*

Proof. Set $v = \sum_{i=1}^{l} v_i$, then

$$0 < (v, v) = \sum_{i=1}^{l} (v_i, v_i) + 2 \sum_{i<j} (v_i, v_j) = l + \sum_{i<j} 2(v_i, v_j).$$

Now if $(v_i, v_j) \neq 0$, then $2(v_i, v_j) \leq -1$. It follows that there are less than l such pairs. $\qquad\square$

Lemma 5.9.3 *Let Γ be the Coxeter diagram of the admissible configuration A. Then Γ contains no cycle (where a cycle is a sequence of points $v_{i_1}, \dots, v_{i_k}$ such that v_{i_j} is connected to $v_{i_{j+1}}$ and v_{i_k} to v_{i_1}).*

Proof. By Lemma 5.9.1 a cycle is the Coxeter graph of an admissible configuration (drop all points not in the cycle). But this Coxeter graph violates Lemma 5.9.2. $\qquad\square$

Lemma 5.9.4 *Let Γ be a Coxeter diagram and v a point of Γ. Let k be the number of edges incident with v. Then $k \leq 3$.*

Proof. Let $u_1, \dots, u_k$ be the points connected to v. Then $(u_i, u_j) = 0$ for $i \neq j$ because of Lemma 5.9.3. Let W be the space spanned by $v, u_1, \dots, u_k$. Extend the set $\{u_1, \dots, u_k\}$ to an orthonormal basis $\{u_0, u_1, \dots, u_k\}$ of W. If v is orthogonal to u_0, then v is in the space spanned by $\{u_1, \dots, u_k\}$ which is not the case. Hence $(v, u_0) \neq 0$. Furthermore $v = \sum_{i=0}^{k}(v, u_i)u_i$ so that

$$1 = (v, v) = \sum_{i=0}^{k}(v, u_i)^2 = (v, u_0)^2 + \sum_{i=1}^{k}(v, u_i)^2.$$

From this it follows that $\sum_{i=1}^{k}(v, u_i)^2 < 1$ and hence $\sum_{i=1}^{k}4(v, u_i)^2 < 4$. We observe that $4(v, u_i)^2$ is the number of lines connecting v and u_i. The result follows. $\qquad\square$

Lemma 5.9.5 *The only connected Coxeter diagram that contains a triple edge is*

G_2: ⊂══════⊃

Proof. This is immediate from Lemma 5.9.4. $\qquad\square$

Lemma 5.9.6 *Let A be an admissible configuration with Coxeter diagram Γ. Let $u_1, \dots, u_k$ be elements from A that form a simple chain in Γ (that is u_i is connected to u_{i+1} by a single edge for $1 \leq i \leq k-1$). Set $A' = \{\sum_{i=1}^{k}u_i\} \cup A \setminus \{u_1, \dots, u_k\}$. Then A' is an admissible configuration. Furthermore, the Coxeter diagram of A' is obtained from Γ by shrinking the simple chain $u_1, \dots, u_k$ to a point.*

Proof. From Lemma 5.9.3 we have that $(u_i, u_j) = 0$ for $1 \leq i < j \leq k$ unless $j = i + 1$ when $2(u_i, u_{i+1}) = -1$. Set $u = \sum_{i=1}^{k} u_i$, then

$$(u, u) = \sum_{i=1}^{k} (u_i, u_i) + \sum_{i<j} 2(u_i, u_j) = k - (k - 1) = 1.$$

Also an element $w \in A'$ is connected to at most one element u_i from the set $\{u_1, \dots, u_k\}$ because by Lemma 5.9.3 there are no cycles. Hence $(u, w) = (u, u_i)$ and $4(u, w)^2 = 4(u_i, w^2) = 0, 1, 2, 3$. The last statement is immediate. □

Lemma 5.9.7 *A Coxeter diagram does not contain a subdiagram of one of the following forms:*

1.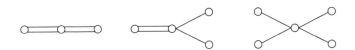

2.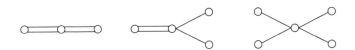

3.

Proof. Suppose one of these is a subdiagram of a Coxeter diagram. Then by Lemma 5.9.1 this is a Coxeter diagram in its own right. Now we use Lemma 5.9.6 to shrink the simple chains to a point and obtain that one of

is a Coxeter diagram. However, none of these can be a Coxeter diagram due to Lemma 5.9.4. □

Lemma 5.9.8 *Let Γ be a Coxeter diagram of an admissible configuration. If Γ is connected, then Γ is one of*

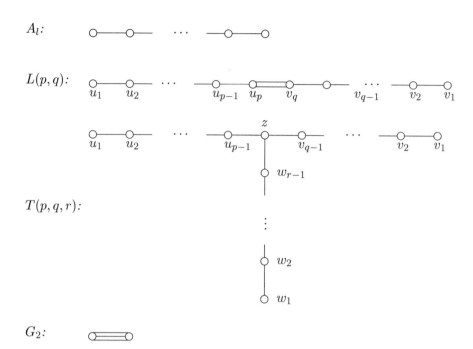

A_l:

$L(p,q)$:

$T(p,q,r)$:

G_2:

Proof. If Γ contains a triple edge, then Γ must be G_2 by Lemma 5.9.5. So we may suppose that Γ does not contain a triple edge.

Suppose that Γ contains a double edge. Then it contains no other double edge by Lemma 5.9.7. Also by the same lemma it does not contain a "node",

i.e., a subdiagram of the form ○———○———○. Hence Γ must be an $L(p,q)$.

Now we suppose that Γ does not contain triple nor double edges. If Γ contains a node, then by Lemma 5.9.5 it contains only one node. Hence Γ is a $T(p,q,r)$.

On the other hand, if Γ does not contain a node, then it must be an A_l by Lemma 5.9.3.   □

Lemma 5.9.9 *Let Γ be a Coxeter diagram of type $L(p,q)$. Then Γ is one of*

- $B_l = C_l$: ○———○—— $\cdots$ ——○══○

- F_4: ○———○══○———○

Proof. Set $u = \sum_{i=1}^{p} iu_i$ and $v = \sum_{j=1}^{q} jv_j$ (where u_i, v_j are as in the diagram of $L(p,q)$). For $i < j$ we have that $(u_i, u_j) = 0$ unless $j = i+1$ when $2(u_i, u_{i+1}) = -1$. Hence

$$(u,u) = \sum_{i=1}^{p} i^2 (u_i, u_i) + \sum_{i<j} ij2(u_i, u_j) =$$

$$\sum_{i=1}^{p} i^2 - \sum_{i=1}^{p-1} i(i+1) = p^2 - \frac{p(p-1)}{2} = \frac{p(p+1)}{2}.$$

And similarly

$$(v,v) = \frac{q(q+1)}{2}.$$

Also $4(u_p, v_q)^2 = 2$, whence

$$(u,v)^2 = p^2 q^2 (u_p, v_q)^2 = \frac{p^2 q^2}{2}.$$

Now we use the inequality of Cauchy-Schwarz, $(u,v)^2 < (u,u)(v,v)$ to obtain

$$\frac{p^2 q^2}{2} < \frac{p(p+1)}{2} \frac{q(q+1)}{2}.$$

After dividing by pq and moving things around we see that this is equivalent to $(p-1)(q-1) < 2$. Since both p and q are positive, this gives that either $p = q = 2$ (leading to F_4) or one of p, q is equal to 1 and the other may be chosen arbitrarily (all yielding a diagram of type $B_l = C_l$). □

Lemma 5.9.10 *Let Γ be a Coxeter diagram of type $T(p,q,r)$, where p, q, r are integers ≥ 2 (if one of them equals 1, then Γ is of type A_l). Then Γ is one of*

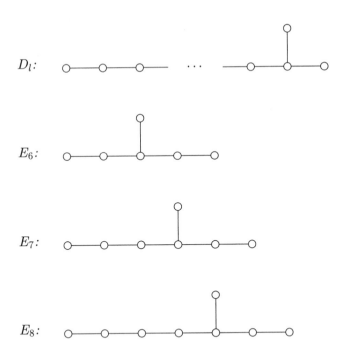

D_l:

E_6:

E_7:

E_8:

Proof. Set

$$u = \sum_{i=1}^{p-1} iu_i, \quad v = \sum_{j=1}^{q-1} jv_j, \quad w = \sum_{k=1}^{r-1} kw_k.$$

Then u, v, w are linearly independent and z is not in the span of u, v, w (since the set of all u_i, all v_j, all w_k together with z is linearly independent). Let W be the space spanned by z, u, v, w and chose a $z_0 \in W$ such that $\{z_0, u, v, w\}$ is an orthogonal basis for W. Expressing z as a linear combination of basis elements we find

$$z = \frac{(z_0, z)}{(z_0, z_0)} z_0 + \frac{(u, z)}{(u, u)} u + \frac{(v, z)}{(v, v)} v + \frac{(w, z)}{(w, w)} w.$$

So that

$$1 = (z, z) = \frac{(z_0, z)^2}{(z_0, z_0)} + \frac{(u, z)^2}{(u, u)} + \frac{(v, z)^2}{(v, v)} + \frac{(w, z)^2}{(w, w)}.$$

From which it follows that

$$\frac{(u, z)^2}{(u, u)} + \frac{(v, z)^2}{(v, v)} + \frac{(w, z)^2}{(w, w)} < 1. \tag{5.5}$$

Now by an analogous argument as the one used in the proof of Lemma 5.9.9, we have that $(u, u) = p(p-1)/2$. Also from the Coxeter diagram $T(p, q, r)$ we have that $(u, z)^2 = (p-1)^2/4$. Hence

$$\frac{(u, z)^2}{(u, u)} = \frac{1}{2}(1 - \frac{1}{p}),$$

and similarly

$$\frac{(v, z)^2}{(v, v)} = \frac{1}{2}(1 - \frac{1}{q}) \quad \text{and} \quad \frac{(w, z)^2}{(w, w)} = \frac{1}{2}(1 - \frac{1}{r}).$$

Combining this with (5.5) we get

$$\frac{1}{p} + \frac{1}{q} + \frac{1}{r} > 1. \tag{5.6}$$

Without loss of generality we may assume that $p \geq q \geq r \geq 2$. Then $\frac{1}{p} \leq \frac{1}{q} \leq \frac{1}{r}$ and hence (5.6) implies $\frac{3}{r} > 1$, which implies $r = 2$. Now (5.6) reduces to

$$\frac{1}{p} + \frac{1}{q} > \frac{1}{2}.$$

Since $\frac{1}{q} \geq \frac{1}{p}$ we have $\frac{2}{q} > \frac{1}{2}$. From this it follows that $q = 2, 3$.

If $q = 2$, then we see that $\frac{1}{p} > 0$ which holds for all $p \geq 2$. In this case Γ is of type D_l.

If $q = 3$, then $\frac{1}{p} > \frac{1}{6}$ and $p = 3, 4, 5$, leading to E_6, E_7, E_8. $\qquad \square$

The Coxeter diagrams A_l, together with those of Lemmas 5.9.9 and 5.9.10 and G_2 are all possible Coxeter diagrams. To obtain the list of all possible connected Dynkin diagrams, we have to put the arrows back on in the cases B_l, C_l, F_4, G_2. Since the diagrams of F_4 and G_2 are symmetric, it does not matter in which direction we point the arrow (both directions will yield equivalent Cartan matrices). However, in the case of B_l, C_l, the two directions are not equivalent. The conclusion is that Table 5.2 contains the list of all possible connected Dynkin diagrams. We have labelled the points of the diagrams in Table 5.2 in order to fix a Cartan matrix corresponding to each diagram.

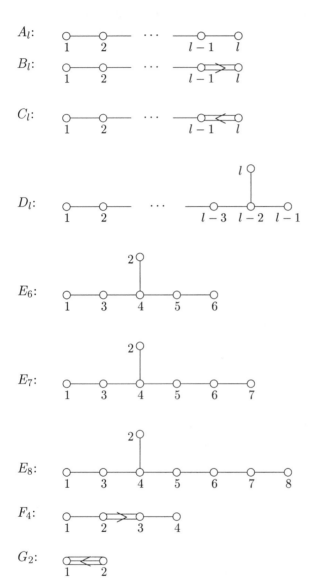

Table 5.2: The connected Dynkin diagrams.

5.10   Constructing the root systems

In the previous section we determined a list of possible root systems. That
is to say: we took the set of all possible Dynkin diagrams and weeded out

those that by some elementary observations could not exist. It is however not yet clear that for every remaining Dynkin diagram D there exists a root system having D as Dynkin diagram. It is the objective of this section to establish this.

Let D be a Dynkin diagram from Table 5.2, and let C be the corresponding Cartan matrix. The obvious way to show that there is a root system having D as its Dynkin diagram is to apply the algorithm CartanMatrixTo-RootSystem to C and prove that the output is a root system. However, it is a somewhat tedious job to describe the output of this algorithm for each Cartan matrix. Instead of this we will for each type describe a Euclidean space and a root system inside this space of that type. From this it will follow that by the algorithm CartanMatrixToRootSystem we can construct the root systems.

Let X be one of A, B, C, D, E, F, G and $l \geq 1$ an integer (and $l = 6, 7, 8$ if $X = E$, $l = 4$ if $X = F$ and $l = 2$ if $X = G$). Then we say that X_l is a *simple type*. Let V be a Euclidean space with inner product $(\ ,\ )$. A linearly independent set of vectors $\Delta = \{\alpha_1, \ldots, \alpha_l\} \subset V$ is said to be a *simple system of type* X_l if

$$\frac{2(\alpha_i, \alpha_j)}{(\alpha_j, \alpha_j)} = C_{ij} \quad \text{for } 1 \leq i, j \leq l,$$

where C is the Cartan matrix corresponding to the Dynkin diagram of type X_l.

Proposition 5.10.1 *Let* X_l *be a simple type. Then there exists a root system* Φ *of type* X_l.

Proof. For each simple type X_l we give an explicit construction of a root system Φ inside a Euclidean space W having a Dynkin diagram of type X_l. First we construct the root systems of types A_l, B_l, D_l and G_2. In each case we work in a real Euclidean space $V = \mathbb{R}^n$ with basis $v_1, \ldots, v_n$ and the usual inner product (defined by $(v_i, v_j) = \delta_{ij}$). In each case we define a *lattice* $\mathcal{L}$ in V (i.e., the $\mathbb{Z}$-span of a basis of V). Then a root system Φ in a subspace W of V is constructed as the set of all elements of $\mathcal{L}$ having prescribed lengths (and maybe satisfying some auxiliary conditions). First we give some general considerations that will help to show in each case that the conditions (R_1)-(R_4) of Definition 5.3.2 hold for Φ.

Since $\mathcal{L}$ is a lattice in V there can be only a finite number of elements of $\mathcal{L}$ at a given distance to the origin. Hence Φ is finite. It will always be clear that Φ spans W and does not contain 0; hence we have (R_1).

Let $\alpha \in \Phi$ and suppose that $\beta = \lambda\alpha \in \Phi$ for a certain $\lambda \in \mathbb{R}$. Then $(\beta,\beta) = \lambda^2(\alpha,\alpha)$. From the requirements on the lengths of the elements of Φ it will follow that λ is either ± 1 or irrational. If λ is irrational, then $\lambda\alpha \notin \mathcal{L}$. So the only multiples of α that also lie in Φ are $\pm\alpha$, and (R_2) is satisfied.

In each case we will prove (R_4) by individual considerations. But then the reflection r_α maps Φ into $\mathcal{L}$. Furthermore, the image of an element of Φ under a reflection will again have the prescribed length by (5.1). Then (R_3) will easily follow.

A_l: we set $V = \mathbb{R}^{l+1}$ and $\mathcal{L} = \{\sum_{i=1}^{l+1} k_i v_i \mid k_i \in \mathbb{Z}\}$. And $\Phi = \{v = \sum_i k_i v_i \in \mathcal{L} \mid (v,v) = 2 \text{ and } \sum_i k_i = 0\}$. Then it is easily seen that Φ consists of the elements $\pm(v_i - v_j)$ for $1 \le i < j \le l+1$. A simple system is given by $\Delta = \{v_1 - v_2, v_2 - v_3, \dots, v_l - v_{l+1}\}$. For the space W we simply take the span of Φ. Since the inner products of elements of Φ are integral and $(v,v) = 2$ for $v \in \Phi$ we have (R_4). Set $v_0 = v_1 + v_2 + \cdots + v_{l+1}$, then $v \in \Phi$ if and only if $(v,v) = 2$ and $(v,v_0) = 0$. Now suppose $v,w \in \Phi$, and let r_v be the reflection with respect to v. Then as seen above $r_v(w)$ has the required length. Now since $r_v(w)$ is a linear combination of v and w, also $(r_v(w),v_0) = 0$. So we also have (R_3).

B_l: we set $V = \mathbb{R}^l$ and $W = V$. The lattice $\mathcal{L}$ is spanned by the basis vectors $v_1, \dots, v_l$. And $\Phi = \{v \in \mathcal{L} \mid (v,v) = 1, 2\}$. Then Φ consists of the vectors $\pm(v_i \pm v_j)$ for $1 \le i < j \le l$ together with $\pm v_i$ for $1 \le i \le l$. A simple system is $\Delta = \{v_1 - v_2, \dots, v_{l-1} - v_l, v_l\}$. In this case (R_3) and (R_4) are clear.

D_l: here $V = \mathbb{R}^l$ and $W = V$. The lattice $\mathcal{L}$ is defined as for B_l. In this case $\Phi = \{v \in \mathcal{L} \mid (v,v) = 2\}$, consisting of $\pm(v_i \pm v_j)$ for $1 \le i < j \le l$. A simple system is given by $\Delta = \{v_1 - v_2, \dots, v_{l-1} - v_l, v_{l-1} + v_l\}$. Again (R_3) and (R_4) are clear.

G_2: put $V = \mathbb{R}^3$ and $v_0 = v_1 + v_1 + v_3$. The lattice is

$$\mathcal{L} = \left\{ \sum_{i=1}^{3} k_i \frac{1}{\sqrt{3}} v_i \mid k_i \in \mathbb{Z} \right\}.$$

Furthermore, $\Phi = \{v \in \mathcal{L} \mid (v,v) = \frac{2}{3}, 2 \text{ and } (v,v_0) = 0\}$. The space W will be the span of Φ. Then

$$\Phi = \pm\left\{ \frac{1}{\sqrt{3}}(v_1 - v_2), \frac{1}{\sqrt{3}}(v_2 - v_3), \frac{1}{\sqrt{3}}(v_1 - v_3) \right\} \cup$$

$$\pm\left\{ \frac{1}{\sqrt{3}}(v_i + v_j - 2v_k) \mid i, j, k \text{ distinct} \right\}.$$

A simple system is $\Delta = \left\{ \frac{1}{\sqrt{3}}(v_1 - v_2), \frac{1}{\sqrt{3}}(-v_1 + 2v_2 - v_3) \right\}$. In this case (R_4) is easily checked and (R_3) follows as in the case of A_l.

Now we turn our attention towards E_8 and F_4. (Since $E_{6,7}$ are subsystems of E_8 in a natural way (cf. Lemma 5.9.1), this also gives a construction of these root systems.) In both cases we construct a set Q that is the $\mathbb{Z}$-span of a linearly dependent set of vectors. So in these cases Q is not a lattice. The root systems will again be defined as subsets of Q consisting of vectors of given lengths. Then analogous considerations to the ones applied for A_l, B_l, D_l, G_2 will prove that Φ is a root system.

E_8: here $V = \mathbb{R}^8$ and $W = V$. Set $v_0 = \frac{1}{2}(v_1 + v_2 + \cdots + v_8)$, and

$$Q = \left\{ k_0 v_0 + \sum_{i=1}^{8} k_i v_i \mid k_i \in \mathbb{Z} \text{ and } \sum_{i=1}^{8} k_i \text{ is even} \right\}.$$

And $\Phi = \{v \in Q \mid (v,v) = 2\}$. Let $v = \sum_{i=0}^{8} k_i v_i \in \Phi$, then

$$2 = (v,v) = \sum_{i=1}^{8} (k_i + \frac{1}{2}k_0)^2.$$

Set $m_i = k_i + k_0/2$ for $1 \le i \le 8$. If k_0 is even then all m_i are integers. Hence the possibilities for v are $\pm(v_i \pm v_j)$ for $1 \le i < j \le 8$. (It is straightforward to check that these elements lie in Q.) Furthermore, if k_0 is odd, then all $m_i \in \mathbb{Z} + \frac{1}{2}$. It is seen that $m_i = \pm\frac{1}{2}$ for $1 \le i \le 8$. This means that $k_i = \frac{1}{2}(\epsilon_i - k_0)$ where $\epsilon_i = \pm 1$. Hence

$$\sum_{i=1}^{8} k_i = -4k_0 + \frac{1}{2}\sum_{i=1}^{8} \epsilon_i.$$

So in order that v lie in Q we must have $\sum_{i=1}^{8} \epsilon_i = 0, \pm 4, \pm 8$. But this is the same as saying that the number of $+1$'s must be even. So Φ contains all vectors $\pm(v_i \pm v_j)$ for $1 \le i < j \le 8$ together with the vectors $\pm\frac{1}{2}(\pm v_1 \pm v_2 \pm \cdots \pm v_8)$ where the number of pluses is even (and hence so is the number of minuses). A simple system is given by $\Delta = \{\frac{1}{2}(v_1 - v_2 - v_3 - v_4 - v_5 - v_6 - v_7 + v_8), v_1 + v_2, v_2 - v_1, v_3 - v_2, v_4 - v_3, v_5 - v_4, v_6 - v_5, v_7 - v_6\}$. It is straightforward to check that all inner products of elements of Δ are integral. Hence the same holds for the elements of Φ. So we have (R_4). Then also (R_3) is immediate.

F_4: we set $V = \mathbb{R}^4$ and $W = V$. Analogously to the previous case we put $v_0 = \frac{1}{2}(v_1 + v_2 + v_3 + v_4)$ and now

$$Q = \{k_0 v_0 + \sum_{i=1}^{4} k_i v_i \mid k_i \in \mathbb{Z}\}.$$

Furthermore $\Phi = \{v \in Q \mid (v,v) = 1,2\}$. It is easily seen that Φ consists of the vectors $\pm v_i$, $\pm v_i \pm v_j$ and $\frac{1}{2}(\pm v_1 \pm v_2 \pm v_3 \pm v_4)$. As simple system we can take $\Delta = \{v_2 - v_3, v_3 - v_4, v_4, \frac{1}{2}(v_1 - v_2 - v_3 - v_4)\}$. Then (R_4) and (R_3) are proved in the same way as for E_8.

Finally we construct C_l. For that let Ψ be the root system of type B_l as constructed above. Then we set

$$\Phi = \{\alpha = \frac{\sqrt{2}}{(\bar{\alpha}, \bar{\alpha})}\bar{\alpha} \mid \bar{\alpha} \in \Psi\}.$$

Properties (R_1) and (R_2) are immediate for Φ. Also by a straightforward calculation we see that for $\alpha, \beta \in \Phi$,

$$\frac{2(\beta, \alpha)}{(\alpha, \alpha)} = \frac{2(\bar{\beta}, \bar{\alpha})}{(\bar{\beta}, \bar{\beta})}.$$

So (R_4) is satisfied. Finally (R_3) is also settled by a short calculation. A simple system of Φ is $\Delta = \{\frac{1}{\sqrt{2}}(v_1 - v_2), \dots, \frac{1}{\sqrt{2}}(v_{l-1} - v_l), v_{l-1} + v_l\}$.   □

For algorithmic purposes we want to be able to construct root systems abstractly. This means that given a type X_l we want to be able to construct a vector space W and a set of vectors in W that form a root system in W, without every time having to perform the explicit construction of Proposition 5.10.1. The next two Corollaries show that the only thing we have to do is call CartanMatrixToRootSystem.

Corollary 5.10.2 *Let X_l be a simple type. Let W be a vector space over $\mathbb{R}$ of dimension l. Let Δ be a basis of W; then there is a bilinear form on W that makes W into a Euclidean space and such that Δ is a simple system of type X_l.*

Proof. Suppose $\Delta = \{\alpha_1, \dots, \alpha_l\}$. A bilinear form f on W is determined by its values $f(\alpha_i, \alpha_j)$ for $1 \leq i, j \leq l$. Now by Proposition 5.10.1 there exists a Euclidean space W_0 and a root system $\Phi_0 \subset W_0$ of type X_l. Let $(,)_0$ be the positive definite bilinear form of W_0 and let $\Delta_0 = \{\beta_1, \dots, \beta_l\}$ be a simple system of Φ_0. Then we define a bilinear form $(,)$ in W by setting $(\alpha_i, \alpha_j) = (\beta_i, \beta_j)_0$. Since the form $(,)_0$ is positive definite, this also holds for the form we defined. Hence W is a Euclidean space and Δ is a simple system of type X_l.   □

Corollary 5.10.3 *Let $X^i_{l_i}$ for $1 \leq i \leq m$ be simple types. Let Φ be a root system that is the direct sum of m irreducible root systems Φ_i of types $X^i_{l_i}$.*

Let C be the Cartan matrix of Φ and set $l = \sum l_i$. Let W be a vector space of dimension l over $\mathbb{R}$ with basis $\Delta = \{\alpha_1, \ldots, \alpha_l\}$. Then the procedure CartanMatrixToRootSystem *finishes on the input C, Δ. Furthermore, there is an inner product on W such that $\langle \alpha_i, \alpha_j \rangle = C(i, j)$.*

Proof. The fact that CartanMatrixToRootSystem(C, Δ) finishes follows from the fact that it finishes when Δ is replaced by a simple system of Φ (having Cartan matrix C). The second statement follows from Corollary 5.10.2. $\square$

5.11 Constructing isomorphisms

In this section we clear up a point that has been left in the dark since Section 5.3; we prove that two semisimple Lie algebras having isomorphic root systems are isomorphic. The results of this section will provide a straightforward algorithm for constructing such an isomorphism.

An isomorphism of two Lie algebras L_1 and L_2 maps a basis of L_1 onto a basis of L_2. Moreover, the isomorphism is determined by the images of the basis elements of L_1. So instead of constructing a linear map between L_1 and L_2 that is an isomorphism, we may as well construct bases B_1, B_2 of L_1 and L_2 such that the linear map that sends B_1 onto B_2 is an isomorphism. Our plan is first to give a method for constructing a canonical basis of a semisimple Lie algebra, and then to prove that canonical bases provide an isomorphism.

Let L be a semisimple Lie algebra with root system Φ relative to the split Cartan subalgebra H. Let $\Delta = \{\alpha_1, \ldots, \alpha_l\}$ be a simple system of Φ. Then by Root facts 12 and 13 there are elements $h_i = (2/(\alpha_i, \alpha_i))h_{\alpha_i}$ (where h_{α_i} is defined by (4.10)), and $x_i \in L_{\alpha_i}$, $y_i \in L_{-\alpha_i}$ such that $[x_i, y_i] = h_i$, $[h_i, x_i] = 2x_i$ and $[h_i, y_i] = -2y_i$. The elements h_i, x_i, y_i for $1 \le i \le l$ satisfy the following commutation relations:

$$
\begin{aligned}
&[h_i, h_j] = 0 \\
&[x_i, y_j] = \delta_{ij} h_i \\
&[h_j, x_i] = \langle \alpha_i, \alpha_j \rangle x_i \\
&[h_j, y_i] = \langle -\alpha_i, \alpha_j \rangle y_i \quad \text{for } 1 \le i, j \le l.
\end{aligned}
\tag{5.7}
$$

The first relation follows from Root fact 6. The second one follows from the fact that $\alpha_i - \alpha_j$ lies in $\Phi^0 = \Phi \cup \{0\}$ only when $i = j$ (Lemma 5.5.3). The last two relations follow from the definition of h_i together with the definition of the bilinear form $(,)$.

Definition 5.11.1 *Let L be a semisimple Lie algebra with root system Φ. Let $\Delta = \{\alpha_1, \ldots, \alpha_l\}$ be a simple system of Φ. A set of non-zero elements $\{x_i \in L_{\alpha_i}, y_i \in L_{-\alpha_i}, h_i \in H \mid 1 \leq i \leq l\}$ satisfying the commutation relations of (5.7) is called a set of* canonical generators *of L.*

By the discussion above we see that a set of canonical generators is guaranteed to exist. The following algorithm provides a straightforward method for constructing such a set.

Algorithm CanonicalGenerators
 Input: a semisimple Lie algebra L, together with a simple system $\Delta = \{\alpha_1, \ldots, \alpha_l\}$ of the root system of L.
Output: a canonical set of generators of L.

Step 1 For $1 \leq i \leq l$ do the following

    Step 1.1 Choose a non-zero $x_i \in L_{\alpha_i}$.

    Step 1.2 Determine an element $y_i \in L_{-\alpha_i}$ such that $[[x_i, y_i], x_i] = 2x_i$.

    Step 1.3 Set $h_i = [x_i, y_i]$.

Step 2 Return $\{x_i, y_i, h_i \mid 1 \leq i \leq l\}$.

Comments: By Root fact 12 we have that there is a $y_i \in L_{-\alpha_i}$ satisfying the requirement of Step 1.2. Furthermore such an element can be found by solving a single linear equation in one variable. Consequently the algorithm terminates and the output consists of non-zero $x_i \in L_{\alpha_i}$, $y_i \in L_{-\alpha_i}$ and $h_i \in H$ such that $[x_i, y_i] = h_i$ and $[h_i, x_i] = 2x_i$. We have to show that the other relations of (4.14) are satisfied as well. For this let $\bar{x}_i \in L_{\alpha_i}, \bar{y}_i \in L_{-\alpha_i}, \bar{h}_i \in H$ be a canonical set of generators. Then since all root spaces are 1-dimensional, there are $\lambda_i, \mu_i \in F$ such that $\bar{x}_i = \lambda_i x_i$, $\bar{y}_i = \mu_i y_i$ and consequently $\bar{h}_i = \lambda_i \mu_i h_i$. By (5.7) we know that $[\bar{h}_j, \bar{x}_i] = \langle \alpha_i, \alpha_j \rangle \bar{x}_j$ for $1 \leq i, j \leq l$. But this is equivalent to

$$\lambda_j \mu_j [h_j, x_i] = \langle \alpha_i, \alpha_j \rangle x_i.$$

Setting $i = j$ we see that $\lambda_j \mu_j = 1$. But then all relations of (5.7) are automatically satisfied by the x_i, y_i, h_i. Hence the output is a set of canonical generators.

Now we have a lemma that will imply that a set of canonical generators $\{h_i, x_i, y_i \mid 1 \leq i \leq l\}$ generates the whole Lie algebra L.

Lemma 5.11.2 *Let* $\alpha, \beta \in \Phi$ *be such that* $\alpha + \beta \in \Phi$. *Choose* $h, x, y \in L$ *(where* $x \in L_\alpha$, $y \in L_{-\alpha}$, $h \in H$*) such that* $[h, x] = 2x$, $[h, y] = -2y$, $[x, y] = h$ *(cf. Root fact 12). Let* $x_\beta \in L_\beta$ *be a non-zero root vector. Then* $[x, x_\beta] \neq 0$. *Furthermore, let* $\beta - r\alpha, \ldots, \beta + q\alpha$ *be the* α*-string containing* β. *Then* $[y, [x, x_\beta]] = q(r+1)x_\beta$ *and* $[x, [y, x_\beta]] = (q+1)rx_\beta$.

Proof. Let K_α be the Lie algebra spanned by h, x, y. Let V be the space spanned by all $L_{\beta + k\alpha}$ for $-r \leq k \leq q$. Then as seen in the proof of Root fact 17, V is an irreducible K_α-module. Choose a non-zero $v_0 \in L_{\beta + q\alpha}$ and set $v_i = (\mathrm{ad}y)^i v_0$ for $i \geq 0$. Then v_0 is a highest weight vector of weight $(\beta + q\alpha)(h) = 2q + \beta(h) = q + r$ (where the last equality follows from Root fact 17). Also v_q spans L_β and hence x_β is a non-zero multiple of v_q. By Lemma 5.1.2 we see that $\mathrm{ad}x(v_q) = q(r+1)v_{q-1}$, which is non-zero because $q > 0$. So also $\mathrm{ad}x(x_\beta) \neq 0$. Furthermore $\mathrm{ad}y(\mathrm{ad}x(v_q)) = q(r+1)v_q$ and $\mathrm{ad}x(\mathrm{ad}y(v_q)) = \mathrm{ad}x(v_{q+1}) = (q+1)rv_q$. Since x_β is a multiple of v_q the same holds with x_β in place of v_q. □

Now we are in a position to construct our basis. Fix a set of canonical generators h_i, x_i, y_i of L. Let $\beta \in \Phi^+$ be a positive root. By Corollary 5.5.6 together with a straightforward induction on $\mathrm{ht}(\beta)$ we see that β can be written as a sum of simple roots

$$\beta = \alpha_{i_1} + \cdots + \alpha_{i_k}$$

such that $\alpha_{i_1} + \cdots + \alpha_{i_m}$ is a root for $1 \leq m \leq k$. We abbreviate the repeated commutator $[x_{i_k}, [x_{i_{k-1}}, [\cdots [x_{i_2}, x_{i_1}] \cdots]]]$ by $[x_{i_k}, \ldots, x_{i_1}]$. Then by Lemma 5.11.2, $[x_{i_k}, \ldots, x_{i_1}]$ is non-zero and hence spans L_β. Now for every positive root $\beta \in \Phi^+$ we fix a sequence $i_1, \ldots, i_k$ such that $\beta = \alpha_{i_1} + \cdots + \alpha_{i_k}$ and $\alpha_{i_1} + \cdots + \alpha_{i_m}$ is a root for $1 \leq m \leq k$. Then it follows that the elements

$$h_1, \ldots, h_l, [x_{i_k}, \ldots, x_{i_1}], [y_{i_k}, \ldots, y_{i_1}]$$

span L. We say that these elements form a *canonical basis* of L with respect to the simple system Δ and the choice of sequences $i_1, \ldots, i_k$.

We introduce the following notation. If $I = (i_1, \ldots, i_k)$ is a sequence of integers, then we set

$$x_I = [x_{i_k}, \ldots, x_{i_1}], \quad \text{and} \quad y_I = [y_{i_k}, \ldots, y_{i_1}].$$

Also we set $\alpha_I = \alpha_{i_1} + \cdots + \alpha_{i_k}$. Now suppose that $I = (i_1, \ldots, i_k)$ and $x \in L$ commutes with x_{i_k}, i.e., $[x, x_{i_k}] = 0$. Write $[x, x_I] = [x, [x_{i_k}, x_J]]$ where $J = (i_1, \ldots, i_{k-1})$. Then by a straightforward application of the

Jacobi identity we see that $[x, x_I] = [x_{i_k}, [x, x_J]]$. And in general, if x commutes with $x_{i_k}, x_{i_{k-1}}, \dots, x_{i_s}$, then

$$[x, x_I] = [x, x_{i_k}, \dots, x_{i_1}] = [x_{i_k}, \dots, x_{i_s}, x, x_{i_{s-1}}, \dots, x_{i_1}]. \qquad (5.8)$$

Lemma 5.11.3 *Let C be the Cartan matrix of Φ with respect to Δ. Let $I = (i_1, \dots, i_k)$ be a sequence such that $\alpha_{i_1} + \dots + \alpha_{i_m}$ is a root for $1 \leq m \leq k$. Let $J = (j_1, \dots, j_k)$ be a second sequence such that $\alpha_I = \alpha_J$. Then $x_I = \lambda x_J$, where $\lambda \in \mathbb{Q}$ depends on C only. And a similar statement holds for y_I and y_J.*

Proof. First we remark that since the expression of α as a linear combination of the simple roots is unique, the sequence J can be obtained from I by permuting the entries.

We prove the result by induction on k. If $k = 1$, then we can take $\lambda = 1$. So suppose that $k > 1$. If $\alpha_{j_1} + \dots + \alpha_{j_m}$ is not a root for a certain m between 1 and k, then we set $\lambda = 0$. This can be checked from the knowledge of the Cartan matrix only (by using the algorithm CartanMatrixToRootSystem). Hence the lemma also holds in this case.

Now we suppose that all $\alpha_{j_1} + \dots + \alpha_{j_m}$ are roots and we set $s = i_k$. If also $j_k = s$, then we set $I' = (i_1, \dots, i_{k-1})$ and likewise $J' = (j_1, \dots, j_{k-1})$. By induction we know that $x_{I'} = \lambda x_{J'}$ where $\lambda \in \mathbb{Q}$ depends on C only. And therefore also $x_I = \lambda x_J$ and we are done with this case.

Now suppose that $j_k \neq s$. The index s must appear in J and we select the right-most j_n such that $j_n = s$. Then because $[y_s, x_i] = 0$ if $i \neq s$, by (5.8) we have that $[y_s, x_J] = [x_{j_k}, \dots, x_{j_{n+1}}, [y_s, [x_s, x_K]]]$ where $K = (j_{n-1}, \dots, j_1)$. Since α_K is a root, x_K is a non-zero root vector in L_{α_K}. Let $\alpha_K - r\alpha_s, \dots, \alpha_K + q\alpha_s$ be the α_s-string containing α_K. We note that $q > 0$ since $j_n = s$. Then by Lemma 5.11.2 we have $[y_s, [x_s, x_K]] = q(r+1)x_K$ and therefore, $[y_s, x_J] = q(r + 1)[x_{j_k}, \dots, x_{j_{n+1}}, x_{j_{n-1}}, \dots, x_{j_1}]$. Furthermore, $[x_s, [y_s, x_J]] = (q' + 1)r'x_J$, where $\alpha_J - r'\alpha_s, \dots, \alpha_J + q'\alpha_s$ is the α_s-string containing α_J. Now $r' > 0$ since $\alpha_J - \alpha_s = \alpha_I - \alpha_s$ is a root. Hence

$$x_J = \frac{1}{(q' + 1)r'}[x_s, [y_s, x_J]] =$$

$$\frac{q(r + 1)}{(q' + 1)r'}[x_s, x_{j_k}, \dots, x_{j_{n+1}}, x_{j_{n-1}}, \dots, x_{j_1}] = \nu x_M$$

where $M = (j_1, \dots, j_{n-1}, j_{n+1}, \dots, j_k, s)$. Note that the integers q, r, q', r' can be determined from the Cartan matrix (cf. the algorithm CartanMatrixToRootSystem), so we have the same for ν. Now $\alpha_M = \alpha_I$ and the last index in M equals s. Hence by the first case above $x_I = \mu x_M$ where μ depends on C only. It follows that $x_I = \lambda x_J$ with $\lambda = \mu/\nu$.

For y_I and y_J we can proceed with exactly the same arguments.   □

Proposition 5.11.4 *Let L be a semisimple Lie algebra with root system Φ. Let $\Delta = \{\alpha_1, \dots, \alpha_l\}$ be a simple system of Φ with Cartan matrix C. Let $h_1, \dots, h_l, x_I, y_I$ be a canonical basis of L. Then the structure constants of L with respect to this basis are rational numbers depending on C only.*

Proof. First $[h_i, h_j] = 0$ and for $I = (i_1, \dots, i_k)$ we have $[h_j, x_I] = \sum_{m=1}^{k} \langle \alpha_{i_m}, \alpha_j \rangle x_I$, which is easily proved by induction on k. Similarly $[h_j, y_I] = -\sum_{m=1}^{k} \langle \alpha_{i_m}, \alpha_j \rangle y_I$.

We consider the remaining commutators $[x_I, x_J]$, $[x_I, y_J]$ and $[y_I, y_J]$. The first and the third are similar so we deal with the first two only. Since

$$\mathrm{ad}x_I = [\mathrm{ad}x_{i_k}, [\mathrm{ad}x_{i_{k-1}}, [\cdots [\mathrm{ad}x_{i_2}, \mathrm{ad}x_{i_1}] \cdots]]]$$

we see that $\mathrm{ad}x_I$ is a linear combination of terms of the form $\mathrm{ad}x_{s_k} \cdots \mathrm{ad}x_{s_1}$. Hence it is enough to prove that $\mathrm{ad}x_s(x_J)$ and $\mathrm{ad}x_s(y_J)$ are linear combinations of basis elements, where the coefficients are in $\mathbb{Q}$ and depend on C only.

Concerning the first commutator, we determine from C whether $\alpha_J + \alpha_s$ is a root. If yes, then we set $K = (J, s)$ and $[x_s, x_J] = \lambda x_K$ where $\lambda \in \mathbb{Q}$ depends only on C by Lemma 5.11.3. Otherwise $[x_s, x_J] = 0$.

For the second commutator suppose $J = (j_1, \dots, j_n)$ and we use induction on n. If $n = 1$, then we know from (5.7) that $[x_s, y_{j_1}] = \delta_{s, j_1} h_s$. Now suppose $n > 1$ and let r be such that $j_r = s$ and $j_{r+1}, \dots, j_n \neq s$ (if there is no such r, then $[x_s, y_J] = 0$). Set $K = (j_1, \dots, j_{r-1})$, then using (5.8) and the Jacobi identity we see that

$$[x_s, y_J] = [y_{j_n}, \dots, y_{j_{r+1}}, [x_s, [y_s, y_K]]]$$
$$= [y_{j_n}, \dots, y_{j_{r+1}}, [y_s, [x_s, y_K]]] + [y_{j_n}, \dots, y_{j_{r+1}}, [[x_s, y_s], y_K]].$$

The first term of this sum is dealt with by the induction hypothesis. For the second term we note that $[x_s, y_s] = h_s$ and $[h_s, y_K]$ is a linear combination of y_M with rational coefficients determined by the Cartan matrix.   □

Algorithm IsomorphismOfSSLieAlgebras
Input: semisimple Lie algebras L_1, L_2 with root systems Φ_1 and Φ_2 respectively, and an isomorphism $\psi : \Phi_1 \to \Phi_2$.
Output: an isomorphism $\phi : L_1 \to L_2$.

Step 1 Compute a simple system $\Delta_1 = \{\alpha_1, \dots, \alpha_l\}$ of Φ_1.

Step 2 For every positive root $\beta \in \Phi_1^+$ fix a sequence $I_\beta = (i_1, \dots, i_k)$ such that $\alpha_{i_1} + \cdots + \alpha_{i_m}$ is a root for $1 \leq m \leq k$ and $\beta = \alpha_I$. Compute a set of canonical generators by CanonicalGenerators(L_1, Δ_1). Let $h_1, \dots, h_l, x_{I_\beta}, y_{I_\beta}$ for $\beta \in \Phi_1^+$ be the corresponding canonical basis.

Step 3 Set $\Delta_2 = \{\psi(\alpha_1), \dots, \psi(\alpha_l)\}$. Then Δ_2 is a simple system of Φ_2. For $\gamma \in \Phi_2^+$ set $I_\gamma = I_{\psi^{-1}(\gamma)}$. By CanonicalGenerators$(L_2, \Delta_2)$ calculate a set of canonical generators of L_2. Let $h'_1, \dots, h'_l, x'_{I_\gamma}, y'_{I_\gamma}$, for $\gamma \in \Phi_2^+$ be the canonical basis of L_2 corresponding to Δ_2.

Step 4 Return the linear map $\phi : L_1 \to L_2$ that sends h_i to h'_i, x_{I_β} to $x'_{I_{\psi(\beta)}}$ and y_{I_β} to $y'_{I_{\psi(\beta)}}$ for $\beta \in \Phi_1^+$.

Comments: Since ψ is an isomorphism of root systems it maps a simple system Δ_1 onto a simple system Δ_2. Also the Cartan matrix of Φ_1 relative to Δ_1 is exactly the same as the Cartan matrix of Φ_2 relative to Δ_2. Hence by Proposition 5.11.4, the structure constants of L_1 with respect to the canonical basis are exactly the same as the structure constants of L_2 with respect to the canonical basis. Hence ϕ is an isomorphism.

Corollary 5.11.5 *Let L_1 and L_2 be semisimple Lie algebras with root systems Φ_1 and Φ_2 of rank l. Let Δ_1 and Δ_2 be simple systems of Φ_1 and Φ_2 respectively. Suppose that the Cartan matrices of Φ_1 with respect to Δ_1 and Φ_2 with respect to Δ_2 are identical. Then there is a unique isomorphism $\phi : L_1 \to L_2$ such that $\phi(h_i) = h'_i$, $\phi(x_i) = x'_i$ and $\phi(y_i) = y'_i$ for $1 \leq i \leq l$, where $\{h_i, x_i, y_i \mid 1 \leq i \leq l\}$ and $\{h'_i, x'_i, y'_i \mid 1 \leq i \leq l\}$ are canonical generators, respectively of L_1 with respect to Δ_1 and L_2 with respect to Δ_2.*

Proof. Since the set $\{h_i, x_i, y_i \mid 1 \leq i \leq l\}$ generates L_1 there can only be one isomorphism satisfying the stated relations. The existence follows from applying IsomorphismOfSSLieAlgebras. $\qquad \square$

Corollary 5.11.6 *The root system is a complete invariant of semisimple Lie algebras.*

Remark. For an example of the use of the algorithm IsomorphismOfSSLieAlgebras we refer to Example 5.15.11.

5.12   Constructing the semisimple Lie algebras

In Sections 5.9 and 5.10 we showed that the indecomposable root systems are exactly those with a Dynkin diagram occurring in Table 5.2. In the previous section we proved that the root system of a simple Lie algebra is a complete structural invariant. The next question is whether for each indecomposable root system there exists a simple Lie algebra. For this several constructions have been proposed. First of all it is possible to argue case-by-case, i.e., to go through the list of Table 5.2 and construct a simple Lie algebra for each diagram. It is for example possible to prove that $\mathfrak{sl}_{l+1}(F)$ is a simple Lie algebra with a root system of type A_l, and $\mathfrak{o}_{2l+1}(F)$, $\mathfrak{sp}_{2l}(F)$, $\mathfrak{o}_{2l}(F)$ are of types B_l, C_l and D_l respectively. For proofs of these results we refer to [48]. A second idea is to give a uniform construction of all simple Lie algebras by generators and relations. This will be described in Section 7.11.

A third approach to the problem is to construct a Lie algebra starting from the root system. From the root system it is easy to see what the dimension of the simple Lie algebra L must be and many products of basis elements are determined by the root system. Root facts 1 and 11, for example, determine commutation relations in L. After collecting as many obvious commutation relations as possible one can try to complete this to a multiplication table of L. This is the approach that we take here.

The idea is straightforward: given a root system Φ of rank l we define a Lie algebra L that is the sum of a Cartan subalgebra H of dimension l and 1-dimensional root spaces. More precisely, let $\Delta = \{\alpha_1, \dots, \alpha_l\}$ be a simple system of Φ. Let H be an l-dimensional vector space with basis $\{h_1, \dots, h_l\}$. For $\alpha \in \Phi$ we set $h_\alpha = \sum_{i=1}^{l} k_i h_i$ if $\alpha = \sum_{i=1}^{l} k_i \alpha_i$. Furthermore, for $\alpha \in \Phi$ let L_α be a 1-dimensional vector space spanned by x_α. Then we set

$$L = H \oplus \bigoplus_{\alpha \in \Phi} L_\alpha.$$

The multiplication table of L relative to the basis $\{h_1, \dots, h_l\} \cup \{x_\alpha \mid \alpha \in \Phi\}$ is given by

$$
\begin{aligned}
&[h_i, h_j] = 0 \quad \text{for } 1 \leq i, j \leq l, \\
&[h_i, x_\alpha] = (\alpha, \alpha_i) x_\alpha \quad \text{for } 1 \leq i \leq l \text{ and } \alpha \in \Phi, \\
&[x_\alpha, x_{-\alpha}] = -\frac{2}{(\alpha, \alpha)} h_\alpha \quad \text{for } \alpha \in \Phi, \\
&[x_\alpha, x_\beta] = 0 \quad \text{for } \alpha, \beta \in \Phi \text{ such that } \alpha + \beta \notin \Phi \text{ and } \beta \neq -\alpha, \\
&[x_\alpha, x_\beta] = N_{\alpha,\beta} x_{\alpha+\beta} \quad \text{for } \alpha, \beta \in \Phi \text{ such that } \alpha + \beta \in \Phi.
\end{aligned}
\tag{5.9}
$$

Where $N_{\alpha,\beta}$ are constants that are to be determined. We note that if there is a semisimple Lie algebra L having Φ as root system, then L has basis relative to which the multiplication table is as (5.9). Indeed, let L be a semisimple Lie algebra with a split Cartan subalgebra H and root system Φ. Let $\Delta = \{\alpha_1, \ldots, \alpha_l\}$ be a simple system of Φ. For $1 \le i \le l$ set $h_i = h_{\alpha_i}$ (where h_{α_i} is defined by (4.10)). Furthermore, by Root fact 12 together with Root fact 13, we can choose $x_\alpha \in L_\alpha$ and $x_{-\alpha} \in L_{-\alpha}$ such that $[x_\alpha, x_{-\alpha}] = (2/(\alpha, \alpha))h_\alpha$ and after taking $-x_\alpha$ instead of x_α we get the relation of (5.9). Then Root facts 1, 6 and Equation (4.10) together with the definition of the bilinear form on H^* imply the other brackets.

In the sequel we determine a choice for the numbers $N_{\alpha,\beta}$ for each root system of simple type X_l. Then we check that the multiplication table (5.9) satisfies the Jacobi identity. This leads to the conclusion that (5.9) defines a Lie algebra. We then still need to show that this Lie algebra is simple. For that we use the following criterion for semisimplicity.

Proposition 5.12.1 *Let L be a finite-dimensional Lie algebra and let $H \subset L$ be a split Abelian subalgebra such that $\mathrm{ad}h$ is a semisimple linear transformation for $h \in H$. For $\alpha \in H^*$ set*

$$L_\alpha = \{x \in L \mid \mathrm{ad}h(x) = \alpha(h)x \quad \text{for } h \in H\}.$$

Let $\Phi^0 = \{\alpha \in H^* \mid L_\alpha \neq 0\}$. Then

$$L = \bigoplus_{\alpha \in \Phi^0} L_\alpha.$$

Set $\Phi = \Phi^0 \setminus \{0\}$. Suppose that $\Phi = -\Phi$, L_α is 1-dimensional for $\alpha \in \Phi$, $[L_\alpha, L_{-\alpha}] \neq 0$ for $\alpha \in \Phi$ and $H = L_0$. Then L is semisimple and H is a Cartan subalgebra of L.

Proof. Since $H = L_0 = L_0(H)$, we have that H is a Cartan subalgebra by definition. Now let $N \subset L$ be the nilradical of L. Then N is invariant under $\mathrm{ad}H$. Hence N splits as a direct sum of simultaneous eigenspaces of the $\mathrm{ad}h$ for $h \in H$. So

$$N = (H \cap N) \oplus \bigoplus_{\alpha \in \Phi}(L_\alpha \cap N).$$

First we note that $\mathrm{ad}_L x$ is nilpotent for all $x \in N$ (Proposition 2.2.2). Hence, because $\mathrm{ad}h$ is semisimple for $h \in H$ we have that $H \cap N = 0$. Since L_α is 1-dimensional, we have $L_\alpha \cap N = 0$ or $L_\alpha \subset N$. Suppose that $L_\alpha \subset N$ for an $\alpha \in \Phi$. By a straightforward application of the Jacobi identity we see that $[[L_\alpha, L_{-\alpha}], H] = 0$, implying $[L_\alpha, L_{-\alpha}] \subset H$. And since $[L_\alpha, L_{-\alpha}] \neq 0$ this implies that $H \cap N \neq 0$, which is a contradiction. It follows that $N = 0$, and as seen in Section 2.6, the solvable radical of L is 0 as well. □

Corollary 5.12.2 *Let Φ be a root system and let L be a Lie algebra with a multiplication table of the form (5.9). Then L is semisimple. Furthermore, the root system of L is isomorphic to Φ.*

Proof. The fact that L is semisimple follows immediately from Proposition 5.12.1. As before $\Delta = \{\alpha_1, \dots, \alpha_l\}$ is a simple system of Φ. Let H be the Cartan subalgebra spanned by $\{h_1, \dots, h_l\}$. Then the x_α are root vectors relative to H and the corresponding root is the function given by $h_i \mapsto (\alpha, \alpha_i)$. Denote this function by α'. Then the set of all α' for $\alpha \in \Phi$ is a root system because L is semisimple. Furthermore, the map $\alpha \mapsto \alpha'$ is linear and maps root strings to root strings. So it is an isomorphism of root systems. □

5.13 The simply-laced case

In this section we determine a choice for the constants $N_{\alpha,\beta}$ in the case where the root system has a Dynkin diagram that contains only simple edges. Such root systems are said to be *simply-laced*.

Let Φ be an irreducible simply-laced root system; i.e., Φ is of type A_l, D_l, E_6, E_7 or E_8. Let Δ be a simple system of Φ. Then because the Dynkin diagram of Φ is connected we have that all $\alpha \in \Delta$ have the same length. Since multiplying the inner product by a positive scalar changes nothing, we may assume that $(\alpha, \alpha) = 2$ for all $\alpha \in \Delta$.

Lemma 5.13.1 *Let Φ be a simply-laced root system with simple system Δ. Suppose that $(\alpha, \alpha) = 2$ for $\alpha \in \Delta$. Then $(\beta, \beta) = 2$ for all $\beta \in \Phi$.*

Proof. This follows immediately from Lemma 5.7.5 together with (5.1). □

Proposition 5.13.2 *Let Φ be a simply-laced root system with a simple system Δ such that $(\alpha, \alpha) = 2$ for all $\alpha \in \Delta$. Let $\alpha, \beta \in \Phi$ be such that $\alpha \neq \pm\beta$. Then $\alpha + \beta \in \Phi$ (respectively $\alpha - \beta \in \Phi$) if and only if $(\alpha, \beta) = -1$ (respectively $(\alpha, \beta) = 1$).*

Proof. If $(\alpha, \beta) = -1$ then by Proposition 5.4.1 we see that $\alpha + \beta \in \Phi$. For the other direction suppose that $\alpha + \beta \in \Phi$. Let $\beta - r\alpha, \ldots, \beta + q\alpha$ be the α-string containing β. Then $q \geq 1$. Set $\gamma = \beta - r\alpha$. Then $\gamma, \gamma + \alpha, \ldots, \gamma + (r+q)\alpha$ is the α-string containing γ. By Proposition 5.4.2 we infer that $-(r+q) = \langle \gamma, \alpha \rangle$. If $\langle \gamma, \alpha \rangle = \pm 2, \pm 3$, then γ and α have unequal length, which is excluded by Lemma 5.13.1. Hence by Table 5.1, we see that $\langle \gamma, \alpha \rangle = \pm 1, 0$. But $r + q \geq 1$. It follows that $\langle \gamma, \alpha \rangle = -1$ and $r = 0$ and $q = 1$. Hence by Proposition 5.4.2 we have that $\langle \beta, \alpha \rangle = -1$. Since by Lemma 5.13.1, $\langle \beta, \alpha \rangle = (\beta, \alpha)$, the result follows.

The case where $\alpha - \beta \in \Phi$ can be treated similarly. $\qquad\square$

Now we define a function that will give us the desired structure constants $N_{\alpha,\beta}$. We assume that Φ is a simply-laced root system with simple system $\Delta = \{\alpha_1, \ldots, \alpha_l\}$ such that $(\alpha_i, \alpha_i) = 2$. By Q we denote the $\mathbb{Z}$-span of Δ, i.e.,

$$Q = \left\{ \sum_{i=1}^{l} k_i \alpha_i \mid k_i \in \mathbb{Z} \right\}.$$

The set Q is called the *root lattice* of Φ.

A function $\varepsilon : Q \times Q \to \{1, -1\}$ is called an *asymmetry function* if it satisfies

$$\begin{aligned}
\varepsilon(\alpha + \beta, \gamma) &= \varepsilon(\alpha, \gamma)\varepsilon(\beta, \gamma) \\
\varepsilon(\alpha, \gamma + \delta) &= \varepsilon(\alpha, \gamma)\varepsilon(\alpha, \delta) \quad \text{for } \alpha, \beta, \gamma, \delta \in Q,
\end{aligned} \tag{5.10}$$

and

$$\varepsilon(\alpha, \alpha) = (-1)^{\frac{1}{2}(\alpha,\alpha)} \quad \text{for } \alpha \in Q. \tag{5.11}$$

Remark. For $\alpha \in \Phi$, (5.11) boils down to $\varepsilon(\alpha, \alpha) = -1$.

Lemma 5.13.3 *Asymmetry functions exist.*

Proof. We start by assigning an arrow to each edge of the Dynkin diagram. (We say that we choose an *orientation* of the Dynkin diagram.) Then we set

$$\varepsilon(\alpha_i, \alpha_j) = \begin{cases} -1 & \text{if } i = j, \\ -1 & \text{if } \alpha_i \text{ and } \alpha_j \text{ are connected and} \\ & \text{the arrow points from } \alpha_i \text{ to } \alpha_j, \\ 1 & \text{if } \alpha_i \text{ and } \alpha_j \text{ are not connected or} \\ & \text{the arrow points from } \alpha_j \text{ to } \alpha_i, \end{cases}$$

and we extend ε to the whole of $Q \times Q$ using condition (5.10). This means that

$$\varepsilon\left(\sum_{i=1}^{l} k_i\alpha_i, \sum_{j=1}^{l} m_j\alpha_j\right) = \prod_{i,j=1}^{l} \varepsilon(\alpha_i,\alpha_j)^{k_i m_j}.$$

So (5.10) is automatically satisfied. It remains to prove (5.11) for ε. For that arrange the labelling of the Dynkin diagram in such a way that α_i, α_j connected and the arrow points from α_i to α_j implies that $i < j$. This is always possible because the Dynkin diagram contains no loops. Then $\varepsilon(\alpha_i,\alpha_j) = 1$ if $i > j$ and $\varepsilon(\alpha_i,\alpha_j) = (-1)^{(\alpha_i,\alpha_j)}$ if $i < j$. Let $\alpha = \sum_{i=1}^{l} k_i\alpha_i$ be an element of Q. Then

$$\varepsilon(\alpha,\alpha) = \prod_{i,j=1}^{l} \varepsilon(\alpha_i,\alpha_j)^{k_i k_j}$$

$$= \prod_{1\le i<j\le l} \varepsilon(\alpha_i,\alpha_j)^{k_i k_j} \cdot \prod_{i=1}^{l} \varepsilon(\alpha_i,\alpha_i)^{k_i^2}$$

$$= \prod_{1\le i<j\le l} (-1)^{k_i k_j(\alpha_i,\alpha_j)} \cdot \prod_{i=1}^{l} (-1)^{\frac{1}{2}k_i^2(\alpha_i,\alpha_i)}$$

$$= (-1)^{\sum_{i<j}(k_i\alpha_i,k_j\alpha_j)} \cdot (-1)^{\sum_i \frac{1}{2}(k_i\alpha_i,k_i\alpha_i)}.$$

Now the observation that

$$\frac{1}{2}(\alpha,\alpha) = \sum_{1\le i<j\le l} (k_i\alpha_i,k_j\alpha_j) + \sum_{i=1}^{l} \frac{1}{2}(k_i\alpha_i,k_i\alpha_i)$$

finishes the proof.   □

Replacing α by $\alpha + \beta$ in (5.11) we get

$$\varepsilon(\alpha,\beta)\varepsilon(\beta,\alpha) = (-1)^{(\alpha,\beta)}. \tag{5.12}$$

Also some easy consequences of (5.10) are

$$\varepsilon(0,\beta) = \varepsilon(\alpha,0) = 1$$
$$\varepsilon(-\alpha,\beta) = \varepsilon(\alpha,\beta)^{-1} = \varepsilon(\alpha,\beta) \quad \text{and similarly for } \varepsilon(\alpha,-\beta). \tag{5.13}$$

Proposition 5.13.4 *Let Φ be a simply-laced irreducible root system with simple system $\Delta = \{\alpha_1,\dots,\alpha_l\}$ and such that $(\alpha,\alpha) = 2$ for $\alpha \in \Phi$. For $\alpha,\beta \in \Phi$ such that $\alpha + \beta \in \Phi$ set $N_{\alpha,\beta} = \varepsilon(\alpha,\beta)$. Then (5.9) defines a Lie algebra.*

Proof. In this case the multiplication table for L reads

$$
\begin{aligned}
&[h_i, h_j] = 0 \quad \text{for } 1 \leq i, j \leq l, \\
&[h_i, x_\alpha] = (\alpha, \alpha_i) x_\alpha \quad \text{for } 1 \leq i \leq l \text{ and } \alpha \in \Phi, \\
&[x_\alpha, x_{-\alpha}] = -h_\alpha \quad \text{for } \alpha \in \Phi, \\
&[x_\alpha, x_\beta] = 0 \quad \text{for } \alpha, \beta \in \Phi \text{ such that } \alpha + \beta \notin \Phi \text{ and } \beta \neq -\alpha, \\
&[x_\alpha, x_\beta] = \varepsilon(\alpha, \beta) x_{\alpha+\beta} \quad \text{for } \alpha, \beta \in \Phi \text{ such that } \alpha + \beta \in \Phi.
\end{aligned}
\tag{5.14}
$$

We have to check the Jacobi identity for L. As seen in Lemma 1.3.1 it is enough to do this for basis elements. So let x, y, z be basis elements of L. If all three of them are from H then the Jacobi identity is trivially satisfied. So suppose that two elements are from H, and the other element is a root vector, i.e., $x = h_i$, $y = h_j$ and $z = x_\alpha$. Then the Jacobi identity reads:

$$[[h_i, h_j], x_\alpha] + [[h_j, x_\alpha], h_i] + [[x_\alpha, h_i], h_j] = 0,$$

and this is equivalent to $-(\alpha, \alpha_i)(\alpha, \alpha_j) + (\alpha, \alpha_j)(\alpha, \alpha_i) = 0$.

Now we suppose that one element is from H and the other two are root vectors. Then the Jacobi identity is

$$[[h_i, x_\alpha], x_\beta] + [[x_\alpha, x_\beta], h_i] + [[x_\beta, h_i], x_\alpha] = 0.$$

If $\alpha + \beta \notin \Phi^0$ then this holds. On the other hand, if $\alpha + \beta \in \Phi$, then the Jacobi identity is equivalent to

$$\varepsilon(\alpha, \beta)(\alpha, \alpha_i) - (\alpha + \beta, \alpha_i)\varepsilon(\alpha, \beta) + (\beta, \alpha_i)\varepsilon(\alpha, \beta) = 0.$$

And if $\beta = -\alpha$ then the Jacobi identity boils down to $(\alpha, \alpha_i)[x_\alpha, x_{-\alpha}] + (-\alpha, \alpha_i)[x_\alpha, x_{-\alpha}] = 0$.

Consequently the Jacobi identity holds if at least one element is from H. Let $\alpha, \beta, \gamma \in \Phi$; it remains to show the Jacobi identity for x_α, x_β, x_γ. We may assume that no pair of these are equal because the Jacobi identity is trivially satisfied in that case. We set $a = [[x_\alpha, x_\beta], x_\gamma]$, $b = [[x_\beta, x_\gamma], x_\alpha]$ and $c = [[x_\gamma, x_\alpha], x_\beta]$.

We remark that if none of the sums $\alpha + \beta$, $\alpha + \gamma$, $\beta + \gamma$ are elements of Φ^0, then $a + b + c$ is trivially 0. So we may assume that at least one of these sums lies in Φ^0. And since permuting the vectors $x_\alpha, x_\beta, x_\gamma$ does not change anything we may assume that $\alpha + \beta \in \Phi^0$.

First we deal with the case where $\alpha + \beta = 0$. This together with the second relation of (5.14) implies that $a = -(\gamma, \alpha)x_\gamma$. We consider a few cases:

1. $\alpha \pm \gamma \notin \Phi^0$; in view of Proposition 5.4.2 this implies that $(\alpha, \gamma) = 0$ and $a = 0$. Also $\beta + \gamma = -\alpha + \gamma \notin \Phi^0$, so that $b = 0$. And since also $c = 0$ the Jacobi identity is verified in this case.

2. $\alpha + \gamma = 0$ or $\alpha - \gamma = 0$; the first possibility means that $\beta = \gamma$ and the second means that $\alpha = \gamma$. Hence in both cases a pair of root vectors are equal and this was excluded.

3. $\alpha + \gamma \in \Phi$; this implies that $\gamma - \beta = \gamma + \alpha \in \Phi$. Hence by Proposition 5.13.2 we see that $\gamma + \beta \notin \Phi^0$. Hence $b = 0$. Also $c = \varepsilon(\gamma, \alpha)\varepsilon(\alpha + \gamma, -\alpha)x_\gamma$, and $a = -(\gamma, \alpha)x_\gamma = x_\gamma$ (by Proposition 5.13.2). So the Jacobi identity is equivalent to $\varepsilon(\gamma, \alpha)\varepsilon(\alpha, -\alpha)\varepsilon(\gamma, -\alpha) = -1$. And since by (5.13), $\varepsilon(\mu, -\nu) = \varepsilon(\mu, \nu)^{-1}$ this is easily seen to hold.

4. $\alpha - \gamma \in \Phi$; by Proposition 5.13.2 this implies that $\alpha + \gamma \notin \Phi$ so that $c = 0$. But $\beta + \gamma = -\alpha + \gamma \in \Phi$ forcing $b = \varepsilon(-\alpha, \gamma)\varepsilon(-\alpha + \gamma, \alpha)x_\gamma$. It follows that in this case the Jacobi identity is equivalent to $\varepsilon(-\alpha, \gamma)\varepsilon(-\alpha + \gamma, \alpha) = 1$. Now using $\varepsilon(\mu, \nu)^{-1} = \varepsilon(\mu, \nu)$ and (5.12) it is seen that this is verified.

The conclusion is that the Jacobi identity holds if $\alpha + \beta = 0$. Now we deal with the last case where $\alpha + \beta \in \Phi$. Then $a = \varepsilon(\alpha, \beta)[x_{\alpha+\beta}, x_\gamma]$. Also $(\alpha + \beta, \gamma) = (\alpha, \gamma) + (\beta, \gamma)$; the two inner products on the right hand side can take the values $\pm 1, 0$ so that the left hand side can be $\pm 2, \pm 1, 0$. (A consequence of Lemma 5.13.1 and Proposition 5.13.2 is that $(\delta, \epsilon) = 0, \pm 1$ for $\delta \neq \epsilon \in \Phi$ and $(\delta, \epsilon) = \pm 2$ if $\delta = \pm \epsilon$.) Again we distinguish a few cases:

1. $(\alpha + \beta, \gamma) = 2$; then $(\alpha, \gamma) = (\beta, \gamma) = 1$. By Proposition 5.13.2 this implies that $\alpha + \gamma$, $\beta + \gamma$ are no elements of Φ and $\alpha + \beta = \gamma$. Hence $a = b = c = 0$.

2. $(\alpha + \beta, \gamma) = 1$. Interchanging α and β changes nothing so we may assume that $(\alpha, \gamma) = 1$ and $(\beta, \gamma) = 0$. Again this forces $a = b = c = 0$ since $\alpha + \beta + \gamma \notin \Phi^0$.

3. $(\alpha + \beta, \gamma) = 0$. If $(\alpha, \gamma) = (\beta, \gamma) = 0$ then $a = b = c = 0$. So after (maybe) interchanging α and β we may assume that $(\alpha, \gamma) = 1$ and $(\beta, \gamma) = -1$. This means that $\alpha + \gamma \notin \Phi^0$ and $c = 0$. Also $\beta + \gamma \in \Phi$, but $\alpha + \beta + \gamma \notin \Phi^0$. So $a = b = 0$.

4. $(\alpha + \beta, \gamma) = -1$. Here we may assume that $(\alpha, \gamma) = -1$ and $(\beta, \gamma) = 0$. This immediately implies that $b = 0$. Furthermore, $\alpha + \beta + \gamma \in \Phi$ and $a = \varepsilon(\alpha, \beta)\varepsilon(\alpha + \beta, \gamma)x_{\alpha+\beta+\gamma}$. Also $c = \varepsilon(\gamma, \alpha)\varepsilon(\alpha + \gamma, \beta)x_{\alpha+\beta+\gamma}$.

After dividing by $\varepsilon(\alpha, \beta)$ and multiplying by $\varepsilon(\gamma, \alpha)\varepsilon(\beta, \gamma)$ the Jacobi identity amounts to

$$\varepsilon(\gamma, \alpha)\varepsilon(\alpha, \gamma) + \varepsilon(\beta, \gamma)\varepsilon(\gamma, \beta) = 0.$$

Using (5.12) and the values of (α, γ) and (β, γ) this is seen to hold.

5. $(\alpha+\beta, \gamma) = -2$. Here $\alpha+\beta+\gamma = 0$ and $(\alpha, \beta) = (\alpha, \gamma) = (\beta, \gamma) = -1$. Now $a = -\varepsilon(\alpha, \beta)h_{\alpha+\beta}$, $b = -\varepsilon(\beta, \gamma)h_{\beta+\gamma}$, $c = -\varepsilon(\gamma, \alpha)h_{\alpha+\gamma}$. Using the facts $h_{\mu+\nu} = h_\mu + h_\nu$ and $\gamma = -\alpha - \beta$ the Jacobi identity is equivalent to

$$\varepsilon(\alpha, \beta)(h_\alpha + h_\beta) - \varepsilon(\beta, -\alpha - \beta)h_\alpha - \varepsilon(-\alpha - \beta, \alpha)h_\beta = 0.$$

By (5.12) we have that $\varepsilon(\alpha, \beta)\varepsilon(\beta, \alpha) = -1$ so that $\varepsilon(\beta, \alpha) = -\varepsilon(\alpha, \beta)$. This implies that $\varepsilon(\beta, -\alpha - \beta) = \varepsilon(\alpha, \beta)$ and $\varepsilon(-\alpha - \beta, \alpha) = \varepsilon(\alpha, \beta)$ and the Jacobi identity is verified.

The conclusion is that the multiplication table of L satisfies the Jacobi identity.   □

5.14   Diagram automorphisms

Before constructing the other simple Lie algebras we need a short intermezzo on certain automorphisms of semisimple Lie algebras, called diagram automorphisms.

Let Φ be a root system with simple system $\Delta = \{\alpha_1, \ldots, \alpha_l\}$. Let D be the Dynkin diagram of Φ with respect to Δ. A bijection $\phi : \Delta \to \Delta$ is called a *diagram automorphism* if

1. the number of lines connecting α_i and α_j in D is equal to the number of lines connecting $\phi(\alpha_i)$ and $\phi(\alpha_j)$,

2. if α_i and α_j are connected in D then there is an arrow pointing from α_i to α_j if and only if the same is true for $\phi(\alpha_i)$ and $\phi(\alpha_j)$, for $1 \le i, j \le l$.

If Φ is irreducible, then it is easily seen that there are diagram automorphisms only in the cases where Φ is simply-laced.

Example 5.14.1 Let Φ be a root system of type D_4 with Dynkin diagram

Define a map ϕ by $\phi(\alpha_1) = \alpha_3$, $\phi(\alpha_2) = \alpha_2$, $\phi(\alpha_3) = \alpha_4$, $\phi(\alpha_4) = \alpha_1$. Then ϕ is a diagram automorphism.

Let $\phi : \Delta \to \Delta$ be a diagram automorphism. Then by $\mathbb{Z}$-linearity we can extend ϕ to a map $\phi : \Phi \to \Phi$, i.e.,

$$\phi\left(\sum_{i=1}^{l} k_i \alpha_i\right) = \sum_{i=1}^{l} k_i \phi(\alpha_i).$$

Let L be a simple Lie algebra with root system Φ, and let ϕ be a diagram automorphism. Then we have that the Cartan matrix of Φ relative to $\alpha_1, \dots, \alpha_l$ is exactly the same as the Cartan matrix of Φ relative to $\phi(\alpha_1), \dots, \phi(\alpha_l)$. Hence by Corollary 5.11.5, ϕ extends to a unique automorphism of L. Let h_i, x_i, y_i be a set of canonical generators of L. Let σ be the permutation of $(1, \dots, l)$ such that $\phi(\alpha_i) = \alpha_{\sigma(i)}$. Then the automorphism ϕ of L is determined by the relations

$$\phi(h_i) = h_{\sigma(i)}, \quad \phi(x_i) = x_{\sigma(i)}, \quad \phi(y_i) = y_{\sigma(i)}.$$

5.15 The non simply-laced case

In this section we construct simple Lie algebras for the non-simply-laced root systems, i.e., for the root systems of types B_l, C_l, F_4 and G_2. The idea is to construct those Lie algebras as subalgebras of simple Lie algebras with a simply-laced root system.

Here we consider simply-laced root systems Ψ of type D_{l+1}, A_{2l-1}, E_6, and D_4 with Dynkin diagram labelled as in Table 5.2. Furthermore we assume that Ψ has a simple system Π such that $(\alpha, \alpha) = 2$ for all $\alpha \in \Pi$. As seen in Section 5.13, this implies that $(\alpha, \alpha) = 2$ for all $\alpha \in \Psi$.

These root systems have the following diagram automorphisms ϕ:

1. D_{l+1}: $\phi(\alpha_i) = \alpha_i$ for $1 \le i \le l-1$, and $\phi(\alpha_l) = \alpha_{l+1}$ and $\phi(\alpha_{l+1}) = \alpha_l$.

2. A_{2l-1}: $\phi(\alpha_i) = \alpha_{2l-i}$ and $\phi(\alpha_{2l-i}) = \phi(\alpha_i)$ for $1 \le i \le l - 1$, and $\phi(\alpha_l) = \alpha_l$.

3. E_6: $\phi(\alpha_1) = \alpha_6$, $\phi(\alpha_2) = \alpha_2$, $\phi(\alpha_3) = \alpha_5$, $\phi(\alpha_4) = \alpha_4$, $\phi(\alpha_5) = \alpha_3$, $\phi(\alpha_6) = \alpha_1$.

4. D_4: $\phi(\alpha_1) = \alpha_3$, $\phi(\alpha_2) = \alpha_2$, $\phi(\alpha_3) = \alpha_4$, $\phi(\alpha_4) = \alpha_1$.

In the sequel we let d be the order of ϕ (i.e., $d = 2$ in the first three cases and $d = 3$ in the fourth case).

Now in each case we choose an orientation of the Dynkin diagram that is invariant under ϕ (i.e., if α_i and α_j are connected and the arrow points from α_i towards α_j, then the arrow belonging to the edge connecting $\phi(\alpha_i)$ and $\phi(\alpha_j)$ points towards $\phi(\alpha_j)$). It is straightforward to see that in each case it is possible to choose such an orientation. Let Q be the root lattice of Ψ and $\varepsilon : Q \times Q \to \{1, -1\}$ be the asymmetry function corresponding to the orientation chosen, as constructed in the proof of Lemma 5.13.3. Furthermore, let K be the Lie algebra with root system Ψ as constructed in Proposition 5.13.4. We fix a basis $h_1, \ldots, h_l, x_\alpha$ (for $\alpha \in \Psi$) relative to which K has multiplication table (5.14). Then as seen in Section 5.14, ϕ extends to a unique automorphism of K. We remark that the automorphism ϕ of K also has order d (this follows from Corollary 5.11.5 together with the observation that $\phi^d(x_\alpha) = x_\alpha$ for all simple roots α).

Lemma 5.15.1 *We have $(\phi(\alpha), \phi(\beta)) = (\alpha, \beta)$ and $\varepsilon(\phi(\alpha), \phi(\beta)) = \varepsilon(\alpha, \beta)$ for $\alpha, \beta \in \Psi$. Furthermore $\phi(x_\alpha) = x_{\phi(\alpha)}$ for all $\alpha \in \Psi$.*

Proof. Because ϕ is a diagram automorphism we have that $\langle \phi(\alpha), \phi(\beta) \rangle = \langle \alpha, \beta \rangle$ for all $\alpha, \beta \in \Pi$. Because $(\alpha, \alpha) = 2$ for all $\alpha \in \Pi$ this implies that $(\phi(\alpha), \phi(\beta)) = (\alpha, \beta)$ for $\alpha, \beta \in \Pi$. Since the simple roots span Ψ this holds for all elements of Ψ. The statement about ε follows from the fact that the orientation is ϕ-invariant. Finally, let α be a positive root. We argue by induction on $\mathrm{ht}(\alpha)$. If this number is 1, then $\phi(x_\alpha) = x_{\phi(\alpha)}$ by definition of ϕ. Furthermore, if $\mathrm{ht}(\alpha) > 1$ then $\alpha = \beta + \gamma$, where β, γ are positive roots of smaller height. But

$$\phi([x_\beta, x_\gamma]) = \varepsilon(\beta, \gamma)\phi(x_{\beta+\gamma})$$

and by induction

$$[\phi(x_\beta), \phi(x_\gamma)] = [x_{\phi(\beta)}, x_{\phi(\gamma)}] = \varepsilon(\phi(\beta), \phi(\gamma))x_{\phi(\beta+\gamma)}.$$

The desired conclusion follows. If α is negative, then we use a similar argument. $\qquad\square$

From the fact that ϕ is an automorphism it follows that the set $K_1(\phi) = \{x \in K \mid \phi(x) = x\}$ is a subalgebra of K. In the rest of this section we will show that $K_1(\phi)$ is simple of type B_l, C_l, F_4 and G_2 if Ψ is of type D_{l+1}, A_{2l-1}, E_6 and D_4 respectively.

Put

$$\Phi_l = \{\alpha \in \Psi \mid \phi(\alpha) = \alpha\},$$

$$\Phi_s = \{\frac{1}{d}(\alpha + \phi(\alpha) + \cdots + \phi^{d-1}(\alpha)) \mid \alpha \in \Psi \quad \text{such that} \quad \phi(\alpha) \neq \alpha\},$$

and $\Phi = \Phi_l \cup \Phi_s$. Note that $\phi(\alpha) = \alpha$ for all $\alpha \in \Phi$. Also for $\alpha \in \Phi$ we define $\alpha' = \alpha$ if $\alpha \in \Phi_l$ and in case $\alpha \in \Phi_s$ we set $\alpha' = \beta$, where $\beta \in \Psi$ is such that $\alpha = \frac{1}{d}(\beta + \cdots + \phi^{d-1}(\beta))$. Note that α' is not uniquely defined for $\alpha \in \Phi_s$ (if $\beta = \alpha'$ then also $\phi(\beta) = \alpha'$ et cetera).

For $\alpha \in \Phi$ we set

$$y_\alpha = \begin{cases} x_{\alpha'} & \text{if } \alpha \in \Phi_l, \\ x_{\alpha'} + x_{\phi(\alpha')} + \cdots + x_{\phi^{d-1}(\alpha')} & \text{if } \alpha \in \Phi_s. \end{cases}$$

(Note that although the choice of α' is ambiguous, there is no ambiguity in the definition of y_α.) Furthermore L_α will be the 1-dimensional subspace of K spanned by y_α for $\alpha \in \Phi$. Set

$$\Delta_l = \{\beta \in \Pi \mid \phi(\beta) = \beta\}$$

$$\Delta_s = \{\frac{1}{d}(\beta + \phi(\beta) + \cdots + \phi^{d-1}(\beta)) \mid \beta \in \Pi \quad \text{such that} \quad \phi(\beta) \neq \beta\},$$

and put $\Delta = \Delta_l \cup \Delta_s$. Let $H' \subset K$ be the Cartan subalgebra spanned by $h_1, \ldots, h_l$. Write $\Pi = \{\beta_1, \ldots, \beta_l\}$. We recall that $h_\beta = \sum_i k_i h_i$ if $\beta = \sum_i k_i \beta_i \in \Psi$. So for $\alpha \in \Phi$ we have

$$h_\alpha = \begin{cases} h_{\alpha'} & \text{if } \alpha \in \Phi_l \\ \frac{1}{d}(h_{\alpha'} + \cdots + h_{\phi^{d-1}(\alpha')}) & \text{if } \alpha \in \Phi_s. \end{cases}$$

We let $H \subset K$ be the space spanned by all h_α for $\alpha \in \Delta$, and we set

$$L = H \oplus \bigoplus_{\alpha \in \Phi} L_\alpha.$$

Lemma 5.15.2 *We have $L = K_1(\phi)$; in particular L is a Lie algebra.*

Proof. We note that the inclusion $L \subset K_1(\phi)$ is clear. We prove the other inclusion. First we deal with the case where $d = 2$. For $\alpha \in \Phi_s$ set $z_\alpha = x_{\alpha'} - x_{\phi(\alpha')}$. Then $\phi(z_\alpha) = -z_\alpha$. Furthermore, the span of $\{y_\alpha \mid \alpha \in \Phi\} \cup \{z_\alpha \mid \alpha \in \Phi_s\}$ is equal to the span of all x_α for $\alpha \in \Psi$. Also if we set $g_\alpha = h_{\alpha'} - h_{\phi(\alpha')}$ for $\alpha \in \Delta_s$, then the span of $\{h_\alpha \mid \alpha \in \Delta\} \cup \{g_\alpha \mid \alpha \in \Delta_s\}$ is equal to H'. The conclusion is that $K = L \oplus K_{-1}(\phi)$ where $K_{-1}(\phi)$ is the eigenspace of ϕ corresponding to the eigenvalue -1, which is spanned by the z_α and g_α. It follows that L is exactly the eigenspace of ϕ corresponding to the eigenvalue 1.

In the case where $d = 3$ we can apply a similar reasoning, but here we have three eigenvalues: 1, ζ and ζ^2, where ζ is a root of $X^2 + X + 1$. So for $\alpha \in \Phi_s$ we set $z_\alpha = x_{\alpha'} + \zeta^2 x_{\phi(\alpha')} + \zeta x_{\phi^2(\alpha')}$ and $w_\alpha = x_{\alpha'} + \zeta^4 x_{\phi(\alpha')} + \zeta^2 x_{\phi^2(\alpha')}$. Then z_α and w_α are eigenvectors with eigenvalues ζ and ζ^2 respectively. We do the same for the h_α and again we reach the conclusion that $L = K_1(\phi)$. $\square$

Lemma 5.15.3 *Let $\alpha' \in \Psi$ be such that $\phi(\alpha') \neq \alpha'$. Then $\alpha' - \phi^i(\alpha') \notin \Psi$ for $i \geq 1$.*

Proof. We note that ϕ maps positive roots to positive roots and negative roots to negative roots. Suppose first that $d = 2$, then we must show that $\beta = \alpha' - \phi(\alpha') \notin \Psi$. However, this follows immediately from $\phi(\beta) = -\beta$.

Now suppose $d = 3$ and set $\beta = \alpha' - \phi(\alpha')$ and $\gamma = \alpha' - \phi^2(\alpha')$. Suppose that $\beta \in \Psi$, then $\gamma = -\phi^2(\beta)$ also lies in Ψ. Suppose further that β is a positive root. Then $\gamma = -\phi^2(\beta)$ is negative. Also $\phi(\beta) = \gamma - \beta$ is positive. However, $\gamma - \beta$ is a sum of negative roots, and hence negative. We have reached a contradiction. If β is negative then we proceed in the same way. It follows that $\beta \notin \Psi$. Finally, suppose that $\gamma \in \Psi$. Then also $\beta = -\phi(\gamma) \in \Psi$ and we have a contradiction. $\square$

Lemma 5.15.4 *For $\alpha \in \Phi$ we have*

$$[y_\alpha, y_{-\alpha}] = \begin{cases} -h_\alpha & \text{if } \alpha \in \Phi_l \\ -dh_\alpha & \text{if } \alpha \in \Phi_s. \end{cases}$$

Proof. If $\alpha \in \Phi_l$ then this is clear from the corresponding commutation relation in K. From Lemma 5.15.3 it follows that for $\alpha' \in \Psi$ such that $\phi(\alpha') \neq \alpha'$ we have $[x_{\phi^i(\alpha')}, x_{\phi^j(-\alpha')}] = 0$. Using this we see that for $\alpha \in \Phi_s$,

$$[y_\alpha, y_{-\alpha}] = [x_{\alpha'} + x_{\phi(\alpha')} \cdots + x_{\phi^{d-1}(\alpha')}, x_{-\alpha'} + x_{\phi(-\alpha')} \cdots + x_{\phi^{d-1}(-\alpha')}]$$
$$= -h_{\alpha'} - h_{\phi(\alpha')} - \cdots - h_{\phi^{d-1}(\alpha')}$$
$$= -dh_{\frac{1}{d}(\alpha' + \phi(\alpha') + \cdots + \phi^{d-1}(\alpha'))}.$$

$\square$

Lemma 5.15.5 *For $\alpha \in \Delta$ and $\beta \in \Phi$ we have $[h_\alpha, y_\beta] = (\beta, \alpha)y_\beta$, where $(\ ,\ )$ is the inner product from Ψ.*

Proof. If $\beta \in \Phi_l$ then there is nothing to prove. If $\beta \in \Phi_s$, then

$$
\begin{aligned}
[h_\alpha, y_\beta] &= [h_\alpha, x_{\beta'} + \cdots + x_{\phi^{d-1}(\beta')}] \\
&= (\alpha, \beta') x_{\beta'} + (\alpha, \phi(\beta')) x_{\phi(\beta')} + \cdots + (\alpha, \phi^{d-1}(\beta')) x_{\phi^{d-1}(\beta')} \\
&= \left(\alpha, \frac{1}{d}(\beta' + \phi(\beta') + \cdots + \phi^{d-1}(\beta'))\right)(x_{\beta'} + \cdots + x_{\phi^{d-1}(\beta')}),
\end{aligned}
$$

where the last equality follows from the fact that $\phi(\alpha) = \alpha$ and hence $(\alpha, \beta') = (\alpha, \phi(\beta')) = (\alpha, \phi^2(\beta')) = \ldots$ (Lemma 5.15.1). $\quad\square$

Lemma 5.15.6 *Let $\beta \in \Phi^0 = \Phi \cup \{0\}$ and suppose that $(\alpha, \beta) = 0$ for all $\alpha \in \Delta$. Then $\beta = 0$.*

Proof. We prove that $(\alpha', \beta) = 0$ for all $\alpha' \in \Pi$. This and the non-degeneracy of $(\,,\,)$ imply the result.

Let $\alpha' \in \Pi$ be such that $\phi(\alpha') \neq \alpha'$. Then $\alpha = \frac{1}{d}(\alpha' + \cdots + \phi^{d-1}(\alpha')) \in \Delta_s$. First note that $\phi(\beta) = \beta$, and hence $(\alpha', \beta) = (\phi^k(\alpha'), \beta)$ for $k \geq 0$ (Lemma 5.15.1). Now

$$
0 = (\alpha, \beta) = \frac{1}{d}\left((\alpha', \beta) + (\phi(\alpha'), \beta) + \cdots + (\phi^{d-1}(\alpha'), \beta)\right) = (\alpha', \beta).
$$

The conclusion is that $(\beta, \alpha') = 0$ for all $\alpha' \in \Pi$. $\quad\square$

Corollary 5.15.7 *L is a semisimple Lie algebra.*

Proof. From Lemma 5.15.4 it follows that $[L_\alpha, L_{-\alpha}] \neq 0$ for $\alpha \in \Phi$. Also by Lemmas 5.15.5 and 5.15.6 we have that $H = L_0(H)$. Finally $\Phi = -\Phi$ and $\dim L_\alpha = 1$ for $\alpha \in \Phi$. The result now follows from Proposition 5.12.1. $\square$

From Corollary 5.15.7 together with Theorem 5.3.5 it follows that Φ (viewed inside some Euclidean space) is a root system. We determine the type of Φ.

Lemma 5.15.8 *The set Δ is a simple system of Φ.*

Proof. By Proposition 5.5.8 it is enough to show that every element of Φ can be written as a $\mathbb{Z}$-linear combination of elements of Δ, where the coefficients are either all non-negative, or all non-positive.

First write $\Pi = \Pi_1 \cup \Pi_2$, where $\Pi_1 = \{\beta \in \Pi \mid \phi(\beta) = \beta\}$ and $\Pi_2 = \{\beta \in \Pi \mid \phi(\beta) \neq \beta\}$. Let $\alpha \in \Phi_l$, then also $\alpha \in \Psi$ and hence

$$\alpha = \sum_{\beta \in \Pi_1} k_\beta \beta + \sum_{\beta \in \Pi_2} k_\beta \beta.$$

Now from $\phi(\alpha) = \alpha$ it follows that

$$\sum_{\beta \in \Pi_2} k_\beta \beta = \sum_{\beta \in \Pi_2} k_\beta \phi(\beta) = \sum_{\beta \in \Pi_2} k_\beta \phi^2(\beta) = \cdots$$

Hence

$$\sum_{\beta \in \Pi_2} k_\beta \beta = \sum_{\beta \in \Pi_2} k_\beta \frac{1}{d}(\beta + \phi(\beta) + \cdots + \phi^{d-1}(\beta)).$$

It follows that α is a $\mathbb{Z}$-linear combination of elements of Δ.

Now let $\alpha \in \Phi_s$, then $\alpha = \frac{1}{d}(\alpha' + \cdots + \phi^{d-1}(\alpha'))$, where $\alpha' \in \Psi$. In particular we have that $\alpha' = \sum_{\beta \in \Pi} m_\beta \beta$, so that

$$\alpha = \sum_{\beta \in \Pi_1} m_\beta \beta + \sum_{\beta \in \Pi_2} m_\beta \frac{1}{d}(\beta + \phi(\beta) + \cdots + \phi^{d-1}(\beta)).$$

And we arrive at the same conclusion. $\qquad\qquad\qquad\qquad\qquad\qquad$ $\square$

Let $\Pi = \{\beta_1, \ldots, \beta_m\}$, where m is the rank of Ψ; then the simple systems Δ of Φ are listed below.

1. D_{l+1}: $\Delta = \{\beta_1, \ldots, \beta_{l-1}, \frac{1}{2}(\beta_l + \beta_{l+1})\}$,

2. A_{2l-1}: $\Delta = \{\frac{1}{2}(\beta_1 + \beta_{2l-1}), \ldots, \frac{1}{2}(\beta_{l-1} + \beta_{l+1}), \beta_l\}$,

3. E_6: $\Delta = \{\beta_2, \beta_4, \frac{1}{2}(\beta_3 + \beta_5), \frac{1}{2}(\beta_1 + \beta_6)\}$,

4. D_4: $\Delta = \{\frac{1}{3}(\beta_1 + \beta_3 + \beta_4), \beta_2\}$.

It is straightforward to see that the Dynkin diagrams of these systems are of types B_l, C_l, F_4 and G_2. The conclusion is that L is a simple Lie algebra of type B_l, C_l, F_4, G_2 respectively. In particular, these Lie algebras exist.

We end this section by giving a multiplication table of L. This will then lead to an algorithm for constructing the simple Lie algebras. First we remark that in Δ there occur roots of two different lengths. If $\alpha \in \Delta_l$, then $(\alpha, \alpha) = 2$ and if $\alpha \in \Delta_s$ then $(\alpha, \alpha) = 1$ if $d = 2$ and $(\alpha, \alpha) = \frac{2}{3}$ if $d = 3$. Using Lemma 5.7.5 we infer that the roots of Φ occur in two different lengths. For $\alpha \in \Phi_l$ we have $(\alpha, \alpha) = 2$ since in that case $\alpha \in \Psi$. Now assume that $d = 2$, then for $\alpha \in \Phi_s$ we calculate

$$(\alpha, \alpha) = (\frac{1}{2}(\alpha' + \phi(\alpha')), \frac{1}{2}(\alpha' + \phi(\alpha'))) = 1 + \frac{1}{2}(\alpha', \phi(\alpha')).$$

Now $(\alpha', \phi(\alpha')) \leq 1$ so that $(\alpha, \alpha) < 2$. Hence by Lemma 5.7.5 we see that

$$(\alpha, \alpha) = 1 \quad \text{and} \quad (\alpha', \phi(\alpha')) = 0 \quad \text{for } \alpha \in \Phi_s. \tag{5.15}$$

Now we deal with the case where $d = 3$. Then for $\alpha \in \Phi_s$ we have

$$(\alpha, \alpha) = (\frac{1}{3}(\alpha' + \phi(\alpha') + \phi^2(\alpha')), \frac{1}{3}(\alpha' + \phi(\alpha') + \phi^2(\alpha'))) =$$
$$\frac{2}{3} + \frac{1}{3}(\alpha', \phi(\alpha')) + \frac{1}{3}(\alpha', \phi^2(\alpha')).$$

And again by Lemma 5.7.5 we see that

$$(\alpha, \alpha) = \frac{2}{3} \quad \text{and} \quad (\alpha', \phi(\alpha')) = (\alpha', \phi^2(\alpha')) = 0 \quad \text{for } \alpha \in \Phi_s. \tag{5.16}$$

The longer roots in Φ are called *long* roots and the shorter ones are called *short* roots.

Lemma 5.15.9 *Let $\alpha, \beta \in \Phi$ be such that $\alpha + \beta \in \Phi$, then $[y_\alpha, y_\beta] = \varepsilon(\alpha', \beta')(r+1)y_{\alpha+\beta}$, where r is the largest integer such that $\alpha - r\beta \in \Phi$, and where $\alpha', \beta' \in \Psi$ are chosen such that $\alpha' + \beta' \in \Psi$.*

Proof. There are three cases to consider: both α and β are long, one of them is short while the other is long and both are short. If both are long, then $(\alpha + \beta, \alpha + \beta) = 4 + 2(\alpha, \beta)$ from which we see that $(\alpha, \beta) = -1$ and $\alpha + \beta$ lies in Ψ (by Proposition 5.13.2) and is long. Then by Proposition 5.13.2 $\alpha - \beta \notin \Psi$ and hence $r = 0$. So we are done in this case.

For the other two cases we first suppose that $d = 2$. Let α be short and β long. Then $\alpha + \beta$ is short (for suppose that $\gamma = \alpha + \beta$ is long, then as seen above $\gamma - \beta$ is also long which is absurd). In this case $[y_\alpha, y_\beta] = [x_{\alpha'} + x_{\phi(\alpha')}, x_{\beta'}]$. By Lemma 5.11.2 we have that $[y_\alpha, y_\beta] \neq 0$ so at least one of $\alpha' + \beta'$, $\phi(\alpha') + \beta'$ must lie in Ψ. But since $\phi(\beta') = \beta'$ we see that if one of them lies in Ψ so does the other. Hence both are elements of Ψ. Now since $\alpha + \beta = \frac{1}{2}(\alpha' + \beta') + \frac{1}{2}\phi(\alpha' + \beta')$,

$$[y_\alpha, y_\beta] = \varepsilon(\alpha', \beta')x_{\alpha'+\beta'} + \varepsilon(\phi(\alpha'), \phi(\beta'))x_{\phi(\alpha'+\beta')} = \varepsilon(\alpha', \beta')y_{\alpha+\beta}.$$

From Proposition 5.13.2 we see that $(\alpha', \beta') = (\phi(\alpha'), \beta') = -1$. Hence $(\alpha, \beta) = -1$ and $(\alpha - \beta, \alpha - \beta) = 5$. Therefore $\alpha - \beta$ is not contained in Φ;

so $r = 0$ and the statement is verified. If α is long and β short, then we use an analogous argument.

We deal with the case where α and β are both short. Then $[y_\alpha, y_\beta] = [x_{\alpha'} + x_{\phi(\alpha')}, x_{\beta'} + x_{\phi(\beta')}]$ and by Lemma 5.11.2 this is non-zero. Hence at least one of $\alpha' + \beta'$, $\alpha' + \phi(\beta')$, $\phi(\alpha') + \beta'$, $\phi(\alpha') + \phi(\beta')$ must lie in Ψ. But this is the same as saying that at least one of $\alpha' + \beta'$, $\alpha' + \phi(\beta')$ is an element of Ψ. And since interchanging β' and $\phi(\beta')$ changes nothing we may assume that $\alpha' + \beta' \in \Psi$, i.e., $(\alpha', \beta') = -1$. Using (5.15) we calculate

$$(\alpha + \beta, \alpha + \beta) = 1 + (\alpha', \phi(\beta')). \tag{5.17}$$

The root $\alpha + \beta$ can be short or long. If it is short, then $(\alpha', \phi(\beta')) = 0$ so that $\alpha' \pm \phi(\beta') \notin \Psi$ by Proposition 5.13.2. Also $(\alpha - \beta, \alpha - \beta) = 3$ and hence $\alpha - \beta \notin \Phi$ and $r = 0$. Also $\phi(\alpha' + \beta') \neq \alpha' + \beta'$ (otherwise $\alpha + \beta$ would be equal to $\alpha' + \beta'$ and hence long). Furthermore, $\alpha + \beta = \frac{1}{2}(\alpha' + \beta') + \frac{1}{2}\phi(\alpha' + \beta')$, and

$$[y_\alpha, y_\beta] = \varepsilon(\alpha', \beta')x_{\alpha'+\beta'} + \varepsilon(\phi(\alpha'), \phi(\beta'))x_{\phi(\alpha'+\beta')} = \varepsilon(\alpha', \beta')y_{\alpha+\beta}.$$

On the other hand, if $\alpha + \beta$ is long, then $\phi(\alpha' + \beta') = \alpha' + \beta'$ (if not, then $\alpha + \beta = \frac{1}{2}(\alpha' + \beta') + \frac{1}{2}\phi(\alpha' + \beta')$ would be short). Also from (5.17) we see that $(\alpha', \phi(\beta')) = 1$ so that $\alpha' - \phi(\beta') \in \Psi$. From $\phi(\alpha' + \beta') = \alpha' + \beta'$ we easily deduce $\phi(\alpha' - \phi(\beta')) = \alpha' - \phi(\beta')$. Therefore $\alpha - \beta = \alpha' - \phi(\beta') \in \Phi_l$, and consequently $r = 1$. Finally,

$$[y_\alpha, y_\beta] = \varepsilon(\alpha', \beta')x_{\alpha'+\beta'} + \varepsilon(\phi(\alpha'), \phi(\beta'))x_{\phi(\alpha'+\beta')} = 2\varepsilon(\alpha', \beta')y_{\alpha+\beta}.$$

For the case where $d = 3$ similar arguments can be applied. Firstly, if α is short and β is long, then $\alpha' + \beta'$, $\phi(\alpha') + \beta'$ and $\phi^2(\alpha') + \beta'$ are all elements of Ψ and no $\phi^k(\alpha') - \beta'$ is an element of Ψ. So $r = 0$ and

$$[x_{\alpha'} + x_{\phi(\alpha')} + x_{\phi^2(\alpha')}, x_{\beta'}] = \varepsilon(\alpha', \beta')y_{\alpha+\beta}.$$

If α and β are both short then again we may assume that $\alpha' + \beta' \in \Psi$ and we use (5.16) to calculate

$$(\alpha + \beta, \alpha + \beta) = \frac{2}{3}(1 + (\alpha', \phi(\beta')) + (\alpha', \phi^2(\beta'))).$$

Now if $\alpha + \beta$ is short, then $(\alpha', \phi(\beta')) + (\alpha', \phi^2(\beta')) = 0$. If both terms are 0, then $r = 0$ and it is straightforward to see that $[y_\alpha, y_\beta] = \varepsilon(\alpha', \beta')y_{\alpha+\beta}$. If on the other hand $(\alpha', \phi(\beta')) = -1$ and $(\alpha', \phi^2(\beta')) = 1$, then we have $r = 1$ and

$$[y_\alpha, y_\beta] = \big(\varepsilon(\alpha', \beta') + \varepsilon(\alpha', \phi(\beta'))\big)y_{\alpha+\beta}.$$

It is easily verified by inspecting Φ (here we are dealing with one root system only) that α, β short such that $\alpha+\beta$ is short and $\alpha'+\beta' \in \Psi$ and $(\alpha', \phi(\beta')) = -1$ imply $\varepsilon(\alpha', \phi(\beta')) = \varepsilon(\alpha', \beta')$. Hence $[y_\alpha, y_\beta] = 2\varepsilon(\alpha', \beta')y_{\alpha+\beta}$. The case where $(\alpha', \phi(\beta')) = 1$ is analogous.

Finally, if $\alpha+\beta$ is long, then $(\alpha', \phi(\beta')) = (\alpha', \phi^2(\beta')) = 1$ and $\alpha' - \phi(\beta')$, $\alpha' - \phi^2(\beta')$ are both elements of Ψ. Furthermore, $\alpha' + \beta' \in \Phi_l$ and so are $\alpha' - \phi(\beta')$ and $\alpha' - \phi(\beta') - \phi^2(\beta')$ (since $(\alpha' - \phi(\beta'), \phi^2(\beta')) = 1$). Hence $r = 2$ and also

$$[y_\alpha, y_\beta] = 3\varepsilon(\alpha', \beta')y_{\alpha+\beta}.$$

$\square$

Now Lemmas 5.15.5, 5.15.6, 5.15.9 add up to the following multiplication table of L:

$$
\begin{aligned}
&[h_\alpha, h_\beta] = 0 \quad \text{for } \alpha, \beta \in \Delta, \\
&[h_\alpha, y_\beta] = (\beta, \alpha)y_\beta \quad \text{for } \alpha \in \Delta \text{ and } \beta \in \Phi, \\
&[y_\alpha, y_{-\alpha}] = \begin{cases} -h_\alpha & \text{for } \alpha \in \Phi_l, \\ -dh_\alpha & \text{for } \alpha \in \Phi_s, \end{cases} \\
&[y_\alpha, y_\beta] = 0 \quad \text{for } \alpha, \beta \in \Phi \text{ such that } \alpha + \beta \notin \Phi \text{ and } \beta \neq -\alpha, \\
&[y_\alpha, y_\beta] = \varepsilon(\alpha', \beta')(r + 1)y_{\alpha+\beta} \quad \text{for } \alpha, \beta \in \Phi \text{ such that } \alpha + \beta \in \Phi, \\
&\qquad \text{where } r \geq 0 \text{ is the largest integer such that } \alpha - r\beta \in \Phi, \\
&\qquad \text{and where } \alpha', \beta' \text{ are chosen such that } \alpha' + \beta' \in \Psi.
\end{aligned}
$$

$$(5.18)$$

The multiplication tables (5.14) and (5.18) yield an algorithm SimpleLieAlgebra for constructing the multiplication table of a simple Lie algebra of type X_l. If $X_l \in \{A_l, D_l, E_{6,7,8}\}$ then we construct a root system Φ of type X_l (by CartanMatrixToRootSystem, cf. Corollary 5.10.3). We choose an orientation of the Dynkin diagram of Φ and we let ε be the corresponding asymmetry function. We return the multiplication table of (5.14). If on the other hand X_l is B_l, C_l, F_4 or G_2, then we construct a root system Ψ of type D_{l+1}, A_{2l-1}, E_6, D_4 respectively. Let ϕ be the diagram automorphism of Ψ. Choose an orientation of the Dynkin diagram of Ψ that is invariant under ϕ and let ε be the corresponding asymmetry function. We construct the set $\Phi = \Phi_s \cup \Phi_l$ and return the table of (5.18).

Example 5.15.10 We use the multiplication table (5.18) to produce the structure constants of the Lie algebra corresponding to the root system of type B_2. This Lie algebra is a subalgebra of the Lie algebra of type D_3. However, the Dynkin diagram of D_3 is the same as the Dynkin diagram of type A_3. We choose the following orientation of the Dynkin diagram of type A_3:

$$\underset{\alpha_1}{\circ} \!\!\rightarrow\!\! \underset{\alpha_2}{\circ} \!\!\leftarrow\!\! \underset{\alpha_3}{\circ}\,.$$

It is clear that this orientation is invariant under the diagram automorphism of A_3. The corresponding asymmetry function is defined by $\varepsilon(\alpha_i, \alpha_i) = -1$ for $i = 1, 2, 3$, $\varepsilon(\alpha_1, \alpha_2) = \varepsilon(\alpha_3, \alpha_2) = -1$ and $\varepsilon(\alpha_1, \alpha_3) = \varepsilon(\alpha_3, \alpha_1) = \varepsilon(\alpha_2, \alpha_1) = \varepsilon(\alpha_2, \alpha_3) = 1$. The root system Φ is the union of

$$\Phi_l = \pm\{\beta_1 = \alpha_2, \beta_2 = \alpha_1 + \alpha_2 + \alpha_3\}$$

and

$$\Phi_s = \pm\{\gamma_1 = \frac{1}{2}(\alpha_1 + \alpha_3), \gamma_2 = \frac{1}{2}(\alpha_1 + 2\alpha_2 + \alpha_3)\}.$$

A basis of L is given by

$$\{g_1, g_2, y_{\beta_1}, y_{\beta_2}, y_{\gamma_1}, y_{\gamma_2}, y_{-\beta_1}, y_{-\beta_2}, y_{-\gamma_1}, y_{-\gamma_2}\},$$

where $g_1 = h_{\beta_1}$ and $g_2 = h_{\gamma_1}$. By way of example we calculate the commutator $[y_{\gamma_1}, y_{\gamma_2}]$. Note that $\gamma_1 + \gamma_2 = \beta_2 \in \Phi$, so if we take $\gamma_1' = \alpha_1$ and $\gamma_2' = \alpha_2 + \alpha_3$, then $\gamma_1' + \gamma_2' \in \Psi$. Furthermore, $\gamma_1 - \gamma_2 = -\beta_1 \in \Phi$ so that $r = 1$. Hence

$$[y_{\gamma_1}, y_{\gamma_2}] = 2\varepsilon(\alpha_1, \alpha_2 + \alpha_3)y_{\beta_2} = -2y_{\beta_2}.$$

Continuing like this we fill the whole table. The result is displayed in Table 5.3.

	g_1	g_2	y_{β_1}	y_{β_2}	y_{γ_1}	y_{γ_2}	$y_{-\beta_1}$	$y_{-\beta_2}$	$y_{-\gamma_1}$	$y_{-\gamma_2}$
g_1	0	0	$2y_{\beta_1}$	0	$-y_{\gamma_1}$	y_{γ_2}	$-2y_{-\beta_1}$	0	$y_{-\gamma_1}$	$-y_{-\gamma_2}$
g_2	·	0	$-y_{\beta_1}$	y_{β_2}	y_{γ_1}	0	$y_{-\beta_1}$	$-y_{-\beta_2}$	$-y_{-\gamma_1}$	0
y_{β_1}	·	·	0	0	y_{γ_2}	0	$-g_1$	0	0	$-y_{-\gamma_1}$
y_{β_2}	·	·	·	0	0	0	0	$-g_1 - 2g_2$	$-y_{-\gamma_2}$	$y_{-\gamma_1}$
y_{γ_1}	·	·	·	·	0	$-2y_{\beta_2}$	0	$y_{-\gamma_2}$	$-2g_2$	$2y_{-\beta_1}$
y_{γ_2}	·	·	·	·	·	0	y_{γ_1}	$-y_{-\gamma_1}$	$-2y_{\beta_1}$	$-2g_1 - 2g_2$
$y_{-\beta_1}$	·	·	·	·	·	·	0	0	$y_{-\gamma_2}$	0
$y_{-\beta_2}$	·	·	·	·	·	·	·	0	0	0
$y_{-\gamma_1}$	·	·	·	·	·	·	·	·	0	$-2y_{-\beta_2}$
$y_{-\gamma_2}$	·	·	·	·	·	·	·	·	·	0

Table 5.3: Multiplication table of the simple Lie algebra of type B_2.

Example 5.15.11 Now that we have two multiplication tables of the Lie algebra of type B_2 we can use the isomorphism theorem to construct an isomorphism between them. Let L_1 be the Lie algebra of Example 4.9.1 and let L_2 be the one of Example 5.15.10. We use the algorithm IsomorphismOfSSLieAlgebras to construct an isomorphism between them. As seen in Example 4.9.1 the root system of L_1 is

$$\Phi_1 = \{(1,-1),(-1,1),(1,1),(-1,-1),(-1,0),(0,-1),(1,0),(0,1)\}.$$

And a simple system is $\Delta_1 = \{\alpha_1 = (0,1), \alpha_2 = (1,-1)\}$ (Example 5.5.2). The space H_2 spanned by g_1, g_2 is a Cartan subalgebra of L_2 and the root system of L_2 relative to H_2 is

$$\Phi_2 = \{(2,-1),(0,1),(-1,1),(1,0),(-2,1),(0,-1),(1,-1),(-1,0)\}$$

(Again we denote a root $\alpha \in H_2^*$ by the vector $(\alpha(g_1), \alpha(g_2))$). A simple system of Φ_2 is $\Delta_2 = \{\eta_1 = (-1,1), \eta_2 = (2,-1)\}$. The Cartan matrix of Φ_1 relative to Δ_1 is identical to the Cartan matrix of Φ_2 relative to Δ_2. Hence the linear map sending α_i to η_i for $i = 1,2$ is an isomorphism of root systems (cf. the proof of Corollary 5.6.3).

Now we have to construct canonical generators of L_1 and L_2. For this we use the algorithm CanonicalGenerators. First we choose $x_1 = q_2 \in L_{1(\alpha_1)}$. Let $y_1 = \mu_1 p_2 \in L_{1(-\alpha_1)}$, where μ_1 is to be determined. The element y_1 is required to satisfy $[[x_1, y_1], x_1] = 2x_1$ and this is equivalent to $-\mu_1 q_2 = 2q_2$, i.e., $\mu_1 = -2$. Then $h_1 = [x_1, y_1] = 2A_{22}$. Continuing we set $x_2 = A_{12}$. Then $y_2 = \mu_2 A_{21}$ and from $[[x_2, y_2], x_2] = 2x_2$ we see that $\mu_2 = 1$. So $h_2 = A_{11} - A_{22}$. In the same way we find canonical generators of L_2: $x_1' = y_{\gamma_1}$, $y_1' = -y_{-\gamma_1}$, $h_1' = 2g_2$, $x_2' = y_{\beta_1}$, $y_2' = -y_{-\beta_1}$ and $h_2' = g_1$. Now we know already a large part of the isomorphism; namely, we know that it sends x_1 to x_1', y_1 to y_1' and so forth. For each remaining root vector $x_\beta \in L_1$ (where β is a positive root) we have to fix a sequence $\alpha_{i_1}, \ldots, \alpha_{i_k}$ of simple roots such that $\alpha_{i_1} + \cdots + \alpha_{i_m}$ is a root for $1 \leq m \leq k$, and $\beta = \alpha_{i_1} + \cdots + \alpha_{i_k}$. In our case there are only two positive roots left and we can take

$$(1,0) = \alpha_1 + \alpha_2 \quad \text{and} \quad (1,1) = \alpha_2 + \alpha_1 + \alpha_1.$$

This means that $[x_1, x_2] = -q_1$ is mapped to $[x_1', x_2'] = -y_{\gamma_2}$. Secondly, $[x_1, [x_1, x_2]] = -B_{12}$ is mapped to $[x_1', [x_1', x_2']] = 2y_{\beta_2}$. We do a similar thing for the negative roots: $[y_1, y_2] = -2p_1$ is mapped to $[y_1', y_2'] = -y_{-\gamma_2}$ and $[y_1, [y_1, y_2]] = 4C_{12}$ to $[y_1', [y_1', y_2']] = -2y_{-\beta_2}$.

We have constructed a linear map $\phi : L_1 \to L_2$ that is necessarily an isomorphism. It is given by

$$A_{22} \mapsto g_2, \ A_{11} \mapsto g_1 + g_2, \ q_2 \mapsto y_{\gamma_1}, \ A_{12} \mapsto y_{\beta_1}, \ p_2 \mapsto \frac{1}{2} y_{-\gamma_1},$$

$$A_{21} \mapsto -y_{-\beta_1}, \ q_1 \mapsto y_{\gamma_2}, \ B_{12} \mapsto -2y_{\beta_2}, \ p_1 \mapsto \frac{1}{2} y_{-\gamma_2}, \ C_{12} \mapsto -\frac{1}{2} y_{-\beta_2}.$$

5.16   The classification theorem

In this section we summarize our efforts into a theorem.

Definition 5.16.1 *Let L be a simple Lie algebra of characteristic 0 with a split Cartan subalgebra. If the rootsystem of L is of type X_l, then L is said to be of type X_l. Furthermore, as simple Lie algebras with the same type are isomorphic (Corollary 5.11.5), we say that X_l is an* isomorphism class *of simple Lie algebras.*

Theorem 5.16.2 *Let F be a field of characteristic 0. Then the isomorphism classes of simple Lie algebras with a split Cartan subalgebra are exactly A_n, B_n (for $n \geq 2$), C_n (for $n \geq 3$), D_n (for $n \geq 4$), E_6, E_7, E_8, F_4 and G_2.*

Proof. Let L be a simple Lie algebra over F with split Cartan subalgebra H. Let Φ be the root system of L with respect to H, then by Corollary 5.11.6, Φ is upto isomorphism determined by L. Furthermore, as seen in Section 5.9, Φ has one of the types mentioned. Conversely let X_l be one of the listed types. Then by Proposition 5.10.1, there is a root system Φ of type X_l. As seen in Sections 5.13 and 5.15 there is a simple Lie algebra over F with a split Cartan subalgebra, having a root system isomorphic to Φ. $\qquad \square$

5.17   Recognizing a semisimple Lie algebra

In this section we consider the problem of deciding whether two given semisimple Lie algebras L_1 and L_2 over $\mathbb{Q}$ are isomorphic over the algebraic closure $\overline{\mathbb{Q}}$ of $\mathbb{Q}$. If the Cartan subalgebras of L_1 and L_2 are split over $\mathbb{Q}$ then the algorithm IsmorphismOfSSLieAlgebras gives a straightforward method for constructing an isomorphism if the Lie algebras are isomorphic, and the method will break down if they are not. However, if the Cartan subalgebras are not split over $\mathbb{Q}$ then the situation becomes more difficult. As an example we consider the 6-dimensional semisimple Lie algebra L of Example 4.12.3. We know that there is only one semisimple Lie algebra of dimension 6 with a split Cartan subalgebra, namely the direct sum of two copies of the Lie algebra with type A_1. However, over $\mathbb{Q}$ the Cartan subalgebra of L is not split and L is a simple Lie algebra. Over the algebraic extension $\mathbb{Q}(i)$ the Cartan subalgebra splits and L is seen to be isomorphic to the direct sum of two copies of the Lie algebra of type A_1. In this case we manage to split the Cartan subalgebra over a rather small algebraic extension. However, since the degree of the algebraic extension needed to split a polynomial of degree n is "generically" $n!$, we see that for larger examples it can become impossible to take this route. In Section 5.17.1 we describe a method for reducing the Lie algebra modulo a suitable prime p. We get a Lie algebra over a finite field and the algebraic extensions of finite fields are much easier to handle. We prove that (under certain conditions on the prime p) the modularized Lie algebra will provide us with the isomorphism class of the original Lie algebra.

In Section 5.17.2 we approach the problem of deciding isomorphism of semisimple Lie algebras in a different way. We encode the adjoint action of a Cartan subalgebra of a semisimple Lie algebra in a number of polynomials. Then we show that L_1 is isomorphic to L_2 if and only if the polynomials obtained for L_1 can be transformed by a change of variables into the polynomials corresponding to L_2.

5.17.1   Identifying a semisimple Lie algebra

Let L be a semisimple Lie algebra defined over $\mathbb{Q}$. Then by Propositions 4.3.6 and 4.3.7 the Lie algebra $L \otimes \overline{\mathbb{Q}}$ has a unique decomposition as a direct sum $I_1 \oplus \cdots \oplus I_m$ of simple ideals. Furthermore, over $\overline{\mathbb{Q}}$ the Cartan subalgebras of the I_k are split and hence the I_k fall into the classification of Theorem 5.16.2. Suppose that the simple ideal I_k is of type X_{n_k}. Then we call $X_{n_1} + \cdots + X_{n_m}$ the type of L. In this section we describe a method for obtaining the type of a semisimple Lie algebra.

Throughout H will be a (maybe non-split) Cartan subalgebra of L. The idea we pursue here is to avoid working over large number fields by reducing the structure constants of L modulo a prime number p. Note that if we multiply all basis elements by a scalar λ, then the structure constants relative to this new basis are also multiplied by λ, so that we can get all structure constants to be integers. Then, using an algebraic extension of $\mathbb{F}_p$ if necessary, we split the Cartan subalgebra and calculate the root system over the modular field. For this root system we calculate the Cartan matrix. We prove that this is also the Cartan matrix of the root system of L.

In the sequel we fix a basis of L such that the structure constants relative to this basis are integers. Furthermore we assume that this basis contains a basis $\{h_1, \ldots, h_l\}$ of the Cartan subalgebra H. From Section 4.11 we recall that $h_0 \in H$ is called a splitting element if the degree of the minimum polynomial of $\mathrm{ad}_L h_0$ is $\dim L - \dim H + 1$. Now fix a splitting element $h_0 = \sum_{i=1}^{l} m_i h_i$ of H such that $m_i \in \mathbb{Z}$ for $1 \le i \le l$ (such splitting elements exist by Proposition 4.11.2). Let A_0 be the matrix of $\mathrm{ad}_L h_0$ relative to the given basis of L. All entries of this matrix are integers, so we can reduce it modulo a prime p and obtain a matrix A_p with entries in the finite field $\mathbb{F}_p$. If $p \ge 5$ is a prime number not dividing the determinant of the matrix of the Killing form of L and such that the minimum polynomial of A_p is square-free and of degree $\dim L - \dim H + 1$, then we call p *pleasant*. In the sequel we use a fixed pleasant prime number p.

Let F be the smallest number field containing all eigenvalues of $\mathrm{ad}_L h_0$. We recall that an algebraic number $\alpha \in F$ is called *integral* over $\mathbb{Z}$ if α satisfies an equation of the form $\alpha^n + a_{n-1}\alpha^{n-1} + \cdots + a_0 = 0$ where $a_i \in \mathbb{Z}$ for $0 \le i \le n - 1$. The set of all elements $\alpha \in F$ that are integral over $\mathbb{Z}$ is a ring called the ring of algebraic integers of F. We denote it by $\mathcal{O}^F$. Then by [56, Chapter I, §3, Proposition 9] there exists a prime ideal P of $\mathcal{O}^F$ such that $P \cap \mathbb{Z} = (p)$, where (p) denotes the ideal of $\mathbb{Z}$ generated by p. The ring

$$\mathcal{O}_P^F = \left\{ \frac{x}{y} \mid x \in \mathcal{O}^F,\ y \in \mathcal{O}^F \setminus P \right\}$$

is called the localization of $\mathcal{O}^F$ at P. It is a local ring, which means that it has a unique maximal ideal M_P given by

$$M_P = \left\{ \frac{x}{y} \mid x \in P,\ y \in \mathcal{O}^F \setminus P \right\}.$$

Since M_P is a maximal ideal, $\mathcal{O}_P^F / M_P$ is a field. Furthermore M_P contains p and hence $\mathcal{O}_P^F / M_P$ is of characteristic p. It follows that there is an $m > 0$ such that $\mathcal{O}_P^F / M_P = \mathbb{F}_{p^m}$, the finite field of p^m elements.

Let T denote the multiplication table of L relative to the fixed basis of L. So all structure constants in T are integers. Let K be a Lie algebra over $\mathcal{O}_P^F$ with multiplication table T. Set

$$K_p = K \otimes_{\mathcal{O}_P^F} \mathbb{F}_{p^m}.$$

Let $\phi : \mathcal{O}_P^F \to \mathbb{F}_{p^m}$ be the projection map. In the obvious way ϕ carries over to a map from K to K_p. Let $\{x_1, \ldots, x_n\}$ be the basis of K corresponding to the multiplication table T, and set $\bar{x}_i = x_i \otimes 1 \in K_p$ for $1 \le i \le n$. Then we set

$$\phi(\sum_{i=1}^n a_i x_i) = \sum_{i=1}^n \phi(a_i)\bar{x}_i.$$

Let κ be the Killing form of K and let κ_p be the Killing form of K_p. The structure constants of K_p are the images under ϕ of the structure constants of K, and hence they lie in the prime field $\mathbb{F}_p$. From this it follows that

$$\kappa_p(\phi(x), \phi(y)) = \phi(\kappa(x, y)) \quad \text{for } x, y \in K.$$

Because p is pleasant, we have that κ_p is non-degenerate.

Since the structure constants of K are the same as those of L, also the basis of K corresponding to these structure constants contains a basis of a Cartan subalgebra, which we also call H. Furthermore, since the coefficients of the matrix of $\mathrm{ad}_K h_0$ (relative to the basis $\{x_1, \ldots, x_n\}$) are integers, it follows that the eigenvalues of the $\mathrm{ad}_K h_0$ are integral and hence contained in $\mathcal{O}_P^F$. The primary decomposition of K relative to $\mathrm{ad}_K h_0$ is identical to the root space decomposition of K relative to H (because h_0 is a splitting element). So the whole Cartan subalgebra H of K is split.

Let H_p be the image under ϕ of H. Furthermore H^* will be the dual space of H, i.e., the space of all linear maps from H into $\mathcal{O}_P^F$. Since H is split we have that H^* contains the roots of K. Let H_p^* be the dual of H_p. The map ϕ induces a map (which we also call ϕ)

$$\phi : H^* \longrightarrow H_p^*.$$

For $\lambda : H \to \mathcal{O}_P^F$ an element of H^* we set $\phi(\lambda)(\phi(h)) = \phi(\lambda(h))$. Since p is pleasant we have that $\mathrm{ad}_{K_p} \phi(h_0)$ is a semisimple linear transformation. Also it has $\dim L - \dim H + 1$ distinct eigenvalues, one of which is 0. Therefore the eigenspace relative to the eigenvalue 0 is H_p and hence the Fitting null-component $(K_p)_0(H_p)$ is equal to H_p. So H_p is a Cartan subalgebra of K_p. Let Φ be the set of roots of K relative to H and Φ_p the set of roots of K_p relative to H_p.

Lemma 5.17.1 *The map ϕ maps Φ bijectively onto Φ_p.*

Proof. A vector x_α is a root vector (corresponding to the root $\alpha \in \Phi$) of K if and only if the coefficients of x_α with respect to the basis $\{x_1, \ldots, x_n\}$ satisfy a certain system of linear equations over $\mathcal{O}_P^F$, given by $[h_i, x_\alpha] = \alpha(h_i)x_\alpha$ for $1 \le i \le l$, where $\{h_1, \ldots, h_l\}$ is a basis of H. This system has a 1-dimensional solution space over F. In particular it has rank $n - 1$. Now the equation system which a root vector of K_p corresponding to $\phi(\alpha)$ must satisfy is exactly the image under ϕ of the equation system corresponding to K and α. It follows that the rank of this equation system is at most $n - 1$ and hence there are non-zero solutions. The conclusion is that $\phi(\alpha)$ is a root of K_p.

If $\alpha \neq \beta \in \Phi$ then because h_0 is a splitting element, $\alpha(h_0) \neq \beta(h_0)$. Because p is pleasant also $\phi(\alpha(h_0)) \neq \phi(\beta(h_0))$, i.e., $\phi(\alpha) \neq \phi(\beta)$. So the map $\phi : \Phi \to \Phi_p$ is injective; moreover, it is surjective as well since Φ and Φ_p have the same cardinality. $\qquad \square$

Lemma 5.17.2 *Let M be a Lie algebra of characteristic $p \geq 5$ with a non-degenerate Killing form. Let α, β be non-zero roots of M such that $\alpha \pm \beta$ are not roots. Then $\kappa_M(h_\alpha, h_\beta) = 0$.*

Proof. We recall that h_α, h_β are defined by (4.10). Let x_α, x_β and $x_{-\beta}$ be non-zero root vectors corresponding to $\alpha, \beta, -\beta$. Since M_β, $M_{-\beta}$ are 1-dimensional (Root fact 16) we see that $\kappa_M(x_\beta, x_{-\beta}) \neq 0$ by Root fact 2 together with the non-degeneracy of κ_M. By Root fact 11, $[x_\beta, x_{-\beta}] = \kappa_M(x_\beta, x_{-\beta})h_\beta \neq 0$. By an application of the Jacobi identity however, we infer that $[x_\alpha, [x_\beta, x_{-\beta}]] = 0$. But this implies that $\alpha(h_\beta) = 0$, i.e., $\kappa_M(h_\alpha, h_\beta) = 0$. $\qquad \square$

Proposition 5.17.3 *Let M be as in Lemma 5.17.2. By R^0 we denote the set of roots of M together with zero. Let $\alpha, \beta \in R^0$, where $\beta \neq 0$. Then not all of $\alpha, \alpha + \beta, \alpha + 2\beta, \alpha + 3\beta, \alpha + 4\beta$ are elements of R^0.*

Proof. Suppose that all of $\alpha, \alpha+\beta, \alpha+2\beta, \alpha+3\beta, \alpha+4\beta$ lie in R^0. Suppose also that one of them is zero. Note that $\alpha + 2\beta$ and $\alpha + 3\beta$ cannot be zero (otherwise 2β or 3β is a root which is excluded by Root fact 16). Hence $\alpha = 0$ or $\alpha = -\beta$ or $\alpha = \beta$ (the last possibility might occur when $p = 5$). But the first possibility implies $2\beta = \alpha + 2\beta$ is a root. The second implies $2\beta = \alpha + 3\beta$ is a root, and the last possibility entails $2\beta = \alpha + \beta$ is a root. In all cases we reach a contradiction, so all of these roots are non-zero.

Now suppose that all of $\alpha, \alpha + \beta, \alpha + 2\beta, \alpha + 3\beta, \alpha + 4\beta$ are non-zero roots. Then $(\alpha+2\beta)\pm\alpha$ and $(\alpha+4\beta)\pm(\alpha+2\beta)$ are not roots. So by Lemma 5.17.2 we have that $\kappa_M(h_{\alpha+4\beta}, h_{\alpha+2\beta}) = 0$ and $\kappa_M(h_\alpha, h_{\alpha+2\beta}) = 0$. This implies that $4\kappa_M(h_\beta, h_{\alpha+2\beta}) = 0$ and hence $\kappa_M(h_{\alpha+2\beta}, h_{\alpha+2\beta}) = 0$. But this is the same as saying that $(\alpha+2\beta, \alpha+2\beta) = 0$ contradicting Root fact 15. $\qquad \square$

From Section 4.9 we recall that we can identify H and H^* via the Killing form. Let σ be an element of H^*. Then the corresponding element h_σ of H is required to satisfy $\kappa(h_\sigma, h) = \sigma(h)$ for all $h \in H$. If $\{h_1, \dots, h_l\}$ is a basis of H and $h_\sigma = a_1 h_1 + \cdots + a_l h_l$, then we have the system of equations

$$(\kappa(h_i, h_j)) \begin{pmatrix} a_1 \\ \vdots \\ a_l \end{pmatrix} = \begin{pmatrix} \sigma(h_1) \\ \vdots \\ \sigma(h_l) \end{pmatrix}. \tag{5.19}$$

Since the restriction of κ_p to H_p is non-degenerate (Root fact 4) we have that the determinant of the matrix of this system is an integer not divisible by p. Let B denote the matrix $(\kappa(h_i, h_j))$, and for $1 \le k \le l$ let B_k be the matrix obtained from B by replacing its k-th column by the right hand side of (5.19). Now Cramer's rule (see, e.g., [54]) states that the k-th coefficient of the solution of (5.19) is $\det(B_k)/\det(B)$. Hence there exists a unique solution over $\mathcal{O}_P^F$. We denote the map sending $\sigma \in H^*$ to $h_\sigma \in H$ by θ. Also for H_p^* we have a similar map $\theta_p : H_p^* \to H$.

Lemma 5.17.4 *We have the following identity:* $\phi \circ \theta = \theta_p \circ \phi$.

Proof. By $\bar{h}_i$ we denote the image of $h_i \in H$ under ϕ, where $\{h_1, \dots, h_l\}$ is the basis of H contained in the basis $\{x_1, \dots, x_n\}$ of K. Choose $\sigma \in H^*$ and suppose that $\theta_p(\phi(\sigma)) = b_1 \bar{h}_1 + \cdots + b_l \bar{h}_l$, where $b_i \in \mathbb{F}_{p^m}$. Then since $\kappa_p(\phi(x), \phi(y)) = \phi(\kappa(x, y))$, the system of equations for the b_i is just the image under ϕ of the system of equations (5.19). Hence $b_i = \phi(a_i)$ and we are done. $\square$

From Section 4.9 we recall that a bilinear form $(,)$ is defined on H^* by $(\sigma, \rho) = \kappa(\theta(\sigma), \theta(\rho))$. In the same way there is a bilinear form $(,)_p$ on H_p^*.

Lemma 5.17.5 *For* $\rho, \sigma \in H^*$ *we have* $\phi((\sigma, \rho)) = (\phi(\sigma), \phi(\rho))_p$.

Proof. The proof is by straightforward calculation:

$$\begin{aligned} \phi((\sigma, \rho)) &= \phi(\kappa(\theta(\sigma), \theta(\rho))) \\ &= \kappa_p(\phi(\theta(\sigma)), \phi(\theta(\rho))) \\ &= \kappa_p(\theta_p(\phi(\sigma)), \theta_p(\phi(\rho))) \quad \text{by Lemma 5.17.4} \\ &= (\phi(\sigma), \phi(\rho))_p. \end{aligned}$$

$\square$

For $\alpha, \beta \in \Phi_p$ let r, q be the smallest nonnegative integers such that $\alpha - (r+1)\beta$ and $\alpha + (q+1)\beta$ are not roots, and set $C(\alpha, \beta) = r - q$. Then by Proposition 5.17.3 we have that $-3 \leq C(\alpha, \beta) \leq 3$ for all $\alpha, \beta \in \Phi_p$.

Lemma 5.17.6 *Let $\alpha, \beta \in \Phi_p$. Then $2\alpha(h_\beta) = C(\alpha, \beta)\beta(h_\beta)$.*

Proof. By Lemma 5.17.1 there are $\gamma, \delta \in \Phi$ such that $\alpha = \phi(\gamma)$, $\beta = \phi(\delta)$. Furthermore, if r, q are the smallest nonnegative integers such that $\alpha - (r+1)\beta$ and $\alpha + (q+1)\beta$ are not roots, then since ϕ maps sums of roots to sums of roots, $\gamma - r\delta, \dots, \gamma + q\delta$ is the δ-string containing γ. Hence by Proposition 5.4.2 we have that $r - q = \langle \gamma, \delta \rangle$. We calculate $2\alpha(h_\beta) = 2(\alpha, \beta)_p = 2(\phi(\gamma), \phi(\delta))_p = \phi(2(\gamma, \delta))$ (Lemma 5.17.5) $= \phi((r - q)(\delta, \delta)) = C(\alpha, \beta)\beta(h_\beta)$. $\square$

Now we define a total order on Φ_p. Since the Killing form of K_p is nondegenerate, by Proposition 4.3.6 K_p is a direct sum of simple ideals I. Since these ideals are simple they satisfy $[I, I] = I$. This implies that $[K_p, K_p] = K_p$. Also $[H_p, H_p] = 0$ and $[(K_p)_\alpha, (K_p)_\beta] \subset (K_p)_{\alpha+\beta}$ for $\alpha, \beta \in \Phi_p$. So H_p is spanned by the spaces $[(K_p)_\alpha, (K_p)_{-\alpha}]$ for $\alpha \in \Phi_p$. Now by Root fact 16, these spaces are 1-dimensional. So we can choose roots $\beta_1, \dots, \beta_l$ from Φ_p such that the spaces $[(K_p)_{\beta_i}, (K_p)_{-\beta_i}]$ for $1 \leq i \leq l$ span H_p. Let $\alpha \in \Phi_p$ and for $1 \leq i \leq l$ set $c_i(\alpha) = C(\alpha, \beta_i)$. If $\alpha, \beta \in \Phi_p$ then we define $\alpha > \beta$ if the first non-zero $c_i(\alpha) - c_i(\beta)$ is positive. It is immediate that this order is transitive. Suppose that $\alpha \geq \beta$ and $\beta \geq \alpha$, then we have to show that $\alpha = \beta$. But this is the same as asserting that $c_i(\alpha) = c_i(\beta)$ for $1 \leq i \leq l$ implies that $\alpha = \beta$. For $1 \leq i \leq l$ let g_i be a non-zero element of $[(K_p)_{\beta_i}, (K_p)_{-\beta_i}]$. Then g_i is a multiple of h_{β_i} by Root fact 11. Hence by Lemma 5.17.6, $c_i(\alpha) = c_i(\beta)$ implies that $2\alpha(g_i) = c_i(\alpha)\beta_i(g_i) = c_i(\beta)\beta_i(g_i) = 2\beta(g_i)$. As the g_i form a basis of H_p the result follows.

A root $\alpha \in \Phi_p$ is said to be positive if $\alpha > 0$. Furthermore, a positive root α is simple if there are no positive roots β, γ such that $\alpha = \beta + \gamma$. We let Δ_p be the set of all simple positive roots in Φ_p, and let Δ be the inverse image under ϕ of Δ_p. Furthermore, we call a root $\alpha \in \Phi$ positive if $\phi(\alpha)$ is positive. Then in the same way as in the proof of Lemma 5.5.3 we see that for $\alpha \neq \beta \in \Delta$ we have that $\alpha - \beta$ is not a root and hence $(\alpha, \beta) \leq 0$. Then by Proposition 5.4.1 it follows that Δ is linearly independent. Furthermore, in the same way as in the proof of Proposition 5.5.5 it follows that every positive root $\alpha \in \Phi$ can be written as a sum

$$\alpha = \sum_{\beta \in \Delta} k_\beta \beta$$

where the k_β are nonnegative integers. Now by Proposition 5.5.8 it follows that Δ is a simple system of Φ. Finally, suppose that $\Delta_p = \{\beta_1, \ldots, \beta_l\}$. Then since ϕ maps root strings to root strings we have that the matrix $(C(\beta_i, \beta_j))_{1 \leq i,j \leq l}$ is a Cartan matrix of Φ relative to Δ.

The above results lead to the following algorithm:

Algorithm Type
Input: a semisimple Lie algebra L over $\mathbb{Q}$.
Output: the type of L.

Step 1 Calculate a Cartan subalgebra H of L (Section 3.2).

Step 2 Extend a basis of H to a basis of L and multiply by an integer in order to ensure that all structure constants relative to this basis are integers.

Step 3 Calculate a splitting element $h_0 \in H$ having integer coefficients relative to the basis of H used in Step 2. Select a pleasant prime p.

Step 4 Let S be the table of structure constants obtained from the table of structure constants of L by reducing every constant modulo p. Let L_p be the Lie algebra with structure constants table S, defined over $\mathbb{F}_{p^m}$ where m is large enough to ensure that the characteristic polynomial of $\mathrm{ad}h_0$ splits into linear factors.

Step 5 Calculate a simple system Δ_p inside the root system Φ_p of L_p. Compute the Cartan matrix of Δ_p. From this matrix determine the type of L.

5.17.2   Isomorphism of semisimple Lie algebras

Here we present an algorithmic method to decide whether two semisimple Lie algebras are isomorphic. There is no attempt to construct the isomorphism (if it exists) as it may only exist over a large extension of the ground field.

Let L be a semisimple Lie algebra of dimension n with Cartan subalgebra H. Let $\{h_1, \ldots, h_l\}$ be a basis of H. Let $x_1, \ldots, x_l$ be indeterminates and set $h = \sum x_i h_i$ which is an element of $L \otimes F(x_1, \ldots, x_l)$. Let

$$f(T) = T^n + p_1(x_1, \ldots, x_l)T^{n-1} + \cdots + p_n(x_1, \ldots, x_l)$$

be the characteristic polynomial of $\mathrm{ad}h$. Then we call f the *characteristic polynomial of the action of H on L*.

Theorem 5.17.7 *Let L_1 and L_2 be semisimple Lie algebras over an algebraically closed field F of characteristic 0. Let H_1 and H_2 be Cartan subalgebras of L_1 and L_2, respectively. Suppose $\dim H_1 = \dim H_2 = l$ and $\dim L_1 = \dim L_2 = n$. Let*

$$f_1(T) = T^n + p_1(x_1, \ldots, x_l)T^{n-1} + \cdots + p_n(x_1, \ldots, x_l)$$

and

$$f_2(T) = T^n + q_1(y_1, \ldots, y_l)T^{n-1} + \cdots + q_n(y_1, \ldots, y_l)$$

be the characteristic polynomials of the action of H_1 (on L_1) and H_2 (on L_2), respectively. Then L_1 and L_2 are isomorphic if and only if there is a transformation

$$
\left\{
\begin{aligned}
\bar{y}_1 &= a_{11}x_1 + a_{12}x_2 + \cdots + a_{1l}x_l \\
&\;\vdots \\
\bar{y}_l &= a_{l1}x_1 + a_{l2}x_2 + \cdots + a_{ll}x_l
\end{aligned}
\right.
\tag{5.20}
$$

such that $\det(a_{ij}) \neq 0$ and $p_i(x_1, \ldots, x_l) = q_i(\bar{y}_1, \ldots, \bar{y}_l)$ for $1 \leq i \leq n$.

Proof. For $i = 1, 2$ let $L_i = V_i \oplus H_i$ be the Fitting decomposition of L_i, where V_i is the Fitting one-component of L_i with respect to H_i. Then V_i is the sum of the root spaces of L_i with respect to H_i. Suppose that L_1 and L_2 are isomorphic. Because all Cartan subalgebras of L_1, L_2 are conjugate under their respective automorphism groups (Theorem 3.5.1) there exists an isomorphism $L_1 \to L_2$ mapping H_1 onto H_2. Let $\alpha_1, \ldots, \alpha_r$ be the roots of L_1 and let $\{h_1, \ldots, h_l\}$ be a basis of H_1. Then

$$f_1(T) = T^l \prod_{i=1}^{r} (T - \alpha_i(h_1)x_1 - \cdots - \alpha_i(h_l)x_l).$$

A base change of V_1 does not affect the characteristic polynomial of the action of H_1. Hence we consider the effect of a base change in H_1 on the polynomial f_1. Suppose $\{\bar{h}_1, \ldots, \bar{h}_l\}$ is a second basis of H_1, where $\bar{h}_i = \sum a_{ki}h_k$. Then

$$T - \sum_{j=1}^{l} \alpha_i(\bar{h}_j)x_j = T - \alpha_i\Big(\sum_{k=1}^{l} a_{k1}h_k\Big)x_1 - \cdots - \alpha_i\Big(\sum_{k=1}^{l} a_{kl}h_k\Big)x_l$$

$$= T - \alpha_i(h_1)\sum_{k=1}^{l} a_{1k}x_k - \cdots - \alpha_i(h_l)\sum_{k=1}^{l} a_{lk}x_k$$

$$= T - \alpha_i(h_1)\bar{y}_1 - \cdots - \alpha_i(h_l)\bar{y}_l.$$

The conclusion is that a base change of H_1 corresponds exactly to a change of variables in the polynomials p_i. So $L_1 \cong L_2$ implies that there is a transformation of the form (5.20).

Now suppose that there is a transformation of the form (5.20). Let $\{\bar{h}_1, \ldots, \bar{h}_l\}$ be a basis of H_2. We define a linear map $\phi : H_1 \to H_2$ by

$$\phi(h_i) = \sum_{j=1}^{l} a_{ji}\bar{h}_j.$$

Then ϕ induces a map (which we also call ϕ) from H_1^* into H_2^*, given by $\phi(\alpha)(\phi(h)) = \alpha(h)$ for $\alpha \in H_1^*$ and $h \in H_1$. We claim that if α is a root of L_1, then $\phi(\alpha)$ is a root of L_2. To see this, note that a root β of L_2 corresponds to a factor

$$T - \beta(\bar{h}_1)y_1 - \cdots - \beta(\bar{h}_l)y_l$$

in f_2. Now choose β such that by the transformation (5.20) this factor is mapped to

$$T - \alpha(h_1)x_1 - \cdots - \alpha(h_l)x_l.$$

Then we calculate

$$T - \sum_{k=1}^{l} \beta(\bar{h}_k)\bar{y}_k = T - \beta(\bar{h}_1)\sum_{j=1}^{l} a_{1j}x_j - \cdots - \beta(\bar{h}_l)\sum_{j=1}^{l} a_{lj}x_j$$

$$= T - (\sum_{j=1}^{l} a_{j1}\beta(\bar{h}_j))x_1 - \cdots - (\sum_{j=1}^{l} a_{jl}\beta(\bar{h}_j))x_l.$$

It follows that

$$\alpha(h_i) = \sum_{j=1}^{l} a_{ji}\beta(\bar{h}_j) = \beta(\sum_{j=1}^{l} a_{ji}\bar{h}_j) = \beta(\phi(h_i)).$$

So on a basis of H_2 the functions $\phi(\alpha)$ and β have the same values, forcing $\phi(\alpha) = \beta$. The conclusion is that L_1 and L_2 have isomorphic root systems and hence $L_1 \cong L_2$ by Corollary 5.11.6. $\qquad\square$

The algorithm resulting from this is the following:

Algorithm: ArelsomorphicSSLieAlgebras
Input: two semisimple Lie algebras L_1 and L_2 over a field of characteristic 0.
Output: true if L_1 and L_2 are isomorphic over the algebraic closure of the ground field; false otherwise.

Step 1 If $\dim L_1 \neq \dim L_2$ then return false. Otherwise set $n = \dim L_1$.

Step 1 Calculate Cartan subalgebras H_1 and H_2 of L_1 and L_2. If $\dim H_1 \neq \dim H_2$ then return false. Otherwise set $l = \dim H_1$.

Step 2 Calculate the polynomials p_i and q_i (as in Theorem 5.17.7).

Step 3 Introduce the variables a_{jk} for $1 \leq j, k \leq l$ and substitute $\bar{y}_j = \sum a_{jk} x_k$ in the q_i. Require that the resulting polynomials are equal to the p_i. This yields a system of polynomial equations in the variables a_{jk}.

Step 4 Now by a Gröbner basis computation we can check whether there is a solution over the algebraic closure of the ground field to the system of equations we obtained in the preceding step. If there is such a solution then return true, otherwise return false.

Comment: Let R be the polynomial ring containing the indeterminates a_{jk}. Let I be the ideal of R generated by the polynomials obtained in Step 3. By Hilbert's Nullstellensatz these polynomials have common zeros over the algebraic closure of the ground field if and only if $I \neq R$, i.e., if and only if $1 \notin I$. Now Buchberger's algorithm for calculating a Gröbner basis of I presents an algorithmic method for checking whether $1 \in I$ (see, e.g., [17], [23]).

Example 5.17.8 Let L_1 be the Lie algebra of Example 4.12.2. This Lie algebra has basis $\{h_1, x_1, y_1, h_2, x_2, y_2\}$ and a Cartan subalgebra H_1 is spanned by $\{h_1, h_2\}$. The matrix of the restriction of $\zeta_1 \mathrm{ad}h_1 + \zeta_2 \mathrm{ad}h_2$ to the space spanned by $\{x_1, y_1, x_2, y_2\}$ is

$$\begin{pmatrix} 2\zeta_1 + 2\zeta_2 & 0 & 0 & 0 \\ 0 & -2\zeta_1 - 2\zeta_2 & 0 & 0 \\ 0 & 0 & 2\zeta_1 - 2\zeta_2 & 0 \\ 0 & 0 & 0 & -2\zeta_1 + 2\zeta_2 \end{pmatrix}.$$

It follows that the characteristic polynomial of the action of H_1 on L_1 is

$$f_1(T) = T^2(T - 2\zeta_1 - 2\zeta_2)(T + 2\zeta_1 + 2\zeta_2)(T - 2\zeta_1 + 2\zeta_2)(T + 2\zeta_1 - 2\zeta_2)$$
$$= T^6 + (-8\zeta_1^2 - 8\zeta_2^2)T^4 + (16\zeta_1^4 - 32\zeta_1^2\zeta_2^2 + 16\zeta_2^4)T^2.$$

Let L_2 be the Lie algebra of Example 4.12.3. This Lie algebra has basis $\{x_1, \ldots, x_6\}$ and a Cartan subalgebra H_2 is spanned by x_1, x_2. The matrix of the restriction of $\xi_1 \mathrm{ad} x_1 + \xi_2 \mathrm{ad} x_2$ to the span of $\{x_3, x_4, x_5, x_6\}$ is

$$\begin{pmatrix} 2\xi_2 & -2\xi_1 & 0 & 0 \\ 2\xi_1 & 2\xi_2 & 0 & 0 \\ 0 & 0 & -2\xi_2 & 2\xi_1 \\ 0 & 0 & -2\xi_1 & -2\xi_2 \end{pmatrix}.$$

Hence the characteristic polynomial of the action of the Cartan subalgebra H_2 of L_2 is

$$\begin{aligned} f_2(T) &= T^2(T^2 - 4\xi_2 T + 4\xi_1^2 + 4\xi_2^2)(T^2 + 4\xi_2 T + 4\xi_1^2 + 4\xi_2^2) \\ &= T^6 + (8\xi_1^2 - 8\xi_2^2)T^4 + (16\xi_1^4 + 32\xi_1^2\xi_2^2 + 16\xi_2^4)T^2 \end{aligned}$$

It is easily seen that the transformation $\zeta_1 = i\xi_1$, $\zeta_2 = \xi_2$ transports f_1 to f_2. So by Theorem 5.17.7, L_1 and L_2 are isomorphic over the algebraic closure of $\mathbb{Q}$.

Example 5.17.9 Again let L_2 be the Lie algebra of Example 4.12.3. We calculate the type of L_2 using the algorithm Type. As seen in Example 4.12.3, $x_1 + x_2$ is a splitting element with minimum polynomial $X(X^2 - 4X + 8)(x^2 + 4X + 8)$. We calculate a pleasant prime p. For that we first try $p = 5$. Over $\mathbb{F}_5$ the matrix of $\mathrm{ad}(x_1 + x_2)$ has minimum polynomial $X(X^4 - 1)$ which factors as $X(X + 1)(X - 1)(X + 2)(X - 2)$. So it is squarefree and of degree $\dim L_2 - \dim H_2 + 1$. Furthermore 5 does not divide the determinant of the matrix of the Killing form of L_2 (which is -2^{20}). The conclusion is that $p = 5$ is pleasant. Moreover, over $\mathbb{F}_5$ the minimum polynomial of $\mathrm{ad}(x_1 + x_2)$ splits into linear factors, so over this field L_2 splits as a direct sum of two ideals I_1 and I_2, where

$$I_1 = \langle x_5 + 2x_6, x_3 + 2x_4, x_1 + 3x_2 \rangle,$$

and

$$I_2 = \langle x_3 + 3x_4, x_5 + 3x_6, x_1 + 2x_2 \rangle.$$

The roots of I_1 and I_2 are easily calculated; it is seen that the type of L_2 is $A_1 + A_1$.

5.18   Notes

Most of the material in this chapter is fairly standard. In Sections 5.12 to 5.15 we have followed [49]. The idea of proving the existence of the semisimple Lie algebras by exhibiting a choice for the constants $N_{\alpha,\beta}$ was also pursued in [83], where the construction was performed in a uniform manner for all root systems. In [18] a different construction of the $N_{\alpha,\beta}$ is described. There the existence of the simple Lie algebras is assumed. It is shown that once the $N_{\alpha,\beta}$ are chosen for the so-called extraspecial pairs (α, β), then all other constants $N_{\gamma,\delta}$ can be determined.

Lemma 5.17.2 and Proposition 5.17.3 as well as the ordering of Φ_p used in Section 5.17.1, are taken from [76]. The algorithm Type from Section 5.17.1 is taken from [33].

Chapter 6

Universal enveloping algebras

Universal enveloping algebras are a basic tool for studying representations of Lie algebras. Let L be a Lie algebra with basis $\{x_1, \ldots, x_n\}$ and let $\rho : L \to \mathfrak{gl}(V)$ be a representation of L. Let $x, y \in L$ then the product $\rho(x)\rho(y)$ is in general not contained in $\rho(L)$. However on many occasions we compose the mappings $\rho(x)$ and $\rho(y)$ (see for instance Section 5.1). The natural framework for doing this is the associative algebra generated by the identity mapping together with $\rho(x)$ for $x \in L$. This associative algebra is called an *enveloping algebra* of L; it is denoted by $\rho(L)^*$. The algebra $\rho(L)^*$ is generated by the identity together with $\rho(x_i)$ for $1 \le i \le n$. Among others, the generators satisfy the relations $\rho(x_i)\rho(x_j) = \rho(x_j)\rho(x_i) + \rho([x_i, x_j])$. And these relations do not depend on the particular representation ρ (but follow from the definition of the concept of representation).

Now the universal enveloping algebra of L is the associative algebra with 1 generated by n abstract symbols, which we also call $x_1, \ldots, x_n$, subject to the relations $x_i x_j - x_j x_i - [x_i, x_j]$. Then any enveloping algebra of L is a quotient of the universal enveloping algebra. So in this sense it contains all enveloping algebras of L.

In Section 6.1 we study ideals in the free associative algebra. We define a special type of generating set of an ideal called Gröbner basis. A criterion for a set to be a Gröbner basis is derived. Then in Section 6.2 we define universal enveloping algebras and we use the criterion of Section 6.1 to prove the Poincaré-Birkhoff-Witt theorem that provides a convenient basis of the universal enveloping algebra. Also in this section we show that representations of a Lie algebra are in one-to-one correspondence with representations of its universal enveloping algebra.

In Section 6.3 we give a criterion for a set of elements of the universal enveloping algebra of a Lie algebra to be a Gröbner basis. This yields an algorithm for calculating a Gröbner basis of an ideal in the universal enveloping algebra. Also we use this criterion to give a proof of Proposition 1.13.4 (that contains a necessary and sufficient condition for a Lie algebra of characteristic $p > 0$ to be restricted).

In Section 6.5 we give an algorithm for constructing a faithful finite-dimensional representation of a finite-dimensional Lie algebra of characteristic 0. By doing this we obtain a proof of Ado's theorem, which asserts that such a representation always exists. Finally in Section 6.6 we prove the corresponding statement for Lie algebras of characteristic $p > 0$, which is known as Iwasawa's theorem.

6.1   Ideals in free associative algebras

Let $X = \{x_1, x_2, \dots\}$ be a set. Then by X^* we denote the set of all words $w = x_{i_1} x_{i_2} \cdots x_{i_k}$ on the elements of X. This includes the empty word which is denoted by 1. The set X^* is called the *free monoid* on X. It is endowed with a binary operation $\cdot : X^* \times X^* \to X^*$ which is defined by $u \cdot v = uv$ (concatenation). Furthermore, if $w = x_{i_1} x_{i_2} \cdots x_{i_k} \in X^*$ then its *degree* is $\deg(w) = k$.

We say that $u \in X^*$ is a *factor* of $v \in X^*$ if there are $w_1, w_2 \in X^*$ such that $w_1 u w_2 = v$ (i.e., u is a subword of v). Also u is a *left factor* of v if $v = u w_2$ and a *right factor* of v if $v = w_1 u$.

Throughout this section we suppose that X^* has a total order $<$ which is *multiplicative* (i.e., $u < v$ implies $wu < wv$ and $uw < vw$ for all $w \in X^*$) and satisfies the *descending chain condition* (i.e., if $w_1 \geq w_2 \geq \cdots$ is a descending chain of words, then there is a $k > 0$ such that $w_k = w_{k+1} = \cdots$). As an example we mention the *deglex* order $<_{\text{dlex}}$. It is defined by $u <_{\text{dlex}} v$ if and only if $\deg(u) < \deg(v)$ or $\deg(u) = \deg(v)$ and $u = w x_i u'$, $v = w x_j v'$ where $w, u', v' \in X^*$ and $i < j$.

Now let F be a field and let $F\langle X \rangle$ be the vector space spanned by X^*. Extend the operation $\cdot$ bilinearly to $F\langle X \rangle$. Then $F\langle X \rangle$ becomes an associative algebra with 1; it is called the *free associative algebra with* 1 over F. The elements of X^* are called *monomials*. And for $f \in F\langle X \rangle$ we define its *support* to be the set of all monomials that occur in f with non-zero coefficient. Furthermore, the order $<$ yields the notion of *leading monomial* of $f \in F\langle X \rangle$, which is the biggest element of the support of f. It is denoted by $\mathrm{LM}(f)$. Also for a subset $S \subset F\langle X \rangle$ we set $\mathrm{LM}(S) = \{\mathrm{LM}(f) \mid f \in S\}$.

Let I be an ideal of $F\langle X \rangle$. We consider the problem of computing inside the quotient $F\langle X \rangle / I$. We need a basis of $F\langle X \rangle / I$, i.e., a set of representatives of all cosets of $F\langle X \rangle$ modulo I. This means that we are looking for a set $B \subset F\langle X \rangle$ such that B spans a complement to I in $F\langle X \rangle$. Furthermore we need to be able to express products of basis elements as linear combinations of basis elements modulo I. This problem is solved by the set of *normal words* of $F\langle X \rangle$ modulo I, i.e., the set

$$N(I) = \{u \in X^* \mid u \notin \mathrm{LM}(I)\}.$$

We let $C(I)$ be the vector space spanned by $N(I)$.

Proposition 6.1.1 *Let $I \subset F\langle X \rangle$ be an ideal. Then $F\langle X \rangle = C(I) \oplus I$.*

Proof. It is clear that $C(I) \cap I = 0$. Let $f \in F\langle X \rangle$, then we prove that $f = v + p$ for some $v \in C(I)$ and $p \in I$. By induction we may assume the result for all $h \in F\langle X \rangle$ such that $\mathrm{LM}(h) < \mathrm{LM}(f)$ (by the descending chain condition there are only a finite number of monomials smaller than $\mathrm{LM}(f)$, so induction is allowed). Write $f = \lambda \mathrm{LM}(f) + h$ where $\lambda \in F$ and $\mathrm{LM}(h) < \mathrm{LM}(f)$. Then by induction $h = v + p$ for a $v \in C(I)$ and a $p \in I$. So if $\mathrm{LM}(f) \in N(I)$ then we are done. However, if this is not the case then $\mathrm{LM}(f) = \mathrm{LM}(g)$ for some $g \in I$. Write $g = \mu \mathrm{LM}(g) + h'$ and by induction the element $f - \frac{\lambda}{\mu} g$ is equal to $u + q$ where $u \in C(I)$ and $q \in I$. Therefore $f = u + q + \frac{\lambda}{\mu} g$ and we are done.   □

Now let $f \in F\langle X \rangle$, then by Proposition 6.1.1 f has a unique expression as $f = v + p$ for $v \in C(I)$ and $p \in I$. The element $v \in C(I)$ is called the *normal form* of f modulo I. It is denoted by $\mathrm{Nf}_I(f)$, or if from the context it is clear which ideal we mean, by $\mathrm{Nf}(f)$.

Let $u, v \in C(I)$ and set $u * v = \mathrm{Nf}(uv)$. Then $C(I)$ together with $*$ becomes an algebra. It is immediate that this algebra is isomorphic to $F\langle X \rangle / I$. So if we have a method for computing normal forms, then we can conveniently calculate in $F\langle X \rangle / I$.

Let $G \subset F\langle X \rangle$ be a set generating an ideal I of $F\langle X \rangle$. An element f of $F\langle X \rangle$ is said to be in *normal form* modulo G if for all $g \in G$, $\mathrm{LM}(g)$ is not a factor of any monomial occurring in f. As a first attempt at calculating normal forms of elements of $F\langle X \rangle$ modulo I we consider the following algorithm for calculating normal forms modulo the generating set G.

Algorithm NormalForm
Input: a generating set G of an ideal $I \subset F\langle X \rangle$, and an element $f \in F\langle X \rangle$.

Output: an element $\phi \in F\langle X \rangle$ that is in normal form modulo G and such that $f = \phi \bmod I$.

Step 1 Set $\phi := 0$, $h := f$.

Step 2 If $h = 0$ then return ϕ. Otherwise set $u := \mathrm{LM}(h)$ and let λ be the coefficient of u in h.

Step 3 Let $g \in G$ be such that $\mathrm{LM}(g)$ is a factor of u. If there is no such g then set $h := h - \lambda u$, $\phi := \phi + \lambda u$ and return to Step 2.

Step 4 Let μ be the coefficient of $\mathrm{LM}(g)$ in g, and let $v, w \in X^*$ be such that $v\mathrm{LM}(g)w = u$. Set $h := h - \frac{\lambda}{\mu}vgw$. Return to Step 2.

Comments: We note that $\mathrm{LM}(h)$ decreases every step. So because $<$ satisfies the descending chain condition, the algorithm terminates. Only monomials that do not have a $\mathrm{LM}(g)$ as a factor for $g \in G$ are added to ϕ. So it is clear that upon termination ϕ is in normal form with respect to G. Furthermore, at any stage during the algorithm we have that $f = h + \phi \bmod I$, so at termination $f = \phi \bmod I$.

We can reformulate the algorithm NormalForm as follows. Let $h \in F\langle X \rangle$ and let $w \in X^*$ be an element of the support h such that $\mathrm{LM}(g)$ is a factor of w for some $g \in G$. Let $u, v \in X^*$ be such that $u\mathrm{LM}(g)v = w$. Then we say that h *reduces* to h', where $h' = h - \frac{\lambda}{\mu}ugv$ and λ and μ are the coefficients of w and $\mathrm{LM}(g)$ in h and g respectively. More generally, if $h_1, \ldots, h_k$ are such that h_i reduces to h_{i+1} for $1 \leq i \leq k - 1$, then we also say that h_1 reduces to h_k, and we write $h_1 \to h_k$. Now the algorithm NormalForm basically performs a series of reduction steps, and if no further reductions are possible it returns what is left.

As the following example demonstrates, this algorithm may not return a normal form modulo the ideal I.

Example 6.1.2 Let $X = \{x, y\}$ and $G = \{xy - x^2, x^3 - yx, y^3\}$ and let the order be $<_{\mathrm{dlex}}$, with $x<_{\mathrm{dlex}}y$. So $\mathrm{LM}(G) = \{xy, x^3, y^3\}$. We consider calculating the normal form of $f = x^2y^2$. Since $f = x(xy)y$ we see that $f \to x^3y$. Continuing like this we find $x^3y \to yxy \to yx^2$. This last monomial is in normal form with respect to G. Hence we output it. However,

$$-(xy - x^2)y^2 + x(y^3) = x^2y^2.$$

So x^2y^2 lies in the ideal I generated by G, so that $\mathrm{Nf}_I(f) = 0$. Also in the algorithm we often have a choice of $g \in G$ such that $\mathrm{LM}(g)$ is a factor of $\mathrm{LM}(h)$. In this particular case we could have made the following series

$x^3y = x^2(xy) \to x^4 = x(x^3) \to xyx \to x^3 \to yx$ and the output is yx.
We see that the algorithm does not give a unique output. However, in the
following we show that if the set G has the property of being a Gröbner
basis, then the output of $\mathsf{NormalForm}(G, f)$ is unique (and equals $\mathrm{Nf}(f)$).

Definition 6.1.3 *Let I be an ideal of $F\langle X \rangle$. Then $G \subset I$ is called a
Gröbner basis of I if for all $f \in I$ there is a $g \in G$ such that $\mathrm{LM}(g)$ is a
factor of $\mathrm{LM}(f)$.*

Theorem 6.1.4 *Let $G \subset F\langle X \rangle$ be a Gröbner basis of the ideal I. Then
$N(I)$ is the set of all words $w \in X^*$ such that for all $g \in G$, $\mathrm{LM}(g)$ is
not a factor of w. Furthermore, $\mathsf{NormalForm}(G, f)$ returns $\mathrm{Nf}_I(f)$ for all
$f \in F\langle X \rangle$.*

Proof. The first assertion is a direct consequence of the definition of
Gröbner basis. Let $f \in F\langle X \rangle$ and set $h = \mathsf{NormalForm}(G, f)$. Let $w \in X^*$
be a monomial occurring in h. Then there is no $g \in G$ such that $\mathrm{LM}(g)$
is a factor of w. Hence by definition of Gröbner basis, w is not a leading
monomial of an element of I. So $w \in N(I)$. Now from $f = h \bmod I$ it
follows that $h = \mathrm{Nf}(I)$.   □

Definition 6.1.5 *Let $G \subset F\langle X \rangle$, then G is called self-reduced if firstly for
$g \neq h \in G$ we have that $\mathrm{LM}(g)$ is not a factor of $\mathrm{LM}(h)$ and secondly the
coefficient of $\mathrm{LM}(g)$ in g is 1 for all $g \in G$.*

The second condition is added to avoid cumbersome notation. This
condition does not amount to much: if we divide all $g \in G$ by a suitable
element $\lambda_g \in F$, then we obtain a set G' satisfying the second condition and
generating the same ideal as G.

Let I be an ideal of $F\langle X \rangle$ generated by $G = \{g_1, g_2, \dots\}$. Then for
$w \in X^*$ we set

$$I_{<w} = \left\{ \sum_i \lambda_i u_i g_{k_i} v_i \mid u_i, v_i \in X^* \text{ such that } u_i \mathrm{LM}(g_{k_i}) v_i < w \right\},$$

i.e., the vector space spanned by all elements of the form ugv for $u, v \in X^*$
and $g \in G$ such that $\mathrm{LM}(ugv) < w$. Note that $I_{<w}$ depends on the particular
generating set G. However, it will always be clear what generating set we
mean.

Theorem 6.1.6 *Let I be an ideal of $F\langle X\rangle$ generated by a self-reduced set G. Then G is a Gröbner basis of I if and only if for all $g_1, g_2 \in G$ and $u, v \in X^*$ such that $\mathrm{LM}(g_1)u = v\mathrm{LM}(g_2)$ we have that $g_1 u - v g_2 \in I_{<t}$, where $t = \mathrm{LM}(g_1)u$.*

Proof. First suppose that G is a Gröbner basis and set $f = g_1 u - v g_2$. Then $f \in I$, so $\mathrm{Nf}(f) = 0$. Furthermore $\mathrm{LM}(f) < t$ and in the algorithm Normal-Form we subtract elements $\frac{\lambda}{\mu} w_1 g w_2$ from h until we reach 0 (cf. Theorem 6.1.4). And for each such element we have $\mathrm{LM}(w_1 g w_2) \leq \mathrm{LM}(f)$; hence $f \in I_{<t}$.

Now we prove the other direction. Let $f \in I$; then f can be written as

$$f = \sum_{i=1}^{n} \lambda_i u_i g_i v_i$$

for some $u_i, v_i \in X^*$ and $g_i \in G$. We prove that there is a $g \in G$ such that $\mathrm{LM}(g)$ divides $\mathrm{LM}(f)$. Set $s_i = \mathrm{LM}(u_i g_i v_i)$ and suppose that the summands are ordered such that

$$s_1 = s_2 = \ldots = s_k > s_{k+1} \geq \cdots \geq s_n.$$

If $k = 1$, then $\mathrm{LM}(f) = u_1 \mathrm{LM}(g_1) v_1$ and we are done. So suppose that $k > 1$ and write

$$f = \lambda_1(u_1 g_1 v_1 - u_2 g_2 v_2) + (\lambda_1 + \lambda_2) u_2 g_2 v_2 + \sum_{i=3}^{n} \lambda_i u_i g_i v_i. \qquad (6.1)$$

Set $w_i = \mathrm{LM}(g_i)$; then $u_1 w_1 v_1 = u_2 w_2 v_2$. Now if $u_1 = u_2$, then $w_1 v_1 = w_2 v_2$ and it is seen that the shorter of w_1, w_2 is a factor of the longer one. But this is excluded since G is self-reduced. It follows that $u_1 \neq u_2$.

Now we suppose that u_1 is shorter than u_2. In this case $u_2 = u_1 u_2'$ where $u_2' \neq 1$. Hence $w_1 v_1 = u_2' w_2 v_2$. Here v_1 must be longer than v_2 because otherwise w_2 is a factor of w_1. So $v_1 = v_1' v_2$ where $v_1' \neq 1$. Now by assumption $g_1 v_1' - u_2' g_2 \in I_{<w_1 v_1'}$ and hence $u_1(g_1 v_1' - u_2' g_2) v_2 = u_1 g_1 v_1 - u_2 g_2 v_2 \in I_{<u_1 w_1 v_1}$. So we can rewrite (6.1) and obtain an expression of the same form where $\mathrm{LM}(u_1 g_1 v_1)$ has decreased (in the case where $k = 2$ and $\lambda_1 + \lambda_2 = 0$) or k has decreased.

If u_1 is longer than u_2, then we reach the same conclusion by similar arguments. So because $<$ satisfies the descending chain condition after a finite number of steps we reach an expression for of of the form (6.1) where $k = 1$ and we are done. $\qquad\square$

Let G be a self-reduced generating set for an ideal I of $F\langle X\rangle$. By Theorem 6.1.6 we see that in order to check that G is a Gröbner basis we need to verify whether $g_1 u - v g_2 \in I_{<t}$ for all $g_1, g_2 \in G$ and $u, v \in X^*$ such that $t = \mathrm{LM}(g_1) u = v \mathrm{LM}(g_2)$. It is easily seen that there are infinitely many such u, v. Indeed, set $w_i = \mathrm{LM}(g_i)$ and $u = s w_2$, $v = w_1 s$ for arbitrary $s \in X^*$. Then $w_1 u = v w_2$. However, if we write $g_i = w_i + \tilde{g}_i$, then

$$g_1 u - v g_2 = \tilde{g}_1 s w_2 - w_1 s \tilde{g}_2 = \tilde{g}_1 s g_2 - g_1 s \tilde{g}_2 \qquad (6.2)$$

which lies in $I_{<t}$ (where $t = w_1 s w_2$). As a consequence we need not be bothered about u and v such that u is longer than or equal to w_2 and v is longer than or equal to w_1. The pairs (u, v) that remain are such that u is a proper right factor of w_2 and v is a proper left factor of w_1. And of those there are finitely many.

Definition 6.1.7 *Let* $g_1, g_2 \in F\langle X\rangle$ *and let* $w_i = \mathrm{LM}(g_i)$ *for* $i = 1, 2$. *Suppose that the coefficients of* w_1, w_2 *in* g_1, g_2 *respectively are* 1. *Suppose further that* w_1 *is not a factor of* w_2 *and* w_2 *is not a factor of* w_1. *Let* $u, v \in X^*$ *be such that* $w_1 u = v w_2$ *and* u *is a proper right factor of* w_2 *and* v *is a proper left factor of* w_1. *Then the element* $g_1 u - v g_2$ *is called a* composition *of* g_1 *and* g_2.

Corollary 6.1.8 *Let* I *be an ideal of* $F\langle X\rangle$ *generated by a self-reduced set* G. *Then* G *is a Gröbner basis of* I *if and only if* $\mathsf{NormalForm}(G, f)$ *is zero for all compositions* f *of the elements of* G.

Proof. If G is a Gröbner basis of I then $\mathsf{NormalForm}(G, f)$ is zero for all compositions f because all such f are in I. On the other hand, let $g_1, g_2 \in G$ and set $f = g_1 u - v g_2$ where $u, v \in X^*$ are such that $\mathrm{LM}(g_1) u = v \mathrm{LM}(g_2)$. Set $t = \mathrm{LM}(g_1) u = v \mathrm{LM}(g_2)$. If u is longer than $\mathrm{LM}(g_2)$ (and hence v is longer than $\mathrm{LM}(g_1)$), then by (6.2), $f \in I_{<t}$. On the other hand, if u is shorter than $\mathrm{LM}(g_2)$ and v is shorter than $\mathrm{LM}(g_1)$, then by assumption $\mathsf{NormalForm}(G, f)$ is zero. Now in Step 4 of the algorithm we subtract elements of the form $\frac{\lambda}{\mu} w_1 g w_2$ from the element h. But $\mathrm{LM}(w_1 g w_2) = \mathrm{LM}(h) \leq \mathrm{LM}(f) < t$. After a finite number of steps this reaches zero, hence $f \in I_{<t}$. Now by Theorem 6.1.6 we have that G is a Gröbner basis. $\qquad\square$

Example 6.1.9 Corollary 6.1.8 yields a straightforward algorithm for computing a Gröbner basis of an ideal I generated by a finite set G. First of all we remark that because G is finite it is straightforward to compute a finite self-reduced set G' generating the same ideal as G. So we may suppose that

G is self-reduced. Then we consider all compositions f of elements of G. If for such an f we have that $\mathsf{NormalForm}(G, f) \neq 0$, then we add this normal form to G. If necessary we perform some reductions to keep G self-reduced. If this procedure finishes, then we have a Gröbner basis. We do not describe this algorithm in greater detail, but illustrate it with an example instead. Let $X = \{x, y\}$ and $G = \{g_1 = xy - x^2, g_2 = x^3 - yx, g_3 = y^3\}$. As order we take the deglex order $<_{\mathrm{dlex}}$, where $x <_{\mathrm{dlex}} y$. First g_1 has no composition with itself, and neither with g_2; but it has a composition with g_3:

$$g_1 \cdot y^2 - x \cdot g_3 = -x^2 y^2.$$

As seen in Example 6.1.2, $\mathsf{NormalForm}(G, f) = yx$; so we add $g_4 = yx$ to G. Now it is not difficult to check that all compositions of the elements $g_1, \ldots, g_4$ reduce to zero modulo G. Hence G is a Gröbner basis.

Unfortunately however, it is by no means guaranteed that this procedure finishes, because a Gröbner basis might be infinite. To illustrate this let $X = \{x, y\}$ and I the ideal of $F\langle X \rangle$ generated by $f_1 = xyx - yx$. Here f_1 has a composition with itself

$$f_1 yx - xy f_1 = xyyx - yxyx$$

which modulo f_1 reduces to $xyyx - yyx$. Set $f_n = xy^n x - y^n x$, then the composition of f_k with f_l reduces to f_{k+l} (modulo f_k, f_l). Hence the Gröbner basis relative to $<_{\mathrm{dlex}}$ is equal to the set of all f_n for $n \geq 1$.

6.2   Universal enveloping algebras

Let L be a Lie algebra over the field F with basis $B = \{x_1, x_2, \ldots\}$ (so L is not necessarily finite-dimensional). Let X be a set of symbols in bijection with B and let $\phi : B \to X$ realize the bijection. Extend ϕ to a map $\phi : L \to F\langle X \rangle$ by linearity. Denote the image $\phi(x_i)$ by $\bar{x}_i$. Now let I be the ideal of $F\langle X \rangle$ generated by the elements

$$g_{ij} = \bar{x}_j \bar{x}_i - \bar{x}_i \bar{x}_j - \phi([x_j, x_i]) \quad \text{for } 1 \leq i < j.$$

Then the *universal enveloping algebra* $U(L)$ of L is defined to be the quotient $F\langle X \rangle / I$. Let $\pi : F\langle X \rangle \to U(L)$ be the projection map, then the map

$$i : L \xrightarrow{\phi} F\langle X \rangle \xrightarrow{\pi} U(L)$$

maps L into $U(L)$.

Theorem 6.2.1 (Poincaré-Birkhoff-Witt) *A basis of $U(L)$ is formed by the monomials*

$$\bar{x}_{i_1}^{m_1} \cdots \bar{x}_{i_k}^{m_k} \quad \text{for } k \geq 0 \text{ and } i_1 < i_2 < \cdots < i_k,$$

(where such a monomial is understood to be 1 if $k = 0$).

Proof. Let G be the generating set of I consisting of the elements g_{ij} for $1 \leq i < j$. We choose the order on X^* to be $<_{\text{dlex}}$ where $\bar{x}_i <_{\text{dlex}} \bar{x}_j$ if $i < j$. Then $\text{LM}(g_{ij}) = \bar{x}_j \bar{x}_i$. We note that G is self-reduced. We consider the possible compositions of elements g_{ij} and g_{kl}. It is straightforward to see that these elements can only have a composition when $\bar{x}_i = \bar{x}_l$. Then for $u = \bar{x}_k$ and $v = \bar{x}_j$ we have

$$g_{ij}u - vg_{ki} = \bar{x}_j\bar{x}_k\bar{x}_i - \bar{x}_i\bar{x}_j\bar{x}_k - \phi([x_j, x_i])\bar{x}_k + \bar{x}_j\phi([x_i, x_k]).$$

Now we calculate the normal form of this element modulo the elements of G. We note that $k < i < j$; so $\bar{x}_j\bar{x}_k$ is the leading monomial of g_{kj}. Hence $\bar{x}_j\bar{x}_k\bar{x}_i$ reduces to $\bar{x}_k\bar{x}_j\bar{x}_i + \phi([x_j, x_k])\bar{x}_i$. The first monomial of this expression has $\bar{x}_j\bar{x}_i$ as factor so it reduces again. As a consequence in two steps $\bar{x}_j\bar{x}_k\bar{x}_i$ reduces to

$$\bar{x}_k\bar{x}_i\bar{x}_j + \bar{x}_k\phi([x_j, x_i]) + \phi([x_j, x_k])\bar{x}_i.$$

In similar fashion $\bar{x}_i\bar{x}_j\bar{x}_k$ reduces to

$$\bar{x}_k\bar{x}_i\bar{x}_j + \bar{x}_i\phi([x_j, x_k]) + \phi([x_i, x_k])\bar{x}_j.$$

It follows that $g_{ij}u - vg_{ki}$ reduces to

$$\bar{x}_k\phi([x_j, x_i]) + \phi([x_j, x_k])\bar{x}_i - \bar{x}_i\phi([x_j, x_k]) -$$
$$\phi([x_i, x_k])\bar{x}_j - \phi([x_j, x_i])\bar{x}_k + \bar{x}_j\phi([x_i, x_k]). \quad (6.3)$$

Let V be the span of X inside $F\langle X \rangle$. Let $v \in V$, then modulo G the element $\bar{x}_r v - v\bar{x}_r$ reduces to $\phi([x_r, \phi^{-1}(v)])$. This is clear if $v = \bar{x}_s$, and for general v it follows by linearity. In particular we see that (6.3) reduces modulo G to

$$\phi([x_k[x_j, x_i]]) + \phi([[x_j, x_k], x_i]) + \phi([x_j, [x_i, x_k]]),$$

but this is zero by the Jacobi identity. The conclusion is that all compositions of elements of G reduce to 0 modulo G. Hence by Corollary 6.1.8, G is a Gröbner basis of I. So the normal words of $F\langle X \rangle$ modulo I are those monomials that do not contain a $\text{LM}(g_{ij})$ as a factor. Now the theorem follows by Proposition 6.1.1. $\qquad\square$

Corollary 6.2.2 *The map* $i : L \to U(L)$ *is injective.*

Proof. By Theorem 6.2.1 the elements $\bar{x}_i$ for $i \geq 1$ are linearly independent modulo I. Hence i is injective. $\qquad\qquad\qquad\qquad\qquad\qquad\qquad\qquad\square$

By the last corollary we may identify L with its image in $U(L)$. To ease notation a little we denote the symbol $\bar{x}_i \in X$ also by x_i. Since L can be viewed as a subspace of $U(L)$ there can be no confusion as to what we mean by this. Furthermore, since $U(L)$ is generated by X it is also generated by L.

Definition 6.2.3 *Let L be a Lie algebra with basis $\{x_1, x_2, \dots\}$. Then an element $x_{i_1}^{k_1} \cdots x_{i_n}^{k_n}$ of $U(L)$ is called a* standard monomial.

Let $\rho : L \to \mathfrak{gl}(V)$ be a representation of L and let $A = \rho(L)^*$ be the corresponding enveloping algebra. We extend ρ to a map $\rho : U(L) \to A$ by setting

$$\rho(x_{i_1}^{m_1} \cdots x_{i_k}^{m_k}) = \rho(x_{i_1})^{m_1} \cdots \rho(x_{i_k})^{m_k}.$$

Then ρ is a surjective algebra homomorphism. So $A \cong U(L)/J$ where $J = \ker \rho$; i.e., any enveloping algebra of L is a quotient of the universal enveloping algebra of L. Furthermore ρ makes V into an $U(L)$-module by $a \cdot v = \rho(a)v$ for $a \in U(L)$ and $v \in V$. Also any algebra homomorphism $U(L) \to A$ whose restriction to L equals ρ must be equal to ρ since L generates $U(L)$. It follows that there is one and only one way to extend a representation from L to $U(L)$. Moreover, if $\rho : U(L) \to \mathfrak{gl}(V)$ is a representation of $U(L)$ then the restriction of ρ to L is a representation of L. It follows that the representations of L are in bijection with the representations of $U(L)$.

Example 6.2.4 When calculating inside $U(L)$ we often compute the normal form of products of basis elements, i.e., we rewrite a product of the form

$$x_{i_1}^{m_1} \cdots x_{i_k}^{m_k} \cdot x_{j_1}^{n_1} \cdots x_{j_l}^{n_l}$$

as a linear combination of standard monomials. We do this using the algorithm NormalForm. However in this case this algorithm boils down to a simple procedure which we call CollectionInUEA (in analogy with the collection algorithm for polycyclic groups, see [80]). The proof of Theorem 6.2.1 already shows how it goes: whenever we encounter in a monomial w some factor of the form $x_j x_i$ with $i < j$, then we replace it by $x_i x_j$ and add the terms obtained from w by replacing the occurrence of $x_j x_i$ by $[x_j, x_i]$.

We do not formulate the algorithm in great detail, but illustrate it with an example. Let L be the 3-dimensional Lie algebra with basis $\{x, y, h\}$ and multiplication table

$$[x, y] = h, \ [h, x] = 2x, \ [h, y] = -2y.$$

(See Example 1.3.6). Then in $U(L)$, $hy = yh - 2y$, and $hy^2 = (hy)y = (yh - 2y)y = y(hy) - 2y^2 = y(yh - 2y) - 2y^2 = y^2h - 4y^2$. By induction we find that $hy^i = y^ih - 2iy^i$. Now let $\rho : L \to \mathfrak{gl}(V)$ be a finite-dimensional representation of L and let $v_0 \in V$ be a highest-weight vector of weight λ (see Definition 5.1.1). Set $v_i = \rho(y)^iv_0$. Then since V is also a $U(L)$-module we can write $h \cdot v_i = hy^i \cdot v_0 = y^ih \cdot v_0 - 2iy^i \cdot v_0 = (\lambda - 2i)v_i$. Which is in accordance with Lemma 5.1.2.

6.3   Gröbner bases in universal enveloping algebras

Throughout this section L will be a finite-dimensional Lie algebra over the field F with basis $\{x_1, \ldots, x_t\}$. In similar fashion as in Section 6.1 we define the notion of a Gröbner basis of an ideal of $U(L)$ and we show how to calculate a Gröbner basis. We remark that although we formulate all results for finite-dimensional Lie algebras, the same proofs hold in the infinite-dimensional case. However in the latter case the indexing of the basis elements is a little tedious.

By the Poincaré-Birkhoff-Witt theorem, a basis of $U(L)$ is formed by the standard monomials $x_1^{k_1} \cdots x_t^{k_t}$. We define the *degree* of a standard monomial $m = x_1^{k_1} \cdots x_t^{k_t}$ to be the number $\deg(m) = k_1 + \cdots + k_t$. Throughout this section we suppose that $<$ is a total order of the set of standard monomials. With respect to this order we define the *leading monomial* LM(f) of $f \in U(L)$ to be the largest standard monomial occurring in f with non-zero coefficient. Furthermore we suppose that the order is *multiplicative* (i.e., if $m_1 < m_2$ then LM$(nm_1p) <$ LM(nm_2p) for all standard monomials n, p) and satisfies the *descending chain condition* (i.e., there are no infinite strictly decreasing chains of standard monomials) and is *degree-compatible* (meaning that $\deg(m) < \deg(n)$ implies $m < n$). An example of such an order is the *deglex* order $<_{\text{dlex}}$. Set $m = x_1^{k_1} \cdots x_t^{k_t}$ and $n = x_1^{l_1} \cdots x_t^{l_t}$. Then $m <_{\text{dlex}} n$ if $\deg(m) < \deg(n)$. And if the degrees of m, n are equal then $m <_{\text{dlex}} n$ if the first non-zero entry in $(k_1 - l_1, \ldots, k_t - l_t)$ is positive. This order is clearly degree compatible and satisfies the descending chain condition. The fact that it is multiplicative follows from the following lemma.

Lemma 6.3.1 *Let the standard monomials be ordered by a degree compatible order* $<$. *Let* $m = x_1^{k_1} \cdots x_n^{k_t}$ *and* $n = x_1^{l_1} \cdots x_n^{l_t}$ *be standard monomials. Then* $\mathrm{LM}(mn) = x_1^{k_1+l_1} \cdots x_n^{k_t+l_t}$.

Proof. This follows from the fact that $<$ is degree compatible. Indeed, in the collection process we substitute occurrences of $x_j x_i$ where $j > i$ by $x_i x_j + [x_j, x_i]$. So the output will consist of the monomial $x_1^{k_1+l_1} \cdots x_t^{k_t+l_t}$ plus terms of lower degree.   □

We say that a standard monomial m is a *factor* of a standard monomial n if there are standard monomials p, q such that $\mathrm{LM}(pmq) = n$.

Example 6.3.2 Let L be the 3-dimensional Lie algebra with basis $\{x, y, h\}$ considered in Example 6.2.4. Let the order be $<_{\mathrm{dlex}}$. Then xh is a factor of xyh because $xh \cdot y = xyh - 2xy$ and the leading monomial of the last expression is xyh.

Let m, n be given standard monomials. The next lemma gives a criterion by which we can decide whether m is a factor of n. It is a direct consequence of Lemma 6.3.1.

Lemma 6.3.3 *Let* $m = x_1^{k_1} \cdots x_t^{k_t}$ *and* $n = x_1^{l_1} \cdots x_t^{l_t}$ *be standard monomials. Then* m *is a factor of* n *if and only if* $k_i \leq l_i$ *for* $1 \leq i \leq t$.

Let $G \subset U(L)$ generate the ideal I of $U(L)$. An element $f \in U(L)$ is said to be in *normal form* with respect to G if for all monomials m occurring in f there is no $g \in G$ such that $\mathrm{LM}(g)$ is a factor of m. A *normal form* of f modulo G is an element $h \in U(L)$ such that h is in normal form with respect to G and $f - h \in I$. The following is an algorithm for calculating a normal form of $f \in U(L)$ modulo a finite set $G \subset U(L)$. We assume that the elements of G are monic (i.e., the coefficient of $\mathrm{LM}(g)$ in g is 1 for all $g \in G$). It is clear that we can do that without loss of generality.

Algorithm NormalForm
Input: a set $G \subset U(L)$ consisting of monic elements, and an element $f \in U(L)$.
Output: a normal form of f modulo G.

Step 1 Set $\phi := 0$, $h := f$.

Step 2 If $h = 0$ then return ϕ. Otherwise set $m := \mathrm{LM}(h)$ and let λ be the coefficient of m in h.

Step 3 Let $g \in G$ be such that $\text{LM}(g)$ is a factor of m. If there is no such g then set $h := h - \lambda m$, $\phi := \phi + \lambda m$ and return to Step 2.

Step 4 Let p, q be standard monomials such that $\text{LM}(p\text{LM}(g)q) = m$. Set $h := h - \lambda pgq$, and return to Step 2.

Comments: We note that by Lemma 6.3.1 it is straightforward to find p, q such that $\text{LM}(p\text{LM}(g)q) = m$ in Step 4. The algorithm terminates since $<$ satisfies the descending chain condition and $\text{LM}(h)$ decreases every round of the iteration. Let I be the ideal of $U(L)$ generated by G. Then throughout the algorithm we have that $\phi + h = f \bmod I$. Furthermore ϕ is always in normal form modulo G. Hence at termination we have that ϕ is a normal form of f modulo G.

We note that we can reformulate the algorithm NormalForm in the same fashion as in Section 6.1. Let $G \subset U(L)$ be a set of monic elements and let $f \in U(L)$. Let m be a monomial occurring in f and let $g \in G$ be such that $\text{LM}(g)$ is a factor of m. Let λ be the coefficient of m in f and let p, q be standard monomials such that $\text{LM}(p\text{LM}(g)q) = m$. Then we say that f *reduces* to f' modulo G, where $f' = f - \lambda pgq$. The algorithm NormalForm executes a series of reductions until no further reductions are possible and outputs the remainder.

Definition 6.3.4 *Let $G \subset U(L)$ and let I be the ideal of $U(L)$ generated by G. Then G is called a Gröbner basis if for all $f \in I$ there is a $g \in G$ such that $\text{LM}(g)$ is a factor of $\text{LM}(f)$.*

As in Section 6.1 we define the set of normal words $N(I)$ as the set of all standard monomials that do not occur as a leading monomial of an element of I. If we let $C(I)$ be the span of $N(I)$ then $U(L) = C(I) \oplus I$ (cf. Proposition 6.1.1). Let $f \in U(L)$ and write $f = v + g$ where $v \in C(I)$ and $g \in I$. Then v is called the *normal form* of f modulo I; it is denoted by $\text{Nf}(f)$. The proof of the following proposition is completely analogous to the proof of Theorem 6.1.4.

Proposition 6.3.5 *Let G be a Gröbner basis of the ideal I of $U(L)$. Then $N(I)$ is the set of all standard monomials m such that $\text{LM}(g)$ is not a factor of m for all $g \in G$. Furthermore NormalForm(G, f) returns $\text{Nf}(f)$ for all $f \in U(L)$.*

Now we consider calculating a Gröbner basis of an ideal I generated by a finite set G. We prove some results analogous to those of Section 6.1. However, in this case we prove that we can restrict our attention to a set of special compositions of the elements of G. To describe those we let $A = F[X_1, \ldots, X_t]$ be the (commutative) polynomial ring in t variables. For $m = x_1^{k_1} \cdots x_t^{k_t}$ we set $\tau(m) = X_1^{k_1} \cdots X_t^{k_t}$ and we extend τ to a map $\tau : U(L) \to A$ by linearity. Since the standard monomials form a basis of $U(L)$ we have that τ is bijective. Now let $f_1, f_2 \in U(L)$ be monic and set $u_i = \tau(\mathrm{LM}(f_i))$ for $i = 1, 2$. Let s be the least common multiple of u_1, u_2 and set $m_i = \tau^{-1}(\frac{s}{u_i})$ for $i = 1, 2$. Then the S-element of f_1, f_2 is defined to be

$$S(f_1, f_2) = m_1 f_1 - m_2 f_2.$$

Note that since the f_i are assumed to be monic the leading monomials cancel in this expression (Lemma 6.3.1).

Example 6.3.6 Let L be the 3-dimensional Lie algebra with basis $\{x, y, h\}$ of Example 6.2.4. Let the order be $<_{\mathrm{dlex}}$ and set $A = F[X, Y, H]$. Put $f_1 = xy - h$ and $f_2 = xh - y$. Then $u_1 = \tau(\mathrm{LM}(f_1)) = XY$ and $u_2 = \tau(\mathrm{LM}(f_2)) = XH$. Furthermore XYH is the least common multiple of u_1, u_2. So $m_1 = h$ and $m_2 = y$ and

$$S(f_1, f_2) = hf_1 - yf_2 = hxy - h^2 - yxh + yx = xy - h.$$

Now let $G = \{g_1, g_2, \ldots\} \subset U(L)$ and let I be the ideal of $U(L)$ generated by G. Let m be a standard monomial. Then we let $I_{<m}$ be the set of elements of the form

$$\sum_{k=1}^{r} \lambda_k p_k g_{i_k} q_k,$$

where p_k, q_k are standard monomials such that $\mathrm{LM}(p_k g_{i_k} q_k) < m$. Note that $I_{<m}$ depends on the generating set G. However, it will always be clear what generating set we mean. Let $f \in I_{<m}$ and $h_1, h_2 \in U(L)$, then by the multiplicativity of $<$ it follows that $h_1 f h_2 \in I_{<m'}$ where $m' = \mathrm{LM}(h_1 m h_2)$. In the sequel we will use this frequently.

Lemma 6.3.7 Let $G \subset U(L)$ generate the ideal I of $U(L)$. Let $g \in G$ and suppose that for $1 \leq i \leq t$ we have that $gx_i - x_i g \in I_{<m_i}$, where $m_i = \mathrm{LM}(gx_i)$. Then for all standard monomials q we have $gq - qg \in I_{<n}$ where $n = \mathrm{LM}(gq)$.

Proof. The proof is by induction on $\deg(q)$. If $\deg(q) = 0$ then the statement is trivial, so suppose that $\deg(q) > 0$. Then $q = x_i q'$ for some index i and some standard monomial q'. By hypothesis $gx_i = x_i g + h'$ for some $h' \in I_{<m_i}$. Hence $gx_i q' = x_i g q' + h' q'$ where $h' q' \in I_{<\mathrm{LM}(m_i q')}$. But since

$m_i = \mathrm{LM}(gx_i)$ we have $\mathrm{LM}(m_i q') = \mathrm{LM}(gq)$ so that $h'q' \in I_{<n}$. Furthermore, by induction, $gq' = q'g + h''$, for some $h'' \in I_{<r}$ where $r = \mathrm{LM}(gq')$. Hence $gq = qg + x_i h'' + h'q'$ and $x_i h'' + h'q' \in I_{<n}$. $\square$

Proposition 6.3.8 *Let* $G \subset U(L)$ *consist of monic elements and let* I *be the ideal of* $U(L)$ *generated by* G. *Then* G *is a Gröbner basis of* I *if and only if*

1. *for all* $g_1, g_2 \in G$ *we have that* $S(g_1, g_2) = m_1 g_1 - m_2 g_2$ *lies in* $I_{<m}$ *where* $m = \mathrm{LM}(m_1 g_1)$ *and*

2. *for all* $g \in G$ *and* $1 \le i \le t$ *we have that* $gx_i - x_i g \in I_{<m}$ *where* $m = \mathrm{LM}(gx_i)$.

Proof. First of all, if G is a Gröbner basis then all these elements reduce to zero modulo G because they are elements of I. In particular they lie in $I_{<m}$ where m is the appropriate standard monomial.

For the other direction we first consider an element $f = p_1 g_1 q_1 - p_2 g_2 q_2$ where $g_1, g_2 \in G$ and p_1, p_2, q_1, q_2 are standard monomials such that $\mathrm{LM}(p_1 g_1 q_1) = \mathrm{LM}(p_2 g_2 q_2)$. Set $m = \mathrm{LM}(p_1 g_1 q_1)$. By Lemma 6.3.7 we see that $p_i g_i q_i = p_i q_i g_i + h_i$ for some $h_i \in I_{<m}$. Now set $f_i = \mathrm{LM}(p_i q_i)$, then it follows that $f = f_1 g_1 - f_2 g_2 + h$, for some $h \in I_{<m}$. Let $A = F[X_1, \ldots, X_t]$ be the commutative polynomial ring in t variables. And let $\tau : U(L) \to A$ be the linear map given by $\tau(x_1^{k_1} \cdots x_t^{k_t}) = X_1^{k_1} \cdots X_t^{k_t}$. Set $u_i = \tau(\mathrm{LM}(g_i))$ and $w_i = \tau(f_i)$ for $i = 1, 2$, and let s be the least common multiple of u_1, u_2. Then $w_1 u_1 = w_2 u_2 = vs$ for some $v \in F[X_1, \ldots, X_t]$. Set $v_i = s/u_i$, then $w_i = v v_i$. Now put $n_i = \tau^{-1}(v_i)$ and $p = \tau^{-1}(v)$. Then $n_1 g_1 - n_2 g_2$ is the S-element of g_1 and g_2. Furthermore, by Lemma 6.3.1 we see that $p n_i = f_i + f_i'$ where $\mathrm{LM}(f_i') < f_i$. It follows that

$$f_1 g_1 - f_2 g_2 = (p n_1 - f_1') g_1 - (p n_2 - f_2') g_2 = p(n_1 g_1 - n_2 g_2) + (f_2' g_2 - f_1' g_1).$$

The first summand of the right hand side is in $I_{<m}$ by the hypothesis of the theorem. And since the second summand also lies in $I_{<m}$, it follows that $f \in I_{<m}$.

Now let $f \in I$, then we must prove that there is a $g \in G$ such that $\mathrm{LM}(g)$ is a factor of $\mathrm{LM}(f)$. Write

$$f = \lambda_1 p_1 g_1 q_1 + \cdots + \lambda_r p_r g_r q_r$$

where $g_i \in G$ for $1 \le i \le r$ and p_i, q_i are standard monomials. We remark that the g_i are not required to be different. Set $m_i = \mathrm{LM}(p_i g_i q_i)$. We suppose that the summands have been ordered so that

$$m_1 = m_2 = \ldots = m_k > m_{k+1} \geq \cdots \geq m_r.$$

Now if $k = 1$, then $\mathrm{LM}(g_1)$ is a factor of $\mathrm{LM}(f)$. So in this case we are done. On the other hand, if $k > 1$ then we write

$$f = \lambda_1(p_1 g_1 q_1 - p_2 g_2 q_2) + (\lambda_1 + \lambda_2)p_2 g_2 q_2 + \sum_{i=3}^{r} \lambda_i p_i g_i q_i.$$

By the discussion above we see that $p_1 g_1 q_1 - p_2 g_2 q_2 \in I_{<m_1}$. Hence by substituting an expression for this summand as element of $I_{<m_1}$ we obtain a different expression for f where k has decreased, or m_1 has decreased (in case $k = 2$ and $\lambda_1 + \lambda_2 = 0$). Because $<$ satisfies the descending chain condition we are done.   □

Corollary 6.3.9 *Let $G \subset U(L)$ consist of monic elements and let I be the ideal of $U(L)$ generated by G. Then G is a Gröbner basis of I if and only if $S(g_1, g_2)$ for all $g_1, g_2 \in G$ and $gx_i - x_i g$ for $g \in G$ and $1 \leq i \leq t$ all reduce to zero modulo G.*

Proof. If G is a Gröbner basis then all these elements reduce to 0 modulo G because they are elements of I. On the other hand let $g_1, g_2 \in G$ and suppose that $S(g_1, g_2) = m_1 g_1 - m_2 g_2$ reduces to zero modulo G. Set $m = \mathrm{LM}(m_1 g_1)$. Then $\mathrm{LM}(S(g_1, g_2)) < m$. In the reduction process we subtract elements of the form $\lambda p g q$ (where $g \in G$) from $S(g_1, g_2)$ such that $\mathrm{LM}(pgq) \leq \mathrm{LM}(S(g_1, g_2)) < m$. Now since $S(g_1, g_2)$ reduces to zero modulo G we have that $S(g_1, g_2) \in I_{<m}$. For the elements of the form $gx_i - x_i g$ we use a similar reasoning. By Proposition 6.3.8 it now follows that G is a Gröbner basis.   □

Now we have the following algorithm for calculating a Gröbner basis of an ideal I generated by a finite set S.

Algorithm GröbnerBasis
Input: a finite set $S = \{g_1, \ldots, g_r\} \subset U(L)$ consisting of monic elements.
Output: a Gröbner basis of the ideal of $U(L)$ generated by S.

Step 1 Set $G := S$ and $D := \{(g_i, g_j) \mid 1 \leq i < j \leq r\} \cup \{(g, i) \mid g \in G,\ 1 \leq i \leq t\}$.

Step 2 If $D = \emptyset$ then return G. Otherwise let p be an element of D and set
$D := D \setminus \{p\}$.

Step 3 If $p = (g_1, g_2)$ for some $g_1, g_2 \in G$, then set $h := S(g_1, g_2)$. Otherwise
$p = (g, i)$ for some $g \in G$ and $1 \le i \le t$; in this case we set $h :=$
$gx_i - x_i g$. Set $h' := \mathsf{NormalForm}(G, h)$.

Step 4 If $h' \neq 0$ then do the following:

Step 4a Divide h' by the coefficient of $\mathrm{LM}(h')$ in h'.

Step 4b Add to D all pairs (g, h') for $g \in G$ and (h', i) for $1 \le i \le t$.

Step 4c Add h' to G.

Return to Step 2.

Theorem 6.3.10 *Let S be a finite subset of $U(L)$ consisting of monic elements. Then $\mathsf{Gr\ddot{o}bnerBasis}(S)$ terminates in a finite number of steps and outputs a Gröbner basis of the ideal generated by S.*

Proof. The fact that the algorithm produces a Gröbner basis (if it terminates) follows immediately from Corollary 6.3.9. Now suppose that the algorithm does not terminate. Then the set G is enlarged infinitely often, and we obtain an infinite sequence $G_1 \subsetneq G_2 \subsetneq G_3 \subsetneq \cdots$. Furthermore G_{i+1} is obtained from G_i by adding an element h_i that is in normal form with respect to G_i. In particular $\mathrm{LM}(g)$ is not a factor of $\mathrm{LM}(h_i)$ for all $g \in G_i$. For $i \ge 1$ we set $\overline{G}_i = \{\tau(\mathrm{LM}(g)) \mid g \in G_i\}$ and we let J_i be the ideal of $F[X_1, \dots, X_t]$ generated by $\overline{G}_i$. Then by Lemma 6.3.3 it follows that $\tau(\mathrm{LM}(h_i))$ is not divisible by any element of $\overline{G}_i$. Now because $\overline{G}_i$ consists of monomials this implies that $\tau(\mathrm{LM}(h_i)) \notin J_i$. As a consequence $J_{i+1} \supsetneq J_i$ and we have obtained an infinite strictly ascending chain of ideals in the ring $F[X_1, \dots, X_t]$. But this is not possible by Hilbert's basis theorem (see for instance [89]). □

One of the main practical problems of the algorithm for calculating a Gröbner basis is that the set D often grows very quickly. A huge number of pairs is checked, which can make the algorithm very time consuming. However, many of these checks do not lead to new elements of G. The next lemmas exhibit some pairs for which we can be certain beforehand that checking them would be superfluous.

Lemma 6.3.11 *Let $G \subset U(L)$ and I be as in Proposition 6.3.8. Suppose that for $1 \le i \le t$ and $g \in G$ we have that $gx_i - x_i g \in I_{<m}$, where $m =$*

$\mathrm{LM}(gx_i)$. Let $g_1, g_2 \in G$ be such that $S(g_1, g_2) = m_2 g_1 - m_1 g_2$, where $m_j = \mathrm{LM}(g_j)$ for $j = 1, 2$. Then $S(g_1, g_2) \in I_{<n}$, where $n = \mathrm{LM}(m_2 g_1)$.

Proof. Write $g_j = m_j + \bar{g}_j$ for $j = 1, 2$. By Lemma 6.3.7 we have that $S(g_1, g_2) = g_1 m_2 - m_1 g_2 + h$ where $h \in I_{<n}$. Furthermore,

$$g_1 m_2 - m_1 g_2 = \bar{g}_1 m_2 - m_1 \bar{g}_2 = \bar{g}_1 g_2 - g_1 \bar{g}_2.$$

Therefore, $g_1 m_2 - m_1 g_2 \in I_{<n}$ and we are done. $\qquad \square$

Lemma 6.3.12 Let $G \subset U(L)$ and I be as in Proposition 6.3.8. Let $g_1, g_2, g_3 \in G$ and write $S(g_1, g_2) = m_1 g_1 - m_2 g_2$, $S(g_1, g_3) = n_1 g_1 - n_3 g_3$ and $S(g_2, g_3) = p_2 g_2 - p_3 g_3$. Suppose that $S(g_1, g_3) \in I_{<n}$ where $n = \mathrm{LM}(n_1 g_1)$ and $S(g_2, g_3) \in I_{<p}$ where $p = \mathrm{LM}(p_2 g_2)$. Suppose further that n and p are factors of $m = \mathrm{LM}(m_1 g_1)$. Then $S(g_1, g_2) \in I_{<m}$.

Proof. Since n and p are factors of m there are standard monomials t_1, t_2 such that $m = \mathrm{LM}(t_1 n)$ and $m = \mathrm{LM}(t_2 p)$. By the hypothesis of the lemma we know that $t_1 S(g_1, g_3) - t_2 S(g_2, g_3) \in I_{<m}$. From $m = \mathrm{LM}(m_1 g_1) = \mathrm{LM}(t_1 n) = \mathrm{LM}(t_1 n_1 g_1)$ it follows that $m_1 = t_1 n_1 + h_1$, where $\mathrm{LM}(h_1) < m_1$. Similarly $m = \mathrm{LM}(m_2 g_2) = \mathrm{LM}(t_2 p_2 g_2)$ implies that $m_2 = t_2 p_2 + h_2$ where $\mathrm{LM}(h_2) < m_2$. We also know that $m = \mathrm{LM}(t_1 n_3 g_3)$ and $m = \mathrm{LM}(t_2 p_3 g_3)$. Hence $(t_2 p_3 - t_1 n_3) g_3 \in I_{<m}$. Now we calculate

$$
\begin{aligned}
t_1 S(g_1, g_3) - t_2 S(g_2, g_3) &= t_1 n_1 g_1 - t_1 n_3 g_3 - t_2 p_2 g_2 + t_2 p_3 g_3 \\
&= m_1 g_1 - m_2 g_2 - h_1 g_1 + h_2 g_2 + (t_2 p_3 - t_1 n_3) g_3.
\end{aligned}
$$

It follows that $S(g_1, g_2) \in I_{<m}$. $\qquad \square$

By Lemma 6.3.11 we do not have to consider S-elements of the form $m_2 g_1 - m_1 g_2$, where $m_i = \mathrm{LM}(g_i)$. Furthermore, if g_1, g_2, g_3 are as in Lemma 6.3.12, and we already dealt with $S(g_1, g_3)$ and $S(g_2, g_3)$, then we can dispense with checking $S(g_1, g_2)$.

Example 6.3.13 Let L be the Lie algebra with basis $\{x, y, h\}$ and multiplication table

$$[x, y] = h, \quad [h, x] = 2x, \quad [h, y] = -2y.$$

Let $G = \{g_1 = x^2, g_2 = y^2, g_3 = h^2 - 1\}$. The standard monomials are ordered by the deglex order, where $x <_{\mathrm{dlex}} y <_{\mathrm{dlex}} h$. We calculate a Gröbner basis of the ideal of $U(L)$ generated by G. All possible S-elements of g_1, g_2, g_3 are of the form considered in Lemma 6.3.11. Therefore we do not have

to check them. So we look at elements of the form $gx_i - x_i g$. First we
have $x^2 y - yx^2 = 2xh + 2x$ which is in normal form modulo G. So we add
$g_4 = xh + x$ to G. The next element is $x^2 h - hx^2 = -4x^2$, which reduces to
0 modulo G. Now we deal with g_2: $y^2 x - xy^2 = -2yh + 2y$ which cannot
be reduced, so we add $g_5 = yh - y$ to G. The element $y^2 h - hy^2 = 4y^2$
reduces to 0 modulo G. Also $(h^2 - 1)x - x(h^2 - 1)$ is a multiple of g_4 and
$(h^2 - 1)y - y(h^2 - 1)$ is a multiple of g_5. Now we check the S-elements containing g_4. First $S(g_1, g_4) = hx^2 - x(xh + x) = 3x^2$ reduces to 0. The S-element
$S(g_2, g_4)$ is of the form considered in Lemma 6.3.11 and $S(g_3, g_4) = -3g_4$.
Also the elements $g_4 x - xg_4$ and $g_4 h - hg_4$ give nothing new. However,
$g_4 y - yg_4 = -2xy + h^2 + h$ reduces to $g_6 = xy - \frac{1}{2}h - \frac{1}{2}$, which we add. Now
by checking the remaining pairs it can be shown that the set $\{g_1, \dots, g_6\}$
is a Gröbner basis. We note that we do not need to check the S-element
$S(g_5, g_6)$ once the S-elements $S(g_4, g_5)$ and $S(g_4, g_6)$ have been dealt with
(Lemma 6.3.12).

We close this section with a proof of Proposition 1.13.4. For this we use
Corollary 6.3.9. Let L be a Lie algebra over the field F of characteristic
$p > 0$ with basis $\{x_1, x_2, \dots\}$. Suppose that there are $y_i \in L$ for $i \geq 1$ such
that $(\mathrm{ad}x_i)^p = \mathrm{ad}y_i$. We let I be the ideal of $U(L)$ generated by the set
$G = \{x_i^p - y_i \mid 1 \leq i\}$.

Lemma 6.3.14 *The set G is a Gröbner basis of I.*

Proof. For $i \geq 1$ set $g_i = x_i^p - y_i$, and define a linear map $\mathrm{ad}x_i : U(L) \to U(L)$ by $\mathrm{ad}x_i(f) = x_i f - f x_i$ for $f \in U(L)$. Then for $x \in L$ we have
$\mathrm{ad}x_i(x) = [x_i, x]$ (where the product on the right hand side is taken in
L). Furthermore, $(\mathrm{ad}x_i)^p(x) = x_i^p x - xx_i^p$ by (1.14) with $n = p$. But also
$(\mathrm{ad}x_i)^p(x) = [y_i, x] = y_i x - xy_i$. This means that $x_i^p x - xx_i^p = y_i x - xy_i$, or
$g_i x = xg_i$ for all $x \in L$. In particular $g_i x_j - x_j g_i = 0$. Also for $i \neq j$,

$$S(g_i, g_j) = x_j^p g_i - x_i^p g_j = g_i x_j^p - x_i^p g_j$$
$$= x_i^p y_j - y_i x_j^p.$$

But modulo G this reduces to $y_i y_j - y_i y_j = 0$. Now by Corollary 6.3.9 we
see that G is a Gröbner basis of I. $\square$

By the preceding lemma we have that the set of normal words is

$$N(I) = \{x_{i_1}^{k_1} \cdots x_{i_r}^{k_r} \mid r \geq 1 \text{ and } 0 \leq k_i < p\}.$$

And a basis of the algebra $A = U(L)/I$ is formed by the cosets of the
normal words. Let $\pi : U(L) \to A$ be the projection map and consider the

linear map $\phi : L \xrightarrow{i} U(L) \xrightarrow{\pi} A_{Lie}$. Since the basis elements x_i are linearly independent modulo I (they are normal words) we have that ϕ is injective. It is also a morphism of Lie algebras so ϕ is an isomorphism of L onto the subalgebra $\tilde{L}$ of A_{Lie} spanned by the $\phi(x_i)$. As seen in Example 1.13.3, A_{Lie} is restricted and the map $a \mapsto a^p$ is a p-th power mapping. Furthermore $\phi(x_i)^p = \phi(y_i)$ and by Definition 1.13.2 (items 2. and 3.) it follows that $\tilde{L}$ is closed under the p-th power mapping of A_{Lie}. So $\tilde{L}$ is restricted and as a consequence so is L, and there is a p-th power mapping in L satisfying $x_i^p \mapsto y_i$.

6.4   Gröbner bases of left ideals

Let L be a Lie algebra. A subspace $I \subset U(L)$ is said to be a *left ideal* if $au \in I$ for all $a \in U(L)$ and all $u \in I$. We can view $U(L)$ as an L-module by setting $x \cdot a = xa$ for $a \in U(L)$ and $x \in L$. Then the left ideals of $U(L)$ are the L-submodules of $U(L)$. Hence if I is a left ideal of $U(L)$, then the quotient space $U(L)/I$ becomes an L-module in a natural way. It is the objective of this section to sketch an approach to Gröbner bases for left ideals in $U(L)$. This will provide a method for constructing a basis of the quotient module $U(L)/I$. All constructions and results are analogous to the results of the previous section.

Throughout we let $<$ be a total order on the standard monomials of $U(L)$. The leading monomial LM(f) of an element $f \in U(L)$ relative to the order $<$ is defined to be the largest monomial occurring in f. We assume that $<$ is left-multiplicative (i.e., $m < n$ implies LM(pm) $<$ LM(pn) for all standard monomials p), satisfies the descending chain condition and is degree compatible. The deglex order of the previous section serves as an example of such an order.

A standard monomial m is called a *left factor* of a standard monomial n if there is a standard monomial p such that LM(pm) $= n$. By Lemma 6.3.1, m is a left factor of n if and only if m is a factor of n.

An element $f \in U(L)$ is said to be in *left-normal form* with respect to a set $G \subset U(L)$ if there is no standard monomial m occurring in f having LM(g) as a left factor for a $g \in G$. Furthermore, $h \in U(L)$ is a *left-normal form* of f modulo G if h is in left-normal form with respect to G and $f - h \in I$, where I is the left ideal of $U(L)$ generated by G. We have an algorithm LeftNormalForm for calculating a left-normal form of an element $f \in U(L)$ modulo a set G of monic elements of $U(L)$. It is the same as the algorithm NormalForm of the previous section, except for Step 4 that is replaced by

Step 4' Let p be a standard monomial such that $\text{LM}(p\text{LM}(g)) = m$. Set $h := h - \lambda pg$.

Let $I \subset U(L)$ be a left ideal. A set $G \subset I$ is called a *left-Gröbner basis* of I if for every $f \in I$ there is a $g \in G$ such that $\text{LM}(g)$ is a factor of $\text{LM}(f)$. If G is a left-Gröbner basis of I, then the algorithm $\mathsf{LeftNormalForm}(f, G)$ returns the normal form of f with respect to I. Also a standard monomial m is said to be left-normal with respect to G if there is no $g \in G$ such that $\text{LM}(g)$ is a left factor of m. If G is a left-Gröbner basis of I, then the cosets of the left-normal monomials with respect to G form a basis of the quotient $U(L)/I$.

Now for calculating a left-Gröbner basis of a left ideal generated by a set of monic elements of $U(L)$ we use the same algorithm as for calculating a Gröbner basis of the ideal of $U(L)$ generated by G, except that we only consider the S-elements (and leave out the elements of the form $gx_i - x_ig$). The algorithm we get is the following.

Algorithm LeftGröbnerBasis
Input: a finite set $S = \{g_1, \dots, g_r\} \subset U(L)$ consisting of monic elements.
Output: a left-Gröbner basis of the left ideal of $U(L)$ generated by S.

Step 1 Set $G := S$ and $D := \{(g_i, g_j) \mid 1 \le i < j \le r\}$.

Step 2 If $D = \emptyset$ then return G. Otherwise let $p = (g_1, g_2)$ be an element of D and set $D := D \setminus \{p\}$.

Step 3 Set $h := S(g_1, g_2)$ and $h' := \mathsf{LeftNormalForm}(G, h)$.

Step 4 If $h' \ne 0$ then do the following:

Step 4a Divide h' by the coefficient of $\text{LM}(h')$ in h'.

Step 4b Add to D all pairs (g, h') for $g \in G$.

Step 4c Add h' to G.

Return to Step 2.

The proof that this algorithm terminates and gives a left-Gröbner basis is analogous to the proof of the corresponding facts concerning the algorithm GröbnerBasis in the previous section. We leave the details to the reader. We note that we can use Lemma 6.3.12 to reduce the number of S-elements that need to be checked. However, Lemma 6.3.11 cannot be used in this case.

6.5   Constructing a representation of a Lie algebra of characteristic 0

Ado's theorem states that any finite-dimensional Lie algebra of characteristic 0 has a faithful finite-dimensional representation. In this section we describe a method for constructing such a representation for a given Lie algebra of characteristic 0. As a byproduct we obtain a proof of Ado's theorem.

A first idea is to look at the adjoint representation of L. The kernel of ad is the centre $C(L)$ of L. So for Lie algebras with a trivial centre the problem is solved by the adjoint representation. Therefore in the rest of this section we will be concerned with Lie algebras that have a nontrivial centre.

This section is divided into several subsections. In Section 6.5.1 we give an algorithm for constructing a tower (with certain properties) of Lie algebra extensions (where every term is an ideal in the next one) with final term L. A representation of the first element of the tower is easily constructed. Then this representation is successively extended to representations of the higher terms of the tower and finally to L itself. Sections 6.5.2 and 6.5.3 focus on a single extension step. In Section 6.5.2 the vector space underlying the extension is described. Then in Section 6.5.3 we derive the algorithm for extending a representation. In Section 6.5.4 an algorithm for the construction of a faithful finite-dimensional representation of L is given and Ado's theorem is obtained as a corollary.

6.5.1   Calculating a series of extensions

From Section 1.10 we recall that $K \rtimes H$ denotes the semidirect sum of K and H. In this Lie algebra K is an ideal and H is a subalgebra. Let L be a finite-dimensional Lie algebra of characteristic 0. Here we describe how a series of subalgebras

$$K_1 \subset K_2 \subset \cdots \subset K_j = \mathrm{NR}(L) \subset K_{j+1} \subset \cdots \subset K_{r-1} = \mathrm{SR}(L) \subset K_r = L$$

can be constructed such that $K_{i+1} = K_i \rtimes H_i$, where H_i is a subalgebra of K_{i+1}. We do this in three steps: first we deal with the case where the Lie algebra is nilpotent, then with the case where the Lie algebra is solvable. The final step consists of calculating a Levi subalgebra. For the first two steps we have the following algorithms.

Algorithm ExtensionSeriesNilpotentCase

Input: a nilpotent Lie algebra L over a field of characteristic 0.
Output: subalgebras $K_1, K_2, \ldots, K_n = L$ and $H_1, \ldots, H_{n-1}$ such that

1. K_1 is commutative,

2. $K_{i+1} = K_i \rtimes H_i$,

3. $\dim H_i = 1$ for $1 \le i \le n-1$.

Step 1 Calculate the lower central series of L. Let K_1 be the final term of this series. Set $n := 1$.

Step 2 If $K_n = L$ then return $K_1, \ldots, K_n, H_1, \ldots, H_{n-1}$. Otherwise go to Step 3.

Step 3 Let I be the unique term of the lower central series of L such that $[L, I] \subset K_n$, but I is not contained in K_n.

Step 4 Let y_n be an element from $I \setminus K_n$ and let H_n be the subalgebra spanned by y_n. Set $K_{n+1} := K_n \rtimes H_n$. Set $n := n+1$ and go to Step 2.

Algorithm ExtensionSeriesSolvableCase
Input: a solvable Lie algebra L over a field of characteristic 0.
Output: subalgebras $K_1, K_2, \ldots, K_m = L$ and $H_1, \ldots, H_{m-1}$ such that

1. K_1 is equal to the nilradical of L,

2. $K_{i+1} = K_i \rtimes H_i$,

3. $\dim H_i = 1$ for $1 \le i \le m-1$.

Step 1 Calculate the nilradical K_1 of L.

Step 2 Let $y_1, \ldots, y_{m-1}$ span a complement to K_1 in L. For $1 \le i \le m-1$ let H_i be the subalgebra spanned by y_i and set $K_{i+1} = K_i \rtimes H_i$. Return $K_1, \ldots, K_m, H_1, \ldots, H_{m-1}$.

Comments: It is straightforward to see that the algorithms terminate. Let L be a solvable Lie algebra of characteristic 0. Since $[L, L] \subset \mathrm{NR}(L)$ (Corollary 2.3.5) the elements y_i constructed in Step 2 of the algorithm for the solvable case satisfy $[y_i, L] \subset \mathrm{NR}(L)$. So we can construct the semidirect sums.

6.5.2   The extension space

In this section we assume that we are given a Lie algebra L over a field F of characteristic 0 together with an ideal K and a subalgebra H such that $L = K \rtimes H$. Starting with a finite-dimensional representation $\rho : K \to \mathfrak{gl}(V)$ of K we try to find a finite-dimensional representation σ of L. Under some conditions we succeed in doing this.

Let $U(K), U(L)$ be the universal enveloping algebras of K, L respectively. Then $U(K)$ is a subalgebra of $U(L)$. Let $a \in U(K)$ and $y \in H$. Then by applying the collection process (CollectionInUEA) we see that $ay = ya + b$ for some $b \in U(K)$. Hence $ay - ya = b \in U(K)$.

We describe the space on which L is to be represented. This will be a finite-dimensional subspace of the dual space

$$U(K)^* = \{ f : U(K) \to F \mid f \text{ is linear} \}.$$

The Lie algebra L acts on $U(K)^*$ in the following way. Let f be an element of $U(K)^*$ and let $x \in K$ and $y \in H$. Then for $a \in U(K)$ we set

$$
\begin{aligned}
(x \cdot f)(a) &= f(ax) \\
(y \cdot f)(a) &= f(ay - ya).
\end{aligned}
\tag{6.4}
$$

Lemma 6.5.1 *The equations (6.4) make $U(K)^*$ into an L-module.*

Proof. We have to prove that $[x, y] \cdot f = x \cdot (y \cdot f) - y \cdot (x \cdot f)$ for all $x, y \in L$ and $f \in U(K)^*$. It is enough to prove this for x, y in a basis of L. So put together a basis of K and a basis of H to obtain a basis of L. Let x, y be elements from this basis. If both $x, y \in K$, then $[x, y] \cdot f(a) = f(a[x, y]) = f(axy - ayx) = x \cdot (y \cdot f)(a) - y \cdot (x \cdot f)(a)$. So in this case we are done. Secondly suppose that $x \in K$ and $y \in H$. Then

$$
\begin{aligned}
x \cdot (y \cdot f)(a) - y \cdot (x \cdot f)(a) &= f(axy - yax) - f(ayx - yax) \\
&= f(axy - ayx) = [x, y] \cdot f(a).
\end{aligned}
$$

(Note that $[x, y] \in K$.) Thirdly suppose that $x, y \in H$. Then

$$
\begin{aligned}
x \cdot (y \cdot f)(a) - y \cdot (x \cdot f)(a) &= f(axy - xay - yax + yxa) - \\
&\quad\, f(ayx - yax - xay + xya) \\
&= f(axy - ayx + yxa - xya) = [x, y] \cdot f(a).
\end{aligned}
$$

$\square$

Now let $\{x_1, \ldots, x_t\}$ be a basis of K. From Section 6.2 we recall that the representation ρ of K extends to a representation of the universal enveloping algebra $U(K)$, by

$$\rho(x_1^{k_1} \cdots x_t^{k_t}) = \rho(x_1)^{k_1} \cdots \rho(x_t)^{k_t}.$$

Consider the map

$$c : V \times V^* \longrightarrow U(K)^*$$

defined by $c(v, w^*)(a) = w^*(\rho(a)v)$ for $v \in V$, $w^* \in V^*$ and $a \in U(K)$. Let $\{e_1, \ldots, e_m\}$ be a basis of V and let e_i^* be the element of V^* defined by $e_i^*(e_j) = \delta_{ij}$. Then $c(e_i, e_j^*)(a)$ is the coefficient on position (j, i) of the matrix of $\rho(a)$ with respect to the basis $\{e_1, \ldots, e_m\}$. Therefore we call an element $c(v, w^*)$ a *coefficient* of the representation ρ. By C_ρ we denote the image of c in $U(K)^*$; it is called the *coefficient space* of ρ.

Let $S_\rho \subset U(K)^*$ be the L-submodule of $U(K)^*$ generated by C_ρ (i.e., the smallest L-submodule of $U(K)^*$ that contains C_ρ). Let $\sigma : L \to \mathfrak{gl}(S_\rho)$ be the corresponding representation. We will call σ the *extension* of ρ to L. Proposition 6.5.3 states some conditions that ensure that S_ρ is finite-dimensional. In the proof we need a lemma that is of independent interest. We recall that the codimension of an ideal I of an algebra A is the number $\dim A/I$.

Lemma 6.5.2 *Let K be a Lie algebra over a field F with basis $\{x_1, \ldots, x_t\}$. An ideal I of $U(K)$ is of finite codimension if and only if for $1 \le i \le t$ there is a univariate polynomial $p_i \in F[X]$ such that $p_i(x_i) \in I$.*

Proof. Suppose that I is of finite codimension. By $\bar{x}_i$ we denote the image of x_i in $U(K)/I$. For $1 \le i \le t$ let p_i be the minimum polynomial of $\bar{x}_i$. Then $p_i(x_i) \in I$. For the other direction suppose that $\deg(p_i) = d_i$. Then modulo I any standard monomial can be reduced to a standard monomial $x_1^{k_1} \cdots x_t^{k_t}$ such that $k_i < d_i$ for $1 \le i \le t$. But there is only a finite number of these; hence the set of normal words $N(I)$ is finite and I is of finite codimension. $\square$

Also we need the concept of nilpotency ideal. Let K be a finite dimensional Lie algebra, and let $\rho : K \to \mathfrak{gl}(V)$ be a finite-dimensional representation of K. If I is an ideal of K such that $\rho(x)$ is a nilpotent linear transformation for all $x \in I$, then I is called a *nilpotency ideal* of K with respect to ρ. Let $0 = V_0 \subset V_1 \subset \cdots \subset V_{s+1} = V$ be a composition series of V with respect to the action of K. Set $I = \{x \in K \mid \rho(x)V_{i+1} \subset V_i$ for $0 \le i \le s\}$. Then I is an ideal of L; furthermore, by Proposition 2.1.4, I is a nilpotency ideal of K with respect to ρ. Now let J be a nilpotency ideal of K with respect to ρ. Then by Proposition 2.1.4, $J \subset I$. It follows that I contains all nilpotency ideals of K with respect to ρ. For this reason I is called the *largest nilpotency ideal* of K with respect to ρ. We denote it by $N_\rho(K)$.

Proposition 6.5.3 *Let* $L = K \rtimes H$ *and let* $\rho : K \to \mathfrak{gl}(V)$ *be a finite-dimensional representation of* K *such that* $[H, K] \subset N_\rho(K)$. *Let* $\sigma : L \to \mathfrak{gl}(S_\rho)$ *be the extension of* ρ *to* L. *Then* S_ρ *is finite-dimensional and* $N_\rho(K) \subset N_\sigma(L)$. *Furthermore, if* L *is solvable and* $\mathrm{ad}_K y$ *is nilpotent for all* $y \in H$, *then* $H \subset N_\sigma(L)$.

Proof. Let $0 = V_0 \subset V_1 \subset \cdots \subset V_{s+1} = V$ be a composition series of V with respect to the action of K. Let $I = \{a \in U(K) \mid \rho(a) = 0\}$ be the kernel of ρ (viewed as a representation of $U(K)$), and

$$J = \{a \in U(K) \mid \rho(a)V_{i+1} \subset V_i \text{ for } 0 \leq i \leq s\}.$$

Then J is an ideal of $U(K)$ and $I \subset J$. Now I has finite codimension (since $U(L)/I \cong \rho(K)^*$) and hence so has J. Now from Lemma 6.5.2 it follows that J^{s+1} also has finite codimension. Set

$$W = \{f \in U(K)^* \mid f(J^{s+1}) = 0\}$$

then W is finite-dimensional as J^{s+1} has finite codimension. We claim that W is stable under the action of L. First let $x \in K$, then for $f \in W$ and $a \in J^{s+1}$ we have $x \cdot f(a) = f(ax) = 0$ since J^{s+1} is an ideal of $U(K)$. Second let $y \in H$. For $a \in U(K)$ we define $\mathrm{ad}y(a) = ya - ay$. Then $\mathrm{ad}y$ is a derivation of $U(K)$ extending the usual map $\mathrm{ad}y : K \to K$. Now $\mathrm{ad}y(K) \subset N_\rho(K) \subset J$; so since K generates $U(K)$ we see that $\mathrm{ad}y(U(K)) \subset J$, whence $\mathrm{ad}y(J^{s+1}) \subset J^{s+1}$. Consequently for $f \in W$ and $a \in J^{s+1}$ we have that $y \cdot f(a) = f(ay - ya) = 0$. Our claim is proved.

Now let $c(v, w^*)$ be a coefficient of ρ, then $c(v, w^*)(I) = 0$. But $J^{s+1} \subset I$ so that $c(v, w^*) \in W$. Therefore $C_\rho \subset W$ and because W is an L-module also $S_\rho \subset W$. And since W is finite-dimensional the same holds for S_ρ.

We prove that $N_\rho(K)$ is an ideal in L. By the definition of largest nilpotency ideal we see that $N_\rho(K) = J \cap K$. But $[H, J \cap K] \subset [H, K] \subset N_\rho(K)$. And this implies that $N_\rho(K)$ is an ideal in L. Now let $x \in N_\rho(K)$, then also $x \in J$ so that $x^{s+1} \in J^{s+1}$. Let $f \in S_\rho$ then $\sigma(x)^{s+1} \cdot f(a) = f(ax^{s+1})$ which is zero because $f \in W$ and $ax^{s+1} \in J^{s+1}$. Therefore $N_\rho(K)$ is a nilpotency ideal of L with respect to σ, i.e., $N_\rho(K) \subset N_\sigma(L)$.

Suppose that L is solvable and $\mathrm{ad}_K y$ is nilpotent for all $y \in H$. Let $y \in H$. Again we consider the derivation $\mathrm{ad}y$ of $U(K)$. Since $\mathrm{ad}y$ acts nilpotently on K, by the Leibniz formula (1.11) we see that $\operatorname{ad} y$ is locally nilpotent on $U(K)$ (meaning that for every $a \in U(K)$ there is a $k_a > 0$ such that $(\operatorname{ad} y)^{k_a}(a) = 0$). This implies that $\operatorname{ad} y$ acts nilpotently on the finite-dimensional space $U(K)/J^{s+1}$, i.e., there is an $N > 0$ such that $(\operatorname{ad} y)^N (U(K)) \subset J^{s+1}$. Then $\sigma(y)^N \cdot f(a) = 0$ for all $f \in S_\rho$ and $a \in U(K)$. The conclusion is that $\sigma(x)$ is nilpotent for all $x \in N_\rho(K) \cup H$. Let M be the Lie algebra spanned by $N_\rho(K)$ along with H. We note that M is an ideal of L. In particular M is a solvable Lie algebra. Suppose that the ground field is algebraically closed. Then by Lie's theorem (Theorem 2.4.4) there is a basis of S_ρ relative to which the matrices of $\sigma(x)$ are upper triangular for all $x \in M$. Hence $\sigma(x)$ is strictly upper triangular for all $x \in N_\rho(K) \cup H$. The conclusion is that $\sigma(x)$ is strictly upper triangular for all $x \in M$. If the ground field is not algebraically closed, then we tensor everything with the algebraic closure of the ground field and arrive at the same conclusion. It follows that M is a nilpotency ideal of L with respect to σ, so that $M \subset N_\sigma(L)$. $\qquad\square$

6.5.3   Extending a representation

Throughout this section $L = K \rtimes H$ and $\rho : K \to \mathfrak{gl}(V)$ is a finite-dimensional representation of K such that $[H, K] \subset N_\rho(K)$. Furthermore, $\sigma : L \to \mathfrak{gl}(S_\rho)$ will be the extension of ρ to L. By Proposition 6.5.3, S_ρ is finite-dimensional.

In this section we give an algorithm to construct a finite-dimensional representation of L. If ρ is faithful and H is 1-dimensional, then the representation of L will also be faithful. The key to the algorithm will be the following proposition.

Proposition 6.5.4 *Suppose that ρ is a faithful representation of K. Then σ is faithful on K. Furthermore, if H is 1-dimensional, then σ is a faithful representation of L or there is an element $\tilde{y} \in K$ such that $y - \tilde{y}$ lies in the centre $C(L)$ of L, where y is an element spanning H.*

Proof. Let x be a non-zero element of K. Then $\rho(x) \neq 0$ and hence there are $v \in V$ and $w^* \in V^*$ such that

$$0 \neq w^*(\rho(x)v) = \sigma(x) \cdot c(v, w^*)(1).$$

Hence $\sigma(x) \neq 0$ for all $x \in K$ so that σ is faithful on K.

Suppose that H is 1-dimensional and let $y \in H$ span H. Suppose further that σ is not faithful on L. This means that there is a relation

$$\sigma(y) - \sigma(\tilde{y}) = 0,$$

for some $\tilde{y} \in K$, i.e., $\sigma(y) = \sigma(\tilde{y})$. Then for all $x \in K$ we have

$$\sigma([\tilde{y}, x]) = [\sigma(\tilde{y}), \sigma(x)] = [\sigma(y), \sigma(x)] = \sigma([y, x]).$$

Since σ is faithful on K, this implies that $[\tilde{y}, x] = [y, x]$; but this means that $[y - \tilde{y}, K] = 0$. Also $\sigma([y, y - \tilde{y}]) = 0$ and because $[y, y - \tilde{y}] \in K$ we have that it is 0. The conclusion is that $y - \tilde{y} \in C(L)$. $\square$

Now the idea of the algorithm is straightforward. Suppose that $H = \langle y \rangle$ is 1-dimensional. If there is an element $\tilde{y} \in K$ such that $y - \tilde{y} \in C(L)$ then we can easily construct a representation of L. If on the other hand, there is no such $\tilde{y}$, then σ will be a faithful representation of L. If H is not 1-dimensional then we have no guarantee that σ is faithful. However in the next section it will become clear that this does not bother us too much.

We consider the problem of calculating a basis of S_ρ. We do this by letting the elements of L act on C_ρ. By the next lemma we only have to consider the action of the elements of H for this.

Lemma 6.5.5 *Let $y_1, \ldots, y_s$ be a basis of H. Then S_ρ is spanned by the elements*

$$y_1^{k_1} \cdots y_s^{k_s} \cdot f$$

where $k_q \geq 0$ $(1 \leq q \leq s)$ and $f \in C_\rho$.

Proof. Let $\{x_1, \ldots, x_t\}$ be a basis of K. Then by the theorem of Poincaré-Birkhoff-Witt, S_ρ is spanned by all elements of the form

$$y_1^{k_1} \cdots y_s^{k_s} \cdot x_1^{l_1} \cdots x_t^{l_t} \cdot f$$

for $f \in C_\rho$. Now for $x \in K$ we have $x \cdot c(v, w^*) = c(\rho(x)v, w^*)$ and hence C_ρ is a $U(K)$-module. This implies that S_ρ is spanned by elements of the form

$$y_1^{k_1} \cdots y_s^{k_s} \cdot f.$$

$\square$

The next problem is how to do linear algebra inside S_ρ. Since the ambient space $U(K)^*$ is infinite-dimensional this is not immediately straightforward. We need to represent each element of S_ρ as a row vector (of finite length); then we can use Gaussian elimination to calculate a basis of S_ρ and calculate matrices of endomorphisms of S_ρ and so on. We solve this problem by taking a finite set B of standard monomials and representing an element $f \in S_\rho$ by the vector $(f(b))_{b \in B}$. We must choose B such that linearly independent elements of S_ρ yield linearly independent row vectors. We shall call such a set B a *discriminating set* for S_ρ. Since S_ρ is finite-dimensional discriminating sets for S_ρ exist and a minimal discriminating set consists of $\dim S_\rho$ standard monomials.

Now we formulate an algorithm for extending the representation ρ to L. There are two cases to be considered; the general case and the case where $H = \langle y \rangle$ and there is a $\tilde{y} \in K$ such that $y - \tilde{y} \in C(L)$. In the second case we can easily construct a faithful representation of L. In the general case we construct the extension of ρ to L. Then by Proposition 6.5.4 we always obtain a faithful representation of L in the case where H is 1-dimensional. For greater clarity we formulate the algorithm using a subroutine that treats the general case. We first state the subroutine.

Algorithm GeneralExtension
Input: $L = K \rtimes H$ and a representation $\rho : K \to \mathfrak{gl}(V)$ such that $[H, K] \subset N_\rho(K)$.
Output: the extension $\sigma : L \to \mathfrak{gl}(S_\rho)$.

Step 1 Calculate a set of standard monomials $B = \{m_1, \dots, m_r\}$ that form a basis of a complement to $\ker \rho$ in $U(K)$.

Step 2 Calculate a basis of C_ρ using B as discriminating set and set $d := \max \deg m_i$.

Step 3 Let B_d be the set of all standard monomials m such that $\deg(m) \le d$.

Step 4 Calculate a basis of S_ρ, using B_d as discriminating set.

Step 5 Calculate the action of the elements of a basis of L on S_ρ. If this yields a representation of L, then return that representation. Otherwise set $d := d + 1$; and go to Step 4.

Comments: The algorithm is straightforward. It calculates a basis of S_ρ and the matrices of the corresponding representation. For want of a better method we use a rather crude way of finding a discriminating set B_d. First we consider the space C_ρ. We have

$$c(v, w^*)(a) = w^*(\rho(a)v),$$

so we can describe a function in C_ρ by giving its values on the monomials m_i constructed in Step 1. Now we let B_d be the set of all monomials of degree $\leq d$. Initially we set d equal to the maximum degree of a monomial m_i, ensuring that all these elements will be contained in B_d. Using Lemma 6.5.5 we calculate a basis of S_ρ, representing each function on the set B_d. Then we calculate the matrices of the action of the elements of a basis of L. If this yields a representation of L then we are done. Otherwise we apparently did not calculate all of S_ρ in the preceding step. This means that B_d is not a discriminating set for S_ρ. So in this case we set $d := d + 1$ and go through the process again. Since S_ρ is finite-dimensional, the procedure will terminate. (Note that once we find in Step 5. a space that is an L-module it must be equal to S_ρ since it contains C_ρ.) We can refine this a little by calculating a much smaller set of monomials in Step 5. In that step we know a basis of our space and we can restrict to a set of standard monomials of cardinality equal to the dimension of this space.

Now we state the routine that also treats the special case. We recall that E_{ij}^n denotes the $n \times n$-matrix with a 1 on position (i, j) and zeros elsewhere.

Algorithm ExtendRepresentation

Input: $L = K \rtimes H$ and a representation $\rho : K \to \mathfrak{gl}(V)$ such that $[H, K] \subset N_\rho(K)$.

Output: a representation $\sigma : L \to \mathfrak{gl}(W)$.

Step 1 If H is spanned by a single element y *and* there is an element $\tilde{y} \in K$ such that $y - \tilde{y} \in C(L)$ then do the following:

Step 1a Set $n := \dim(V)$. Set $\sigma(y - \tilde{y}) := E_{n+1,n+2}^{n+2}$.

Step 1b For x in a basis of K let $\sigma(x)$ be the $n + 2 \times n + 2$-matrix of which the $n \times n$-submatrix in the top left corner is $\rho(x)$, and the rest of the entries are 0.

Step 1c Return σ.

Otherwise return GeneralExtension(L, ρ).

Proposition 6.5.6 *Let $L = K \rtimes H$ and let $\rho : K \to \mathfrak{gl}(V)$ be a representation of K such that $[H, K] \subset N_\rho(K)$. Then* ExtendRepresentation(L, ρ) *returns a finite-dimensional representation σ of L such that $N_\rho(K) \subset N_\sigma(L)$. Also if ρ is faithful and H is 1-dimensional, then σ is a faithful representation of L.*

Proof. First we suppose that σ is constructed in Steps 1a., 1b. of the algorithm. We remark that finding a $\tilde{y}$ such that $y - \tilde{y} \in C(L)$ amounts to solving a system of linear equations. Also since $y - \tilde{y} \in C(L)$ we have that the map σ is a representation of L. It is obviously finite-dimensional and $\sigma(x)$ is nilpotent for all $x \in N_\rho(K)$. Since $[y, N_\rho(K)] = [\tilde{y}, N_\rho(K)] \subset N_\rho(K)$ we see that $N_\rho(K)$ is an ideal of L. Therefore, $N_\rho(K) \subset N_\sigma(L)$. On the other hand, if σ is constructed by GeneralExtension then these properties follow from Proposition 6.5.3.

Finally suppose that ρ is faithful and H is 1-dimensional. If σ is constructed in Steps 1a., 1b., then σ is obviously faithful on L. And if σ is constructed by GeneralExtension then σ is faithful by Proposition 6.5.4.   □

6.5.4   Ado's theorem

Here we formulate algorithms for calculating a faithful finite-dimensional representation of a Lie algebra L of characteristic 0. We treat the cases where L is nilpotent and solvable separately.

Algorithm RepresentationNilpotentCase
Input: a nilpotent Lie algebra L of characteristic 0.
Output: a finite-dimensional faithful representation σ of L, such that $\sigma(x)$ is nilpotent for all $x \in L$.

Step 1 Let $[K_1, \ldots, K_n, H_1, \ldots, H_{n-1}]$ be the output of ExtensionSeries-NilpotentCase with input (L).

Step 2 Let $\{x_1, \ldots, x_s\}$ be a basis of K_1 and set $\rho_1(x_i) := E_{1,i+1}^{s+1}$ for $1 \leq i \leq s$.

Step 3 For $2 \leq i \leq n$ set $\rho_i :=$ ExtendRepresentation(K_i, ρ_{i-1}).

Step 4 Return ρ_n.

Lemma 6.5.7 *Let L be a nilpotent Lie algebra of characteristic 0, then* RepresentationNilpotentCase *on input L returns a faithful finite-dimensional representation σ of L such that $\sigma(x)$ is nilpotent for all $x \in L$.*

Proof. By induction we show that for $1 \leq i \leq n$ we have that ρ_i is a faithful finite-dimensional representation of K_i such that $\rho_i(x)$ is nilpotent for all $x \in K_i$. This is certainly true for $i = 1$. Now assume that $i > 1$. Then $K_i = K_{i-1} \rtimes H_{i-1}$; and since by induction, $N_{\rho_{i-1}}(K_{i-1}) = K_{i-1}$ we have that $[H_{i-1}, K_{i-1}] \subset N_{\rho_{i-1}}(K_{i-1})$. By Proposition 6.5.6 we see that $K_{i-1} = N_{\rho_{i-1}}(K_{i-1}) \subset N_{\rho_i}(K_i)$. Let y_{i-1} be an element spanning H_{i-1} (note that the H_j are all 1-dimensional in this case). We show that $\rho_i(y_{i-1})$

is nilpotent. If ρ_i is constructed in Steps 1a., 1b. of the algorithm ExtendRepresentation, then $\rho_i(y_{i-1}) = \rho_i(y_{i-1} - \tilde{y}_{i-1}) + \rho_i(\tilde{y}_{i-1})$. Both summands on the right hand side are nilpotent and commute, hence $\rho_i(y_{i-1})$ is nilpotent. On the other hand, if ρ_i is constructed by GeneralExtension, then $\rho_i(y_{i-1})$ is nilpotent by Proposition 6.5.3. Now in exactly the same way as in the last part of the proof of Proposition 6.5.3 it follows that $K_{i-1} + H_{i-1} \subset N_{\rho_i}(K_i)$, i.e., $\rho_i(x)$ is nilpotent for all $x \in K_i$. Also ρ_i is faithful by Proposition 6.5.6. $\square$

Algorithm RepresentationSolvableCase
Input: a solvable Lie algebra L of characteristic 0.
Output: a finite-dimensional faithful representation σ of L, such that $\sigma(x)$ is nilpotent for all $x \in$ NR(L).

Step 1 Let $[K_1, \dots, K_m, H_1, \dots, H_{m-1}]$ be the output of ExtensionSeriesSolvableCase with input (L).

Step 2 Let σ_1 be the output of RepresentationNilpotentCase(K_1).

Step 3 For $2 \le i \le m$ set $\sigma_i :=$ExtendRepresentation(K_i, σ_{i-1}).

Step 4 Return σ_m.

Lemma 6.5.8 *Let L be a solvable Lie algebra of characteristic 0, then* RepresentationSolvableCase *on input L returns a faithful finite-dimensional representation σ of L such that $\sigma(x)$ is nilpotent for all $x \in$ NR(L).*

Proof. By induction we show that for $1 \le i \le m$ we have that σ_i is a faithful finite-dimensional representation of K_i such that $\sigma_i(x)$ is nilpotent for all $x \in$ NR(L). For $i = 1$ this follows from Lemma 6.5.7 as $K_1 =$ NR(L). Now suppose that $i > 1$, then $K_i = K_{i-1} \rtimes H_{i-1}$, and by Corollary 2.3.5 we have that $[H_{i-1}, K_{i-1}] \subset$ NR(L). But by induction, NR$(L) \subset N_{\sigma_{i-1}}(K_{i-1})$. Hence the hypothesis of Proposition 6.5.6 is satisfied. It follows that NR$(L) \subset N_{\sigma_i}(K_i)$, and that σ_i is faithful. $\square$

Algorithm Representation
Input: a Lie algebra L of characteristic 0.
Output: a finite-dimensional faithful representation σ of L, such that $\sigma(x)$ is nilpotent for all $x \in$ NR(L).

Step 1 (Catch nilpotent and solvable cases.) If L is nilpotent, then return the output of RepresentationNilpotentCase on input L. If L is solvable, then return the output of RepresentationSolvableCase with input (L).

Step 2 Calculate the solvable radical R and a Levi subalgebra S of L.

Step 3 Let η be output of RepresentationSolvableCase(R). Let ρ be the output of ExtendRepresentation with input L and η. Return the direct sum of ρ and the adjoint representation of L.

Theorem 6.5.9 *Let L be a a finite-dimensional Lie algebra of characteristic 0. Then* Representation(L) *returns a faithful representation σ of L, such that $\sigma(x)$ is nilpotent for all $x \in$ NR(L).*

Proof. Firstly, if σ is output in Step 1, then this follows from Lemmas 6.5.7 and 6.5.8. If L is not solvable, then $L = R \rtimes S$. We note that by Proposition 2.3.6, NR(L) = NR(R). So by Lemma 6.5.8 we have that NR(L) $\subset N_\eta(R)$. Using Corollary 2.3.5, we now see that $[S, R] \subset N_\eta(R)$. Therefore, by Proposition 6.5.6, ρ is a finite-dimensional representation of L. In this case S is not 1-dimensional, so ρ is constructed by GeneralExtension. Therefore, by Proposition 6.5.4, ρ is faithful on R. In particular, ρ is faithful on the centre $C(L)$ of L. But then it follows that the direct sum of ρ and the adjoint representation of L is faithful on L. Furthermore, by Proposition 6.5.6, $\rho(x)$ is nilpotent for all $x \in$ NR(L). Also ad$x$ is nilpotent for all $x \in$ NR(L). Hence $\sigma(x)$ is nilpotent for all $x \in$ NR(L). $\square$

Corollary 6.5.10 (Ado's theorem) *Let L be a finite-dimensional Lie algebra over a field of characteristic zero. Then L has a faithful finite-dimensional representation. Moreover, a representation can be constructed such that the elements of the nilradical of L are represented by nilpotent endomorphisms.*

Example 6.5.11 Let L be the 5-dimensional Lie algebra of characteristic 0 with basis $\{x_1, \dots, x_5\}$ and multiplication table

$$[x_3, x_4] = x_2, \ [x_3, x_5] = x_1, \ [x_4, x_5] = x_3.$$

This is a nilpotent Lie algebra and the final term of its lower central series is spanned by x_1, x_2, x_3. So we set $K_1 = \langle x_1, x_2, x_3 \rangle$, $K_2 = K_1 \rtimes \langle x_4 \rangle$ and $K_3 = K_2 \rtimes \langle x_5 \rangle$. A representation ρ_1 of K_1 is easily constructed by setting $\rho_1(x_1) = E_{12}^4$, $\rho_1(x_2) = E_{13}^4$ and $\rho_1(x_3) = E_{14}^4$. Then the only standard monomials p for which $\rho_1(p)$ is non-zero are $1, x_1, x_2, x_3$. So any coefficient will yield zero when applied to any other monomial. Denoting the function that assigns to a matrix the coefficient on position (i, j) by c_{ij} we see that there are four coefficient functions. First there is $c_{11} : U(L) \to F$ defined by $c_{11}(1) = 1$ and $c_{11}(p) = 0$ if p is a standard monomial not equal to 1. Second we have c_{12} given by $c_{12}(x_1) = 1$ and $c_{12}(p) = 0$ if $p \neq x_1$. Third and fourth we have c_{13} and c_{14} that satisfy $c_{13}(x_2) = 1$ and $c_{13}(p) = 0$ if $p \neq x_2$ and $c_{14}(x_3) = 1$ and $c_{14} = 0$ if $p \neq x_3$. So C_{ρ_1} is 4-dimensional.

Now we let x_4 act on C_{ρ_1}. By induction it is straightforward to establish that $x_4 x_3^m = x_3^m x_4 - m x_2 x_3^{m-1}$ and hence

$$x_1^k x_2^l x_3^m x_4 - x_4 x_1^k x_2^l x_3^m = m x_1^k x_2^{l+1} x_3^{m-1}.$$

Set $a_{k,l,m} = m x_1^k x_2^{l+1} x_3^{m-1}$, then for $f \in U(K_1)^*$ we have $x_4 \cdot f(x_1^k x_2^l x_3^m) = f(a_{k,l,m})$. Since $a_{k,l,m}$ is never 1 we see that $x_4 \cdot c_{11} = 0$. Furthermore $a_{k,l,m}$ is also never x_1 or x_3 implying that $x_4 \cdot c_{12} = x_4 \cdot c_{14} = 0$. However, if we set $k = l = 0$ and $m = 1$, then $a_{k,l,m} = x_2$. This means that $x_4 \cdot c_{13}(x_3) = c_{13}(x_2) = 1$. As a consequence $x_4 \cdot c_{13} = c_{14}$. So C_{ρ_1} is stable under the action of x_4, i.e., $S_{\rho_1} = C_{\rho_1}$. This yields a representation ρ_2 of K_2. By the above we see that $\rho_2(x_4) = E_{4,3}^4$. Also for $f \in U(K_1)^*$ we have $x_1 \cdot f(x_1^k x_2^l x_3^m) = f(x_1^{k+1} x_2^l x_3^m)$ from which it follows that $x_1 \cdot c_{11} = x_1 \cdot c_{13} = x_1 \cdot c_{14} = 0$ and $x_1 \cdot c_{12} = c_{11}$. For x_2, x_3 we use a similar argument and we conclude that $\rho_2(x_1) = E_{12}^4$, $\rho_2(x_2) = E_{13}^4$ and $\rho_2(x_3) = E_{14}^4$ (the same matrices as for ρ_1).

We perform the final extension. The standard monomials $p \in U(K_2)$ such that $\rho_2(p)$ is non-zero are $1, x_1, x_2, x_3, x_4, x_3 x_4$ ($\rho_2(x_3 x_4) = \rho_2(x_2)$). So the coefficients are $c_{11}, c_{12}, c_{13}, c_{14}, c_{43}$ where

$$c_{11}(1) = 1 \text{ and } 0 \text{ for all other monomials}$$
$$c_{12}(x_1) = 1 \text{ and } 0 \text{ for all other monomials}$$
$$c_{13}(x_2) = c_{13}(x_3 x_4) = 1 \text{ and } 0 \text{ for all other monomials}$$
$$c_{14}(x_3) = 1 \text{ and } 0 \text{ for all other monomials}$$
$$c_{43}(x_4) = 1 \text{ and } 0 \text{ for all other monomials}.$$

We let x_5 act on the 5-dimensional space C_{ρ_2} spanned by the coefficients. First by some induction arguments we see that

$$x_1^k x_2^l x_3^m x_4^n x_5 - x_5 x_1^k x_2^l x_3^m x_4^n =$$
$$n x_1^k x_2^l x_3^{m+1} x_4^{n-1} - \binom{n}{2} x_1^k x_2^{l+1} x_3^m x_4^{n-2} + m x_1^{k+1} x_2^l x_3^{m-1} x_4^n.$$

We denote the expression on the right hand side by $b_{k,l,m,n}$. Then for $f \in U(K_2)^*$ we have $x_5 \cdot f(x_1^k x_2^l x_3^m x_4^n) = f(b_{k,l,m,n})$. Since $b_{k,l,m,n}$ is never 1, $x_5 \cdot c_{11} = 0$. Furthermore $b_{k,l,m,n} = x_1$ if and only if $k = l = n = 0$ and $m = 1$ so $x_5 \cdot c_{12}(x_3) = c_{12}(x_1) = 1$ and $x_5 \cdot c_{12} = c_{14}$. In similar fashion $x_5 \cdot c_{14} = c_{43}$.

Also $b_{k,l,m,n}$ is x_2 or x_3x_4 if and only if $k = l = m = 0$ and $n = 2$, in which case $b_{k,l,m,n} = 2x_3x_4 - x_2$. So $x_5 \cdot c_{13}(x_4^2) = c_{13}(2x_3x_4 - x_2) = 1$. This yields a function $g : U(K_2) \to F$ not contained in C_{ρ_2}; it is defined by $g(x_4^2) = 1$ and $g(p) = 0$ for all standard monomials $p \neq x_4^2$. Since $b_{k,l,m,n}$ is never x_4 or x_4^2 we have that $x_5 \cdot c_{43} = x_5 \cdot g = 0$. It follows that S_{ρ_2} is 6-dimensional. We leave it to the reader to write down the matrices of the corresponding representation of L.

6.6   The theorem of Iwasawa

In the preceding section we proved Ado's theorem, stating that any finite-dimensional Lie algebra of characteristic 0 can be obtained as a linear Lie algebra. In this section we prove the corresponding result for Lie algebras of characteristic $p > 0$, which is known as Iwasawa's theorem.

Throughout F will be a field of characteristic $p > 0$. We recall that a polynomial of the form

$$\alpha_0 X^{p^m} + \alpha_1 X^{p^{m-1}} + \cdots + \alpha_m X$$

is called a *p-polynomial*.

Lemma 6.6.1 *Let L be a finite-dimensional Lie algebra over F. Let $x \in L$, then there is a monic p-polynomial $f \in F[X]$ such that $f(\mathrm{ad}x) = 0$.*

Proof. Let g be the minimum polynomial of $\mathrm{ad}x$. Suppose that the degree of g is d. Then for $0 \leq i \leq d$ we write

$$X^{p^i} = gq_i + r_i$$

where $\deg(r_i) < d$. Now since the space of polynomials of degree $< d$ is d, the polynomials r_i are linearly dependent. So there are $\lambda_0, \ldots, \lambda_d$ such that $\sum_{i=0}^d \lambda_i r_i = 0$. This means that

$$\sum_{i=0}^d \lambda_i X^{p^i} = g \sum_{i=0}^d \lambda_i q_i.$$

So g is a factor of a p-polynomial f. After dividing by a suitable scalar we may assume that f is monic. Finally, because g divides f, we have that $f(\mathrm{ad}x) = 0$.   □

Now let L be a finite-dimensional Lie algebra over the field F with basis $\{x_1, \ldots, x_n\}$. For $1 \leq i \leq n$ let $f_i \in F[X]$ be a monic p-polynomial such that $f_i(\mathrm{ad}x_i) = 0$. Set $G = \{f_1(x_1), \ldots, f_n(x_n)\}$ which is a subset of $U(L)$. Let I be the ideal of $U(L)$ generated by G.

Lemma 6.6.2 *The set G is a Gröbner basis of I.*

Proof. We claim that $f_i(x_i)$ commutes with all elements of $U(L)$. For this it is enough to show that $f_i(x_i)$ commutes with all elements of L. For $a \in U(L)$ let $\mathrm{ad}a$ be the derivation of $U(L)$ given by $\mathrm{ad}a(b) = ab - ba$ for $b \in U(L)$. By (1.14) with $n = p$ it follows that $(\mathrm{ad}a)^p(b) = a^p b - b a^p = \mathrm{ad}a^p(b)$ for $b \in U(L)$. So $(\mathrm{ad}a)^p = \mathrm{ad}a^p$. By a straightforward induction this implies $(\mathrm{ad}a)^{p^k} = \mathrm{ad}a^{p^k}$. In particular, for $y \in L$ we have $\mathrm{ad}f_i(x_i)(y) = f_i(\mathrm{ad}x_i)(y) = 0$, i.e. $f_i(x_i)$ commutes with every $y \in L$ and our claim is proved.

Now we show that G is a Gröbner basis. Firstly, $f_i(x_i)x_j - x_j f_i(x_i) = 0$ by the claim above. Write $f_i(x_i) = x_i^{p^{m_i}} + \bar{f}_i(x_i)$. Then

$$S(f_i, f_j) = x_j^{p^{m_j}} f_i - x_i^{p^{m_i}} f_j = f_i x_j^{p^{m_j}} - x_i^{p^{m_i}} f_j$$
$$= \bar{f}_i(x_i) x_j^{p^{m_j}} - x_i^{p^{m_i}} \bar{f}_j(x_j).$$

But modulo G this reduces to $-\bar{f}_i(x_i)\bar{f}_j(x_j) + \bar{f}_i(x_i)\bar{f}_j(x_j) = 0$. By Corollary 6.3.9 it now follows that G is a Gröbner basis of I. □

As in the proof of the preceding lemma write $f_i(x_i) = x_i^{p^{m_i}} + \bar{f}_i(x_i)$. Then by Lemma 6.6.2 it follows that the set of normal words of $U(L)$ modulo I is

$$N(I) = \{x_{i_1}^{k_1} \cdots x_{i_r}^{k_r} \mid r \geq 1 \text{ and } 0 \leq k_i < p^{m_i}\}.$$

And a basis of the algebra $A = U(L)/I$ is formed by the cosets of the normal words. Let $\pi : U(L) \to A$ be the projection map and consider the linear map $\phi : L \xrightarrow{i} U(L) \xrightarrow{\pi} A_{Lie}$. Then by the same argument as we used at the end of Section 6.3 we have that ϕ is an isomorphism of L onto its image in A_{Lie}. We identify L with its image and we let $\rho : L \to \mathfrak{gl}(A)$ be the map given by $\rho(x)(b) = xb$ for $x \in L$, $b \in A$ (where the product on the right hand side is taken in A). Then $\rho([x,y])(b) = [x,y]b = (xy - yx)b = (\rho(x)\rho(y) - \rho(y)\rho(x))b$. Hence ρ s a representation of L. Also since A has an identity this representation is faithful. So we have proved the following theorem.

Theorem 6.6.3 (Iwasawa) *Let L be a finite-dimensional Lie algebra of characteristic $p > 0$. Then L has a faithful finite-dimensional representation.*

Remark. We can easily transform the proof of Iwasawa's theorem into an algorithm Representation for Lie algebras L of characteristic $p > 0$. For every basis element x_i of L we calculate a p-polynomial f_i such that $f_i(\mathrm{ad}x_i) = 0$. Then we form the algebra A spanned by the normal words $N(I)$ modulo the ideal I of $U(L)$ generated by the elements $f_i(x_i)$. Finally we calculate the matrices of the left multiplication of x_i on A. However, since the dimension of A will generally be huge, this is not a practical algorithm. To the best of our knowledge the problem of finding an algorithm for constructing a finite-dimensional faithful representation of manageable dimension for a Lie algebra of characteristic $p > 0$ is still open.

6.7   Notes

Gröbner bases for commutative polynomial rings were invented by B. Buchberger ([15], [16], [17]). In [9] the concept of Gröbner basis was extended to noncommutative associative algebras (see also [64], [65]). The proof of the Poincaré-Bikhoff-Witt theorem by the use of Gröbner bases was given in [9]. For a different approach we refer to [42], [48].

Gröbner bases for left ideals in universal enveloping algebras were introduced in [1]. In [50] Gröbner bases for so-called solvable polynomial rings were described. Universal enveloping algebras are an example of solvable polynomial rings. This last paper also describes Gröbner bases for two-sided ideals. Lemmas 6.3.11 and 6.3.12 are to the best of our knowledge not contained elsewhere in the literature, they are however, straightforward translations of the corresponding statements for commutative polynomial rings (see, e.g., [65]). Proposition 1.13.4 and Theorem 6.6.3 are contained in [48]. However, the approach to these results via Gröbner bases, is as far as we know not described elsewhere.

The algorithms of Section 6.5 are taken from [34]. The proof of Proposition 6.5.3 follows [13]. In general we do not know any bounds on the dimension of the representation output by Representation(L). In [34] it is proved that for a nilpotent Lie algebra of dimension n and nilpotency class c the dimension of the resulting representation is bounded above by $\binom{n+c}{c}$.

Chapter 7

Finitely presented Lie algebras

Let L be a Lie algebra and $X \subset L$ a set of elements of L. The Lie algebra generated by X is defined to be the smallest subalgebra of L containing X.

Example 7.0.1 Let $L = \mathfrak{sl}_3(F)$ (see Example 1.2.1). Set

$$x = \begin{pmatrix} 1 & 0 & 0 \\ 0 & 1 & 0 \\ 0 & 0 & -2 \end{pmatrix} \qquad y = \begin{pmatrix} 0 & 1 & 0 \\ 0 & 0 & 1 \\ 0 & 0 & 0 \end{pmatrix}.$$

Let K be the subspace of L spanned by $\{x, y, [x, y], [y, [x, y]]\}$. Then a small calculation shows that K is closed under the Lie bracket, so that K is a subalgebra of L. Furthermore, K contains x, y and it is clear that any subalgebra of L that contains x and y must also contain K. Hence K is the Lie algebra generated by x and y.

Let $X = \{x_1, x_2, x_3, \dots\}$ be a subset of the Lie algebra L. Then the Lie algebra K generated by X is spanned by elements of the form

$$x_i, \ [x_i, x_j], \ [[x_i, x_j], x_k], \ [x_i, [x_j, x_k]], \ [[[x_i, x_j], x_k], x_l], \cdots$$

i.e., by all bracketed expressions in the $x_i \in X$. It follows that the multiplicative structure of K is described by a set of relations that are satisfied by the x_i. For example, in Example 7.0.1 we have the relation $[x, [x, y]] - 3[x, y] = 0$. The form and number of such relations will depend on the Lie algebra L and the subset X. However, in all cases we have the identities

(L_1) $[a, a] = 0$ for all $a \in K$,

(L_2) $[a, [b, c]] + [b, [c, a]] + [c, [a, b]] = 0$ for all $a, b, c \in K$.

In this chapter we study Lie algebras generated by a set of elements. For this we want to treat the relations that depend on the particular Lie algebra L separately form the relations following from (L_1) and (L_2) that hold in any Lie algebra. If we concentrate on the latter relations and forget about the first, then we get the concept of free Lie algebra. In Section 7.1 we formally define the concept of free Lie algebra. If we take a free Lie algebra and impose some relations on the generators, then we get a so-called finitely presented Lie algebra. This is the subject of Section 7.2. A large part of the rest of the chapter is devoted to describing algorithms for finding a basis of a finitely presented Lie algebra. In Sections 7.3 and 7.4 an algorithm for this task is described that works inside the free algebra. It enumerates a basis of a finitely presented Lie algebra, and therefore it will only terminate if this Lie algebra is finite-dimensional.

It is also possible to work directly inside the free Lie algebra. For that we need a basis of the free Lie algebra. This is the subject of Sections 7.5, 7.6 and 7.7, where we describe the so-called Hall sets, which form a class of bases of the free Lie algebra. In Section 7.8 we give two examples of Hall sets. These are then used in Section 7.9 to give an algorithm for reducing elements of the free Lie algebra modulo other elements of the free Lie algebra. In Section 7.10 we prove a theorem due to A. I. Shirshov, giving an algorithm for calculating a Gröbner basis of an ideal in a free Lie algebra. This algorithm is only guaranteed to terminate if a finite Gröbner basis exists. This gives us one more algorithm for calculating a basis of a finitely-presented Lie algebra. Section 7.11 contains an application of the theory of finitely presented Lie algebras. We prove a theorem by J.-P. Serre, describing the semisimple Lie algebras of characteristic 0 as finitely presented Lie algebras.

7.1   Free Lie algebras

Let X be a set. We construct a Lie algebra generated by X, satisfying no relations other than (L_1) and (L_2).

Let $M(X)$ be the set inductively defined as follows:

1. $X \subset M(X)$,

2. if $m, n \in M(X)$, then also the pair $(m, n) \in M(X)$.

The set $M(X)$ is called the *free magma* on X. It consists of all bracketed expressions on the elements of X. On $M(X)$ we define a binary operation $\cdot : M(X) \times M(X) \to M(X)$ by $m \cdot n = (m, n)$ for all $m, n \in M(X)$.

For $m \in M(X)$ we define its *degree* recursively: $\deg(m) = 1$ if $m \in X$ and $\deg(m) = \deg(m') + \deg(m'')$ if $m = (m', m'')$. So the degree of an element $m \in M(X)$ is just the number of elements of X that occur in m (counted with multiplicities). For an integer $d \geq 1$ we let $M_d(X)$ be the subset of $M(X)$ consisting of all $m \in M(X)$ of degree d. Then

$$M(X) = \bigcup_{d \geq 1} M_d(X).$$

Let F be a field and let $A(X)$ be the vector space over F spanned by $M(X)$. If we extend the binary operation on $M(X)$ bilinearly to $A(X)$, then $A(X)$ becomes a (non-associative) algebra; it is called the *free algebra* over F on X. Let $f \in A(X)$; if also $f \in M(X)$, then f is said to be a *monomial*. For $f \in A(X)$ we define $\deg(f)$ to be the maximum of the degrees of m where m runs over all $m \in M(X)$ that occur in f with non-zero coefficient.

Let I_0 be the ideal of $A(X)$ generated by all elements

$$(m, m) \text{ for } m \in M(X),$$
$$(m, n) + (n, m) \text{ for } m, n \in M(X), \tag{7.1}$$
$$(m, (n, p)) + (n, (p, m)) + (p, (m, n)) \text{ for } m, n, p \in M(X).$$

Set $L(X) = A(X)/I_0$. Let B be a basis of $L(X)$ consisting of (images of) elements of $M(X)$. Then it is immediate that we have $(x, x) = 0$, $(x, y) + (y, x) = 0$ and $(x, (y, z)) + (y, (z, x)) + (z, (x, y)) = 0$ for all $x, y, z \in B$. By Lemma 1.3.1 it follows that the relations (L_1) and (L_2) hold for all elements of $L(X)$ so that $L(X)$ is a Lie algebra. Therefore we will use the bracket to denote the product in $L(X)$, i.e., if $\bar{a}, \bar{b}$ are two elements of $L(X)$ (representing the cosets of $a, b \in A(X)$), then $[\bar{a}, \bar{b}]$ will represent the coset of $a \cdot b$.

Definition 7.1.1 *The Lie algebra $L(X)$ is called the* free *Lie algebra on* X.

Example 7.1.2 Let $X = \{x\}$ be a set consisting of one element. Then because $(x, x) \in I_0$, we have that $L(X)$ is 1-dimensional, with basis $\{\bar{x}\}$.

Let $\pi : A(X) \to L(X)$ be the projection map. Since X is naturally contained in $A(X)$, the map

$$i : X \longrightarrow A(X) \overset{\pi}{\longrightarrow} L(X)$$

maps X into $L(X)$.

Lemma 7.1.3 *Let K be a Lie algebra and let $\psi : X \to K$ be a map from X into K. Then there exists a unique morphism of Lie algebras $\phi : L(X) \to K$ such that $\phi \circ i = \psi$.*

Proof. Let $\phi : A(X) \to K$ be the linear map defined by $\phi(x) = \psi(x)$ for $x \in X$ and $\phi(a \cdot b) = [\phi(a), \phi(b)]_K$, for $a, b \in A(X)$, (where $[\ ,\ ]_K$ denotes the product in K). Then ϕ is a morphism of algebras. Furthermore, for $a \in A(X)$ we have $\phi(a \cdot a) = [\phi(a), \phi(a)]_K = 0$ and similarly for the Jacobi identity. Hence $\phi(I_0) = 0$ so that ϕ induces a morphism from $L(X)$ into K, which we also denote by ϕ. Suppose that there is a second morphism $\phi' : L(X) \to K$ such that $\phi' \circ i = \psi$. Then ϕ and ϕ' coincide on $i(X)$ and because $i(X)$ generates $L(X)$, we must have $\phi = \phi'$. $\square$

Lemma 7.1.4 *The set $i(X)$ is linearly independent. In particular i is injective.*

Proof. Suppose that there are $x_1, \dots, x_m \in X$ such that

$$\lambda_1 i(x_1) + \cdots + \lambda_m i(x_m) = 0 \tag{7.2}$$

for some $\lambda_1, \dots, \lambda_m \in F$. Let K be the 1-dimensional Lie algebra, with underlying vector space F. For $1 \le k \le m$ let $\psi_k : X \to K$ be the map that sends x_k to 1 and all other elements of X to 0. Then by Lemma 7.1.3, there are morphisms $\phi_k : L(X) \to K$ such that $\phi_k \circ i = \psi_k$. By applying ϕ_k to (7.2) we see that $\lambda_k = 0$. As a consequence $i(X)$ is linearly independent. $\square$

Proposition 7.1.5 *Let K be a Lie algebra generated by $X \subset K$. Let $L(X)$ be the free Lie algebra on X. Then there is a surjective morphism $\phi : L(X) \to K$.*

Proof. Let $\psi : X \to K$ be the identity mapping. Then by Lemma 7.1.3 there is a morphism of Lie algebras $\phi : L(X) \to K$ such that $\phi \circ i = \psi$. Furthermore, ϕ is surjective because X generates K. $\square$

By Lemma 7.1.4 we may identify X with its image in $L(X)$. Therefore we will view X as a subset of $L(X)$. Then X generates the free Lie algebra $L(X)$. Furthermore, it is the most general Lie algebra generated by X in the sense that every Lie algebra generated by X is a homomorphic image of $L(X)$ (Proposition 7.1.5).

7.2   Finitely presented Lie algebras

In Section 1.5 we discussed two ways of representing a Lie algebra on a computer: by matrices and by an array of structure constants. Here we describe a third way, namely by generators and relations.

Let L be a Lie algebra generated by a set $X \subset L$. Then by Proposition 7.1.5 there is a surjective morphism $\phi : L(X) \to L$. Set $I = \ker(\phi)$, then by Lemma 1.8.1 we have that $L \cong L(X)/I$. The conclusion is that any Lie algebra is isomorphic to a free Lie algebra modulo an ideal.

Now let X be a finite set and let $I \subset L(X)$ be an ideal generated by a finite set $R \subset L(X)$. Then the Lie algebra $L = L(X)/I$ is said to be *finitely presented*. The pair (X, R) is called a *finite presentation* of L, and we write $L = \langle X \mid R \rangle$.

It is easily seen that any finite-dimensional Lie algebra is finitely presented. Indeed, let L be a finite-dimensional Lie algebra and let $X = \{x_1, \ldots, x_n\}$ be a basis of L. Then L is generated by X. Let

$$[x_i, x_j] = \sum_{k=1}^{n} c_{ij}^k x_k$$

be the multiplication table of L. Set $R = \{[x_i, x_j] - \sum_{k=1}^{n} c_{ij}^k x_k \mid 1 \le i, j \le n\} \subset L(X)$. Then we have that $L = \langle X \mid R \rangle$.

In Section 1.5 we elected the representation by structure constants as our preferred way of representing a Lie algebra. The algorithms given subsequently operate on Lie algebras that are given by an array of structure constants. However, sometimes the natural way to represent a Lie algebra is by a finite presentation. So the question presents itself whether we can produce a table of structure constants for a finitely presented Lie algebra. First of all we note that a finitely presented Lie algebra can be infinite-dimensional (indeed, any free Lie algebra on a finite set is finitely presented). So the best we can hope for is to have an algorithm that constructs a multiplication table for a finitely presented Lie algebra L whenever L happens to be finite-dimensional. The objective of the next sections is to give such an algorithm.

7.3   Gröbner bases in free algebras

Let $L = L(X)/I$ be a finitely presented Lie algebra. Then L is also equal to the quotient $A(X)/J$ where J is the ideal of $A(X)$ generated by the pre-images in $A(X)$ of the generators of I together with the generators of I_0

(7.1). In analogy with Sections 6.1, 6.3 we use an order $<$ on the monomials of $A(X)$ and a Gröbner basis to calculate a set of normal monomials which forms a set of coset representatives for the elements of $A(X)/J$.

In this section we study ideals of the free algebra $A(X)$ in general. Throughout J will be an ideal of $A(X)$ generated by $a_i \in A(X)$ for $i \geq 1$ (so we may have an infinite generating set). We first introduce some machinery that allows us to describe elements of J. Let $\sigma = (m_1, \dots, m_k)$ be a sequence of elements of the free magma $M(X)$, and let $\delta = (d_1, \dots, d_k)$ be a sequence (of length equal to the length of σ) of letters $d_i \in \{l, r\}$. Set $\alpha = (\sigma, \delta)$, then we call α an *appliance*. The integer k is called the *length* of α.

For an appliance α we define a map $P_\alpha : M(X) \to M(X)$ in the following way. If $k = 0$, then $P_\alpha(m) = m$ for all $m \in M(X)$. On the other hand, if $k > 0$, then we set $\beta = ((m_2, \dots, m_k), (d_2, \dots, d_k))$. If $d_1 = l$, then $P_\alpha(m) = P_\beta((m_1, m))$, and $P_\alpha(m) = P_\beta((m, m_1))$ if $d_1 = r$. So the map corresponding to an appliance consists of a series of multiplications and the letter d_i determines whether we multiply from the left or the right by m_i. Finally we extend the map P_α to a map $P_\alpha : A(X) \to A(X)$ by linearity.

Example 7.3.1 Let $X = \{x, y, z\}$ and set $\sigma = (x, (y, z), z)$ and $\delta = (l, r, r)$, and $\alpha = (\sigma, \delta)$. Then $P_\alpha(m) = (((x, m), (y, z)), z)$ for $m \in M(X)$.

Now a general element f of J will be of the form

$$f = \lambda_1 P_{\alpha_1}(a_{i_1}) + \cdots + \lambda_r P_{\alpha_r}(a_{i_r}). \tag{7.3}$$

An element $m_1 \in M(X)$ is called a *factor* of $m_2 \in M(X)$ if there is an appliance α such that $m_2 = P_\alpha(m_1)$.

Throughout this section $<$ will be a total order on $M(X)$ that is *multiplicative* (i.e., $m < n$ implies $(m, p) < (n, p)$ and $(p, m) < (p, n)$ for all $p \in M(X)$). Furthermore we assume that $<$ satisfies the *descending chain condition*, i.e., if $m_1 \geq m_2 \geq m_3 \geq \dots$ is a descending chain of monomials then there is a $k > 0$ such that $m_k = m_j$ for $j \geq k$. Such an order can for instance be constructed by fixing it arbitrarily on X, and by postulating that $\deg(m) < \deg(n)$ implies $m < n$. Furthermore the elements of equal degree are ordered lexicographically, i.e., if $\deg(m) = \deg(n) > 1$ (so that $m = (m', m'')$ and $n = (n', n'')$) then $m < n$ if $m' < n'$ or ($m' = n'$ and $m'' < n''$).

As usual, the leading monomial $\mathrm{LM}(f)$ of $f \in A(X)$ is the largest monomial occurring in f with non-zero coefficient. Since $<$ is multiplicative we have that $\mathrm{LM}(P_\alpha(f)) = P_\alpha(\mathrm{LM}(f))$ for any appliance α and $f \in A(X)$.

Set $\mathrm{LM}(J) = \{\mathrm{LM}(f) \mid f \in J\}$. Then $N(J) = \{m \in M(X) \mid m \notin \mathrm{LM}(J)\}$ is called the set of *normal monomials* of $A(X)$ modulo J. Let $C(J)$ be the span of $N(J)$ inside $A(X)$. Then by a completely analogous argument to the one used in the proof of Proposition 6.1.1 we have that $A(X) = C(J) \oplus J$. And this means that the cosets of the elements of $N(J)$ form a basis of $A(X)/J$. Also for $f \in A(X)$ there are unique $v \in C(J)$ and $g \in J$ such that $f = v + g$. The element v is called the *normal form* of f modulo J; it is denoted by $\mathrm{Nf}_J(f)$, or by $\mathrm{Nf}(f)$ if it is clear which ideal we mean.

As a first attempt at calculating the normal form of $f \in A(X)$ modulo J we reduce f modulo the generators of J. Let $G \subset A(X)$ and $f \in A(X)$ then f is called *reduced* modulo G if for all $g \in G$ we have that $\mathrm{LM}(g)$ is not a factor of any monomial occurring in f with non-zero coefficient. If f is not reduced modulo G, then there is a monomial m in occurring in f, and an element $g \in G$ and an appliance α such that $P_\alpha(\mathrm{LM}(g)) = m$. Let λ be the coefficient of m in f and μ the coefficient of $\mathrm{LM}(g)$ in g. Set $h = f - \frac{\lambda}{\mu}P_\alpha(g)$. We say that f reduces modulo G in one step to h. More generally, we say that f reduces to f' modulo g if there exist $f_1, \ldots, f_k$ such that $f = f_1$, $f' = f_k$ and f_i reduces modulo g in one step to f_{i+1} for $1 \leq i \leq k - 1$. Because $<$ satisfies the descending chain condition we have that any maximal sequence of reduction steps finishes in a finite number of steps with an element that is reduced modulo G.

Definition 7.3.2 *A set $G \subset A(X)$ generating an ideal J is called a Gröbner basis if for all $f \in J$ there is a $g \in G$ such that $\mathrm{LM}(g)$ is a factor of $\mathrm{LM}(f)$.*

Proposition 7.3.3 *Let G be a Gröbner basis for the ideal J of $A(X)$. Then $N(J)$ is the set of all monomials $m \in M(X)$ such that for all $g \in G$, $\mathrm{LM}(g)$ is not a factor of m. Also if $f \in A(X)$ reduces to $f' \in A(X)$ modulo G, where f' is reduced modulo G, then $f' = \mathrm{Nf}(f)$.*

Proof. The first statement is a direct consequence of the definition of Gröbner basis. Since G is a Gröbner basis, f' is reduced modulo J, i.e., $f' \in C(I)$. Furthermore, $f' = f \bmod J$ and therefore $f' = \mathrm{Nf}(f)$. □

Definition 7.3.4 *A set $G = \{a_1, a_2, \ldots\} \subset A(X)$ is said to be self-reduced if $\mathrm{LM}(a_i)$ is no factor of $\mathrm{LM}(a_j)$ for $i \neq j$.*

We remark that if G is a finite subset of $A(X)$, then by successively reducing the elements of G modulo each other we can compute a finite set

G' such that G' is self-reduced and generates the same ideal as G. The rest of this section is devoted to showing that a self-reduced set is a Gröbner basis.

Lemma 7.3.5 *Let $m_1, m_2 \in M(X)$ be such that m_1 is not a factor of m_2 and m_2 is not a factor of m_1. Suppose that there are appliances α, β such that $P_\alpha(m_1) = P_\beta(m_2)$. Then m_2 is a factor of $P_\alpha(n)$ and m_1 is a factor of $P_\beta(n)$ for all $n \in M(X)$.*

Proof. First we note that the lengths of α and β are at least 1, since otherwise one of m_1, m_2 would be a factor of the other one. Let $\beta = (\sigma, \delta)$, where $\sigma = (p_1, \dots, p_k)$ and $\delta = (d_1, \dots, d_k)$. For $0 \le i \le k$ we set

$$\beta_i = ((p_1, \dots, p_i), (d_1, \dots, d_i))$$

and $n_i = P_{\beta_i}(m_2)$ (so that $n_0 = m_2$). Now m_1 is not a factor of n_0, but it is a factor of n_k. So there is an index $i > 0$ such that m_1 is not a factor of n_{i-1} but m_1 is a factor of n_i. We have that $n_i = (n_{i-1}, p_i)$ or $n_i = (p_i, n_{i-1})$. And since $n_i \ne m_1$ (otherwise m_2 would be a factor of m_1) we have that m_1 must be a factor of p_i and hence m_1 is a factor of $P_\beta(n)$ for all n in $M(X)$. The proof for P_α is similar. $\square$

Lemma 7.3.6 *Let $m \in M(X)$ and let $\alpha \ne \beta$ be appliances such that $P_\alpha(m) = P_\beta(m)$. Then m is a factor of $P_\alpha(n)$ and of $P_\beta(n)$ for all $n \in M(X)$.*

Proof. Let k and l be the lengths of α and β respectively. If one of them is zero, then the other must be zero as well and $\alpha = \beta$. So k and l are both non-zero. We prove the statement by induction on $k + l$. Suppose that $\alpha = (\sigma_1, \delta_1)$ and $\beta = (\sigma_2, \delta_2)$, where $\sigma_1 = (m_1, \dots, m_k)$ and $\sigma_2 = (n_1, \dots, n_l)$. Write $P_\alpha(m) = (a, b)$; then $m_k = a$ or $m_k = b$. First suppose that $m_k = a$. If also $n_l = a$ then we can erase the last element from α and β and obtain two appliances α', β' of smaller length that also satisfy the hypothesis of the lemma. So we can conclude by induction. On the other hand, if $n_l = b$, then $a = P_{\beta'}(m)$ where β' is obtained from β by erasing n_l. Hence m is a factor of m_k and in the same way, m is a factor of n_l. As a consequence m is a factor of $P_\alpha(n)$ and of $P_\beta(n)$ for all $n \in M(X)$. The proof for the case where $m_k = b$ is completely analogous. $\square$

Now we turn to our ideal J generated by a_i for $i \ge 1$. If $m \in M(X)$ then by $J_{<m}$ we denote the subspace of J consisting of all elements of the form $\lambda_1 P_{\alpha_1}(a_{i_1}) + \cdots + \lambda_r P_{\alpha_r}(a_{i_r})$ such that $\mathrm{LM}(P_{\alpha_j}(a_{i_j})) < m$ for $1 \le i \le r$.

Lemma 7.3.7 *Let b_1, b_2 be generators of J and set $m_i = \mathrm{LM}(b_i)$ for $i = 1, 2$. If $b_1 \neq b_2$ then we assume that m_1 is not a factor of m_2 and m_2 is not a factor of m_1. Let α_1, α_2 be two appliances such that $m = P_{\alpha_1}(m_1) = P_{\alpha_2}(m_2)$. Then $P_{\alpha_1}(b_1) - P_{\alpha_2}(b_2) \in J_{<m}$.*

Proof. First we deal with the case where $b_1 \neq b_2$. Then by Lemma 7.3.5 we have that m_2 is a factor of $P_{\alpha_1}(n)$ and m_1 is a factor of $P_{\alpha_2}(n)$ for all $n \in M(X)$. So m is a bracketed expression containing m_1 and m_2 as subexpressions. Furthermore, $P_{\alpha_1}(n)$ is obtained from m by replacing one occurrence of m_1 in m at position p_1 by n. Similarly, there is a position p_2 in m such that $P_{\alpha_2}(n)$ is obtained from m by replacing the m_2 at p_2 by n. Now we define a function $P_m : A(X) \times A(X) \to A(X)$. Let $n_1, n_2 \in M(X)$; then $P_m(n_1, n_2)$ is obtained from m by replacing m_1 at position p_1 in m by n_1, and replacing the m_2 at position p_2 by n_2. Furthermore the function P_m is extended to the whole of $A(X) \times A(X)$ by bilinearity.

To ease notation a little we suppose that the coefficient of m_i in b_i is 1 (it is obvious that we can do this without loss of generality). Set $\bar{b}_i = b_i - m_i$ for $i = 1, 2$. And let $P_1, P_2 : A(X) \to A(X)$ be functions defined by $P_1(n) = P_m(n, -\bar{b}_2)$ and $P_2(n) = P_m(-\bar{b}_1, n)$. Then P_1 and P_2 are linear combinations of functions P_γ for appliances γ. Also we have that $P_{\alpha_1}(n) = P_m(n, m_2)$ and $P_{\alpha_2}(n) = P_m(m_1, n)$. So

$$
\begin{aligned}
P_{\alpha_1}(b_1) - P_{\alpha_2}(b_2) &= P_m(m_1, m_2) + P_m(\bar{b}_1, m_2) - P_m(m_1, m_2) - P_m(m_1, \bar{b}_2) \\
&= P_m(\bar{b}_1, m_2) - P_m(m_1, \bar{b}_2).
\end{aligned}
$$

And

$$
\begin{aligned}
P_1(b_1) - P_2(b_2) &= P_m(m_1, -\bar{b}_2) + P_m(\bar{b}_1, -\bar{b}_2) - P_m(-\bar{b}_1, m_2) - P_m(-\bar{b}_1, \bar{b}_2) \\
&= P_m(m_1, -\bar{b}_2) - P_m(-\bar{b}_1, m_2).
\end{aligned}
$$

Hence $P_{\alpha_1}(b_1) - P_{\alpha_2}(b_2) = P_1(b_1) - P_2(b_2)$. Furthermore, $P_i(b_i) \in J_{<m}$.

Now we consider the case where $b_1 = b_2$. If in addition $\alpha_1 = \alpha_2$ then there is nothing to prove. So suppose that $\alpha_1 \neq \alpha_2$, and set $q = m_1 = m_2$. Then $P_{\alpha_1}(q) = P_{\alpha_2}(q)$. By Lemma 7.3.6, q is a factor of $P_{\alpha_1}(n)$ and of $P_{\alpha_2}(n)$ for all $n \in M(X)$. Hence we can proceed as above. $\square$

Proposition 7.3.8 *Let $G = \{a_1, a_2, \dots\} \subset A(X)$ be a self-reduced set. Then G is a Gröbner basis.*

Proof. Let J be the ideal of $A(X)$ generated by G. Let $f \in J$ and write

$$
f = \lambda_1 P_{\alpha_1}(a_{i_1}) + \cdots + \lambda_r P_{\alpha_r}(a_{i_r}),
$$

where $\alpha_1, \ldots, \alpha_r$ are certain appliances. We show that there is a $g \in G$ such that $\mathrm{LM}(g)$ is a factor of $\mathrm{LM}(f)$. Set $n_j = \mathrm{LM}(P_{\alpha_j}(a_{i_j}))$. We suppose that the summands have been ordered such that the leading monomials are in decreasing order, i.e.,

$$m = n_1 = \ldots = n_k > n_{k+1} \geq \cdots \geq n_r.$$

If $k = 1$, then $\mathrm{LM}(f) = \mathrm{LM}(P_{\alpha_1}(a_{i_1})) = P_{\alpha_1}(\mathrm{LM}(a_{i_1}))$ and there is nothing to prove. So assume that $k > 1$. Then by Lemma 7.3.7 we have that $P_{\alpha_1}(a_{i_1}) - P_{\alpha_2}(a_{i_2}) \in J_{<m}$. Furthermore

$$f = \lambda_1(P_{\alpha_1}(a_{i_1}) - P_{\alpha_2}(a_{i_2})) + (\lambda_1 + \lambda_2)P_{\alpha_2}(a_{i_2}) + \sum_{k=3}^{r} \lambda_k P_{\alpha_k}(a_{i_k}).$$

Since the first term of this expression is in $J_{<m}$ we can write it as a linear combination of $P_{\gamma_k}(a_{i_k})$ such that $\mathrm{LM}(P_{\gamma_k}(a_{i_k})) < m$. Hence we find a new expression for f where the term m has decreased (in the case where $k = 2$ and $\lambda_1 + \lambda_2 = 0$), or the number k has decreased. Now because $<$ satisfies the descending chain condition, we can conclude by induction. $\square$

7.4 Constructing a basis of a finitely presented Lie algebra

In this section we use the results of the preceding section to describe an algorithm for calculating a basis of a finitely presented Lie algebra, that is finite-dimensional.

Let X be a finite set and let $R = \{r_1, \ldots, r_s\}$ be a finite subset of the free Lie algebra $L(X)$. Let I be the ideal of $L(X)$ generated by R and set $L = L(X)/I$. Let $\pi : A(X) \to L(X)$ be the projection map. For $1 \leq i \leq s$ we let $h_i \in A(X)$ be such that $\pi(h_i) = r_i$. Then we let $J \subset A(X)$ be the ideal generated by the set G consisting of the generators of I_0 (7.1) together with the h_i. Then $A(X)/J \cong L$. Hence we are in the situation considered in Section 7.3. We have an ideal J of $A(X)$ generated by an infinite number of generators, and supposing that the quotient $A(X)/J$ is finite-dimensional, we want to construct a basis of it. By Propositions 7.3.3 and 7.3.8 this is easy as soon as we have found a self-reduced generating set for J. This is not immediately straightforward as J is generated by an infinite number of elements.

In the sequel we suppose that the monomials in $M(X)$ are ordered by an order $<$ that is not only multiplicative and satisfies the descending chain condition, but is also degree compatible, i.e., $\deg(m) < \deg(n)$ implies $m < n$ for all $m, n \in M(X)$.

We recall that $M_k(X)$ is the set of all $m \in M(X)$ such that $\deg(m) = k$, and we set

$$G_k = \{g \in G \mid \deg(g) \leq k\}.$$

We let $J_k \subset A(X)$ be the ideal generated by G_k. Note that G_k is a finite set. So by doing successive reductions of the elements of G_k we can calculate a self-reduced set G'_k that also generates J_k. Set

$$\mathrm{LM}(J_k) = \{\mathrm{LM}(f) \mid f \in J_k\},$$

and for $k, l > 0$, let $B_{k,l}$ be the set of all $m \in M_k(X)$ that are not contained in $\mathrm{LM}(J_l)$. It is straightforward to see that $J = \bigcup_{l>0} J_l$. Hence, since $A(X)/J$ is finite-dimensional there are $k_0, l_0 \geq 0$ such that $B_{k_0+1,l_0} = \emptyset$. Now since $J_l \subset J_{l+1}$, also $B_{k_0+1,l} = \emptyset$ for $l \geq l_0$. In the sequel we let l be the smallest number such that $l \geq l_0$, $l \geq 2k_0 + 1$ and such that $h_i \in G_l$ for $1 \leq i \leq s$. We set

$$B = \bigcup_{1 \leq i \leq k_0} B_{i,l}.$$

In the sequel we prove that $J_l = J$, thereby proving that J is generated by G_l. Since G_l is finite this allows us to calculate a self-reduced generating set for J.

Lemma 7.4.1 *Let* $m \in M_{d+1}(X)$, *then modulo elements of* J_{d+1}, m *can be written as a linear combination of elements of the form* (n, x) *where* $n \in M_d(X)$ *and* $x \in X$.

Proof. This is trivial if $d = 0$. So suppose that $d \geq 1$ and $m = (m', m'')$. We prove the lemma by induction on $\deg(m'')$. If $m'' \in X$ then there is nothing to prove. So suppose that $m'' = (a, y)$ for $a, y \in M(X)$. Furthermore, by induction (note that $J_k \subset J_{d+1}$ whenever $k \leq d + 1$) we may assume that $y \in X$. Since J_{d+1} contains all relations of the form $(p, q) + (q, p)$ such that $\deg(p) + \deg(q) \leq d + 1$ and all Jacobi identities of degree $\leq d + 1$ we can write modulo J_{d+1},

$$(m', (a, y)) = ((y, m'), a) + ((m', a), y).$$

The second summand is of the required form. Also, since $\deg(a) < \deg(m'')$, by induction the first summand can be written as a linear combination of elements of the required form. $\qquad\qquad\square$

Lemma 7.4.2 *Let $f \in A(X)$; then modulo J_l, f can be written as a unique linear combination of elements from B.*

Proof. We prove the statement by induction on $\mathrm{LM}(f)$; as $<$ satisfies the descending chain condition the induction is correct. We note that B is the set of normal monomials of degree $\leq k_0$ of $A(X)$ modulo J_l. From this follows uniqueness. Set $m = \mathrm{LM}(f)$ and write $f = \lambda m + \bar{f}$. First suppose that $\deg(m) \leq k_0$. By induction $\bar{f}$ is modulo J_l equal to a linear combination of elements of B. So if $m \in B$ then the same holds for f. On the other hand, if $m \notin B$, then m reduces modulo J_l to an element $g \in A(X)$. Since $<$ is degree compatible also $\deg(\mathrm{LM}(g)) \leq k_0$. Hence by induction g can modulo J_l be written as a linear combination of elements of B and therefore so can f.

If $\deg(m) = k_0 + 1$ then $m \in \mathrm{LM}(J_l)$ and again we can reduce m to an element $g \in A(X)$ modulo J_l. Because $<$ is degree compatible, $\deg(\mathrm{LM}(g)) \leq k_0 + 1$ and again we conclude by induction.

If $k_0 + 1 < \deg(m) \leq 2k_0 + 1$, then using Lemma 7.4.1 we can write m modulo J_l as a linear combination of elements of the form (n, x), where $\deg(n) = \deg(m) - 1$ and $x \in X$. By induction, we can write n modulo J_l as a linear combination of elements of B. So m reduces modulo J_l to a linear combination of elements p such that $\deg(p) \leq k_0 + 1$. But above we have seen that such a p can be reduced to a linear combination of elements of B modulo J_l.

Finally suppose that $\deg(m) > 2k_0 + 1$ and $m = (m', m'')$. By induction m' and m'' reduce modulo J_l to linear combinations of elements in B. So m can modulo J_l be rewritten as a linear combination of elements of the form (a, b) where $a, b \in B$. But since $\deg((a, b)) \leq 2k_0$ these expressions all reduce modulo J_l to linear combinations of the elements of B. □

We abbreviate the left hand side of the Jacobi identity on a, b, c by $\mathrm{Jac}(a, b, c)$, i.e.,

$$\mathrm{Jac}(a, b, c) = (a, (b, c)) + (b, (c, a)) + (c, (a, b)).$$

Lemma 7.4.3 *Let K be an ideal of $A(X)$ containing all $(m_1, m_2) + (m_2, m_1)$ for $m_1, m_2 \in M(X)$. Suppose that $\mathrm{Jac}(x, m, n) = 0 \pmod{K}$ for all $m, n \in M(X)$, and all $x \in X$. Then $\mathrm{Jac}(p, m, n) = 0 \pmod{K}$ for all $m, n, p \in M(X)$.*

Proof. The proof is by induction on $\deg(p)$. If $\deg(p) = 1$, then there is nothing to prove. If $\deg(p) > 1$, then $p = (a, b)$. Using induction and relations $(m_1, m_2) = -(m_2, m_1)$ (mod K) we calculate modulo K,

$$
\begin{aligned}
((a,b),(m,n)) &= - \, ((b,(m,n)),a) - (((m,n),a),b) \\
&= ((m,(n,b)),a) + ((n,(b,m)),a) \\
&\quad + (((n,a),m),b) + (((a,m),n),b) \\
&= - \, (((n,b),a),m) - ((a,m),(n,b)) \\
&\quad - (((b,m),a),n) - ((a,n),(b,m)) \\
&\quad - ((m,b),(n,a)) - ((b,(n,a)),m) \\
&\quad - ((n,b),(a,m)) - ((b,(a,m)),n) \\
&= - \, (((n,b),a) + (b,(n,a)),m) - (((b,m),a) + (b,(a,m)),n) \\
&= ((n,(a,b)),m) - ((m,(a,b)),n).
\end{aligned}
$$

Hence $\mathrm{Jac}(p,m,n) = 0$ (mod K). □

Now let $p, q \in M(X)$. Then by Lemma 7.4.2, $(p,q) + (q,p)$ can modulo J_l be written as a linear combination of elements $(p',q') + (q',p')$ where $p', q' \in B$. But since $l \geq 2k_0 + 1$ all these elements lie in J_l. Also by Lemma 7.4.2, modulo J_l all elements $\mathrm{Jac}(x,m,n)$ for $x \in X$ and $m, n \in M(X)$ can be written as linear combinations of elements of the form $\mathrm{Jac}(x, m', n')$, where $m', n' \in B$. But since $l \geq 2k_0 + 1$ all those elements are in J_l. So J_l satisfies the hypothesis of Lemma 7.4.3. It follows that J_l contains all elements $\mathrm{Jac}(p,m,n)$ for all $p, m, n \in M(X)$. Also J_l contains (a pre-image of) R. The conclusion is that all generators of J lie in J_l, i.e., $J_l = J$. So B is the set of normal monomials of $A(X)$ modulo J and hence B is a basis of $A(X)/J$. Furthermore a self-reduced generating set G'_l of J_l will be a Gröbner basis of J (Proposition 7.3.8). It follows that B is the set of all monomials m such that $\deg(m) \leq k_0$ and $\mathrm{LM}(g)$ is not a factor of m for all $g \in G'_l$.

We summarize our findings in the following algorithm.

Algorithm FpLieAlgebra

Input: a finite set X and a finite subset $R \subset L(X)$.

Output: a basis of $L(X)/I$ where I is the ideal generated by R.

Step 1 Let J be the ideal of $A(X)$ generated by a pre-image of R together with the elements (7.1). Set $l := 1$ and repeat the following:

Step 1a Calculate a self-reduced generating set G'_l of J_l.

Step 1b For $1 \leq k \leq l$ calculate the set $B_{k,l}$ of monomials $m \in M(X)$ of degree k such that $\mathrm{LM}(b)$ is not a factor of m for all $b \in G'_l$.

Step 1c If there is a $k \in \{1, \dots, l-1\}$ such that $B_{k+1,l} = \emptyset$ then go to Step 2. Otherwise increase l by 1 and return to Step 1a.

Step 2 Increase l such that J_l contains a pre-image of R in $A(X)$. Set $t = \max(2k+1, l)$.

Step 3 Calculate a self-reduced generating set G'_t of J_t. Calculate the set B of all monomials m in $M(X)$ of degree $\leq k$ such that $\mathrm{LM}(b)$ is not a factor of m for $b \in G'_t$. Return B.

Remark. This algorithm can easily be extended so as to produce a multiplication table of $L(X)/I$. Indeed, any product of elements of B can be reduced (uniquely) modulo the elements of G'_t to a linear combination of the elements of B.

Example 7.4.4 Let $X = \{x, y\}$, $R = \{[x, [x, y]] - [x, y], [y, [y, [x, y]]]\}$ (as usual we use the brackets $[\ ,\ ]$ to denote the product in $L(X)$). While performing the algorithm it is much more convenient to use the anticommutativity relations $(m, m) = 0$ and $(m, n) + (n, m) = 0$ directly to reduce expressions to a convenient form, than to put them into the ideal first. So in the ideal we only consider (pre-images) of the elements o R together with the Jacobi identities.

By anticommutativity we immediately have that $\mathrm{Jac}(m, n, p) = 0$ if any two of m, n, p are equal. It follows that the Jacobi identity of lowest degree is $\mathrm{Jac}(x, y, (x, y))$. So we consider the set G_4 consisting of this Jacobi identity together with the elements of R. We have

$$\mathrm{Jac}(x, y, (x, y)) = (x, (y, (x, y))) + (y, ((x, y), x)) + ((x, y), (x, y)).$$

The last summand is zero by anticommutativity. The second summand is equal to $-(y, (x, (x, y)))$ which is equal to $-(y, (x, y))$ by the first relation in R. So the self-reduced set G'_4 generating the same ideal as G_4 is

$$G'_4 = \{(x, (x, y)) - (x, y), (y, (y, (x, y))), (x, (y, (x, y))) - (y, (x, y))\}.$$

Now we enumerate the basis elements of $A(X)$ upto degree 4 modulo the elements in G'_4. By anticommutativity we see that these are x, y, (x, y), $(y, (x, y))$ (an element of degree 4 is the product of an element of degree 1 and an element of degree 3, but all those elements are leading monomials of elements of G'_4). Therefore $B_{4,4} = \emptyset$. So we set $t = 7$ and reduce all elements of G_7. We only have to consider Jacobi identities of those monomials that do not already reduce modulo G_4'. The first of these is

$$\operatorname{Jac}(x, y, (y, (x, y))) = (x, (y, (y, (x, y)))) + (y, ((y, (x, y)), x))$$
$$+ ((y, (x, y)), (x, y)).$$

The first two summands reduce to zero modulo G_4'. Hence only the last summand remains. Using this, the Jacobi identities $\operatorname{Jac}(x, (x, y), (y, (x, y)))$ and $\operatorname{Jac}(y, (x, y), (y, (x, y)))$ reduce to zero. So G_7' consists of the elements of G_4' together with $((y, (x, y)), (x, y))$. Hence $B = \{x, y, (x, y), (y, (x, y))\}$ is a basis of $L(X)/I$ where I is the ideal of $L(X)$ generated by R.

7.5   Hall sets

Let X be a set. We know that the free Lie algebra $L(X)$ is spanned by (the image of) the set $M(X)$ of all bracketed expressions in the elements of X. However, this set is by no means linearly independent. Indeed, all elements of the form (m, m) are zero and also the Jacobi identity yields linear dependencies. If we want to calculate in $L(X)$ then we need a basis of this algebra (i.e., a set of coset representatives of $A(X)$ modulo the ideal I_0) and a way of rewriting a product of basis elements as a linear combination of basis elements. A solution to this problem is formed by the so-called *Hall sets*. In this section we define the concept of Hall set. Then in subsequent sections we prove that a Hall set yields a basis of the free Lie algebra. An algorithm for rewriting an element of $M(X)$ as a linear combination of the elements of a Hall set will be part of this proof.

Let $<$ be a total order on $M(X)$. Relative to $<$ we define a set $H \subset M(X)$ by

$$X \subset H, \tag{7.4}$$

if $h_1, h_2 \in M(X)$ then $(h_1, h_2) \in H$ if and only if $h_1, h_2 \in H$ and

$$h_1 < h_2, \tag{7.5}$$
$$h_1 \in X \text{ or } h_1 = (a, b) \text{ with } b \geq h_2.$$

A set satisfying these requirements is called a *quasi-Hall* set. The order $<$ is said to be a *Hall order* if $(h_1, h_2) < h_2$ for all $h_1, h_2 \in H$ such that $(h_1, h_2) \in H$. In that case H is called a *Hall set*.

Example 7.5.1 Hall orders are easily constructed. Inside each set $M_d(X)$ the order is chosen arbitrarily. Furthermore, if $m, n \in M(X)$ have different degrees, then $m < n$ if and only if $\deg(m) > \deg(n)$. Then $(h_1, h_2) < h_2$ for all $h_1, h_2 \in M(X)$. So $<$ is a Hall order.

We note that a Hall order defines a unique Hall set. Using conditions (7.4) and (7.5) we formulate an algorithm for recognizing whether a given element of $M(X)$ lies in a given Hall set. We stress that we do not need to know all elements of the Hall set for this; we only need to be able to calculate the order of elements of the Hall set.

Algorithm IsHallElement

Input: a Hall order $\leq$ on $M(X)$ and an element $h \in M(X)$.

Output: true if h is in the Hall set corresponding to $\leq$, false otherwise.

Step 1 If $h \in X$ then return true.

Step 2 Write $h = (h', h'')$. If IsHallElement$(\leq, h')$ or IsHallElement$(\leq, h'')$ is false then return false.

Step 3 If $h' \geq h''$ then return false.

Step 4 If $h' \in X$ then return true. Otherwise write $h' = (a, b)$ and if $b < h''$ then return false. Otherwise return true.

Comments: The two recursive calls in Step 2 finish and give the correct output by induction on the degree. Then the fact that the initial call gives the correct output is a direct consequence of (7.4), (7.5).

We also have an algorithm for calculating the elements of a Hall set upto a given degree.

Algorithm HallSet

Input: a finite set X and a Hall order $\leq$ on $M(X)$ and a number $d > 0$.

Output: all elements of the Hall set corresponding to $\leq$ upto degree d.

Step 1 Set $H := X$ and $k := 2$.

Step 2 If $k = d + 1$ then return H, otherwise go to Step 3.

Step 3 Compute the set S of all pairs $(h, g) \in H \times H$ such that $\deg(h) + \deg(g) = k$ and IsHallElement$(\leq, (h, g))$ =true.

Step 4 Set $H := H \cup S$, $k := k + 1$ and return to Step 2.

Example 7.5.2 Let $X = \{x, y\}$. We construct the Hall set H upto degree 4 corresponding to a Hall order as in Example 7.5.1. That is we fix the order arbitrarily for elements of equal degree and if h, g have unequal degree then

$h < g$ if and only if $\deg(h) > \deg(g)$. First we choose $x < y$. Then (x, y) is the only element of degree 2 in H. And necessarily $(x, y) < x < y$. Continuing, $((x, y), x)$ and $((x, y), y)$ are the elements of degree 3 of H. We choose $((x, y), x) < ((x, y), y)$. Then the elements of degree 4 of H are $(((x, y), x), x)$, $(((x, y), y), x)$ and $(((x, y), y), y)$.

7.6   Standard sequences

In this section we fix a set X and a Hall set $H \subset M(X)$ relative to the Hall order $<$. The elements of H will be called *Hall elements*. If a Hall element $h \in H$ does not lie in X, then by h' and h'' we will denote its left and right factor (so $h = (h', h'')$).

The elements of X are called *letters*. From Section 6.1 we recall that X^* is the set of all words in the elements of X. In this section we make a first step towards proving that a Hall set forms a basis of the free Lie algebra. We study standard sequences of Hall elements and establish a bijection between decreasing standard sequences and words in X^*. In the next section this bijection will be used to prove that a Hall set is linearly independent.

Definition 7.6.1 *A sequence of Hall elements $s = (h_1, \dots, h_n)$, where $n \geq 1$ is called a* standard sequence *if for $1 \leq i \leq n$ we have either $h_i \in X$ or $h_i = (h_i', h_i'')$ with $h_i'' \geq h_{i+1}, \dots, h_n$.*

Example 7.6.2 A sequence of letters is always standard. A sequence $(h_1, \dots, h_n)$ is said to be decreasing if $h_1 \geq h_2 \geq \cdots \geq h_n$. Any decreasing sequence of Hall elements $s = (h_1, \dots, h_n)$ is standard. Indeed, if $h_i = (h_i', h_i'')$ then because $\leq$ is a Hall order we have $h_i'' > h_i \geq h_{i+1}, \dots, h_n$.

If a standard sequence of Hall elements is not decreasing, then we apply a series of rewriting steps in order to make it decreasing.

Definition 7.6.3 *Let $s = (h_1, \dots, h_n)$ be a standard sequence of Hall elements. A* rise *of s is an index i such that $h_i < h_{i+1}$.*

Let $s = (h_1, \dots, h_n)$ be a standard sequence. Suppose that s is not decreasing, and let i be the right-most rise of s. Then we say that s *reduces in one step* to s' where $s' = (h_1, \dots, h_{i-1}, (h_i, h_{i+1}), h_{i+2}, \dots, h_n)$.

Lemma 7.6.4 *The sequence s' is a standard sequence of Hall elements.*

Proof. First we prove that (h_i, h_{i+1}) is a Hall element. Certainly h_i and h_{i+1} are elements of H. Also $h_i < h_{i+1}$ because i is a rise of s. Furthermore, if h_i is not an element of X, then $h_i = (h_i', h_i'')$ and $h_i'' \geq h_{i+1}$ because s is a standard sequence. So, by (7.5), (h_i, h_{i+1}) is a Hall element.

Now we prove that s' is standard. First we have $h_{i+1} \geq h_{i+2}, \ldots, h_n$ because i is the right-most rise of s. Secondly, for $j = 1, \ldots, i-1$ we have that either h_j is a letter, or $h_j'' \geq h_{j+1}, \ldots, h_{i+1}$ (because s is standard) $> (h_i, h_{i+1})$ (because $\leq$ is a Hall order). The conclusion is that

$$h_j'' \geq h_{j+1}, \ldots, h_{i-1}, (h_i, h_{i+1}), h_{i+2}, \ldots, h_n.$$

So s' is standard.   □

Let s and t be two standard sequences of Hall elements, then we say that s *reduces* to t if there are standard sequences $s_1, \ldots, s_k$ such that $s_1 = s$, $s_k = t$ and s_i reduces to s_{i+1} in one step for $1 \leq i \leq k-1$.

Lemma 7.6.5 *Let s be a standard sequence of Hall elements. Then s reduces to a decreasing standard sequence in a finite number of steps.*

Proof. We use induction on the length of s. If s has length 1, then s is obviously decreasing. Now suppose that s has length n. If s has no rises then s is decreasing. On the other hand, if s has rises, then s reduces in one step to a sequence s' of length $n-1$. By induction s' reduces to a decreasing standard sequence, hence so does s.   □

The reduction process makes a sequence shorter and the degree of some of its elements bigger. We consider an inverse process that we call *unpacking*; after a finite number of steps it arrives at a sequence of letters. Let $s = (h_1, \ldots, h_n)$ be a standard sequence such that not all of its elements lie in X. Let h_r be the *left-most* element of s that does not lie in X. Then $h_r = (h_r', h_r'')$ and set $s' = (h_1, \ldots, h_{r-1}, h_r', h_r'', h_{r+1}, \ldots, h_n)$. We say that s *unpacks in one step* to s'.

Lemma 7.6.6 *Let s be a standard sequence of Hall elements that unpacks in one step to s'. Then s' is a standard sequence of Hall elements.*

Proof. Let s and s' be as above. Firstly $h_r' \in X$ or $h_r' = (a, b)$ and $b \geq h_r''$ (by definition of Hall set) $\geq h_{r+1}, \ldots, h_n$ (because s is a standard sequence). Also $h_r'' \in X$ or $h_r'' = (a, b)$ and $b > h_r''$ (because $>$ is a Hall order) $\geq h_{r+1}, \ldots, h_n$.   □

Proposition 7.6.7 *Let s be a decreasing sequence of Hall elements. Denote by t_k the standard sequence of Hall elements obtained from s by k unpacking steps ($t_0 = s$). Then t_k reduces to t_{k-1} in one step for $k \geq 1$.*

Proof. We prove the statement by induction on k. First let $k = 1$. We obtain t_1 from t_0 by replacing a Hall element (h'_r, h''_r) by the two elements h'_r, h''_r. Since t_0 is decreasing we have that the right-most rise of t_1 occurs at the position of h'_r, h''_r. Hence t_1 reduces in one step to t_0.

Now suppose $k \geq 1$. By induction we know that t_k reduces to t_{k-1} in one step. We prove that t_{k+1} reduces to t_k in one step. Write $t_{k-1} = (h_1, \dots, h_r, h_{r+1}, \dots, h_n)$ where $h_r = (h'_r, h''_r)$ is the left-most element not in X. Then

$$t_k = (h_1, \dots, h_{r-1}, h'_r, h''_r, h_{r+1}, \dots, h_n)$$

and $h'_r < h''_r$ is the right-most rise of t_k (by induction), i.e., the subsequence $h_{r+1}, \dots, h_n$ is decreasing. We prove that t_{k+1} reduces to t_k in one step by considering a few cases:

1. $h'_r, h''_r \in X$; here the left-most element in t_k that is not in X, will occur after h''_r and because the subsequence starting with h_{r+1} is decreasing, t_{k+1} reduces to t_k.

2. $h'_r \in X$ but h''_r not. Then $h''_r = (a, b)$ and

$$t_{k+1} = (h_1, \dots, h'_r, a, b, h_{r+1}, \dots, h_n),$$

   and $b \geq h_{r+1}, \dots, h_n$ (because t_k is a standard sequence). So the right-most rise of t_{k+1} occurs at the position of a, b.

3. h'_r is not in X. Then $h'_r = (a, b)$ and

$$t_{k+1} = (h_1, \dots, h_{r-1}, a, b, h''_r, h_{r+1}, \dots, h_n),$$

   and as $((a, b), h''_r)$ is a Hall element, we have $b \geq h''_r \geq h_{r+1}, \dots, h_n$ and the right-most rise in t_{k+1} occurs at a, b.

In all cases we see that t_{k+1} reduces in one step to t_k and we are done. $\square$

Let $\varphi : M(X) \longrightarrow X^*$ be defined as follows: $\varphi(x) = x$ for $x \in X$, and $\varphi((t', t'')) = \varphi(t')\varphi(t'')$ for $(t', t'') \in M(X) \setminus X$. The word $\varphi(t)$ is called the *foliage* of t. Furthermore, if $s = (h_1, \dots, h_n)$ is a sequence of Hall elements, then the foliage of s is defined by $\varphi(s) = \varphi(h_1) \cdots \varphi(h_n)$.

Proposition 7.6.8 *For every word $w \in X^*$ there is a unique decreasing standard sequence of Hall elements s such that $\varphi(s) = w$.*

Proof. Let $w = x_1 \ldots x_p$ be an element of X^* and let $s = (x_1, \ldots, x_p)$ be the corresponding standard sequence of Hall elements. By Lemma 7.6.5, s reduces to a decreasing standard sequence t in a finite number of steps. Furthermore, from the definition of the reduction process, we see that $\varphi(t) = \varphi(s) = w$.

Suppose that there are two decreasing standard sequences t and t' such that $\varphi(t) = \varphi(t') = w$. Then we unpack both t and t' to the sequences of letters s and s' respectively. By the definition of the unpacking process we see that $\varphi(s) = \varphi(t) = \varphi(t') = \varphi(s')$, so that $s = s'$. But by Proposition 7.6.7 this implies that $t = t'$. $\square$

The following corollary is an immediate consequence of Proposition 7.6.8.

Corollary 7.6.9 *Let* $h_1, h_2 \in H$ *then* $h_1 \neq h_2$ *if and only if* $\varphi(h_1) \neq \varphi(h_2)$.

Corollary 7.6.9 implies that every Hall element h corresponds to a unique Hall word $w = \varphi(h)$. If from the context it is clear which Hall set H we mean, then we call a word of the form $u = \varphi(h)$ for $h \in H$ a *Hall word*.

7.7 A Hall set provides a basis

Let F be a field. From Section 6.1 we recall that $F\langle X \rangle$ is the associative algebra spanned by X^*; it is called the free associative algebra on X.

Recall that the free algebra $A(X)$ is the algebra spanned by all elements of $M(X)$ (where $M(X)$ is the free magma on X). We define a linear map

$$P : A(X) \longrightarrow F\langle X \rangle$$

by defining it on the basis $M(X)$: $P(x) = x$ for $x \in X$ and $P((a, b)) = P(a)P(b) - P(b)P(a)$. An element of the image of P is called a *Lie polynomial*.

Example 7.7.1 Let $X = \{x, y\}$ and set $m = (x, ((x, y), y))$. Then the Lie polynomial corresponding to m is $xxyy - 2xyxy + 2yxyx - yyxx$.

We fix a Hall set $H \subset M(X)$. We let P act on sequences of Hall elements in the obvious way; for $s = (h_1, \ldots, h_n)$ a sequence of Hall elements, we set $P(s) = P(h_1) \cdots P(h_n)$.

Theorem 7.7.2 *The set of all* $P(s)$ *where* s *runs through all decreasing standard sequences of Hall elements forms a basis of* $F\langle X \rangle$.

Proof. First we prove that every word $w \in X^*$ can be written as a linear combination of elements of the form $P(s)$ where s is a decreasing standard sequence of Hall elements.

Let $t = (h_1, \ldots, h_n)$ be a sequence of Hall elements. Suppose that t is not decreasing and let i be the right-most rise of t, i.e., $h_i < h_{i+1}$. Then we set

$$\rho(t) = (h_1, \ldots, h_{i-1}, h_{i+1}, h_i, h_{i+2}, \ldots, h_n).$$

Then $\rho(t)$ is again a standard sequence: if $h_{i+1} \notin X$ then $h''_{i+1} > h_{i+1}$ (because $>$ is a Hall order) $> h_i$. Furthermore, if t reduces in one step to t' then we set

$$\lambda(t) = t' = (h_1, \ldots, h_{i-1}, (h_i, h_{i+1}), h_{i+2}, \ldots, h_n).$$

Now we define the *derivation tree* $D(t)$ of t to be the tree with root labelled t and $D(\lambda(t))$ and $D(\rho(t))$ as left and right subtrees. Furthermore, if t is decreasing, then its derivation tree only consists of the root labelled t.

We claim that the derivation tree of t is always finite. We prove this by induction on the pair $v(t) = (n, |\{(i, j) \mid i < j \text{ and } h_i < h_j\}|)$. These pairs are ordered lexicographically: $(n_1, r_1) < (n_2, r_2)$ if $n_1 < n_2$ or $n_1 = n_2$ and $r_1 < r_2$. If t is decreasing, then $D(t)$ only consists of the root labelled t. Now suppose that t is not decreasing. Then $v(\lambda(t)) < v(t)$ and $v(\rho(t)) < v(t)$. So by induction $D(\lambda(t))$ and $D(\rho(t))$ are finite, and hence $D(t)$ is finite.

By the definition of P it is seen that $P(t) = P(\lambda(t)) + P(\rho(t))$. Hence $P(t)$ equals the sum of $P(t')$ where t' runs through the leaves of the derivation tree of t. Furthermore these t' are decreasing standard sequences.

Let $w = x_1 \cdots x_r$ be a word in X^*. Let $t = (x_1, \ldots, x_r)$ be the corresponding standard sequence. Then $w = P(t)$ which by the above is equal to a sum of $P(t')$ where the t' are decreasing standard sequences of Hall elements. So the elements $P(s)$, where s runs through the decreasing standard sequences of Hall elements, span $F\langle X \rangle$.

In order to prove that the elements $P(s)$ are linearly independent we may assume that the set X is finite. Let $F\langle X \rangle_d$ denote the subspace of $F\langle X \rangle$ spanned by all words of degree d. Let $s = (h_1, \ldots, h_n)$ be a standard sequence of Hall elements and set $d(s) = \deg(h_1) + \cdots + \deg(h_n)$. Then it is straightforward to see that $P(s) \in F\langle X \rangle_{d(s)}$. Now by Proposition 7.6.8, the number of decreasing standard sequences s such that $d(s) = d$ is equal to the number of basis elements of $F\langle X \rangle_d$. Furthermore, by the above the elements $P(s)$ where s is a decreasing standard sequence such that $d(s) = d$ span $F\langle X \rangle_d$. Hence this set must be linearly independent. □

From Section 7.1 we recall that $L(X) = A(X)/I_0$, where I_0 is the ideal of $A(X)$ generated by the elements (7.1). It is straightforward to see that $P(I_0) = 0$; hence P can be viewed as a linear map from $L(X)$ into $F\langle X \rangle$. We recall that $\pi : A(X) \to L(X)$ is the projection map.

Corollary 7.7.3 *The set $\pi(H)$ is linearly independent in $L(X)$.*

Proof. By Theorem 7.7.2 we even have that the set $P(\pi(H))$ is linearly independent in $F\langle X \rangle$. So $\pi(H)$ must also be linearly independent in $L(X)$. $\square$

We will write a finite linear combination of Hall elements as

$$\sum_{h \in H} \alpha_h h,$$

where it is understood that $\alpha_h = 0$ for all but a finite number of $h \in H$.

Let m be an element of $M(X)$. We give an algorithm that computes a linear combination of Hall elements $a = \sum_{h \in H} \alpha_h h$ such that $\pi(a) = \pi(m)$. We suppose that the Hall set H is defined relative to the Hall order $\leq$ (and that our routine has access to this order).

Algorithm RewriteMagmaElement

Input: an element $m \in M(X)$.

Output: a linear combination $a = \sum_{h \in H} \alpha_h h$ with such that $\pi(a) = \pi(m)$.

Step 1 If IsHallElement($\leq, m$)=true then return m.

Step 2 If either m' or m'' is no Hall element, then let $\sum_{h \in H} \beta_h h$ be the output of RewriteMagmaElementon input m' and $\sum_{g \in H} \gamma_g g$ the output of RewriteMagmaElement on input m''. For each element $(h, g) \in M(X)$ such that $\beta_h \gamma_g \neq 0$ perform the algorithm RewriteMagmaElement and multiply the coefficients of the output by $\beta_h \gamma_g$; return the sum of the results.

Step 3 If $m', m'' \in H$ then

    3a if $m' > m''$ then let $\sum_{h \in H} \delta_h h$ be the output of RewriteMagmaElement with input (m'', m'). Return $\sum_{h \in H} -\delta_h h$.

    3b if $m' = m''$ then return 0.

    3c if $m' < m''$ then write $m' = (a, b)$ and set $n_1 := ((a, m''), b)$ and $n_2 := (a, (b, m''))$. For n_1 and n_2 perform RewriteMagmaElement; collect the outputs together and return the result.

Example 7.7.4 Before proving that this algorithm terminates with the correct output we give an example. For this let H be the Hall set of Example 7.5.2. We rewrite the element $m = (y, ((x, y), x))$. Set $m' = y$ and $m'' = ((x, y), x)$. Then m' and m'' are Hall elements so we arrive in Step 3. Since $m' > m''$ we are in Step 3a. We rewrite the element $n = (((x, y), x), y)$. Now $n' < n''$, but $n' = (a, b)$ with $b < n''$. So we consider the elements $n_1 = (((x, y), y), x)$ and $n_2 = ((x, y), (x, y))$ and rewrite them. First, since n_2 is of the form (p, p) it is reduced to 0. Secondly n_2 is a Hall element so it is not rewritten. So the output is $-(((x, y), y), x)$ (where the $-$ is left over from Step 3a where we entered the recursion).

Lemma 7.7.5 *Let* $m = (m', m'') \in M(X)$ *be such that* $m', m'' \in H$. *Then on input* m *the algorithm* RewriteMagmaElement *terminates in a finite number of steps and the output is a linear combination of Hall elements* $m_i = (m_i', m_i'')$ *such that* $\deg(m_i) = \deg(m)$ *and* $m_i'' \leq \max(m', m'')$.

Proof. We use induction on the tuple $(\deg(m), \max(m', m''))$; where these tuples are ordered lexicographically: $(d_1, m_1) < (d_2, m_2)$ if and only if either $d_1 < d_2$ or $d_1 = d_2$ and $m_1 < m_2$. This is a well-ordering, so we can apply induction.

We may assume that $(m', m'') \notin H$ since otherwise the lemma is trivial. Because $m', m'' \in H$ we end up in Step 3. First suppose that $m' > m''$, i.e., we are in Step 3a. Then we perform the algorithm on the input (m'', m'). This time we arrive at Step 3c. Below we prove that this step terminates and gives output satisfying the requirements of the lemma. So in this case we are done. Furthermore, if $m' = m''$ then the algorithm returns 0 and the requirements on the output are trivially fulfilled.

Now we assume that $m' < m''$, i.e., we are in Step 3c. We must prove that the algorithm terminates on the inputs n_1 and n_2. First we observe that $b < m''$ because otherwise $m \in H$ which is excluded.

We now investigate what happens if we perform the algorithm on the input $n_1 = ((a, m''), b)$. First if n_1 is a Hall element, then in Step 1 it is returned and there is nothing to prove. Now suppose that it is not a Hall element. If (a, m'') is a Hall element then $(a, m'') < m''$ as $<$ is a Hall order. Also $b < m''$, so that $\max((a, m''), b) < m'' = \max(m', m'')$. Therefore, by induction, RewriteMagmaElement terminates on input n_1 and returns the correct output. On the other hand, if (a, m'') is not a Hall element, then in Step 2 the element (a, m'') is rewritten. Since $\deg((a, m'')) < \deg(m)$, by induction we conclude that (a, m'') is rewritten as a sum $\sum_h \beta_h h$ where $\deg(h) = \deg(a) + \deg(m'')$ and $h'' \leq \max(a, m'') = m''$ for all h such that $\beta_h \neq 0$. Next we perform the algorithm on each element (h, b) where h is such that $\beta_h \neq 0$. Fix such a Hall element h. We have $\deg(h, b) = \deg(m)$. But $h < h''$ (since $\leq$ is a Hall order) $\leq m''$ and also $b < m''$. Hence $\max(h, b) < m''$ and by induction the element (h, b) is rewritten as $\sum_g \gamma_g g$ with $\deg(g) = \deg(h) + \deg(b)$ and $g'' \leq \max(h, b)$ for all g such that $\gamma_g \neq 0$. So $\deg(g) = \deg(a) + \deg(m'') + \deg(b) = \deg(m)$. Furthermore, since $h < m''$ and $b < m''$ also $g'' < m'' = \max(m', m'')$. The conclusion is that the output of the call to RewriteMagmaElement on the input n_1 is a linear combination of Hall elements g such that $\deg(g) = \deg(m)$ and $g'' \leq \max(m', m'')$.

Now we consider $n_2 = (a, (b, m''))$. First suppose that (b, m'') is a Hall element. Then $a < b$ (since $(a, b) \in H$) $< m''$ (since $(b, m'') \in H$). Also $(b, m'') < m''$. So $\max(a, (b, m'')) < m'' = \max(m', m'')$. Hence by induction, on input n_2, RewriteMagmaElement terminates with the correct output. On the other hand, if $(b, m'') \notin H$, then we proceed as with n_1. Collecting the output of the two calls to RewriteMagmaElement together we obtain a linear combination of Hall elements satisfying the requirements of the lemma. $\square$

Theorem 7.7.6 *Let $m \in M(X)$ be arbitrary. Then* RewriteMagmaElement *terminates on input M with output f such that $\pi(f) = \pi(m)$.*

Proof. First, if $\deg(m) = 1$ then the statement is trivial. So suppose that $m = (m', m'')$. If $m', m'' \in H$ then the call terminates by Lemma 7.7.5. On the other hand, if one of m', m'' is not a Hall element then in Step 2 it is rewritten. These calls terminate by induction on the degree. Now for each pair (h, g) the algorithm is performed again (where h is from the output of the rewriting of m' and g from the output of the rewriting of m''). However, in this case $h, g \in H$ so by Lemma 7.7.5 also these calls terminate.

It is straightforward to see that $\pi(f) = \pi(m)$ because all transformations performed in the algorithm leave this value unchanged. Indeed, Step 2 is justified by bilinearity of the product in $A(X)$. In Step 3a we use the relation $\pi((m', m'')) = -\pi((m'', m'))$. In Step 3b we use that $\pi((m', m')) = 0$. And Step 3c is justified by the Jacobi identity: $\pi((a, b), m'') = \pi((a, m''), b) + \pi(a, (b, m''))$. $\square$

Corollary 7.7.7 *Let H be a Hall set in $M(X)$ then $\pi(H)$ is a basis of the free Lie algebra $L(X)$.*

Proof. Corollary 7.7.3 states that $\pi(H)$ is linearly independent. And by Theorem 7.7.6 we see that every element of $L(X)$ is a linear combination of

$\pi(h)$ for $h \in H$. $\square$

7.8   Two examples of Hall orders

Let X be a set. In this section we give examples of Hall orders that can be used in conjunction with the algorithm IsHallElement, i.e., Hall orders which allow methods for deciding whether $h_1 < h_2$ for h_1, h_2 in the corresponding Hall set.

Let $<$ be a total order on X^*. Let $<'$ be any order on $M(X)$ such that $m <' n$ whenever $\varphi(m) < \varphi(n)$ for $m, n \in M(X)$. Then we say that $<'$ is an *extension* of $<$ to $M(X)$. Let $H \subset M(X)$ be the quasi-Hall set relative to $<'$. If $\varphi(h_1)\varphi(h_2) < \varphi(h_2)$ for all $h_1, h_2 \in H$ such that $(h_1, h_2) \in H$, then it is clear that $<'$ is a Hall order and H is a Hall set. But then by Corollary 7.6.9 we see that the function $\varphi : H \to X^*$ is injective. So the order of the elements of H is determined by the order $<$ on X^*. But for constructing the elements of H we only need to know the order of the elements of H (cf. the algorithms HallSet and IsHallElement). It turns out that the order of elements of $M(X)$ whose order is not determined by $<$, is irrelevant for the construction of H. Hence $<$ determines a unique Hall set in $M(X)$. This justifies the following definition.

Definition 7.8.1 *Let $<$ be a total order on X^*. Let $<'$ be an extension of $<$ to $M(X)$ and let H be the quasi-Hall set corresponding to $<'$. If $\varphi(h_1)\varphi(h_2) < \varphi(h_2)$ for all $h_1, h_2 \in H$ such that $(h_1, h_2) \in H$, then $<$ is said to be a Hall order on X^*.*

Our first example of a Hall order on X^* will be the lexicographical order.

Example 7.8.2 (Lexicographical order) Let $X = \{x_1, x_2, \dots\}$ be a set. The lexicographical order $<_{\text{lex}}$ on X^* is defined as follows. For two words $w_1, w_2 \in X^*$ we have $w_1 <_{\text{lex}} w_2$ if and only if either $w_2 = w_1 u$ for some non-empty word $u \in X^*$ (i.e., w_1 is a proper left factor of w_2) or $w_1 = u x_i v_1$ and $w_2 = u x_j v_2$ where $u, v_1, v_2 \in X^*$ and $i < j$ (i.e., on the left-most position where w_1 and w_2 differ, the entry in w_1 is the smaller one). The proof of the following lemma is straightforward.

Lemma 7.8.3 *Let $u, v \in X^*$ such that $u <_{\text{lex}} v$. Then $wu <_{\text{lex}} wv$ for all $w \in X^*$. Furthermore, if u is not a left factor of v, then also $uw <_{\text{lex}} vw$ for all $w \in X^*$.*

Lemma 7.8.4 *The order $<_{\text{lex}}$ is a Hall order on X^*. Let w be a Hall word. If y is a proper non-empty right factor of w (i.e., $w = xy$ for a non-empty $x \in X^*$), then $w <_{\text{lex}} y$.*

Proof. First we remark that the first statement follows from the second. Let $<'$ be any extension of $<_{\text{lex}}$ to $M(X)$. Let H be the quasi-Hall set corresponding to $<'$. Let $h \in H$ be such that $w = \varphi(h)$. We prove by induction on $\deg(h)$ that $w <_{\text{lex}} y$ for all proper non-empty right factors y of w. This is trivial if $\deg(h) = 1$, so suppose that $\deg(h) > 1$, i.e., $h = (h', h'')$. Set $u = \varphi(h')$ and $v = \varphi(h'')$. We claim that $u \neq v$. This is clear if $u \in X$. So suppose that $u \notin X$ and $u = v$. Then $h' = (a, b)$ with $b \geq' h''$, but $\varphi(b) \neq \varphi(h'')$ so $\varphi(b) \geq_{\text{lex}} \varphi(h'') = v$. Now $u = v$ implies that $\varphi(b)$ is a right factor of v, so by induction $v <_{\text{lex}} \varphi(b)$, and we have reached a contradiction. Now from $u \neq v$ it follows that $u <_{\text{lex}} v$ because $h \in H$.

We distinguish three cases. First suppose that y is longer than v. Then $y = u''v$ where u'' is such that $u = u'u''$ for a non-empty $u' \in X^*$. Then by induction $u <_{\text{lex}} u''$ and since u is not a left factor of u'', $w = uv <_{\text{lex}} u''v = y$. In the second case y is equal to v. First suppose that u is a left factor of v, i.e., $v = uv'$. Then by induction $v <_{\text{lex}} v'$ and $w = uv <_{\text{lex}} uv' = v$. Also, if u is not a left factor of v, then from $u <_{\text{lex}} v$ we see that $w = uv <_{\text{lex}} v$. Finally suppose that y is shorter than v. Then $v = v'y$ and by induction $v <_{\text{lex}} y$. Furthermore, above we have shown that $w = uv <_{\text{lex}} v$ and hence $w <_{\text{lex}} y$. $\qquad\qquad\square$

Let H be the Hall set corresponding to $<_{\text{lex}}$, then the elements $\varphi(h)$ for $h \in H$ are called *Lyndon-Shirshov words*.

Lemma 7.8.5 *Let H be the Hall set corresponding to $<_{\text{lex}}$. A non-empty word $w \in X^*$ is a Hall word if and only if for all factorizations $w = xy$ where $x, y \in X^*$ are non-empty, we have $w <_{\text{lex}} y$.*

Proof. Let w be a Hall word and write $w = xy$, where x, y are non-empty. Then by Lemma 7.8.4, we see that $w <_{\text{lex}} y$. For the other direction suppose that $w <_{\text{lex}} y$ for all proper non-empty right factors y of w. By induction on the degree of w we prove that this implies that w is a Hall word. This is trivial if $\deg(w) = 1$, so suppose that $\deg(w) > 1$. Let v be the smallest non-empty proper right factor of w and write $w = uv$. We show that both u and v are Hall words.

Let v_2 be a proper non-empty right factor of v. Then $v <_{\text{lex}} v_2$ because v_2 is also a right factor of w and v is the smallest of those. So by induction we have that v is a Hall word. Now let u_2 be a non-empty proper right factor of u. Suppose that $u_2<_{\text{lex}}v$. This means that $u_2<_{\text{lex}}v<_{\text{lex}}u_2v$ (as v is the smallest right factor of w). This can only happen if u_2 is a left factor of v, i.e., $v = u_2v'$. But then $u_2v'<_{\text{lex}}u_2v$, which implies that $v'<_{\text{lex}}v$, contradicting the fact that v is the smallest right factor of w. The conclusion is that $u_2 \geq_{\text{lex}} v$. Therefore $u<_{\text{lex}}uv = w<_{\text{lex}}v \leq_{\text{lex}} u_2$. Hence by induction u is a Hall word.

Let $h_1, h_2 \in H$ be such that $\varphi(h_1) = u$ and $\varphi(h_2) = v$. Because $u<_{\text{lex}}uv = w<_{\text{lex}}v$ we have that $h_1 < h_2$, where $<$ is any extension of $<_{\text{lex}}$ to $M(X)$. Also, if $h_1 \notin X$, then $h_1 = (h_1', h_1'')$. We write $u = u'u''$ for the corresponding factorization of u. Then as seen above, $u'' \geq_{\text{lex}} v$. It follows that $(h_1, h_2) \in H$ and w is a Hall word. $\square$

Corollary 7.8.6 *Let H be as in Lemma 7.8.5. Let u, v be Hall words with $u<_{\text{lex}}v$, then uv is a Hall word.*

Proof. The proof is straightforward, using Lemma 7.8.5 as well as the proof of Lemma 7.8.4. $\square$

Let H be the Hall set relative to $<_{\text{lex}}$. Let w be a Hall word. The proof of Lemma 7.8.5 suggests a straightforward algorithm to find the corresponding element of H. Let v be the smallest proper non-trivial right factor of w and write $w = uv$. Then as seen in the proof of Lemma 7.8.5, u and v are Hall words. So recursively we can determine $h', h'' \in H$ such that $\varphi(h') = u$ and $\varphi(h'') = v$. Now the proof of Lemma 7.8.5 shows that $(h', h'') \in H$. Furthermore, $\varphi((h', h'')) = w$. We call this algorithm BracketingOfHallWord.

Example 7.8.7 (Reverse lexicographical order) We define the order $<_{\text{Rlex}}$ on X^*. Let $w_1, w_2 \in X^*$ then $w_1 <_{\text{Rlex}} w_2$ if and only if $w_1 = uw_2$ (i.e., w_2 is a proper right factor of w_1), or $w_1 = v_1x_iu$ and $w_2 = v_2x_ju$ where $v_1, v_2, u \in X^*$ and $i < j$. We call $<_{\text{Rlex}}$ the *reverse lexicographical order*. This order is trivially a Hall order on X^*, as $\varphi(h_1)\varphi(h_2)<_{\text{Rlex}}\varphi(h_2)$ for all $h_1, h_2 \in M(X)$.

We note that the reverse lexicographical order in general does not yield the same Hall set as the lexicographical order. Let $X = \{x, y, z\}$ with $x < y < z$. Then $((x, y), (x, z))$ is a Hall element with respect to $<_{\text{lex}}$, but not with respect to $<_{\text{Rlex}}$. Also $(y, (x, (x, z)))$ is a Hall element with respect to $<_{\text{Rlex}}$ but not with respect to $<_{\text{lex}}$.

We now characterize the words $\varphi(h)$ for h in the Hall set corresponding to the order $<_{\text{Rlex}}$. The proof of the following lemma is straightforward.

Lemma 7.8.8 *Let $u, v \in X^*$ be non-empty words. If $u <_{\mathrm{Rlex}} v$ then $uw <_{\mathrm{Rlex}} vw$ for all $w \in X^*$. Also if v is not a right factor of u, then $wu <_{\mathrm{Rlex}} wv$ for all $w \in X^*$.*

Proposition 7.8.9 *Let H be the Hall set corresponding to $<_{\mathrm{Rlex}}$. Let $w \in X^*$ be a non-empty word. Then w is a Hall word if and only if for each factorization $w = uv$ where u, v are non-empty we have $w >_{\mathrm{Rlex}} u$.*

Proof. We say that a word x has property Q if for each factorization $x = yz$ with y, z are non-empty we have $x >_{\mathrm{Rlex}} y$.

We claim that for $x, y \in X^*$ that have Q and are such that $x <_{\mathrm{Rlex}} y$ we have that $a = xy$ has Q. For this let $a = uv$ where u, v are non-empty. First suppose that u is longer than x, i.e., $u = xu'$. Then since y has Q we have $y >_{\mathrm{Rlex}} u'$. By Lemma 7.8.8 together with the fact that y is not a right factor of u' we have $xy >_{\mathrm{Rlex}} xu'$, i.e., $a >_{\mathrm{Rlex}} u$.

Suppose that $u = x$; then $u <_{\mathrm{Rlex}} y$. If $u = u'y$, then because in this case u has Q, $u >_{\mathrm{Rlex}} u'$ and by Lemma 7.8.8, $uy >_{\mathrm{Rlex}} u'y = u$. On the other hand if $u = u_1 x_i z$ and $y = y_1 x_j z$ where $i < j$, then $u <_{\mathrm{Rlex}} uy = a$.

Finally, suppose that u is shorter than x, i.e., $x = ux'$. Then since x has Q, $x >_{\mathrm{Rlex}} u$. But by the above, $xy >_{\mathrm{Rlex}} x$ so also $xy >_{\mathrm{Rlex}} u$. In all three cases we have $a >_{\mathrm{Rlex}} u$. We conclude that a has Q.

Let $w = \varphi(h)$ for some $h \in H$; we prove that w has property Q by induction on $\deg(h)$. If $\deg(h) = 1$ then w trivially has Q. Now suppose $\deg(h) > 1$, i.e., $h = (h', h'')$ and hence $w = xy$ where $x = \varphi(h')$ and $y = \varphi(h'')$. By induction both x and y have Q. Furthermore, since $h \in H$, we have that $x <_{\mathrm{Rlex}} y$. Hence by the above claim, w has Q.

For the other direction suppose that w has Q. Let u be the biggest left factor of w (in the order $<_{\mathrm{Rlex}}$). If u' is a left factor of u then it is also a left factor of w and hence $u >_{\mathrm{Rlex}} u'$. It follows that u has Q. Let $v \in X^*$ be such that $w = uv$. Let y be a proper left factor of v. Then uy is a left factor of w and hence $u >_{\mathrm{Rlex}} uy$. Now suppose that $y >_{\mathrm{Rlex}} u$. This implies that $uy <_{\mathrm{Rlex}} u <_{\mathrm{Rlex}} y$. But this is only possible if $u = u'y$. This means that $uy <_{\mathrm{Rlex}} u'y$, from which we have that $u <_{\mathrm{Rlex}} u'$. But this is not possible because u has Q. So $y \leq_{\mathrm{Rlex}} u$. Now $v >_{\mathrm{Rlex}} uv = w >_{\mathrm{Rlex}} u$, where the second inequality follows from the fact that w has Q. Hence we have that $v >_{\mathrm{Rlex}} y$. The conclusion is that also v has Q.

We prove that $w = \varphi(h)$ for an $h \in H$ by induction on the length of w. The statement is trivial if $w \in X$ so suppose that the length of w is bigger than 1. Then by the above we may write $w = uv$ where both u and v have Q and $u <_{\mathrm{Rlex}} v$. So by induction $u = \varphi(h_1)$ and $v = \varphi(h_2)$ for some $h_1, h_2 \in H$. Now by induction on $\deg(h_1)$ we prove that this, together with $u<_{\mathrm{Rlex}}v$ implies that $uv = \varphi(h)$ for some $h \in H$. If $u \in X$, then $(h_1, h_2) \in H$ by (7.4). If $u \notin X$ then $h_1 = (h'_1, h''_1)$ and if $\varphi(h''_1) \geq_{\mathrm{Rlex}} v$ then also $(h_1, h_2) \in H$. In both cases we are done. So suppose that $\varphi(h''_1)<_{\mathrm{Rlex}}v$ and set $u'' = \varphi(h''_1)$. Then (by the first part of the proposition) also u'' has Q, so by induction $u''v = \varphi(g)$ for some $g \in H$. Furthermore if we set $u' = \varphi(h'_1)$ then $u''v>_{\mathrm{Rlex}}u'u''v = w>_{\mathrm{Rlex}}u'$, where the first inequality follows from the definition of $>_{\mathrm{Rlex}}$ and the second by the fact that w has Q. So again by induction $u'u''v = \varphi(h)$ for some $h \in H$ and we are done. $\square$

Corollary 7.8.10 *Let H be the Hall set corresponding to $<_{\mathrm{Rlex}}$, and let u, v be Hall words with $u<_{\mathrm{Rlex}}v$. Then uv is a Hall word.*

Proof. Let x be a proper left factor of uv. We show that $uv>_{\mathrm{Rlex}}x$. First suppose that $x = u$. If v is not a right factor of u then $v>_{\mathrm{Rlex}}u$ implies $uv>_{\mathrm{Rlex}}u$. On the other hand, if $u = u'v$ for some non-empty u' then $u'<_{\mathrm{Rlex}}u$ by Proposition 7.8.9. Hence by Lemma 7.8.8 we see that $u = u'v<_{\mathrm{Rlex}}uv$; so in this case we are done. Now if x is a proper left factor of u, then $x<_{\mathrm{Rlex}}u$ (by Proposition 7.8.9) $<_{\mathrm{Rlex}}uv$. Also if x is longer than u, then $x = ux'$ where x' is a proper left factor of v. Hence $x'<_{\mathrm{Rlex}}v$ by Proposition 7.8.9, and since v is not a right factor of x', by Lemma 7.8.8 we have $ux'<_{\mathrm{Rlex}}uv$. Now by Proposition 7.8.9 uv is a Hall word. $\square$

Let H be the Hall set relative to $<_{\mathrm{Rlex}}$, and let w be a Hall word. The proof of Proposition 7.8.9 suggests a way of implementing BracketingOfHall-Word in this case. Let u be the biggest left factor of w and write $w = uv$. Then by the proof of Proposition 7.8.9, u and v are both Hall words. So recursively we can determine $h_1, h_2 \in H$ such that $u = \varphi(h_1)$ and $v = \varphi(h_2)$. If $h_1 \in X$, or $h_1 = (h'_1, h''_1)$ with $h''_1 \geq_{\mathrm{Rlex}} h_2$, then $(h_1, h_2) \in H$ and we are done. If $h''_1<_{\mathrm{Rlex}}h_2$, then write $u' = \varphi(h'_1)$ and $u'' = \varphi(h''_1)$. We determine $g_1, g_2 \in H$ such that $\varphi(g_1) = u'$ and $\varphi(g_2) = u''v$, and we return to the beginning, with g_1, g_2 in place of h_2, h_2.

7.9 Reduction in $L(X)$

In this section we describe an algorithm to reduce elements of $L(X)$ modulo a set $G \subset L(X)$. We let H be a Hall set relative to $<_{\mathrm{lex}}$ or to $<_{\mathrm{Rlex}}$. Throughout we also denote the Hall order on H by $<_{\mathrm{lex}}$ or $<_{\mathrm{Rlex}}$. Since the projection $\pi : H \to L(X)$ is injective (Corollary 7.6.9), H can be viewed as a subset of $L(X)$. Moreover H forms a basis of $L(X)$ (Corollary 7.7.7).

The algorithm RewriteMagmaElement yields an algorithmic method for calculating the product $[g, h]$ for $g, h \in H$.

Let $<_R$ be a total order on H. Relative to this order every element $a \in L(X)$ has a *leading monomial*, which is the biggest element of H occurring in a. It is denoted by $\mathrm{LM}(a)$. We say that $<_R$ is a *reduction order* if $<_R$ is multiplicative (i.e., $a <_R b$ for $a, b \in H$ implies $\mathrm{LM}([a, c]) <_R \mathrm{LM}([b, c])$ and $\mathrm{LM}([c, a]) <_R \mathrm{LM}([c, b])$ for all $c \in H$) and satisfies the descending chain condition (i.e., there is no infinite strictly decreasing chain of elements of H). We note that the Hall orders $<_{\mathrm{lex}}$ and $<_{\mathrm{Rlex}}$ are *not* reduction orders. In the Hall set corresponding to $<_{\mathrm{lex}}$ we have, for example, the infinite decreasing chain $(x, y), ((x, y), y), (((x, y), y), y) \dots$.

Now we construct a reduction order on H; we do this separately for the case where H is defined relative to $<_{\mathrm{Rlex}}$ and for the case where H is defined relative to $<_{\mathrm{lex}}$. We recall that $P : A(X) \to F\langle X \rangle$ is the map assigning a Lie polynomial to every element of $A(X)$ as defined in Section 7.7.

Lemma 7.9.1 *Let H be the Hall set corresponding to $<_{\mathrm{Rlex}}$. Let $h \in H$ and set $w = \varphi(h)$. Then $P(h)$ is equal to w plus a linear combination of smaller words having the same degree as w. Also let $h_1 <_{\mathrm{Rlex}} h_2 \in H$ and set $w_i = \varphi(h_i)$ for $i = 1, 2$. Then $P((h_1, h_2))$ is equal to $w_1 w_2$ plus a linear combination of smaller words.*

Proof. The proof is by induction on $\deg(h)$. If $\deg(h) = 1$ then the result is trivial, so suppose that $\deg(h) > 1$. Then $h = (h', h'')$ and we write $w = w' w''$ for the corresponding factorization of w. Then by induction we have

$$P(h') = w' + \sum_{u <_{\mathrm{Rlex}} w'} \alpha_u u \ \text{ and } \ P(h'') = w'' + \sum_{v <_{\mathrm{Rlex}} w''} \beta_v v,$$

where all words occurring in the sums have the same degree as w' and w'' respectively. Hence

$$
\begin{aligned}
P(h) &= P(h') P(h'') - P(h'') P(h') \\
&= w' w'' + \sum_{v <_{\mathrm{Rlex}} w''} \beta_v w' v + \sum_{u <_{\mathrm{Rlex}} w'} \alpha_u u w'' + \sum_{u <_{\mathrm{Rlex}} w', v <_{\mathrm{Rlex}} w''} \alpha_u \beta_v u v \\
&\quad - w'' w' - \sum_{u <_{\mathrm{Rlex}} w'} \alpha_u w'' u - \sum_{v <_{\mathrm{Rlex}} w''} \beta_v v w' - \sum_{u <_{\mathrm{Rlex}} w', v <_{\mathrm{Rlex}} w''} \alpha_u \beta_v v u.
\end{aligned}
$$

First we have that $w' w'' = w$. Let $w' v$ be a term from the second sum, where $v <_{\mathrm{Rlex}} w''$. Since $\deg(v) = \deg(w'')$, w'' cannot be a right factor of v.

Hence by Lemma 7.8.8, we see that $w'v<_{\text{Rlex}}w'w'' = w$. The terms in the second and third sum are dealt with in the same way. Also $w''w'<_{\text{Rlex}}w'$ (by definition of $<_{\text{Rlex}}$) $<_{\text{Rlex}}w'w''$ (by Proposition 7.8.9). Furthermore, by analogous arguments we see that the terms in the remaining sums are all $<_{\text{Rlex}}w''w'$. The first statement of the lemma follows. For the second statement we use the first and proceed by completely analogous arguments. $\square$

Let H be the Hall set corresponding to $<_{\text{Rlex}}$. Then we define the order $<_R$ as follows. If $\deg(h) < \deg(g)$, then $h <_R g$. On the other hand, if $\deg(h) = \deg(g)$ then $h <_R g$ if $\varphi(h)<_{\text{Rlex}}\varphi(g)$. By Corollary 7.6.9 this defines $<_R$ for all elements of H. In the sequel, when we write $\text{LM}(a)$ for some $a \in L(X)$, we mean the leading monomial of a relative to $<_R$.

Lemma 7.9.2 *Let H be the Hall set corresponding to $<_{\text{Rlex}}$. Let u,v be Hall words such that $u<_{\text{Rlex}}v$. Let $h,g \in H$ be such that $\varphi(h) = u$ and $\varphi(g) = v$. Then $\varphi(\text{LM}([h,g])) = uv$.*

Proof. We recall that I_0 is the ideal of $A(X)$ generated by the elements (7.1). Since $P(I_0) = 0$ we can view P as a map $P : L(X) \to F\langle X\rangle$. Furthermore, $P([h,g]) = P((h,g))$. From Lemma 7.9.1 it now follows that $P([h,g])$ is equal to uv plus a linear combination of smaller words. Set $h_0 = \text{LM}([h,g])$. Then by Lemma 7.9.1, and the linearity of P we see that $\varphi(h_0)$ is the biggest word (in the order $<_{\text{Rlex}}$) occurring in $P([h,g])$. Hence $\varphi(h_0) = uv$. $\square$

Proposition 7.9.3 *Let H be the Hall set corresponding to $<_{\text{Rlex}}$. Then $<_R$ is a reduction order on H.*

Proof. Let $h <_R g \in H$ and $f \in H$. We show that $\text{LM}([h,f]) <_R \text{LM}([g,f])$. If $\deg(h) < \deg(g)$ then this is clear. So suppose that $\deg(h) = \deg(g)$. Then $h<_{\text{Rlex}}g$. First suppose that $g<_{\text{Rlex}}f$. Then using Lemma 7.9.2 we see that $\varphi(\text{LM}([h,f])) = \varphi(h)\varphi(f)<_{\text{Rlex}}\varphi(g)\varphi(f) = \varphi(\text{LM}([g,f]))$. Secondly suppose that $h<_{\text{Rlex}}f<_{\text{Rlex}}g$. Then we write $u = \varphi(h)$, $v = \varphi(g)$ and $w = \varphi(f)$. We show that $uw<_{\text{Rlex}}wv$. If v is not a right factor of w then this follows from $w<_{\text{Rlex}}v$. If v is a right factor of w, then we write $w = w'v$. Since w is a Hall word, by Proposition 7.8.9, $w>_{\text{Rlex}}w'$. But w is not a right factor of w' so also $w>_{\text{Rlex}}uw'$. Hence by Lemma 7.8.8, $wv>_{\text{Rlex}}uw'v = uw$. It follows that $\varphi(\text{LM}([h,f])) = uw<_{\text{Rlex}}wv = \varphi(\text{LM}([g,f]))$. Finally, if $f<_{\text{Rlex}}h$, then $\varphi(\text{LM}([h,f])) = \varphi(f)\varphi(h)<_{\text{Rlex}}\varphi(f)\varphi(g) = \varphi(\text{LM}([g,f]))$. In all cases we have $\varphi(\text{LM}([h,f]))<_{\text{Rlex}}\varphi(\text{LM}([g,f]))$, implying $\text{LM}([h,f]) <_R \text{LM}([g,f])$.

Because $[f, h] = -[h, f]$ and $[f, g] = -[g, f]$ we also have $\mathrm{LM}([f, h]) <_R$ $\mathrm{LM}([f, g])$. So $<_R$ is multiplicative. It is evident that $<_R$ satisfies the descending chain condition. The conclusion is that $<_R$ is a reduction order. $\square$

Lemma 7.9.4 *Let H be the Hall set corresponding to $<_{\mathrm{lex}}$. Let $h \in H$ and set $w = \varphi(h)$. Then $P(h)$ is equal to w plus a linear combination of bigger words having the same degree as w. Also let $h_1 <_{\mathrm{lex}} h_2 \in H$ and set $w_i = \varphi(h_i)$ for $i = 1, 2$. Then $P((h_1, h_2))$ is equal to $w_1 w_2$ plus a linear combination of bigger words.*

Proof. Let w be a Hall word and write $w = uv$. Then using Lemma 7.8.5 we see that $uv <_{\mathrm{lex}} v <_{\mathrm{lex}} vu$. Using this we proceed in exactly the same way as in the proof of Lemma 7.9.1. $\square$

Let H be the Hall set corresponding to $<_{\mathrm{lex}}$. Then we define the order $<_R$ as follows. If $\deg(h) < \deg(g)$, then $h <_R g$. On the other hand, if $\deg(h) = \deg(g)$ then $h <_R g$ if $\varphi(h) >_{\mathrm{lex}} \varphi(g)$.

Lemma 7.9.5 *Let H be the Hall set corresponding to $<_{\mathrm{lex}}$. Let u, v be Hall words such that $u <_{\mathrm{lex}} v$. Let $h, g \in H$ be such that $\varphi(h) = u$ and $\varphi(g) = v$. Then $\varphi(\mathrm{LM}([h, g])) = uv$.*

Proof. The proof is analogous to the proof of Lemma 7.9.2. Again set $h_0 = \mathrm{LM}([h, g])$. This time we prove that $\varphi(h_0)$ is the smallest word (in the order $<_{\mathrm{lex}}$) occurring in $P([h, g])$. This implies that h_0 is the biggest Hall element (in the order $<_R$) occurring in $[h, g]$. $\square$

Proposition 7.9.6 *Let H be the Hall set corresponding to $<_{\mathrm{lex}}$. Then $<_R$ is a reduction order on H.*

Proof. The proof is analogous to the proof of Proposition 7.9.3. Let $h <_R g \in H$, and $f \in H$. If $\deg(h) < \deg(g)$, then clearly $\mathrm{LM}([h, f]) <_R$ $\mathrm{LM}([g, f])$. So suppose that $\deg(h) = \deg(g)$. This implies that $g <_{\mathrm{lex}} h$. The difficult case is where $g <_{\mathrm{lex}} f <_{\mathrm{lex}} h$. Write $u = \varphi(h)$, $v = \varphi(g)$ and $w = \varphi(f)$. We have to show that $wu >_{\mathrm{lex}} vw$. This is clear if v is not a left factor of w, as $v <_{\mathrm{lex}} w$. If v is a left factor of w, then we write $w = vw'$. By Lemma 7.8.5, $w <_{\mathrm{lex}} w'$, and therefore $vw <_{\mathrm{lex}} vw'$. But vw is no left factor of vw' and hence $vw <_{\mathrm{lex}} vw'u = wu$. $\square$

Let $G \subset L(X)$, and let $J \subset L(X)$ be the ideal generated by G. We describe elements of J in a fashion analogous to the one used in Section 7.3. Let $\sigma = (h_1, \dots, h_k)$ be a sequence of elements of H, and let $\delta = (d_1, \dots, d_k)$ be a sequence (of length equal to the length of σ) of letters $d_i \in \{l, r\}$. Set $\alpha = (\sigma, \delta)$, then we call α an *appliance*. The integer k is called the *length* of α. For an appliance α we define a map $P_\alpha : L(X) \to L(X)$ in the following way. If $k = 0$, then $P_\alpha(a) = a$ for all $a \in L(X)$. On the other hand, if $k > 0$, then we set $\beta = ((m_2, \dots, m_k), (d_2, \dots, d_k))$. If $d_1 = l$, then $P_\alpha(a) = P_\beta([h_1, a])$, and $P_\alpha(a) = P_\beta([a, h_1])$ if $d_1 = r$.

Let $g, h \in H$. Then g is said to be a *factor* of h if there is an appliance α such that $\mathrm{LM}(P_\alpha(g)) = h$. Let $G \subset L(X)$, and $f \in L(X)$. Then f is said to be in *normal form* modulo G if no Hall element occurring in f has a $\mathrm{LM}(g)$ as a factor for $g \in G$. Furthermore, $u \in L(X)$ is said to be a *normal form* of f modulo G if u is in normal form modulo G and $f = u \bmod I$, where I is the ideal of $L(X)$ generated by G. Now we formulate an algorithm NormalForm, analogous to the ones of Chapter 6. An element $f \in L(X)$ is called *monic* if the coefficient of $\mathrm{LM}(f)$ in f is 1.

Algorithm NormalForm
Input: a set $G \subset L(X)$ consisting of monic elements, and an element $f \in L(X)$.
Output: a normal form of f modulo G.

Step 1 Set $\phi := 0$, $a := f$.

Step 2 If $a = 0$ then return ϕ. Otherwise set $h := \mathrm{LM}(a)$ and let λ be the coefficient of h in a.

Step 3 Let $g \in G$ be such that $\mathrm{LM}(g)$ is a factor of h. If there is no such g then set $a := a - \lambda h$, $\phi := \phi + \lambda h$ and return to Step 2.

Step 4 Let α be an appliance such that $\mathrm{LM}(P_\alpha(\mathrm{LM}(g))) = h$. Set $a := a - \lambda P_\alpha(g)$.

Comments: The algorithm terminates since $<$ satisfies the descending chain condition and $\mathrm{LM}(a)$ decreases every round of the iteration. Let I be the ideal of $L(X)$ generated by G. An invariant of the algorithm is the relation $\phi + a = f \bmod I$. Furthermore ϕ is always in normal form modulo G. Hence at termination we have that ϕ is a normal form of f modulo G.

We reformulate the algorithm NormalForm in the same fashion as in Section 6.1. Let $G \subset L(X)$ be a set consisting of monic elements and let

$a \in L(X)$. Suppose that there is a $g \in G$ such that $\mathrm{LM}(g)$ is a factor of h, where h is a Hall element occurring in a. Let α be an appliance such that $\mathrm{LM}(P_\alpha(\mathrm{LM}(g))) = h$. Let λ be the coefficient of h in a. Then we say that a reduces modulo G to $b = a - \lambda P_\alpha(g)$. More generally we say that a reduces modulo G to b if there are $c_1, \ldots, c_k \in L(X)$ such that $c_1 = a$, $c_k = b$ and c_i reduces modulo G to c_{i+1} for $1 \leq i \leq k-1$. Because $<_R$ satisfies the descending chain condition, any maximal sequence of reduction steps finishes with an element that cannot be reduced further.

The problem with the algorithm NormalForm and the reduction procedure is that in general it is difficult to determine whether a given $g \in H$ is a factor of a given $h \in H$. However, in the case where the Hall set is defined relative to $<_{\mathrm{lex}}$ or $<_{\mathrm{Rlex}}$ we have a criterion for deciding this.

Let H be a Hall set and let u be a Hall word, i.e., $u = \varphi(h)$ for some $h \in H$. Suppose that $h \notin X$, i.e., $h = (h', h'')$. Then the factorization $u = vw$ where $v = \varphi(h')$ and $w = \varphi(h'')$ is called the *standard factorization* of u.

Lemma 7.9.7 *Let H be the Hall set corresponding to $<_{\mathrm{Rlex}}$ or to $<_{\mathrm{lex}}$. Let u, v be two Hall words. If $u \notin X$ then $u = xy$ will be its standard factorization. Suppose that v is a subword of u. Then there are three possibilities: v is a left factor of u, v is a subword of x, v is a subword of y.*

Proof. First we prove the statement for the case where H is defined relative to $<_{\mathrm{Rlex}}$. Suppose that v is not a subword of x or of y. Then $v = v_1 v_2$ where v_1 and v_2 are non-empty words such that $x = w_1 v_1$ and $y = v_2 w_2$ for some $w_1, w_2 \in X^*$. By Proposition 7.8.9 we have $y >_{\mathrm{Rlex}} v_2$ and since y is not a right factor of v_2, also $y >_{\mathrm{Rlex}} v_1 v_2$. Now for x we have that $x = w_1 v_1$ where v_1 is a non-empty left factor of v and if $x = ab$ is the standard factorization of x, then $b \geq_{\mathrm{Rlex}} y$ (the last inequality follows from (7.5)). We prove that from this it follows that w_1 is empty, by induction on the length of x. First if $x \in X$, then w_1 is empty because v_1 is not. So suppose that the length of x is > 1. First suppose that w_1 is longer than or equal to a, i.e., $b = b' v_1$ for some $b' \in X^*$. By the definition of $<_{\mathrm{Rlex}}$ we have $v_1 \geq_{\mathrm{Rlex}} b$ and hence $v_1 \geq_{\mathrm{Rlex}} y$. But because v_1 is a left factor of v, by Proposition 7.8.9 we see that $v >_{\mathrm{Rlex}} v_1$, and as $y >_{\mathrm{Rlex}} v$, also $y >_{\mathrm{Rlex}} v_1$, a contradiction. So w_1 is shorter than a, i.e., $a = w_1 a'$ where a' is a non-empty left factor of v and if $a = cd$ is the standard factorization of a, then $d \geq_{\mathrm{Rlex}} b \geq_{\mathrm{Rlex}} y$. So a has the same property as x. Therefore, by induction we have that w_1 is empty.

If H is defined relative to $<_{\mathrm{lex}}$, then we use the same line of argument. In this case, from $b = b' v_1$ we deduce $y \leq_{\mathrm{lex}} b <_{\mathrm{lex}} v_1$ (by Lemma 7.8.5; b is a Hall word) $<_{\mathrm{lex}} v$ (because v_1 is a left factor of v). But also $y >_{\mathrm{lex}} v_2$

(because v_2 is a left factor of y) $>_{\text{lex}} v$ (by Lemma 7.8.5). So we reach the same contradiction. $\square$

Lemma 7.9.8 *Let H be the Hall set corresponding to $<_{\text{Rlex}}$ or to $<_{\text{lex}}$. Let $g, h \in H$; then g is a factor of h if and only if $\varphi(g)$ is a subword of $\varphi(h)$.*

Proof. First we prove the lemma for the case where H is defined relative to $<_{\text{Rlex}}$. Set $u = \varphi(g)$ and $v = \varphi(h)$. Suppose first that u is a subword of v. By Lemma 7.9.7 we may assume that $v = uv'$. Let $w_1, \ldots, w_s$ be Hall words such that $w_1 \cdots w_s = v'$. (Note that these exist as the letters of v' are Hall words.) By Corollary 7.8.10 we may assume that $w_1 \geq_{\text{Rlex}} w_2 \geq_{\text{Rlex}} \cdots \geq_{\text{Rlex}} w_s$. For $1 \leq i \leq s$ let $h_i \in H$ be such that $\varphi(h_i) = w_i$. Set $f_k = [[[g, h_1], h_2] \cdots, h_k]$. By induction on k we prove that $\varphi(\text{LM}(f_k)) = uw_1 \cdots w_k$. This is clear for $k = 0$. So suppose $k \geq 0$ and $\varphi(\text{LM}(f_k)) = uw_1 \cdots w_k$. Set $u_k = uw_1 \cdots w_k$ and $v_k = w_{k+1} \cdots w_s$. Then $v = u_k v_k$, and $w_{k+1} \geq_{\text{Rlex}} w_s \geq_{\text{Rlex}} v_k$ (because w_s is a right factor of v_k) $>_{\text{Rlex}} u_k v_k >_{\text{Rlex}} u_k$ (by Proposition 7.8.9). Hence $u_k <_{\text{Rlex}} w_{k+1}$ and by Lemma 7.9.2 we have that $\varphi(\text{LM}(f_{k+1})) = uw_1 \cdots w_{k+1}$. Therefore $\varphi(\text{LM}(f_s)) = uv'$ and by Corollary 7.6.9 we see that $\text{LM}(f_s) = h$.

Now suppose that g is a factor of h. Then there is an appliance α such that $\text{LM}(P_\alpha(g)) = h$. Hence $\varphi(\text{LM}(P_\alpha(g))) = \varphi(h)$. We prove by induction on the length of α that $\varphi(\text{LM}(P_\alpha(g)))$ contains $\varphi(g)$ as a subword. This is trivial if the length of α is 0, so suppose that the length of α is > 0. Write $\alpha = ((h_1, \ldots, h_k), (d_1, \ldots, d_k))$, and set $\beta = ((h_1, \ldots, h_{k-1}), (d_1, \ldots, d_{k-1}))$. Set $\tilde{g} = \text{LM}(P_\beta(g))$. By induction $\varphi(\tilde{g})$ contains $\varphi(g)$ as a subword. But $\text{LM}(P_\alpha(g)) = \text{LM}([h_k, \tilde{g}])$, and by Lemma 7.9.2 we see that $\varphi(\text{LM}([h_k, \tilde{g}]))$ contains $\varphi(\tilde{g})$ as a subword and we are done.

If H is defined relative to $<_{\text{lex}}$ then we use the same line of argument. In this case, in the first half of the proof, we deduce $w_{k+1} \geq_{\text{lex}} w_s >_{\text{lex}} v$ (by Lemma 7.8.5) $>_{\text{lex}} u_k$. $\square$

The proof of Lemma 7.9.8 yields an algorithm for computing an appliance α such that $\text{LM}(P_\alpha)(g) = h$.

Algorithm Appliance

Input: $g, h \in H$.

Output: an appliance α such that $\text{LM}(P_\alpha(g)) = h$ if g is a factor of h, fail otherwise.

Step 1 (Catch some trivial cases.) If $g = h$ then return the empty appliance. If $h \in X$, and $g \neq h$ then return fail.

Step 2 Set $u := \varphi(g)$ and $v := \varphi(h)$. If u is not a subword of v then return fail; else go to Step 3.

Step 3 Let $v = ab$ be the standard factorization of v, where $a = \varphi(h')$ and $b = \varphi(h'')$. If u is a subword of a then let $\beta = (\sigma, \delta)$ be the output of Appliance(g, h'). Add h'' to σ and add the letter r to δ and output the result. If u is a subword of v, then we let $\beta = (\sigma, \delta)$ be the output of Appliance(g, h''), and we proceed in the same way, this time adding h' to σ and l to δ. If u is not a subword of u nor a subword of v, then go to Step 4.

Step 4 Write $v = uv'$ and calculate Hall words $w_1 \geq w_2 \geq \cdots \geq w_s$ such that $v' = w_1 \cdots w_s$. Let $h_1, \ldots, h_s$ be elements of H such that $\varphi(h_i) = w_i$. Return $((h_1, \ldots, h_s), (r, \ldots, r))$.

Comments: By Lemma 7.9.7, we have that u is either a subword of a, a subword of b or a left factor of v (longer than a). In the first two cases we use a recursive call. In the second case (Step 4) we perform the construction of the proof of Lemma 7.9.8. The Hall words $w_1, \ldots, w_s$ can be calculated as follows. Write $v' = x_{i_1} \cdots x_{i_t}$ for certain $x_{i_k} \in X$. Start with the standard sequence $s = (x_{i_1}, \ldots, x_{i_t})$, and apply the reduction procedure of Section 7.6 to arrive at a decreasing standard sequence having the same foliage.

7.10   Gröbner bases in free Lie algebras

Let X be a set. Throughout this section we let H be the Hall set in $M(X)$ relative to $<_{\text{lex}}$ or to $<_{\text{Rlex}}$. As in the previous section we also view H as a subset of $L(X)$, of which it forms a basis. As shown in Section 7.9, a reduction order $<_R$ on H exists. In this section we take all leading monomials relative to a fixed reduction order $<_R$.

Let $I \subset L(X)$ be an ideal of $L(X)$. We define the set of *normal monomials* $N(I)$ of $L(X)$ modulo I to be the set of all $h \in H$ such that h is not a leading monomial of an element of I. If we let $C(I)$ be the span of $N(I)$, then $L(X) = C(I) \oplus I$ (cf. Proposition 6.1.1). This implies that the cosets of the normal monomials span $L(X)/I$. Also for $f \in L(X)$ we define $\text{Nf}_I(f) = u$, where u is the unique element of $C(I)$ such that $f = u + g$ for a $g \in I$.

Now let $G \subset L(X)$ and let I be the ideal of $L(X)$ generated by G. Then we say that G is a *Gröbner basis* of I if for all $f \in I$ there is a $g \in G$ such that $\text{LM}(g)$ is a factor of $\text{LM}(f)$. As in Section 6.1, it is straightforward to show that reduction of an element $f \in L(X)$ modulo a Gröbner basis of I always yields a unique result, which is $\mathrm{Nf}_I(f)$.

In this section we establish a sufficient condition for a set $G \subset L(X)$ to be a Gröbner basis of I. This condition is similar to the condition obtained in the associative case (Corollary 6.1.8). The next result will enable us to construct compositions of two elements of $L(X)$.

Proposition 7.10.1 *Let $h_1, h_2 \in H$ and suppose that $\varphi(h_1) = uv$ and $\varphi(h_2) = vw$, where v is non-empty. Then uvw is a Hall word.*

Proof. First we deal with the case where H is defined relative to $<_{\mathrm{Rlex}}$. Let Q be the property of the proof of Proposition 7.8.9. Then by that proposition the words uv and vw both have Q. From this it follows that $u<_{\mathrm{Rlex}}v<_{\mathrm{Rlex}}w$. We show that uvw has Q. For that let x be a left factor of uvw. First we suppose that $x = uv$. Now by Proposition 7.8.9 we have $v<_{\mathrm{Rlex}}vw$ and by Lemma 7.8.8 (vw is not a right factor of v), $uv<_{\mathrm{Rlex}}uvw$. Now suppose that x is shorter than uv. Then by Proposition 7.8.9 and by the previous case, $x<_{\mathrm{Rlex}}uv<_{\mathrm{Rlex}}uvw$. Finally suppose that x is longer than uv, i.e., $x = uvx'$. Then vx' is a left factor of vw and hence $vx'<_{\mathrm{Rlex}}vw$. Now by Lemma 7.8.8 (vw is not a right factor of vx') we see that $uvx'<_{\mathrm{Rlex}}uvw$. It follows that uvw has Q and by Proposition 7.8.9 we are done.

If H is defined relative to $<_{\mathrm{lex}}$ then we use analogous arguments. In this case we show that for all proper non-trivial right factors y of uvw we have $uvw<_{\mathrm{lex}}y$. First suppose that $y = vw$. By Lemma 7.8.5, $uvw<_{\mathrm{lex}}v$, and as uv is not a left factor of v, also $uvw<_{\mathrm{lex}}vw = y$. If y is shorter than vw, then $y>_{\mathrm{lex}}vw$ (by Lemma 7.8.5) $>_{\mathrm{lex}}uvw$ (as seen above). Finally, if y is longer than vw, then $y = y'vw$ and $y'v$ is a proper right factor of uv. Hence by Lemma 7.8.5, $uv<_{\mathrm{lex}}y'v$ and since uv is not a left factor of $y'v$, we have $uvw<_{\mathrm{lex}}y'vw = y$. Now by Lemma 7.8.5, uvw is a Hall word. $\square$

Let $g_1, g_2 \in L(X)$ and set $h_1 = \mathrm{LM}(g_1)$ and $h_2 = \mathrm{LM}(g_2)$. Suppose that the coefficients of h_1, h_2 in g_1, g_2 respectively are 1. Suppose further that $\varphi(h_1) = uv$ and $\varphi(h_2) = vw$, where v is non-empty. Then by Proposition 7.10.1, uvw is a Hall word. Furthermore, it contains both uv and vw as subwords. Let $h \in H$ be such that $\varphi(h) = uvw$. Then by Lemma 7.9.8, h_1, h_2 both are factors of h. Let α_1, α_2 be two appliances, as found by the algorithm Appliance, such that $\mathrm{LM}(P_{\alpha_i}(h_i)) = h$ for $i = 1, 2$. Then the element

$$P_{\alpha_1}(g_1) - P_{\alpha_2}(g_2)$$

is called a *composition* of g_1 and g_2.

Let $I \subset L(X)$ be an ideal of $L(X)$ generated by the set $\{g_i \mid i \geq 1\}$. Let $h \in H$, then by $I_{<_R h}$ we denote the subspace of I spanned by elements of the form $\lambda_1 P_{\alpha_1}(g_{i_1}) + \cdots + \lambda_s P_{\alpha_s}(g_{i_s})$, where the α_k are appliances such that $\mathrm{LM}(P_{\alpha_k}(g_{i_k})) <_R h$. Note that this depends on the particular generating set G. Usually it will be clear what generating set we mean; however, if we want to stress the dependency on G, then we also write $I_{<_R h}(G)$ in place of $I_{<_R h}$.

A set $G \subset L(X)$ is said to be *self-reduced* if for any pair $g_1, g_2 \in G$ we have that $\mathrm{LM}(g_1)$ is not a factor of $\mathrm{LM}(g_2)$, and the coefficient of $\mathrm{LM}(g)$ in g is 1 for all $g \in G$.

Theorem 7.10.2 (Shirshov) *Let $I \subset L(X)$ be an ideal of $L(X)$ generated by the set $G \subset L(X)$. Suppose that G is self-reduced. Suppose that all compositions $P_{\alpha_1}(g_1) - P_{\alpha_2}(g_2)$ of elements g_1, g_2 of G lie in $I_{<_R h}(G)$, where $h = \mathrm{LM}(P_{\alpha_1}(g_1)) = \mathrm{LM}(P_{\alpha_2}(g_2))$. Then G is a Gröbner basis of I.*

Proof. First we prove the theorem for the case where H is defined relative to $<_{\mathrm{Rlex}}$. On X^* we use an order which we also denote by $<_R$. It is defined in an analogous way to the order $<_R$ on H. If $u, v \in X^*$ have different degrees, then $u <_R v$ if and only if $\deg(u) < \deg(v)$. If $\deg(u) = \deg(v)$, then $u <_R v$ if and only if $u <_{\mathrm{Rlex}} v$. We note that $<_R$ is multiplicative (cf. Lemma 7.8.8), and satisfies the descending chain condition. In the sequel leading monomials of elements of $F\langle X \rangle$ are taken relative to this order.

We use the map $P : L(X) \to F\langle X \rangle$ that assigns to each element of $L(X)$ its Lie polynomial. Let J be the ideal of $F\langle X \rangle$ generated by $P(G)$. We show that $P(G)$ is a Gröbner basis of J. For this we use Theorem 6.1.6. Let $f_1, f_2 \in P(G)$ and let $u, v \in X^*$ be such that $\mathrm{LM}(f_1)u = v\mathrm{LM}(f_2) = w$. We have to show that $f_1 u - v f_2 \in J_{<_R w}$. If v is longer than $\mathrm{LM}(f_1)$, or equal to $\mathrm{LM}(f_2)$, then $f_1 u - v f_2 \in J_{<_R w}$ by a calculation analogous to (6.2). Now suppose that v is shorter than $\mathrm{LM}(f_1)$. Then $\mathrm{LM}(f_1) = vv_1$ and $\mathrm{LM}(f_2) = v_1 u$ for a non-empty $v_1 \in X^*$. Let $g_i \in G$ be such that $f_i = P(g_i)$ for $i = 1, 2$. By Lemma 7.9.1 we have $\mathrm{LM}(f_i) = \varphi(\mathrm{LM}(g_i))$. So g_1, g_2 have a composition. Let α_1, α_2 be appliances (found by the algorithm Appliance) such that $P_{\alpha_1}(g_1) - P_{\alpha_2}(g_2)$ is the composition of g_1 and g_2 corresponding to the decomposition $\mathrm{LM}(f_1) = vv_1$ and $\mathrm{LM}(f_2) = v_1 u$.

We claim that $P(g_1)u - P(P_{\alpha_1}(g_1)) = \sum_i \lambda_i a_i P(g_1) b_i$, where the λ_i are certain scalars and $a_i, b_i \in X^*$ are such that $\mathrm{LM}(a_i P(g_1) b_i) <_R w$. Since $\mathrm{LM}(f_1)$ is a left factor of $\mathrm{LM}(f_1)u$ we have $P_{\alpha_1}(g_1) = [[[g_1, h_1], h_2], \cdots, h_k]$ where $h_i \in H$ are such that $\varphi(h_1) \cdots \varphi(h_k) = u$. But this means that $P(P_{\alpha_1}(g_1))$ is a linear combination of elements of the form

$$P(h_{i_1}) \cdots P(h_{i_t}) P(g_1) P(h_{i_{t+1}}) \cdots P(h_{i_k}).$$

The leading monomial of $P(P_{\alpha_1}(g_1))$ is $\mathrm{LM}(f_1)u$. We note that this word only occurs in $P(g_1)P(h_1)\cdots P(h_k)$ (cf. Lemma 7.9.1). So

$$P(g_1)u - P(P_{\alpha_1}(g_1)) = P(g_1)(u - P(h_1)\cdots P(h_k)) + \sum_i \mu_i \tilde{a}_i P(g_1)\tilde{b}_i,$$

where $\tilde{a}_i, \tilde{b}_i \in X^*$ are such that $\mathrm{LM}(\tilde{a}_i P(g_1)\tilde{b}_i) <_R w$. By Lemma 7.9.1 we see that the leading monomial of $P(h_1)\cdots P(h_k)$ is u. Therefore $P(g_1)(u - P(h_1)\cdots P(h_k))$ is a linear combination of elements $P(g_1)a$, where $a \in X^*$ is such that $\mathrm{LM}(P(g_1)a) <_R w$, and our claim follows. In the same way we can show that $vP(g_2) - P(P_{\alpha_2}(g_2)) = \sum_i \nu_i c_i P(g_2)d_i$, where $c_i, d_i \in X^*$ are such that $\mathrm{LM}(c_i P(g_2)d_i) <_R w$.

It follows that

$$P(g_1)u - vP(g_2) = P(P_{\alpha_1}(g_1)) - P(P_{\alpha_2}(g_2)) +$$
$$\sum_i \lambda_i a_i P(g_1)b_i - \sum_i \nu_i c_i P(g_2)d_i,$$

where the last two sums on the right-hand side are in $J_{<_R w}$. Furthermore, $P(P_{\alpha_1}(g_1)) - P(P_{\alpha_2}(g_2)) = P(P_{\alpha_1}(g_1) - P_{\alpha_2}(g_2))$. But by assumption $P_{\alpha_1}(g_1) - P_{\alpha_2}(g_2)$ lies in $I_{<_R h}$, where $h = \mathrm{LM}(P_{\alpha_1}(g_1))$. Therefore the image under P of this element lies in $J_{<_R w}$ by Lemma 7.9.1.

Now by Theorem 6.1.6, $P(G)$ is a Gröbner basis of J. This implies that G is a Gröbner basis of I. Indeed, let $a \in I$, then $P(a) \in J$. Hence there is a $g \in G$ such that $\mathrm{LM}(P(g))$ is a factor (i.e., a subword) of $\mathrm{LM}(P(a))$. But by Lemma 7.9.1, $\mathrm{LM}(P(g)) = \varphi(\mathrm{LM}(g))$ and likewise for $\mathrm{LM}(P(a))$. So by Lemma 7.9.8, $\mathrm{LM}(g)$ is a factor of $\mathrm{LM}(a)$. Hence G is a Gröbner basis of I.

If H is defined relative $<_{\mathrm{lex}}$, then we use analogous arguments. In this case the order $<_R$ is defined as follows. If $u, v \in X^*$ have different degrees, then $u <_R v$ if and only if $\deg(u) < \deg(v)$. Furthermore, if $\deg(u) = \deg(v)$, then $u <_R v$ if and only if $u >_{\mathrm{lex}} v$. $\square$

Now let $G \subset L(X)$ and let $g_1, g_2 \in G$. Suppose that g_1, g_2 have a composition $c = P_{\alpha_1}(g_1) - P_{\alpha_2}(g_2)$ and set $h = \mathrm{LM}(P_{\alpha_1}(g_1)) = \mathrm{LM}(P_{\alpha_2}(g_2))$. Then the composition c is called *useless* if there is a $g_3 \in G$ such that $\mathrm{LM}(g_3)$ is a factor of h. Otherwise the composition c is called *useful*.

Lemma 7.10.3 *Let G, I be as in Theorem 7.10.2. Suppose that G is self-reduced. Suppose that all useful compositions $P_{\alpha_1}(g_1) - P_{\alpha_2}(g_2)$ of elements g_1, g_2 of G lie in $I_{<_R h}$, where $h = \mathrm{LM}(P_{\alpha_1}(g_1)) = \mathrm{LM}(P_{\alpha_2}(g_2))$. Then G is a Gröbner basis of I.*

Proof. We use the notation of the proof of Theorem 7.10.2. Again we prove that the set $P(G)$ is a Gröbner basis of J. For that let $f_1, f_2 \in P(G)$ and suppose that they have the composition $c = f_1 u_1 - u_2 f_2$. Set $w = \mathrm{LM}(f_1 u_1)$. We must show that $c \in I_{<_R w}$. If there is no $f_3 \in P(G)$ such that $\mathrm{LM}(f_3)$ is a factor of w, then this is already done in the proof of Theorem 7.10.2. So suppose that there is such an f_3 in $P(G)$. Write $w_i = \mathrm{LM}(f_i)$ for $i = 1, 2, 3$. Then $w_1 = u_2 v$ and $w_2 = v u_1$ for some $v \neq 1$. Also we know that w_3 is a factor of $u_2 v u_1$, so we may write $u_2 v u_1 = x w_3 y$ where $x, y \in X^*$. If x is longer than, or equal to u_2, then w_3 is a factor of $v u_1 = w_2$, which is excluded as G is self-reduced. So x is shorter than u_2. In the same way we see that y is shorter than u_1. So there are $a, b \neq 1$ such that $w_3 = avb$ and $u_2 = xa$ and $u_1 = by$. Now we calculate

$$
\begin{aligned}
f_1 u_1 - u_2 f_2 &= f_1 by - x a f_2 \\
&= (f_1 b - x f_3) y + x(f_3 y - a f_2).
\end{aligned}
$$

Now $f_1 b - x f_3$ is a composition of f_1 and f_3, and the degree of $z = \mathrm{LM}(f_1 b)$ is strictly less than $\deg(w)$. (Note that $y \neq 1$, as otherwise w_2 is a factor of w_3.) So by induction on the degree we may assume that $f_1 b - x f_3 \in I_{<_R z}$, and similarly for the composition $f_3 y - a f_2$. But then it follows that $f_1 u_1 - u_2 f_2 \in I_{<_R w}$. $\qquad\square$

Corollary 7.10.4 *Let $G \subset L(X)$ be a self-reduced set generating an ideal I of $L(X)$. Then G is a Gröbner basis of I if and only if every useful composition of any pair of elements of G reduces to zero modulo G.*

Proof. Using Lemma 7.10.3 the proof is completely analogous to the proof of Corollary 6.1.8. $\qquad\square$

    These results lead to an algorithm for computing a Gröbner basis of an ideal generated by $G \subset L(X)$. In this algorithm we need a subroutine that given a set $G \subset L(X)$ computes a self-reduced set $M \subset L(X)$ generating the same ideal as G. We first state this subroutine.

Algorithm InterReduce
Input: a finite set $G \subset L(X)$.
Output: a finite set $M \subset L(X)$, where M is self-reduced and generates the same ideal as G.

Step 1 Set $M := G$.

Step 2 Denote the elements of M by $g_1, \ldots, g_s$. For $1 \leq i \leq s$ replace g_i in M by $\mathsf{NormalForm}(M \setminus \{g_i\}, g_i)$ (and if necessary divide by a scalar to ensure that the coefficient of the leading monomial is 1).

Step 3 If M is unchanged in Step 2., then return M. Otherwise return to Step 2.

Comments: We show that the algorithm terminates. Let $(h_1, \ldots, h_s)$ be the s-tuple of leading monomials of the elements of M. We order these s-tuples lexicographically. If M is changed in Step 2., then this s-tuple decreases. However, this cannot happen infinitely often because $<_R$ satisfies the descending chain condition. So after a finite number of rounds the algorithm terminates. It is straightforward to see that at termination the set M is self-reduced. It also generates the same ideal as the initial set G because the replacement operation of Step 2. does not change this.

Lemma 7.10.5 *Let $G \subset L(X)$ be a finite set generating an ideal I of $L(X)$. Set $G' = \mathsf{InterReduce}(G)$. Let $h \in H$; then $I_{<_R h}(G) \subset I_{<_R h}(G')$.*

Proof. Let $M_1 = \{g_1, \ldots, g_s\} \subset L(X)$ and let M_2 be the set obtained from M_1 by replacing g_i by $g_i' = \mathsf{NormalForm}(M_1 \setminus \{g_i\}, g_i)$. Then $g_i' = g_i - \sum_{k=1}^{m} \mu_k P_{\beta_k}(g_{j_k})$, where $\mathrm{LM}(P_{\beta_k}(g_{j_k})) \leq \mathrm{LM}(g_i)$. Let α be an appliance such that $P_\alpha(g_i) \in I_{<_R h}(M_1)$. Now $P_\alpha(g_i) = P_\alpha(g_i') + \sum_{k=1}^{m} \mu_k P_\alpha(P_{\beta_k}(g_{j_k}))$, which lies in $I_{<_R h}(M_2)$. It follows that $I_{<_R h}(M_1) \subset I_{<_R h}(M_2)$, and hence also $I_{<_R h}(G) \subset I_{<_R h}(G')$. $\qquad\square$

Now the algorithm for calculating a Gröbner basis reads as follows.

Algorithm GröbnerBasis
Input: a finite set $G \subset L(X)$.
Output: a Gröbner basis of the ideal of $L(X)$ generated by G.

Step 1 (Initialization.) Set $M_1 := \mathsf{InterReduce}(G)$. Set $D := \{(g_1, g_2) \mid g_1 \neq g_2 \in M_1\}$. Set $k := 1$.

Step 2 (Choose pair from D.) If $D = \emptyset$ then return M_k. Otherwise let $p = (g_1, g_2)$ be an element of D and set $D := D \setminus \{p\}$.

Step 3 (Process compositions.) If g_1 and g_2 do not have a useful composition, then return to Step 2. Otherwise, for all useful compositions $c = P_{\alpha_1}(g_1) - P_{\alpha_2}(g_2)$ of g_1, g_2 do the following:

Step 3a Let g be the result of reducing c modulo M_k. If $g \neq 0$, then divide g by the coefficient of $\mathrm{LM}(g)$ in r, and perform Steps 3b., and 3c.

Step 3b Add to D all pairs (f, g) for $f \in M_k$.

Step 3c Add g to M_k and set $M_{k+1} := \mathsf{InterReduce}(M_k)$. (Where in Step 2. of the algorithm $\mathsf{InterReduce}$ the set D is also changed. If in this step a g_i is replaced by g_i', then all pairs in D containing a g_i are erased and to D are added all pairs (f, g_i') for $f \in M$.) Set $k := k + 1$.

Return to Step 2.

Comments: We remark that an element $f \in L(X)$ can never form a composition with itself. Otherwise $\varphi(\mathrm{LM}(f)) = uvu$ for some non-empty $u \in X^*$. We show that there can be no Hall word of this form. First suppose that H is defined relative to $<_{\mathrm{Rlex}}$. Then u is a right factor of uvu, so that $uvu <_{\mathrm{Rlex}} u$. Now by Proposition 7.8.9, uvu cannot be a Hall word. If H is defined relative to $<_{\mathrm{lex}}$, then we use a similar argument, this time uvu cannot be a Hall word by Lemma 7.8.5. So in the set D we do not have to include pairs of the form (g, g).

Suppose that the algorithm terminates, and denote the output by M. Let $f_1, f_2 \in M$ and suppose that they have a useful composition $c = P_{\alpha_1}(f_1) - P_{\alpha_2}(f_2)$. At some point in Step 3., this composition is considered. So c lies in some $I_{<_R h}(M_k)$, where $h = \mathrm{LM}(P_{\alpha_1}(f_1))$. Therefore, by Lemma 7.10.5, c also lies in $I_{<_R h}(M_l)$ for $l \geq k$ and hence also in $I_{<_R h}(M)$. So by Lemma 7.10.3, M is a Gröbner basis of the ideal I generated by G. However, it is by no means guaranteed that the algorithm terminates, as an ideal may not have a finite Gröbner basis. Indeed, there is no algorithm for deciding whether or not two arbitrary elements of an arbitrary finitely presented Lie algebra are equal (this is a result by G. P. Kukin, [55], see also [85]). In particular there can be no algorithm that computes a finite Gröbner basis of an arbitrary ideal in $L(X)$.

On the other hand, if the ideal generated by G has a finite Gröbner basis, then the question arises whether the algorithm GröbnerBasis will find it on input G. The next lemma guarantees this as long as we choose the pairs from the set D in such a way that every pair that is added to the set will be processed after a finite number of steps. (We could, for instance let D be an ordered list, and take the pairs from the beginning of the list, and add the pairs at the end of it.)

Lemma 7.10.6 *let $G \subset L(X)$, and let $I \subset L(X)$ be the ideal generated by G. Suppose that I has a finite Gröbner basis. Suppose further that the elements p in Step 2. of the algorithm* GröbnerBasis *are chosen in such a way as to ensure that every element that is added to the set D (in Steps 3b. and 3c.) is processed after a finite number of steps. Then the algorithm* GröbnerBasis *terminates.*

Proof. Set $M = \bigcup_{k \geq 1} M_k$, and let $\overline{G}$ be the set of all $g \in M$ such that no LM(f) is a factor of LM(g) for $f \in M \setminus \{g\}$. Let $h \in H$, then we claim that $I_{<_R h}(M_k) \subset I_{<_R h}(\overline{G})$ for $k \geq 1$. Let $g_1, \ldots, g_s \in M_k$ and set

$$f = \lambda_1 P_{\alpha_1}(g_1) + \cdots + \lambda_s P_{\alpha_s}(g_s), \qquad (7.6)$$

where the α_i are appliances such that LM($P_{\alpha_i}(g_i)$) $<_R h$. We show that $f \in I_{<_R h}(\overline{G})$. If $g_i \in \overline{G}$ for $1 \leq i \leq s$, then we are done. On the other hand, suppose that a $g_i \notin I_{<_R h}(\overline{G})$. Then g_i is replaced at least once in the algorithm InterReduce. Suppose that this happens in the transition from M_l to M_{l+1}. By g'_i denote the output of the routine NormalForm that replaces g_i. As at any stage during the algorithm InterReduce the leading monomials of the elements of M differ (the input to the algorithm certainly has this property, and it is preserved during the algorithm), we have that in the first round of NormalForm, g_i is reduced modulo an element a such that LM(a) $<_R$ LM(g_i). This implies that g_i can be written as a sum $g_i = g'_i + \mu_1 P_{\beta_1}(f_1) + \cdots + \mu_r P_{\beta_r}(f_r)$, where the $f_k \in M_{l+1}$ and LM(f_k) $<_R$ LM(g_i) and LM($P_{\beta_k}(f_k)$) $\leq_R$ LM(g_i) for $1 \leq k \leq r$. Now if we substitute this expression into (7.6) we get a different expression for f. However, since $<_R$ satisfies the descending chain condition we have that such reductions cannot happen infinitely often. Therefore after a finite number of steps we reach a stable situation, $f = \sum_j \nu_j P_{\gamma_j}(e_j)$ where the e_j are never reduced. This means that the $e_j \in \overline{G}$ and therefore $f \in I_{<_R h}(\overline{G})$.

Now $I_{<_R h}(M_k) \subset I_{<_R h}(\overline{G})$ for all $h \in H$ implies that the ideal generated by M_k (which is I) is a subset of the ideal generated by $\overline{G}$. This means that the ideal generated by $\overline{G}$ is equal to I. Furthermore, we claim that $\overline{G}$ is a Gröbner basis of I. First of all, it is clear that $\overline{G}$ is self-reduced. Let $g_1, g_2 \in \overline{G}$ and suppose that they have a useful composition $c = P_{\alpha_1}(g_1) - P_{\alpha_2}(g_2)$. Then at some point during the algorithm c is considered. So there is a $k > 0$ such that $c \in I_{<_R h}(M_k)$, where $h = $ LM($P_{\alpha_1}(g_1)$). Therefore $c \in I_{<_R h}(\overline{G})$. So by Lemma 7.10.3, $\overline{G}$ is a Gröbner basis of I. Now let $S = \{f_1, \ldots, f_n\}$ be a finite Gröbner basis of I. Then for $1 \leq j \leq n$ there is a $g_j \in \overline{G}$ such that LM(g_j) divides LM(f_j). Let $l > 0$ be such that $g_j \in M_l$ for $1 \leq j \leq n$. Then for all $f \in I$ there is a $g \in M_l$ such that LM(g)

divides $\mathrm{LM}(f)$. Hence M_l is a finite Gröbner basis. $\qquad\square$

Example 7.10.7 Let $X = \{x, y\}$ and $r_1 = [x, [x, y]] - [x, y]$ and $r_2 = [[x, y], y], y]$, and $R = \{r_1, r_2\}$ (note that this presentation equivalent to the one used in Example 7.4.4). Let $I \subset L(X)$ be the ideal generated by R. We use the Hall set relative to $<_{\mathrm{lex}}$ with $x <_{\mathrm{lex}} y$. We see that r_1 and r_2 have a composition, namely

$$[[r_1, y], y] - [x, r_2].$$

Using the algorithm RewriteMagmaElement, $[[[x, [x, y]], y], y]$ is rewritten as $[x, [[[x, y], y], y]] + [[x, y], [[x, y], y]] - [[[x, y], y], y], y]$. So the composition equals $[[x, y], [[x, y], y]] - [[[x, y], y], y], y]$. The last term of this element reduces to zero modulo r_2. The first term cannot be reduced so we add $r_3 = [[x, y], [[x, y], y]]$. Now r_1 and r_3 have the composition $[r_1, [[x, y], y]] - [x, r_3]$, which rewrites to $[[x, [[x, y], y]], [x, y]] - [[x, y], [[x, y], y]]$. The last term reduces to zero modulo r_3. The first term has $[x, [x, y]]$ as a factor. We have $[[r_1, y], [x, y]] = [[x, [[x, y], y]], [x, y]] + [[x, y], [[x, y], y]]$ so that $[[x, [[x, y], y]], [x, y]]$ reduces to $-[[x, y], [[x, y], y]]$, which reduces to zero modulo r_3. The elements r_1, r_2, r_3 have no further compositions, and hence $G = \{r_1, r_2, r_3\}$ is a Gröbner basis of the ideal generated by r_1, r_2. Now we enumerate the basis elements of the quotient, i.e., the normal monomials of $L(X)$ modulo I. These are exactly those Hall elements that do not have a leading monomial of an element of G as a factor. It is straightforward to see that $\{x, y, [x, y], [[x, y], y]\}$ are the normal monomials upto degree 3. In degree 4 we have the monomials $[x, [[x, y], y]]$ and $[[[x, y], y], y]$. The first one has $\mathrm{LM}(r_1)$ as a factor and the second one has $\mathrm{LM}(r_2)$ as factor. We also do not get any monomials of degree 5 and 6. Now by induction we show that all Hall elements of degree at least 7 have a $\mathrm{LM}(r_i)$ as a factor. Let $h = (h', h'')$ be a Hall element of degree at least 7. Then either h' or h'' has degree between 4 and $\deg(h) - 1$. But all monomials of those degrees have a $\mathrm{LM}(r_i)$ as a factor. So we have the same for h. The conclusion is that the quotient is 4-dimensional. Reducing products of basis elements modulo G we can write down a multiplication table of it.

7.11 Presentations of the simple Lie algebras of characteristic zero

The objective of this section is to prove a theorem by J.-P. Serre that gives a construction of any semisimple Lie algebra with a split Cartan subalgebra of characteristic zero as a finitely presented Lie algebra.

Let C be a Cartan matrix (cf. Section 5.6). Suppose that C is an $l \times l$-matrix and let

$$X = \{h_1, \ldots, h_l, x_1, \ldots, x_l, y_1, \ldots, y_l\}$$

be a set of $3l$ symbols. We use the lexicographic order $<_{\text{lex}}$ where

$$h_1 <_{\text{lex}} \cdots <_{\text{lex}} h_l <_{\text{lex}} x_1 <_{\text{lex}} \cdots <_{\text{lex}} x_l <_{\text{lex}} y_1 \cdots <_{\text{lex}} y_l.$$

As seen in Section 7.8 this order defines a Hall set in $M(X)$.

Throughout we will denote the entry of C on position (i, j) by c_{ij}. In the free Lie algebra $L(X)$ we consider the following elements:

$$p_{ij} = [h_i, h_j] \text{ for } i < j,$$
$$q_i = [x_i, y_i] - h_i \text{ for } 1 \leq i \leq l,$$
$$r_{ij} = [x_i, y_j] \text{ for } i \neq j,$$
$$s_{ij} = [h_i, x_j] - c_{ji}x_j \text{ for } 1 \leq i, j \leq l,$$
$$t_{ij} = [h_i, y_j] + c_{ji}y_j \text{ for } 1 \leq i, j \leq l.$$

Note that if L is a semisimple Lie algebra having a root system Φ with Cartan matrix C then the relations above are satisfied by the canonical generators of L (see Section 5.11). Let R be the set of all elements $p_{ij}, q_i, r_{ij}, s_{ij}, t_{ij}$.

Lemma 7.11.1 *The set R is a Gröbner basis in $L(X)$ with respect to the order $<_{\text{lex}}$.*

Proof. It is evident that R is a self-reduced set. So we must prove that all compositions of the elements of R reduce to zero modulo R. First we see that p_{ij} and s_{jk} have the composition $[p_{ij}, x_k] - [h_i, s_{jk}]$. Rewriting this as a linear combination of Hall elements we have that this is equal to $[[h_i, x_k], h_j] + c_{kj}[h_i, x_k]$. But modulo R this reduces to $-c_{ki}c_{kj}x_k + c_{kj}c_{ki}x_k = 0$. Also p_{ij} and t_{jk} form a composition, but this one reduces to zero in a similar way.

Now s_{ij} and q_j have the composition $[s_{ij}, y_j] - [h_i, q_j]$. After rewriting this is equal to $[[h_i, y_j], x_j] - c_{ji}[x_j, y_j] + [h_i, h_j]$ which reduces to $-c_{ji}[y_j, x_j] - c_{ji}[x_j, y_j] = 0$. The elements s_{ij} and r_{jk} (where $j \neq k$) have the composition $[s_{ij}, y_k] - [h_i, r_{jk}]$, which after rewriting equals $[[h_i, y_k], x_j] - c_{ji}[x_j, y_k]$. This reduces to $c_{ki}[x_j, y_k] - c_{ji}[x_j, y_k]$ which reduces to zero modulo R as $j \neq k$. There are no further compositions between the elements of R, so by Corollary 7.10.4 we conclude that R is a Gröbner basis. $\square$

Now let I be the ideal of $L(X)$ generated by R and set $K = L(X)/I$.

Theorem 7.11.2 *We have $K = K_y \oplus K_h \oplus K_x$ (direct sum of vector spaces), where K_x is isomorphic to the free Lie algebra generated by $x_1, \ldots, x_l$ and K_y is isomorphic to the free Lie algebra generated by $y_1, \ldots, y_l$. Furthermore, K_h is an Abelian subalgebra spanned by the images of h_i in K.*

Proof. As before we use the Hall set corresponding to $<_{\text{lex}}$. The cosets of the normal Hall elements modulo I form a basis of K. Furthermore, since R is a Gröbner basis of I (Lemma 7.11.1), the normal Hall elements are exactly those that do not have a LM(r) as a factor for $r \in R$. We claim that a normal Hall element m is either composed entirely of letters x_i or entirely of letters y_i or entirely of letters h_i. We prove this by induction on $\deg(m)$. It is certainly true if $\deg(m) = 1$. So suppose that $\deg(m) \geq 2$, i.e., $m = (m', m'')$ and the result holds by induction for m' and m''. First suppose that m' is composed entirely from letters h_i and m'' entirely from letters x_i. Let h_i be the last letter from $\varphi(m')$ and x_j the first letter from $\phi(m'')$. Then by Lemma 7.9.8 $[h_i, x_j]$ is a factor of m and m is not normal; so we have reached a contradiction. For the other cases we use a similar argument and the claim follows.

Now any Hall element m consisting entirely of letters h_i reduces to 0 modulo R if $\deg(m) > 1$. It follows that the only normal monomials of this type are the generators $h_1, \ldots, h_l$. Furthermore, no Hall element consisting entirely of letters x_i or entirely of letters y_i has a LM(r) for an $r \in R$ as a factor. As a consequence they are all normal, and the theorem follows. $\square$

By Theorem 7.11.2, the images of the h_i, x_i, y_i are linearly independent in K, and we denote these images also by h_i, x_i, y_i. For $1 \leq i \neq j \leq l$ set

$$\theta_{ij}^+ = (\text{ad} x_i)^{-c_{ji}+1}(x_j)$$
$$\theta_{ij}^- = (\text{ad} y_i)^{-c_{ji}+1}(y_j).$$

Then $\theta_{ij}^+ \in K_x$ and $\theta_{ij}^- \in K_y$.

Lemma 7.11.3 *We have $[y_k, \theta_{ij}^+] = 0$ and $[x_k, \theta_{ij}^-] = 0$ for $1 \leq k \leq l$.*

Proof. We prove $[y_k, \theta_{ij}^+] = 0$; the proof of the other identity is similar. First suppose that $k \neq i$, then $[y_k, x_i] = 0$ by relation r_{ik}. Hence also $[\text{ad} y_k, \text{ad} x_i] = 0$, so that

$$\text{ad} y_k (\text{ad} x_i)^{-c_{ji}+1}(x_j) = (\text{ad} x_i)^{-c_{ji}+1} \text{ad} y_k (x_j). \tag{7.7}$$

If $k \neq j$ then (7.7) is zero by relation r_{jk}. On the other hand, if $k = j$, then using relations q_j and s_{ij} we see that (7.7) is equal to

$$-(\text{ad} x_i)^{-c_{ji}+1}(h_j) = c_{ij}(\text{ad} x_i)^{-c_{ji}}(x_i).$$

Now if $c_{ji} = 0$ then also $c_{ij} = 0$ (because C is a Cartan matrix) and hence this is zero. If $c_{ji} \neq 0$, then $-c_{ji} > 0$ and hence $(\text{ad}x_i)^{-c_{ji}}(x_i) = 0$.

Now suppose $k = i$. By induction the following identity is seen to hold in the universal enveloping algebra $U(K)$

$$ y_i x_i^n = x_i^n y_i - n x_i^{n-1} h_i - \frac{n(n-1)}{2} c_{ii} x_i^{n-1}. $$

The adjoint representation of K extends to a representation of $U(K)$ (cf. Section 6.2). So using $c_{ii} = 2$ we have

$$ \text{ad}y_i(\text{ad}x_i)^{-c_{ji}+1}(x_j) = $$

$$ \left((\text{ad}x_i)^{-c_{ji}+1}\text{ad}y_i - (-c_{ji}+1)(\text{ad}x_i)^{-c_{ji}}\text{ad}h_i + (-c_{ji}+1)c_{ji}(\text{ad}x_i)^{-c_{ji}} \right)(x_j). $$

The first term of this expression gives zero as $i \neq j$. So using $\text{ad}h_i(x_j) = c_{ji}x_j$ we get

$$ \text{ad}y_i(\theta_{ij}^+) = \left(-c_{ji}(-c_{ji}+1) + (-c_{ji}+1)c_{ji} \right)(\text{ad}x_i)^{-c_{ji}}(x_j) = 0. $$

$\square$

Set $H = K_h$, the Abelian subalgebra of K spanned by $h_1, \ldots, h_l$. Let H^* denote the dual space of H. For $\sigma \in H^*$ set $K_\sigma = \{z \in K \mid [h,z] = \sigma(h)z$ for all $h \in H\}$. If $K_\sigma \neq 0$ then σ is said to be a *weight* and K_σ is a *weight space*. Furthermore an element $0 \neq z \in K_\sigma$ is said to be a *weight vector* of weight σ.

Let $\alpha_i \in H^*$ be given by $\alpha_i(h_j) = c_{ij}$. Then α_i is a weight of K and x_i is a weight vector of weight α_i. Furthermore, the y_i are weight vectors of weight $-\alpha_i$.

Lemma 7.11.4 *Let K_x, K_y be as in Theorem 7.11.2. Let m be a monomial in K_x (i.e., m is a bracketed expression in the x_i) and n a monomial in K_y (i.e., n is a bracketed expression in the y_i). Then m, n are weight vectors of certain weights σ_m, σ_n. Furthermore σ_m, σ_n are linear combinations of the α_i where the coefficients are non-negative integral numbers in the case of σ_m and non-positive integral numbers in the case of σ_n.*

Proof. The generators x_i of K_x are weight vectors of weight α_i. Furthermore, if u, v are weight vectors of weights σ and ρ respectively, then $[u, v]$ is a weight vector of weight $\sigma + \rho$. This follows by a straightforward application of the Jacobi identity. From this the statements for m follow. The argument for n is analogous. $\square$

Let J^+ be the ideal of K_x generated by the θ_{ij}^+, and J^- the ideal of K_y generated by the θ_{ij}^-.

Lemma 7.11.5 J^+, J^- *are ideals of* K.

Proof. Let M^+ be the ideal of K generated by the θ_{ij}^+. It is clear that $J^+ \subset M^+$; we prove that $M^+ \subset J^+$. The adjoint representation of K extends to a representation of $U(K)$. So M^+ is spanned by the spaces $U(K) \cdot \theta_{ij}^+$. Using the Poincaré-Birkhoff-Witt theorem (Theorem 6.2.1), we see that M^+ is spanned by the elements $\tilde{x}\tilde{h}\tilde{y} \cdot \theta_{ij}^+$, where $\tilde{x} \in U(K_x)$, $\tilde{h} \in U(H)$ and $\tilde{y} \in U(K_y)$. Now by Lemma 7.11.3 we see that $\tilde{y} \cdot \theta_{ij}^+ = 0$ for all standard monomials $\tilde{y} \in U(K_y)$ except $\tilde{y} = 1$. Furthermore $\tilde{h} \cdot \theta_{ij}^+$ is proportional to θ_{ij}^+ by Lemma 7.11.4. The conclusion is that M^+ is spanned by the elements $\tilde{x} \cdot \theta_{ij}^+$ for $\tilde{x} \in U(K_x)$. But this implies $M^+ \subset J^+$. The argument for J^- is similar. $\square$

Now set $J = J^+ \oplus J^-$; then J is an ideal of K by Lemma 7.11.5. Put $L = K/J$. Since $H \cap J = 0$ we can identify the space $H \subset K$ with a subspace of L, which we also denote by H. So $L = N^+ \oplus H \oplus N^-$, where $N^+ = K_x/J^+$ and $N^- = K_y/J^-$. If $\sigma \in H^*$ then we set $L_\sigma = \{z \in L \mid [h, z] = \sigma(h)z$ for all $h \in H\}$. A function σ such that $L_\sigma \neq 0$ is called a weight, in which case L_σ is a weight space. In the same way as for K we have that N^+ is the sum of weight spaces corresponding to weights that are $\mathbb{Z}$-linear combinations of the α_i where the coefficients are non-negative, and similarly for N^- where the coefficients are always non-positive. In the sequel we denote the images of x_i, h_i, y_i in L also by x_i, h_i, y_i.

Lemma 7.11.6 *The elements* $x_i, h_i, y_i \in L$ *for* $1 \leq i \leq l$ *are linearly independent.*

Proof. As $L = N^+ \oplus H \oplus N^-$ and $x_i \in N^+$, $h_i \in H$ and $y_i \in N^-$, it is enough to show that the h_i are linearly independent, the x_i are linearly independent and the y_i are linearly independent. First of all the h_i are linearly independent in L since $J \cap H = 0$. Also $[h_i, x_i] = 2x_i$, $[h_i, y_i] = -2y_i$ and $[x_i, y_i] = h_i$. So the subalgebra of L spanned by the three elements x_i, y_i, h_i is a homomorphic image of $\mathfrak{sl}_2(F)$. The kernel of this homomorphism is not all of $\mathfrak{sl}_2(F)$ because $h_i \neq 0$. So since $\mathfrak{sl}_2(F)$ is a simple Lie algebra we have that the kernel is 0 and therefore also x_i, y_i are non-zero in L. Suppose that the x_i are linearly dependent, and let t be minimal such that there are $\lambda_1, \ldots, \lambda_t \in F$ such that not all of them are zero and $\sum_{i=1}^{t} \lambda_i x_i = 0$. Then $\lambda_t \neq 0$ and hence there are $\mu_i \in F$ such that $x_t = \sum_{i=1}^{t-1} \mu_i x_i$. Now let $h \in H$ be such that $\alpha_t(h) \neq \alpha_i(h)$ for some $i \in \{1, \ldots, t-1\}$ (such h exists as $\alpha_i \neq \alpha_j$ for $i \neq j$). Then $[h, x_t] = \alpha_t(h)x_t = \sum_{i=1}^{t-1} \alpha_t(h)\mu_i x_i$. But also $[h, x_t] = \sum_{i=1}^{t-1} \mu_i \alpha_i(h)x_i$. So we get a non-trivial linear combination of the first $t-1$ vectors x_i to be zero. This however contradicts the minimality of t and the x_i are linearly independent. By a similar argument we decide that the y_i are linearly independent, and we are done.   □

We recall that an endomorphism a of a (possibly infinite-dimensional) vector space V is called *locally nilpotent* if for all $v \in V$ there is a $k > 0$ such that $a^k \cdot v = 0$.

Lemma 7.11.7 *The endomorphisms* $\mathrm{ad}x_i$, $\mathrm{ad}y_i$ *of* L *are locally nilpotent.*

Proof. Let V_i be the space of all $z \in L$ such that $(\mathrm{ad}x_i)^k(z) = 0$ for some $k > 0$. By the Leibniz formula (1.11) we have that V_i is a subalgebra of L. Furthermore since $\theta_{ij}^+ = 0$ in L, $x_j \in V_i$ for $1 \leq j \leq l$. Also by relations q_i, r_{ij} and s_{ij} we have $y_j \in V_i$ for $1 \leq j \leq l$. Hence also $h_j = [x_j, y_j] \in V_i$. It follows that V_i contains the generators of L and hence $V_i = L$. The argument for y_i is similar.   □

Fix a generator x_i; we define a linear map $\exp(\mathrm{ad}x_i) : L \to L$. Let $u \in L$, then by Lemma 7.11.7, there is an integer k_u such that $(\mathrm{ad}x_i)^{k_u+1} = 0$, and we set

$$\exp(\mathrm{ad}x_i)(u) = \sum_{m=0}^{k_u} \frac{1}{m!}(\mathrm{ad}x_i)^m(u).$$

It is obvious that $\exp(\mathrm{ad}x_i)$ is a linear map on L; moreover it is an automorphism of L (cf. Section 1.11).

We recall that $\alpha_i \in H^*$ is given by $\alpha_i(h_j) = c_{ij}$. Since C is nonsingular, $\alpha_1, \ldots, \alpha_l$ are linearly independent. By Corollary 5.10.3, the algorithm CartanMatrixToRootSystem finishes on input C and $\{\alpha_1, \ldots, \alpha_l\}$. Let Φ be the output and let V be the vector space over $\mathbb{R}$ spanned by $\alpha_1, \ldots, \alpha_l$. Then by Corollary 5.10.3, there is an inner product $(\, , \,)$ on V such that $\frac{2(\alpha_i, \alpha_j)}{(\alpha_j, \alpha_j)} = c_{ij}$ for $1 \leq i, j \leq l$. We fix such an inner product, to be used in the sequel. We have that Φ is a root system in V with Cartan matrix C.

From Section 5.3 we recall that for $\alpha \in \Phi$ there is a reflection r_α of V given by $r_\alpha(v) = v - \langle v, \alpha \rangle \alpha$. The group generated by all reflections r_α for $\alpha \in \Phi$ is called the Weyl group $W(\Phi)$ of Φ (Section 5.7). We recall that

W(Φ) leaves (,) invariant. Set $\Delta = \{\alpha_1, \ldots, \alpha_l\}$, then Δ is a simple system of Φ. We recall that W(Φ) is generated by the reflections r_{α_i} (Theorem 5.7.6). We note that for $\sigma \in V$ we have $\sigma(h_j) = \langle \sigma, \alpha_j \rangle$. For $\sigma = \alpha_i$ this is obvious as $\alpha_i(h_j) = c_{ij}$. The general case follows by linearity.

Lemma 7.11.8 *Let* σ, ρ *be weights of* L *and suppose that there is a* $g \in$ W(Φ) *such that* $g(\sigma) = \rho$. *Then* $\dim L_\sigma = \dim L_\rho$.

Proof. We note that since all weights are linear combinations of the α_i with integer coefficients they all lie in the real space spanned by the roots. In particular the Weyl group acts on them. It is enough to show the lemma when $g = r_{\alpha_i}$ for some i. The general statement will then follow by Theorem 5.7.6. Set $\tau_i = (\exp(\mathrm{ad}x_i))(\exp(\mathrm{ad} - y_i))(\exp(\mathrm{ad}x_i))$. Then τ_i is an automorphism of L. Furthermore, a short calculation shows that $\tau_i(h_j) = h_j - c_{ij}h_i$. From this it follows that τ_i stabilizes H and $\tau_i^2(h_j) = h_j$, so that $\tau_i^{-1}(h) = \tau_i(h)$ for all $h \in H$. Now let $z \in L_\rho$, then for $h \in H$, $[h, \tau_i(z)] = \tau_i([\tau_i(h), z]) = \rho(\tau_i(h))\tau_i(z)$. So $\tau_i(z)$ lies in the weight space belonging to $\rho\tau_i$. We know that $r_{\alpha_i}(\sigma) = \rho$. So for $1 \leq j \leq l$, $\rho\tau_i(h_j) = r_{\alpha_i}(\sigma)(h_j - c_{ij}h_i) = \sigma(h_j) - c_{ij}\sigma(h_i) - \langle \sigma, \alpha_i \rangle \alpha_i(h_j) + \langle \sigma, \alpha_i \rangle c_{ij}\alpha_i(h_i) = \sigma(h_j)$ (in the last equality we have used $\sigma(h_i) = \langle \sigma, \alpha_i \rangle$). It follows that τ_i maps L_ρ into L_σ. Similarly τ_i maps L_σ into L_ρ. Therefore $\dim L_\sigma = \dim L_\rho$. $\quad\square$

Lemma 7.11.9 $\dim L_{\alpha_i} = 1$ *and* $\dim L_{k\alpha_i} = 0$ *if* $k \neq \pm 1, 0$.

Proof. The corresponding facts for K are clear since the elements x_i are weight vectors for the α_i and a weight vector for $k\alpha_i$ must be of the form $[[x_i, x_i], \ldots, x_i]$ (k factors x_i), but these are always zero. Now the lemma follows from Lemma 7.11.6. $\quad\square$

Lemma 7.11.10 *If* $\alpha \in \Phi$, *then* $\dim L_\alpha = 1$.

Proof. This follows from Lemmas 5.7.5, 7.11.8, and 7.11.9. $\quad\square$

Lemma 7.11.11 *Let* $\sigma = \sum_{i=1}^{l} \lambda_i \alpha_i$ *where the* $\lambda_i \in \mathbb{R}$. *Suppose that* σ *is not a multiple of a root. Then there is a* $g \in$ W(Φ) *such that* $g(\sigma) = \sum_{i=1}^{l} \mu_i \alpha_i$ *where at least one* $\mu_i > 0$ *and at least one* $\mu_i < 0$.

Proof. As before V is the vector space over $\mathbb{R}$ spanned by Φ. For $u \in V$ we denote the hyperplane perpendicular to u by P_u (i.e., the set of all $w \in V$ such that $(u, w) = 0$). As σ is not a multiple of a root we have

that P_σ is not contained in any P_α for $\alpha \in \Phi$. Choose a $v \in P_\sigma$ not lying in any P_α. Choose a $g \in W(\Phi)$ such that $(g(v), \rho)$ is maximal, where
$\rho = \frac{1}{2} \sum_{\alpha > 0} \alpha$ is the Weyl vector (see Section 5.7). Then exactly in the
same way as in the proof of Theorem 5.7.3 we have that $(g(v), \alpha_i) > 0$ for
$1 \leq i \leq l$. Write $g(\sigma) = \sum_{i=1}^l \mu_i \alpha_i$. Then by the choice of v we see that
$0 = (v, \sigma) = (g(v), g(\sigma)) = \sum_{i=1}^l \mu_i (g(v), \alpha_i)$. This can only happen if at
least one coefficient μ_i is > 0 and at least one μ_i is < 0. $\square$

Lemma 7.11.12 $\dim L_\sigma = 0$ *if* $\sigma \notin \Phi$.

Proof. We may assume that σ is a linear combination of the α_i with integer coefficients (otherwise $\dim L_\sigma = 0$). The case where σ is a multiple of a
root is taken care of by Lemma 7.11.9. Otherwise by Lemma 7.11.11, there
is a $g \in W(\Phi)$ such that $g(\sigma) = \sum_{i=1}^l \mu_i \alpha_i$ where at least one $\mu_i > 0$ and
at least one $\mu_i < 0$. But this implies that $\dim L_{g(\sigma)} = 0$ and therefore by
Lemma 7.11.8, $\dim L_\sigma = 0$. $\square$

Let $\alpha \in \Phi$ be a root and set $\beta = r_{\alpha_i}(\alpha)$. Then β is also a root and the
automorphism τ_i from the proof of Lemma 7.11.8 maps L_α onto L_β. Let
$x_\alpha \in L_\alpha$ and $x_{-\alpha} \in L_{-\alpha}$ and suppose that $[x_\alpha, x_{-\alpha}]$ is a non-zero element
of H. Set $x_{\pm\beta} = \tau_i(x_{\pm\alpha})$ Then $[x_\beta, x_{-\beta}] = \tau_i([x_\alpha, x_{-\alpha}])$ is also a non-zero
element of H. Now $[x_i, y_i] = h_i$ so for the simple roots α_i we have that
$[L_{\alpha_i}, L_{-\alpha_i}]$ is non-zero and is contained in H. By Lemma 5.7.5 we have the
corresponding statement for all roots $\alpha \in \Phi$. By Proposition 5.12.1 it now
follows that L is a semisimple Lie algebra. Furthermore L has root system
Φ. We summarize our findings in the following theorem.

Theorem 7.11.13 (Serre) *Let C be an $l \times l$-Cartan matrix. Let*

$$X = \{h_1, \ldots, h_l, x_1, \ldots, x_l, y_1, \ldots, y_l\}$$

be a set of $3l$ symbols. Let I be the ideal of $L(X)$ generated by the elements
p_{ij}, q_i, r_{ij}, s_{ij}, t_{ij}, θ_{ij}^+, θ_{ij}^-. *Then $L = L(X)/I$ is a finite-dimensional
semisimple Lie algebra with a split Cartan subalgebra and a root system Φ
having Cartan matrix C.*

This theorem together with the algorithms from Section 7.4 and from
Section 7.10 gives one more algorithm for constructing the semisimple Lie
algebras of characteristic 0.

Example 7.11.14 Set

$$C = \begin{pmatrix} 2 & -1 \\ -2 & 2 \end{pmatrix},$$

i.e., the Cartan matrix belonging to the root system of type B_2. Set $X = \{h_1, h_2, x_1, x_2, y_1, y_2\}$. Then the relations $p_{ij}, q_i, r_{ij}, s_{ij}, t_{ij}, \theta_{ij}^+, \theta_{ij}^-$ amount to the following:

$$
\begin{aligned}
&[h_1, h_2], \\
&[x_1, y_1] - h_1, && [x_1, y_2], \\
&[x_2, y_2] - h_2, && [x_2, y_1], \\
&[h_1, x_1] - 2x_1, && [h_1, y_1] + 2y_1, \\
&[h_1, x_2] + 2x_2, && [h_1, y_2] - 2y_2, \\
&[h_2, x_1] + x_1, && [h_2, y_1] - y_1, \\
&[h_2, x_2] - 2x_2, && [h_2, y_2] + 2y_2, \\
&[x_1, [x_1, [x_1, x_2]]], && [x_2, [x_2, x_1]], \\
&[y_1, [y_1, [y_1, y_2]]], && [y_2, [y_2, y_1]].
\end{aligned}
$$

It is straightforward to see that the basis elements other than those in X can only be $[x_1, x_2]$, $[x_1, [x_1, x_2]]$, $[y_1, y_2]$ and $[y_1, [y_1, y_2]]$. So the only thing we have to do is check the Jacobi identity for these basis elements. Doing this we will complete the multiplication table. For instance

$$\mathrm{Jac}(y_1, x_1, x_2) = [y_1, [x_1, x_2]] + [x_1, [x_2, y_1]] + [x_2, [y_1, x_1]].$$

The second term of this sum reduces to zero and the third term to $-2x_2$. We obtain $[y_1, [x_1, x_2]] = 2x_2$, giving the product of the basis elements y_1 and $[x_1, x_2]$. Also

$$\mathrm{Jac}(h_1, x_1, x_2) = [h_1, [x_1, x_2]] + [x_1, [x_2, h_1]] + [x_2, [h_1, x_1]].$$

The second summand reduces to $2[x_1, x_2]$ and the third to $-2[x_1, x_2]$. So we get $[h_1, [x_1, x_2]] = 0$. Now we see that also $\mathrm{Jac}(h_1, y_1, [x_1, x_2]) = 0$. Continuing like this we fill the multiplication table. There are no Jacobi identities that lead to linear dependencies among the basis elements. Hence we construct a 10-dimensional Lie algebra, isomorphic to the Lie algebras of Examples 4.9.1 and 5.15.10.

7.12   Notes

Sections 7.3 and 7.4 closely follow [38]. In [57] a similar algorithm is described to the one of Section 7.4. In that paper the authors concentrate on finding a multiplication table of a finitely-presented Lie algebra. The proof is combinatorial.

A particular instance of the problem of constructing a multiplication table of a finitely presented Lie algebra, is the problem of constructing a nilpotent quotient of a finitely presented Lie algebra. Here the set of relations R is homogeneous, in the sense that for all $r \in R$ we have that all monomials occurring in r have the same degree d_r. Let X be the generating set and let L be the finitely presented Lie algebra $\langle X \mid R \rangle$. Given an integer $c > 0$ the problem is to construct the Lie algebra L/L^{c+1}, where L^{c+1} is the $c + 1$-st term of the lower central series of L. We note that this can be done using an algorithm like the one of Section 7.4. We have to modify the algorithm given there so that every monomial of degree $> c$ is automatically reduced to 0. We note that all the relations appearing in the ideal I_0 are homogeneous. Furthermore, if we reduce a homogeneous element modulo a set of homogeneous elements, the result will be homogeneous again, and of the same degree as the original element (if it is not zero). Therefore the elements of a Gröbner basis will also be homogeneous. It follows that we can calculate the elements of a Gröbner basis that are of a certain degree d, simply by taking all relations of degree d and inter-reducing them. We refer to [41], [87] for a more elaborate treatment of this problem.

Hall bases originated in a paper of M. Hall Jr., [39]. Our definition of Hall bases follows [72]; it is a generalization of the original definition of Hall. The material of Sections 7.6 and 7.7 is (with a slight variation) taken from [72] (see also [62]).

Gröbner bases and reduction in $L(X)$ were invented by A. I. Shirshov ([79]). Note that this means that Gröbner bases for ideals of $L(X)$ were invented well before the corresponding concept in associative algebras. The proof of Theorem 7.10.2 is based on [4]. Lemma 7.10.6 is a direct translation of the same result in the associative case (see [64], [65]).

The proof of Lemma 7.11.1 follows [11]. For the proof of Serre's theorem we have followed [42] and [78]. Let X_l be a simple type and let C be a corresponding Cartan matrix. Let I be the ideal from Theorem 7.11.13. Then it is of interest to construct a Gröbner basis of I. For the simple types A_l, B_l, C_l, D_l, this is done in [11].

Chapter 8

Representations of semisimple Lie algebras

In this chapter we study finite-dimensional representations of semisimple Lie algebras of characteristic 0 with a split Cartan subalgebra. We fix some notation, to be used throughout this chapter. First of all, L will be a finite-dimensional semisimple Lie algebra over a field F of characteristic 0, with split Cartan subalgebra H. Furthermore, Φ will be the set of roots of L with respect to H. In Φ we fix a simple system $\Delta = \{\alpha_1, \ldots, \alpha_l\}$ and Φ^+, Φ^- are the sets of positive respectively negative roots of Φ. Relative to Δ we have a set of canonical generators $x_1, \ldots, x_l, h_1, \ldots, h_l, y_1, \ldots, y_l$ satisfying the relations (5.7). From Section 4.9 we recall that on H^* we have a non-degenerate symmetric bilinear form $(\ ,\ )$ defined by $(\lambda, \mu) = \kappa_L(h_\lambda, h_\mu)$. For $\lambda, \mu \in H^*$ we set

$$\langle \lambda, \mu \rangle = \frac{2(\lambda, \mu)}{(\mu, \mu)}.$$

Then $\alpha_i(h_j) = \langle \alpha_i, \alpha_j \rangle$ (cf., (5.7)). And because $\langle\ ,\ \rangle$ is linear in its first argument we have more generally, $\lambda(h_j) = \langle \lambda, \alpha_j \rangle$ for $\lambda \in H^*$.

We have that L decomposes as $L = N^- \oplus H \oplus N^+$, where N^-, N^+ are the direct sums of the root spaces corresponding to the negative and positive roots respectively. We note that N^- is generated by $y_1, \ldots, y_l$ and N^+ by $x_1, \ldots, x_l$ (cf. Lemma 5.11.2).

In Section 8.1 we show that a finite-dimensional L-module V is spanned by common eigenvectors of the Cartan subalgebra H. Furthermore, this holds over any field of characteristic 0 as the eigenvalues of the h_j are integers. These eigenvectors are called weight vectors. The function assigning to each element of H the eigenvalue of its restriction to an eigenspace, is called a weight. These functions are of paramount importance throughout the chapter. We define an order on the set of weights, and we show that an irreducible finite-dimensional representation has a unique maximal weight, called the highest weight.

In Section 8.2 we describe Verma modules, which can be viewed as universal highest-weight modules. This gives us a construction of a highest-weight module for every dominant weight.

In Section 8.3, we collect a few facts on weights. These are used in the rest of the chapter. In Section 8.4 we prove that any irreducible highest-weight module over L must be finite-dimensional, thus obtaining a classification of the irreducible finite-dimensional L-modules (and because any L-module is completely reducible (Theorem 4.4.6), a classification of all finite-dimensional L-modules).

In the rest of the chapter we derive methods for calculating with the weights of a representation. In Section 8.5 we describe how we represent weights for the purpose of computing with them. Sections 8.6 to 8.9, and Section 8.11 are devoted to algorithms for computing the weights of an irreducible L-module, and their multiplicities. The subject of Section 8.10 are two formulas due to H. Weyl. The first of these is important primarily for theoretical reasons. The second however gives us an efficient algorithm for calculating the dimension of an irreducible L-module. In Section 8.12 we consider the problem of decomposing the character of the tensor product of two irreducible L-modules into a sum of irreducible characters. We derive three formulas for performing this task. Finally in Section 8.13 we briefly consider the problem of computing so-called branching rules.

8.1   The weights of a representation

Let $\phi : L \to \mathfrak{gl}(V)$ be a finite-dimensional representation of L. Let $h \in H$. Root fact 8 states that $\mathrm{ad}_L h$ is semisimple. So Corollary 4.6.4 implies that also $\phi(h)$ is a semisimple linear transformation. In the sequel we assume that $\phi(H)$ is split over F, i.e., that F contains all eigenvalues of the $\phi(h)$ for $h \in H$. Let $V = V_1 \oplus \cdots \oplus V_s$ be the (collected) primary decomposition of V relative to $\phi(H)$ (cf. Theorem 3.1.10). Let $h \in H$, then the minimum polynomial of $\phi(h)|_{V_i}$ is irreducible (since $\phi(h)$ is semisimple). So as $\phi(H)$ is split, we have that $\phi(h)|_{V_i}$ is multiplication by a scalar $\mu_i(h)$. Now the functions $\mu_i : H \to F$ are called weights.

Definition 8.1.1 *Let $\lambda \in H^*$. Then λ is called a* weight *of V if the space $V_\lambda = \{v \in V \mid \phi(h)v = \lambda(h)v$ for all $h \in H\}$ is non-zero. In that case a*

$v \in V_\lambda$ *is called a* weight vector *of weight* λ *and* V_λ *is called the* weight space *of weight* λ.

Let $P(V)$ denote the set of all weights of V, i.e., $P(V) = \{\lambda \in H^* \mid V_\lambda \neq 0\}$. Then V is the direct sum of the weight spaces V_λ for $\lambda \in P(V)$.

An element $\lambda \in H^*$ is said to be *integral* if $\langle \lambda, \alpha_i \rangle \in \mathbb{Z}$ for $1 \leq i \leq l$ (note that this depends on the choice of the simple system Δ that we fixed at the outset). As $\langle \lambda, \alpha_j \rangle = \lambda(h_j)$, we have that λ is integral if and only if $\lambda(h_j) \in \mathbb{Z}$ for $1 \leq j \leq l$.

Now fix $j \in \{1, \dots, l\}$ and let K_j be the subalgebra of L spanned by x_j, h_j, y_j. Then K_j is isomorphic to $\mathfrak{sl}_2(F)$. By restricting ϕ, V turns into a K_j-module, and the $\lambda(h_j)$ are the eigenvalues of $\phi(h_j)$ for $\lambda \in P(V)$. So by Theorem 5.1.4, $\lambda(h_j)$ is an integer for all $\lambda \in P(V)$. The conclusion is that all $\lambda \in P(V)$ are integral. As a consequence our assumption that F contains all eigenvalues of $\phi(h)$ for $h \in H$ turns out to be unnecessary since this holds for all fields of characteristic 0.

By $P \subset H^*$ we denote the set of all integral functions in H^*. On P we define a partial order $<$ by $\lambda < \mu$ if $\mu - \lambda = \sum_{i=1}^{l} k_i \alpha_i$ with all k_i nonnegative integers. The transitivity of this order is clear; we prove that $\lambda \leq \mu$ and $\mu \leq \lambda$ imply $\lambda = \mu$. From the first inequality we have $\mu - \lambda = \sum_{i=1}^{l} k_i \alpha_i$ and from the second $\lambda - \mu = \sum_{i=1}^{l} m_i \alpha_i$. So $0 = \mu - \lambda + \lambda - \mu = \sum_{i=1}^{l} (k_i + m_i) \alpha_i$. Since the α_i are linearly independent, $k_i + m_i = 0$ for all i, and because $k_i, m_i \geq 0$, we have $k_i = m_i = 0$. We note that not all elements of P are necessarily comparable with the order $<$. The difference of two elements from P might, for example, not be a linear combination of α_i with integral coefficients.

Lemma 8.1.2 *Let* $\alpha \in \Phi$ *and let* x_α *be a non-zero element of the root space* L_α. *Let* $\lambda \in P(V)$. *Then* $\phi(x_\alpha)V_\lambda \subset V_{\lambda+\alpha}$.

Proof. The proof is settled by a short calculation. For $v \in V_\lambda$ and $h \in H$ we have

$$\phi(h)\phi(x_\alpha)v = \phi(x_\alpha)\phi(h)v + \phi([h, x_\alpha])v$$
$$= (\lambda(h) + \alpha(h))\phi(x_\alpha)v.$$

$\square$

From Lemma 8.1.2 it follows that $\phi(x_\alpha)$ maps weight spaces into weight spaces. Furthermore, if α is a positive root and $\phi(x_\alpha)V_\lambda \subset V_\mu$, then $\mu > \lambda$. This leads to the following definition.

Definition 8.1.3 *An element $\lambda \in P(V)$ is called a* highest weight *of V if for all $v \in V_\lambda$ we have that $\phi(x_\alpha)v = 0$ for all $\alpha \in \Phi^+$ and root vectors $x_\alpha \in L_\alpha$. In this case a non-zero vector $v \in V_\lambda$ is called a* highest-weight *vector.*

We remark that V must have highest weights since V is finite-dimensional.

Let K_j again be the subalgebra of L spanned by x_j, h_j, y_j, and let $\lambda \in P(V)$ be a highest weight. Then $\lambda(h_j)$ is a highest weight of V (viewed as K_j-module) in the sense of Definition 5.1.1. So by Theorem 5.1.4, $\lambda(h_j) \geq 0$. Now an element $\lambda \in P$ is called *dominant* if $\langle \lambda, \alpha_j \rangle \geq 0$ for $1 \leq j \leq l$. The set of all dominant weights in P is denoted by P_+. So every highest weight of V is integral and dominant.

In order to study the role played by the highest weights of a representation we broaden our perspective somewhat and we let V be an L-module that is not necessarily finite-dimensional. From Section 6.2 we recall that the representation of L afforded by V can uniquely be extended to $U(L)$, making V into a $U(L)$-module. Now the L-module V is called a *highest-weight module* over L with highest weight λ, if there is a weight vector $v_\lambda \in V$ of weight λ such that

$$x_\alpha \cdot v_\lambda = 0 \text{ for all positive roots } \alpha \text{ and vectors } x_\alpha \in L_\alpha, \text{ and} \qquad (8.1)$$

$$U(L) \cdot v_\lambda = V. \qquad (8.2)$$

Lemma 8.1.4 *Let V be a highest-weight module over L with highest weight λ. Then*

$$V = \bigoplus_{\mu \leq \lambda} V_\mu$$

Furthermore, $\dim V_\lambda = 1$ and $\dim V_\mu$ is finite for all $\mu \in P(V)$.

Proof. Let $v_\lambda \in V$ be a highest-weight vector of weight λ. We recall that L decomposes as $L = N^- \oplus H \oplus N^+$. So by the Poincaré-Birkhoff-Witt theorem, $U(L)$ is spanned by monomials of the form $a^- h a^+$ where a^-, a^+, h are standard monomials in $U(N^-)$, $U(N^+)$ and $U(H)$ respectively. Hence by (8.2) V is spanned by all $a^- h a^+ \cdot v_\lambda$. Now by (8.1), $a^+ \cdot v_\lambda = 0$ unless $a^+ = 1$. Furthermore $h \cdot v_\lambda$ is proportional to v_λ. The conclusion is that V is spanned by vectors of the form $a^- \cdot v_\lambda$. By Lemma 8.1.2, any such vector lies in a weight space V_μ, where $\mu = \lambda - \sum_{i=1}^{l} k_i \alpha_i$, where the k_i are nonnegative integers for $1 \leq i \leq l$. In particular $\mu \leq \lambda$. Consequently V is the sum of its weight spaces V_μ, where $\mu \leq \lambda$. That this sum is direct is proved as follows. Let v_i be a non-zero weight vector of weight μ_i and suppose that $\sum_{i=1}^{s} t_i v_i = 0$ for some $t_i \in F$ such that not all of them are zero. We prove by induction on s that all t_i must be zero. If $s = 1$ then this is trivial. If $s > 1$ then we may suppose that $t_s \neq 0$. Hence we have a relation $v_s = \sum_{i=1}^{s-1} s_i v_i$. Let $h \in H$ be such that $\mu_s(h) \neq \mu_i(h)$ for some $i \in \{1, \dots, s-1\}$. By applying h twice to the relation for v_s we get $\sum_{i=1}^{s-1} (\mu_s - \mu_i)(h) s_i v_i = 0$ and by induction this implies that the s_i are zero.

Also any vector $a^- \cdot v_\lambda$ will never lie in V_λ unless $a^- = 1$, so $\dim V_\lambda = 1$. Finally, for any given μ the number of linear combinations $\sum_{i=1}^{l} k_i \alpha_i$ such that $\mu = \lambda - \sum_{i=1}^{l} k_i \alpha_i$ is finite. Hence $\dim V_\mu < \infty$. $\qquad\square$

Theorem 8.1.5 *Let V be a finite-dimensional irreducible L-module. Then V has a unique highest weight $\lambda \in P_+$. Furthermore, V is a highest-weight module over L with highest weight λ.*

Proof. Choose a $\lambda \in P(V)$ maximal with respect to the order $<$, and let $0 \neq v_\lambda \in V_\lambda$. Then $x_\alpha \cdot v_\lambda = 0$ for all positive α by Lemma 8.1.2. Also $U(L) \cdot v_\lambda = V$ since V is irreducible. So V is a highest-weight module with highest weight λ. Hence by Lemma 8.1.4, all weights $\mu \in P(V)$ are $\leq \lambda$. So if there is one more highest weight, say λ', then $\lambda' \leq \lambda$. By applying the above reasoning to λ' we see that $\lambda \leq \lambda'$. It follows that $\lambda = \lambda'$. $\qquad\square$

8.2   Verma modules

In the previous section we showed that every irreducible finite-dimensional L-module V is a highest-weight module with respect to a unique highest weight $\lambda \in P_+$. In this section we show that to every dominant weight $\lambda \in P_+$ corresponds an irreducible L-module $V(\lambda)$ of highest weight λ. Then in Section 8.4 we prove that such a module is necessarily finite-dimensional, thereby obtaining a one-to-one correspondence between irreducible finite-dimensional representations of L and elements of P_+.

Let $\lambda \in P_+$. Then a highest-weight module $M(\lambda)$ of highest weight λ is called a *Verma module* if for any highest-weight module V over L with highest weight λ there is a surjective homomorphism of L-modules $\tau : M(\lambda) \to V$. In other words, any highest-weight module for L with highest weight λ is a quotient of $M(\lambda)$.

The universal enveloping algebra $U(L)$ is made into an L-module by $x \cdot a = xa$ for $x \in L, a \in U(L)$. For $\alpha \in \Phi$ fix a root vector $x_\alpha \in L_\alpha$. Let $B(\lambda)$ be the submodule generated by the x_α for all positive roots α together with $h - \lambda(h) \cdot 1$ for $h \in H$ (i.e., $B(\lambda)$ is the left ideal of $U(L)$ generated by these elements). Set $M(\lambda) = U(L)/B(\lambda)$ (quotient of L-modules). Let $v_\lambda \in M(\lambda)$ be the image of 1. It is straightforward to see that the generators of $B(\lambda)$ form a left-Gröbner basis of $B(\lambda)$ (see Section 6.4). Therefore $1 \notin B(\lambda)$ and $v_\lambda \neq 0$. Furthermore v_λ is a highest-weight vector of weight λ. Hence $M(\lambda)$ is a highest-weight module for L with highest weight λ. We note that $M(\lambda)$ is infinite-dimensional. Indeed, the generators of $B(\lambda)$ form a left-Gröbner basis for $B(\lambda)$. Therefore the vectors

$$\prod_{\alpha>0} x_{-\alpha}^{k_\alpha} \cdot v_\lambda,$$

where the k_α are non-negative integers, form a basis of $M(\lambda)$.

Now let V be any highest-weight module over L with highest weight λ. Set $A = \{a \in U(L) \mid a \cdot V_\lambda = 0\}$. Then A is an L-submodule of $U(L)$ containing $B(\lambda)$. Now A is the kernel of the surjective morphism of L-modules $U(L) \to V$ that sends $1 \in U(L)$ to a fixed highest-weight vector of V. So $V \cong U(L)/A$ and hence there is a surjective homomorphism of L-modules $\tau : M(\lambda) \to V$. The conclusion is that $M(\lambda)$ is a Verma module of highest weight λ.

Now suppose that there is a second Verma module $M_1(\lambda)$ with highest weight λ. Then there are surjective homomorphisms of L-modules, $\tau_1 : M(\lambda) \to M_1(\lambda)$ and $\tau_2 : M_1(\lambda) \to M(\lambda)$. For all weights μ we have $\tau_1(M(\lambda)_\mu) = M_1(\lambda)_\mu$ and similarly for τ_2. So $\dim M(\lambda)_\mu = \dim M_1(\lambda)_\mu$, and it follows that τ_1 and τ_2 are bijective. As a consequence $M(\lambda)$ and $M_1(\lambda)$ are isomorphic L-modules. So upto isomorphism there exists a unique Verma module $M(\lambda)$ of highest weight λ.

Lemma 8.2.1 $M(\lambda)$ *contains a unique maximal proper L-submodule $M'(\lambda)$.*

Proof. Any proper L-submodule of $M(\lambda)$ does not contain the span of the highest-weight vector of $M(\lambda)$ (otherwise this submodule is equal to $M(\lambda)$ by (8.2)). So the sum of all proper submodules is again proper, and the lemma follows.   □

Now set $V(\lambda) = M(\lambda)/M'(\lambda)$ (quotient of L-modules), where $M'(\lambda)$ is the unique maximal proper L-submodule of $M(\lambda)$. Then $V(\lambda)$ is an irreducible highest-weight module over L of highest weight λ. Let V be an irreducible highest-weight module over L with highest weight λ. Then $V \cong M(\lambda)/W$ for some L-submodule W of $M(\lambda)$. But then $W \subset M'(\lambda)$. If $W \neq M'(\lambda)$, then V contains the image of $M'(\lambda)$ as a non-zero proper submodule. But this is impossible because V is irreducible. Hence $W = M'(\lambda)$ and $V \cong V(\lambda)$. We have proved the following proposition.

Proposition 8.2.2 *Let $\lambda \in P_+$. Then up to isomorphism there exists a unique irreducible highest-weight module $V(\lambda)$ over L with highest weight λ.*

8.3 Integral functions and the Weyl group

In this section we leave the area of representation theory for a moment to collect some facts on the set P of integral functions. These will be of paramount importance in the remainder of this chapter.

Throughout this section V will be the vector space over $\mathbb{R}$ spanned by the set of roots Φ. From Section 5.3 we recall that V is a Euclidean space relative to the bilinear form $(\ ,\ )$.

For $1 \leq i \leq l$ let $\lambda_i \in P$ satisfy the equations

$$\langle \lambda_i, \alpha_j \rangle = \delta_{ij} \text{ for } j = 1, \dots, l. \tag{8.3}$$

If we express λ_i as a linear combination of the α_i, then we see that the matrix of the equation system that the coefficients must satisfy, is the Cartan matrix of Φ. In particular this matrix is nonsingular and (8.3) has a unique solution over $\mathbb{R}$. In particular $\lambda_i \in V$. Note that the λ_i are necessarily dominant. They are called the *fundamental dominant weights*.

Lemma 8.3.1 *Let $\lambda \in P$, and set $m_j = \langle \lambda, \alpha_j \rangle$ for $1 \leq j \leq l$. Then $\lambda = m_1\lambda_1 + \cdots + m_l\lambda_l$. In particular $\lambda \in P_+$ if and only if $m_j \geq 0$ for $1 \leq j \leq l$.*

Proof. We calculate

$$\langle \lambda - \sum_{i=1}^{l} m_i\lambda_i, \alpha_j \rangle = \langle \lambda, \alpha_j \rangle - \sum_{i=1}^{l} m_i \langle \lambda_i, \alpha_j \rangle$$
$$= \langle \lambda, \alpha_j \rangle - m_j = 0.$$

So also $(\lambda - \sum_{i=1}^{l} m_i\lambda_i, \alpha_j) = 0$ for $1 \leq j \leq l$. Beçause $(\ ,\ )$ is non-degenerate, $\lambda - \sum_{i=1}^{l} m_i\lambda_i = 0$. $\square$

By Lemma 8.3.1 we see that P is a lattice in V. It is called the *weight lattice* of V. We recall that the root lattice Q is the set of all integral linear combinations of the simple roots. For all roots $\alpha \in \Phi$ we have that $\langle \alpha, \alpha_i \rangle$ is an integer. So P contains the root lattice Q.

From Section 5.3 we recall that for $\alpha \in \Phi$ we have a reflection $r_\alpha : V \to V$ defined by $r_\alpha(\sigma) = \sigma - \langle \sigma, \alpha \rangle \alpha$. Furthermore, the group generated by all reflections r_α is called the Weyl group; it is denoted by $W(\Phi)$ (Section 5.7). By a straightforward calculation we see that $r_{\alpha_i}(\lambda_j) = \lambda_j - \delta_{ij}\alpha_i$ for $1 \le i, j \le l$. So since the r_{α_i} generate $W(\Phi)$ (Theorem 5.7.6) we have that $W(\Phi)$ leaves P invariant.

We note that $r_\alpha^2 = 1$ for every reflection r_α. So the inverse of a product $r_\alpha r_\beta \cdots r_\gamma$ is $r_\gamma \cdots r_\beta r_\alpha$.

Lemma 8.3.2 *Let $\beta_1, \dots, \beta_t \in \Delta$ be (not necessarily distinct) simple roots. Set $g_i = r_{\beta_i}$. If $g_1 \cdots g_{t-1}(\beta_t)$ is a negative root, then there is an $s \in \{1, \dots, t-1\}$ such that $g_1 \cdots g_t = g_1 \cdots g_{s-1} g_{s+1} \cdots g_{t-1}$.*

Proof. Set $\gamma_i = g_{i+1} \cdots g_{t-1}(\beta_t)$ for $0 \le i \le t-2$ and $\gamma_{t-1} = \beta_t$. Then $\gamma_0 < 0$ and $\gamma_{t-1} > 0$. Now let γ_s for some $s > 0$ be the first positive root in this sequence. Then $g_s(\gamma_s) = \gamma_{s-1} < 0$. So g_s maps a positive root to a negative one. Now Lemma 5.7.1 states that g_s permutes the positive roots other than β_s. So we must have $\gamma_s = \beta_s$. Since $g \in W(\Phi)$ leaves the inner product $(\ ,\ )$ invariant, by a short calculation it can be established that $g r_\alpha g^{-1} = r_{g(\alpha)}$ for $\alpha \in \Phi$. Using this we calculate

$$(g_{s+1} \cdots g_{t-1}) g_t (g_{t-1} \cdots g_{s+1}) = r_{g_{s+1} \cdots g_{t-1}(\beta_t)}$$
$$= r_{\gamma_s} = r_{\beta_s} = g_s.$$

Now if we substitute this expression for g_s into $g_1 \cdots g_t$ we arrive at the desired result. $\qquad\square$

Let $g \in W(\Phi)$. Then by Theorem 5.7.6 we may write $g = g_1 \cdots g_t$, where $g_i = r_{\beta_i}$ for some $\beta_i \in \Delta$. This expression for g is said to be *reduced* if t is minimal among all ways of writing g as a product of simple reflections.

Corollary 8.3.3 *Let $g = g_1 \cdots g_t$ be a product of elements $g_i = r_{\beta_i}$, $\beta_i \in \Delta$, and suppose that this expression is reduced. Then $g(\beta_t)$ is a negative root.*

Proof. Suppose that $g(\beta_t) > 0$. Then $g_1 \cdots g_{t-1}(\beta_t) = -g(\beta_t) < 0$. Now by Lemma 8.3.2 we can write g as a shorter product of simple reflections and we have derived a contradiction. $\qquad\square$

Lemma 8.3.4 *Let $\lambda, \mu \in P_+$ and let $g \in W(\Phi)$ be such that $g(\lambda) = \mu$. Then $\lambda = \mu$.*

Proof. Let $g = g_1 \cdots g_t$ be a reduced expression for g where $g_i = r_{\beta_i}$, $\beta_i \in \Delta$. We prove the lemma by induction on t. If $t = 0$ then $g = 1$ and there is nothing to prove. So suppose that $t > 0$. By Corollary 8.3.3 we have $g(\beta_t) < 0$ so that

$$0 \le (\lambda, \beta_t) = (g(\lambda), g(\beta_t)) = (\mu, g(\beta_t)) \le 0.$$

As a consequence $(\lambda, \beta_t) = 0$; so $r_{\beta_t}(\lambda) = \lambda$. Hence $g r_{\beta_t}(\lambda) = \mu$. But $g r_{\beta_t} = g_1 \cdots g_{t-1}$ and we conclude by induction.   □

Theorem 8.3.5 *Each element $\mu \in P$ is conjugate under $W(\Phi)$ to exactly one element of P_+.*

Proof. First we show that there is a $\lambda \in P_+$ conjugate to μ. For that write $\mu = m_1 \lambda_1 + \cdots + m_l \lambda_l$, where $m_j = \langle \mu, \alpha_j \rangle$ (Lemma 8.3.1). If all $m_j \ge 0$ then $\mu \in P_+$ and there is nothing to prove. On the other hand, if there is a j such that $m_j < 0$, then set $\mu_1 = r_{\alpha_j}(\mu) = \mu - m_j \alpha_j$. We have that $\mu_1 > \mu$ and μ_1 is conjugate to μ under $W(\Phi)$. Continuing like this we construct a series

$$\mu = \mu_0 < \mu_1 < \mu_2 < \cdots$$

of weights conjugate to μ. But $W(\Phi)$ is finite and as a consequence μ only has a finite number of conjugates. It follows that after a finite number of steps we have found a dominant weight conjugate to μ.

Now we prove uniqueness. For that suppose that there are two dominant weights λ_1, λ_2 conjugate to μ, i.e., there are $g_1, g_2 \in W(\Phi)$ such that $g_i(\mu) = \lambda_i$ for $i = 1, 2$. Then $g_2 g_1^{-1}(\lambda_1) = \lambda_2$, and by Lemma 8.3.4 we conclude that $\lambda_1 = \lambda_2$ □

The proof of Theorem 8.3.5 gives us an algorithm for finding a dominant weight conjugate under the Weyl group to a given weight. We call this algorithm ConjugateDominantWeight. It takes as input the Weyl group $W(\Phi)$ and a weight $\mu = m_1 \lambda_1 + \cdots + m_l \lambda_l$. If all $m_j \ge 0$, then μ is returned. Otherwise we determine a j such that $m_j < 0$, replace μ by $\mu - m_j \alpha_j$ and continue.

Lemma 8.3.6 *Let $\alpha_1, \ldots, \alpha_l$ be linearly independent vectors in a Euclidean space W such that $(\alpha_i, \alpha_j) \le 0$ for $i \ne j$. Let $v \in W$ be such that $(v, \alpha_i) \ge 0$ for $1 \le i \le l$. Then $v = \sum_{i=1}^{l} a_i \alpha_i$ where $a_i \ge 0$.*

Proof. The proof is by induction on l. Let $w \in W$ be the unique vector satisfying $(w, \alpha_i) = \delta_{il}$ for $1 \le i \le l$. Then w is perpendicular to $\alpha_1, \ldots, \alpha_{l-1}$ and hence $\{\alpha_1, \ldots, \alpha_{l-1}, w\}$ forms a basis of W. Write $v = \sum_{i=1}^{l-1} b_i \alpha_i + cw$

and set $v_0 = \sum_{i=1}^{l-1} b_i \alpha_i$. For $i = 1, \dots, l-1$ we have $0 \leq (v, \alpha_i) = (v_0, \alpha_i)$. So by induction $b_i \geq 0$ for $1 \leq i \leq l-1$. Also $0 \leq (v, \alpha_l) = (v_0, \alpha_l) + c$ and since $(v_0, \alpha_l) \leq 0$ we have $c \geq 0$. Now write $w = \sum_{i=1}^{l} c_i \alpha_i = u + c_l \alpha_l$. Then $0 < (w, w) = (w, u) + c_l = c_l$, i.e., $c_l > 0$. Also for $i = 1, \dots, l-1$ we have $(w, \alpha_i) = 0$ implying $(u, \alpha_i) + c_l(\alpha_l, \alpha_i) = 0$, from which $(u, \alpha_i) \geq 0$. By induction this implies $c_i \geq 0$ for $1 \leq i \leq l-1$, and we are done. $\quad\square$

Lemma 8.3.7 *Let* $\lambda \in P_+$ *be a dominant weight. Then the number of dominant weights* $\mu \in P_+$ *such that* $\mu < \lambda$ *is finite.*

Proof. Write $\lambda = \sum_{i=1}^{l} s_i \alpha_i$ and $\mu = \sum_{i=1}^{l} t_i \alpha_i$ where $s_i, t_i \in \mathbb{Q}$. Furthermore, s_i, t_i are non-negative by Lemma 8.3.6 (cf. Lemma 5.5.3). However, since $\lambda > \mu$ all $s_i - t_i$ are non-negative integers. Hence there are only finitely many possibilities for the t_i. $\quad\square$

From Section 5.7 we recall that

$$\rho = \frac{1}{2} \sum_{\alpha \in \Phi^+} \alpha$$

is the Weyl vector.

Lemma 8.3.8 $\rho = \sum_{i=1}^{l} \lambda_i$.

Proof. Let α_i be a simple root. Corollary 5.7.2 states that $r_{\alpha_i}(\rho) = \rho - \alpha_i$. Hence $(\rho - \alpha_i, \alpha_i) = (r_{\alpha_i}(\rho), r_{\alpha_i}^2(\alpha_i)) = (\rho, r_{\alpha_i}(\alpha_i)) = (\rho, -\alpha_i)$. This means that $2(\rho, \alpha_i) = (\alpha_i, \alpha_i)$, i.e., $\langle \rho, \alpha_i \rangle = 1$. Now by Lemma 8.3.1 we get the result. $\quad\square$

Proposition 8.3.9 *Let* $\lambda \in P_+$ *be a dominant integral function. Then* $g(\lambda) \leq \lambda$ *for all* $g \in W(\Phi)$. *Let* W_λ *be the stabilizer of* λ *in* $W(\Phi)$, *i.e., the subgroup of* $W(\Phi)$ *consisting of all* $g \in W(\Phi)$ *such that* $g(\lambda) = \lambda$. *Then* W_λ *is generated by the simple reflections* r_{α_i} *for* $\alpha_i \in \Delta$ *such that* $(\lambda, \alpha_i) = 0$.

Proof. Write $g_i = r_{\alpha_i}$ and let $g = g_{i_1} \cdots g_{i_t}$ be a reduced expression for g. We may assume that $g \neq 1$, i.e., that $t > 0$. For $1 \leq s \leq t$ consider the element $\lambda_s = g_{i_s} \cdots g_{i_t}(\lambda)$. We calculate

$$(\lambda_s, \alpha_{i_{s-1}}) = (g_{i_s} \cdots g_{i_t} \lambda, \alpha_{i_{s-1}})$$
$$= (\lambda, g_{i_t} \cdots g_{i_s}(\alpha_{i_{s-1}})).$$

But by Corollary 8.3.3 we have that $g_{i_t} \cdots g_{i_{s-1}}(\alpha_{i_{s-1}})$ is a negative root. So $g_{i_t} \cdots g_{i_s}(\alpha_{i_{s-1}})$ is a positive root and $(\lambda_s, \alpha_{i_{s-1}}) \geq 0$. Therefore $\lambda_{s-1} = \lambda_s - m\alpha_{i_{s-1}}$ where $m \geq 0$. From this it follows that $g(\lambda) \leq \lambda$. Furthermore, $g(\lambda) = \lambda$ implies that $(\lambda, \alpha_{i_s}) = 0$ for $1 \leq s \leq t$. This implies the second statement of the proposition. $\qquad\square$

8.4 Finite dimensionality

In this section we show that any irreducible highest-weight module over L is necessarily finite-dimensional. We start with a lemma.

Lemma 8.4.1 *Let* x_i, y_i, h_i *for* $1 \leq i \leq l$ *be a set of canonical generators of* L *and set*

$$\tau_i = \exp(\mathrm{ad}x_i)\exp(-\mathrm{ad}y_i)\exp(\mathrm{ad}x_i).$$

Then $\tau_i(h_j) = h_j - \langle \alpha_i, \alpha_j \rangle h_i$.

Proof. The proof is by straightforward calculation. First we have that $\exp(\mathrm{ad}x_i)(h_j) = h_j - \alpha_i(h_j)x_i$. Secondly,

$$\exp(-\mathrm{ad}y_i)(h_j - \alpha_i(h_j)x_i) = h_j - \alpha_i(h_j)x_i - \alpha_i(h_j)h_i.$$

And finally

$$\exp(\mathrm{ad}x_i)(h_j - \alpha_i(h_j)x_i - \alpha_i(h_j)) = h_j - \alpha_i(h_j)h_i$$

implying the result. $\qquad\square$

Theorem 8.4.2 *Let* $\lambda \in P_+$ *be a dominant weight. Let* V *be an irreducible highest-weight module over* L *with highest weight* λ. *Then* V *is finite-dimensional. Furthermore, the set of weights* $P(V)$ *is stable under* $\mathrm{W}(\Phi)$ *and* $\dim V_\mu = \dim V_{g(\mu)}$ *for* $\mu \in P(V)$ *and* $g \in \mathrm{W}(\Phi)$.

Proof. Let $v_\lambda \in V$ be a highest-weight vector with weight λ. We recall that $x_1, \ldots, x_l, h_1, \ldots, h_l, y_1, \ldots, y_l$ are the canonical generators of L. Set $m_i = \langle \lambda, \alpha_i \rangle$ and $u_i = y_i^{m_i+1} \cdot v_\lambda$. We claim that $u_i = 0$ for $1 \leq i \leq l$. First, if $j \neq i$, then $[x_i, y_j] = 0$ and hence $x_j \cdot u_i = y_i^{m_i+1} x_j \cdot v_\lambda = 0$. By induction on n it is straightforward to prove the following identity in $U(L)$:

$$x_i y_i^{n+1} = y_i^{n+1} x_i + (n+1)y_i^n h_i - n(n+1)y_i^n.$$

So $x_i \cdot u_i = y_i^{m_i+1} x_i \cdot v_\lambda + (m_i+1)\lambda(h_i)y_i^n \cdot v_\lambda - m_i(m_i+1)y_i^n \cdot v_\lambda$, which is zero because $\lambda(h_i) = m_i$. Consequently $x_j \cdot u_i = 0$ for $1 \le j \le l$, and because the x_j generate N^+, also $N^+ \cdot u_i = 0$. But this implies that $u_i = 0$ because otherwise u_i is a highest-weight vector of weight $\mu_i = \lambda - (m_i + 1)\alpha_i$. By Lemma 8.1.4 this implies that $\lambda \le \mu_i$, a contradiction.

For $1 \le i \le l$ we let K_i be the subalgebra of L spanned by x_i, h_i, y_i. Then K_i is isomorphic to $\mathfrak{sl}_2(F)$ and by restricting the representation of L, V becomes a K_i-module. Now let $W_i \subset V$ be the subspace spanned by $v_\lambda, y_i \cdot v_\lambda, \dots, y_i^{m_i} \cdot v_\lambda$. Then W_i is stable under the action of y_i by our claim above. Furthermore, calculations as those in the proof of Proposition 4.10.1 show that W_i is also stable under x_i and h_i. So W_i is a finite-dimensional K_i-module contained in V.

Now we show that V is a sum of finite-dimensional K_i-submodules of V. Let V' be the sum of all finite-dimensional K_i-submodules of V. Then by the above, $V' \ne 0$. Let W be a finite-dimensional K_i-submodule of V. Let $\overline{W}$ be the span of W together with all subspaces of the form $x_\alpha \cdot W$ for root vectors $x_\alpha \in L_\alpha$, where α ranges over Φ. Then since $x_i x_\alpha \cdot w = x_\alpha x_i \cdot w + [x_i, x_\alpha] \cdot w$ we have that $\overline{W}$ is stable under x_i. Similarly $\overline{W}$ is stable under h_i and y_i, and therefore it is a finite-dimensional K_i-submodule of V. In particular, $\overline{W} \subset V'$. So if $w \in V'$, then also $x_\alpha \cdot w \in V'$, and since the x_α generate L, we have that V' is an L-submodule of V. It follows that $V = V'$.

Now let $w \in V$, then by the above, w lies in some finite-dimensional K_i-submodule of V. But x_i and y_i act nilpotently on such modules (cf. Lemma 5.1.2). Hence x_i and y_i are locally nilpotent on V. We define the endomorphisms $\exp(x_i)$ and $\exp(y_i)$ as follows. Let $u \in V$, then there is an integer $k_u > 0$ such that $x_i^{k_u+1} \cdot u = 0$. Set $\exp(x_i)(u) = \sum_{r=0}^{k_u} \frac{1}{r!} x_i^r \cdot u$. Then $\exp(x_i)$ is a linear map. The map $\exp(y_i)$ is defined similarly. The endomorphisms $\exp(x_i), \exp(y_i)$ are bijective as their inverses are $\exp(-x_i)$, $\exp(-y_i)$.

Now let ϕ denote the representation of L afforded by V. Set

$$\tau_i = \exp \mathrm{ad}\phi(x_i) \exp -\mathrm{ad}\phi(y_i) \exp \mathrm{ad}\phi(x_i)$$

and

$$\eta_i = \exp \phi(x_i) \exp -\phi(y_i) \exp \phi(x_i).$$

Then τ_i is an automorphism of the Lie algebra $\phi(L)$ (cf. Section 1.11). Denote by l_{x_i} and r_{x_i} the left and right multiplication by $\phi(x_i)$, which are endomorphisms of $\mathfrak{gl}(V)$. Then l_{x_i} and r_{x_i} are locally nilpotent (since $\phi(x_i)$ is a locally nilpotent linear transformation) and they commute. Hence $\exp \mathrm{ad}\phi(x_i) = \exp(\mathrm{l}_{x_i} + \mathrm{r}_{-x_i}) = \exp(\mathrm{l}_{x_i})\exp(\mathrm{r}_{-x_i}) = \mathrm{l}_{\exp x_i} \cdot \mathrm{r}_{\exp -x_i}$, from which it follows that $\exp\phi(x_i)\phi(z)(\exp\phi(x_i))^{-1} = \exp(\mathrm{ad}\phi(x_i))(\phi(z))$ for all $z \in L$. This however implies that

$$\tau_i(\phi(z)) = \eta_i\phi(z)\eta_i^{-1} \text{ for all } z \in L. \tag{8.4}$$

By Lemma 8.4.1 we have that $\tau_i(\phi(h_j)) = \phi(h_j) - \langle \alpha_i, \alpha_j \rangle \phi(h_i)$. By (8.4) this implies that $\phi(h_j)\eta_i = \eta_i\phi(h_j) + \langle \alpha_i, \alpha_j \rangle \phi(h_i)\eta_i$, which for $i = j$ boils down to $\phi(h_i)\eta_i = -\eta_i\phi(h_i)$. Using this we calculate for $v \in V_\mu$,

$$\begin{aligned}
\phi(h_j)\eta_i \cdot v &= \big(\eta_i\phi(h_j) + \langle \alpha_i, \alpha_j \rangle \phi(h_i)\eta_i\big) \cdot v \\
&= (\mu(h_j) - \langle \alpha_i, \alpha_j \rangle \mu(h_i))\eta_i \cdot v \\
&= (\mu(h_j) - \langle \mu, \alpha_i \rangle \alpha_i(h_j))\eta_i \cdot v
\end{aligned}$$

(where the last equality follows from $\mu(h_i) = \langle \mu, \alpha_i \rangle$). As a consequence $\eta_i \cdot v$ lies in the weight space corresponding to $r_{\alpha_i}(\mu)$. Since $\mathrm{W}(\Phi)$ is generated by the r_{α_i} (Theorem 5.7.6) we see that $P(V)$ is stable under $\mathrm{W}(\Phi)$. Also since η_i is nonsingular we have $\dim V_\mu = \dim V_{r_{\alpha_i}(\mu)}$ and the same holds for $g \in \mathrm{W}(\Phi)$ in place of r_{α_i} by Theorem 5.7.6.

Let A be the set of all $\mu \in P_+$ such that $\mu \leq \lambda$. Then by Lemma 8.3.7, A is finite. Furthermore let B be the set of all $g(\mu)$ for $g \in \mathrm{W}(\Phi)$ and $\mu \in A$. Then B is finite since $\mathrm{W}(\Phi)$ is. But by Theorem 8.3.5 together with the fact that $P(V)$ is stable under $\mathrm{W}(\Phi)$, $P(V) \subset B$. So $P(V)$ is finite and by Lemma 8.1.4, V is finite-dimensional. □

Corollary 8.4.3 *There is a bijection between the set P_+ and the set of (isomorphism classes of) irreducible finite-dimensional L-modules.*

Proof. By Theorem 8.1.5, every finite-dimensional irreducible L-module has a unique highest weight in P_+. Also if $\lambda \in P_+$, then by Proposition 8.2.2, there is a unique irreducible highest-weight module V having λ as highest weight. And by Theorem 8.4.2 V is finite-dimensional. □

8.5　On representing the weights

We want to calculate with the set of weights of an L-module V. For this we need a compact format for representing a weight $\lambda \in H^*$. We do this by taking the list of coefficients of λ relative to the basis of fundamental weights $\lambda_1, \ldots, \lambda_l$. This means that we represent λ as a vector $(n_1, \ldots, n_l)$ where the n_i are such that $\lambda = \sum_i n_i\lambda_i$. We note that by Lemma 8.3.1, we have $n_i = \lambda(h_i)$, where the h_i come from a set of canonical generators of L that we fixed at the start. We call the set $\{(\lambda(h_1), \ldots, \lambda(h_l)) \mid \lambda \in P(V)\}$ the set of *weight-coordinate vectors* of V relative to $h_1, \ldots, h_l$. This set does not only depend on L and V, but also on various choices that are made. First we choose a Cartan subalgebra $H \subset L$. This Cartan subalgebra gives us a root system Φ. Then in that root system we choose a fundamental system $\Delta = \{\alpha_1, \ldots, \alpha_l\}$. This determines the h_j as

$$h_j = \frac{2h_{\alpha_j}}{(\alpha_j, \alpha_j)}.$$

Finally we list the h_j in a certain order, which gives us the weight-coordinate vectors $(\lambda(h_1), \ldots, \lambda(h_l))$. Now the question presents itself as to what extend the set of weight-coordinate vectors of an L-module V depends on these choices.

The h_i come from a set of canonical generators x_i, y_i, h_i ($1 \leq i \leq l$). Let x_i', y_i', h_i' be a second set of canonical generators. Then by Corollary 5.11.5 there is a $g \in \mathrm{Aut}(L)$ such that $g(h_i) = h_i'$. Now let $\phi : L \to \mathfrak{gl}(V)$ denote the representation of L afforded by V, and suppose that there is a bijective linear map $\tau : V \to V$ such that $\phi(g(x)) \cdot \tau(v) = \tau(\phi(x) \cdot v)$ for all $x \in L$ and $v \in V$ (i.e., the representations ϕ and $\phi \circ g$ of L are equivalent). Let $v_\lambda \in V$ be a weight vector of weight λ relative to the h_i (i.e., $\phi(h_i) \cdot v_\lambda = \lambda(h_i) v_\lambda$). Then $\phi(g(h_i)) \cdot \tau(v_\lambda) = \tau(\phi(h_i) \cdot v_\lambda) = \lambda(h_i) \tau(v_\lambda)$. So $\tau(v_\lambda)$ is a weight vector relative to the h_i'. Denote the corresponding weight by μ (note that μ might lie in the dual space of a different Cartan subalgebra); then $\mu(h_i') = \lambda(h_i)$. The conclusion is that the existence of the map τ implies that the set of weight-coordinate vectors $(\lambda(h_1), \ldots, \lambda(h_l))$ of V with respect to $h_1, \ldots, h_l$ is equal to the set of weight-coordinate vectors of V with respect to $h_1', \ldots, h_l'$. By the next lemma we can always find such a map τ if g is an inner automorphism.

Lemma 8.5.1 *Let $\phi : L \to \mathfrak{gl}(V)$ be a finite-dimensional representation of L. Let g be an inner automorphism of L. Then there is a bijective linear map $\tau : V \to V$ such that $\phi(g(x)) \cdot \tau(v) = \tau(\phi(x) \cdot v)$ for all $x \in L$ and $v \in V$.*

Proof. First suppose that $g = \exp(\mathrm{ad}x)$, where $x \in L$ is such that $\mathrm{ad}_L x$ is nilpotent. Then by Corollary 4.6.4 we have that $\phi(x)$ is also nilpotent. Now set $\tau = \exp(\phi(x))$. Then it is straightforward to see that

$$\exp(\mathrm{ad}\phi(x))(\phi(z)) = \tau\phi(z)\tau^{-1} \quad \text{for } z \in L$$

(cf. the proof of Theorem 8.4.2). Using this we calculate $\phi(\exp(\operatorname{ad}x)(z)) = \exp(\operatorname{ad}\phi(x))(\phi(z)) = \tau\phi(z)\tau^{-1}$. So $\tau(\phi(z)(v)) = \phi(\exp(\operatorname{ad}x)(z))\tau(v) = \phi(g(z))\tau(v)$. Since τ is clearly bijective we are done for the case where $g = \exp(\operatorname{ad}x)$. Now if g_1, g_2 correspond to the maps τ_1, τ_2, then it is straight-forward to see that $g_1 g_2$ corresponds to $\tau_1 \tau_2$. So because $\operatorname{Int}(L)$ is generated by elements of the form $\exp(\operatorname{ad}x)$, we get the result in general. $\qquad\square$

Let H denote the Cartan subalgebra containing the h_i. Suppose that also the $h_i' \in H$, i.e., the automorphism g mapping the h_i to the h_i' maps H into itself. The h_i correspond to the simple system $\Delta = \{\alpha_1, \dots, \alpha_l\}$ of the root system Φ. Let $\Delta' = \{\alpha_1', \dots, \alpha_l'\}$ be the simple system corresponding to the h_i'. Then Δ' is also a simple system of Φ. Furthermore we may suppose that the Cartan matrix of Φ relative to Δ' is *equal* to the Cartan matrix of Φ relative to Δ (cf. Corollary 5.7.4). By Theorem 5.7.3 there is a $\sigma \in W(\Phi)$ such that $\sigma(\Delta) = \Delta'$. We show that this implies that there is an inner automorphism g_σ of L such that $g_\sigma(h_j) = h_j'$. By Theorem 5.7.6 we may assume that $\sigma = r_{\alpha_k}$ for some $\alpha_k \in \Delta$. Set

$$\tau_k = \exp(\operatorname{ad}x_k)\exp(-\operatorname{ad}y_k)\exp(\operatorname{ad}x_k).$$

Then by Lemma 8.4.1, $\tau_k(h_j) = h_j - \langle \alpha_k, \alpha_j \rangle h_k$. Now the h_j' are the unique elements of H satisfying $\alpha_i'(h_j') = \langle \alpha_i', \alpha_j' \rangle$, and this is equal to $\langle \alpha_i, \alpha_j \rangle$ (because of our assumption on the Cartan matrices of Φ relative to Δ and Δ'). We calculate

$$\begin{aligned}
\alpha_i'(\tau_k(h_j)) &= \alpha_i'(h_j - \langle \alpha_k, \alpha_j \rangle h_k) \\
&= \alpha_i'(h_j) - \alpha_k(h_j)\langle \alpha_i', \alpha_k \rangle \\
&= r_{\alpha_k}(\alpha_i')(h_j) = \alpha_i(h_j) = \langle \alpha_i, \alpha_j \rangle.
\end{aligned}$$

Which implies that $\tau_k(h_j) = h_j'$. And therefore, by Lemma 8.5.1 we have that the set of weight-coordinate vectors of V relative to the h_i is equal to the set of weight-coordinate vectors of V relative to the h_i'.

Now suppose that the h_i' belong to a different Cartan subalgebra H'. Suppose further that the ground field is algebraically closed. (This does not make much difference as $\overline{V} = V \otimes_F \overline{F}$ is a module over $\overline{L} = L \otimes_F \overline{F}$, and the set of weight-coordinate vectors of $\overline{V}$ relative to $h_i \otimes 1$ is equal to the set of weight-coordinate vectors of V relative to the h_i.) Then by Theorem 3.5.1 there is a $g \in \operatorname{Int}(L)$ such that $g(H) = H'$. Now by Lemma 8.5.1 the set of weight-coordinate vectors of V relative to the h_i is the same as the set of weight-coordinate vectors of V relative to the $g(h_i)$. Hence we are in the situation considered above; and it follows that if the $g(h_i)$ correspond to a different simple system than the h_i', then the set of weight-coordinate vectors relative to the $g(h_i)$ is equal to the set of coordinate vectors relative to the h_i'.

The conclusion that we draw from this is that the set of weight-coordinate vectors does not depend on the particular Cartan subalgebra H of L which we choose, and neither does it matter which simple system Δ in the corresponding root system we choose. So the only thing that matters is the order in which we list the h_i. If the Dynkin diagram does not have any diagram automorphisms, then we can fix this order by fixing the Cartan matrix. However, if the Dynkin diagram does allow automorphisms, then this is not enough. Indeed, by way of example let L be the simple Lie algebra of type A_2. Let $x_1, x_2, y_1, y_2, h_1, h_2$ be a fixed set of canonical generators. Let $\Delta = \{\alpha_1, \alpha_2\}$ be the corresponding simple system. Let $\mu_i \in H^*$ for $i = 1, 2$ be defined by $\mu_i(h_j) = \delta_{ij}$. So μ_1 is represented by the coordinate vector $(1, 0)$ and μ_2 by $(0, 1)$. Then the L-module V_1 with highest weight μ_1 is not equivalent to the L-module V_2 with highest weight μ_2. However, if we interchange α_1 and α_2 (and hence also h_1 and h_2), then μ_1 is represented by $(0, 1)$ and μ_2 by $(1, 0)$. Furthermore, the Cartan matrix of Φ relative to α_1, α_2 is equal to the Cartan matrix of Φ relative to α_2, α_1. So in this case we also have to fix the order of the simple system Δ in order to give the representation of weights by means of coordinate vectors a meaning.

8.6 Computing orbits of the Weyl group

When computing with the weights of a representation we often want the orbit of a given weight under the action of the Weyl group. This is due to Theorem 8.4.2: if μ is a weight of V, then so are all elements of the orbit $W(\Phi) \cdot \mu$. Furthermore, for all elements of a fixed orbit the dimension of the corresponding weight space is the same.

By Theorem 5.7.6, $W(\Phi)$ is generated by the simple reflections r_{α_i}. The action of these simple reflections on the weight lattice is easily written down: $r_{\alpha_i}(\lambda_j) = \lambda_j - \delta_{ij}\alpha_i$. So by conventional techniques for computing the orbit of a point under the action of a finite group, we can calculate the orbit. However, in this case we can utilize the structure of the Weyl group to give a much more efficient method. For this we need some results on the length of a Weyl group element.

Write $g_i = r_{\alpha_i}$ for $1 \leq i \leq l$. Let $g \in W(\Phi)$ and write $g = g_{i_1} \cdots g_{i_s}$. If s is minimal among all ways of writing g as a product of simple reflections (i.e., if this expression for g is reduced), then we say that the *length* of g is s; and we write $\ell(g) = s$. We define the length of the identity to be 0.

Also for $g \in W(\Phi)$ we let $n(g)$ be the number of positive roots α such that $g(\alpha)$ is a negative root.

Lemma 8.6.1 *Let $g \in W(\Phi)$ and let $\alpha_i \in \Delta$ be a simple root. If $g(\alpha_i) > 0$, then $n(gr_{\alpha_i}) = n(g) + 1$. On the other hand, if $g(\alpha_i) < 0$, then $n(gr_{\alpha_i}) = n(g) - 1$.*

Proof. By Lemma 5.7.1, r_{α_i} permutes the positive roots other than α_i. Furthermore r_{α_i} maps α_i to $-\alpha_i$. The lemma follows from this. $\square$

Proposition 8.6.2 *For $g \in W(\Phi)$ we have $\ell(g) = n(g)$.*

Proof. Suppose $\ell(g) = r$ and write $g = g_{i_1} \cdots g_{i_r}$. We prove the theorem by induction on r. For $r = 0$ the statement is trivial, so we suppose that $r > 0$. By Corollary 8.3.3, $g(\alpha_{i_r}) < 0$; so by Lemma 8.6.1, $n(gr_{\alpha_{i_r}}) = n(g) - 1$. Also $\ell(gr_{\alpha_{i_r}}) = \ell(g_{i_1} \cdots g_{i_{r-1}}) = \ell(g) - 1$. Therefore by induction we have $n(g) - 1 = \ell(g) - 1$; which is the same as $n(g) = \ell(g)$. $\square$

Corollary 8.6.3 *Let $g \in W(\Phi)$ be such that $g(\Delta) = \Delta$. Then $g = 1$.*

Proof. For such a g we have $g(\Phi^+) = \Phi^+$. Hence $n(g) = 0$, and by Proposition 8.6.2, also $\ell(g) = 0$. $\square$

Corollary 8.6.4 *There is a unique element $g_0 \in W(\Phi)$ such that $g_0(\Delta) = -\Delta$. Moreover, $g_0^2 = 1$ and $\ell(g_0) = |\Phi^+|$.*

Proof. By Proposition 5.5.8, $-\Delta$ is a simple system of Φ. So by Theorem 5.7.3 there is a $g_0 \in W(\Phi)$ such that $g_0(\Delta) = -\Delta$. The uniqueness of this g_0 follows from Corollary 8.6.3. This corollary also implies that $g_0^2 = 1$. Finally, the length of g_0 is equal to the number of positive roots by Proposition 8.6.2. $\square$

The unique element g_0 of Corollary 8.6.4 is called the *longest element* of $W(\Phi)$.

Now let $\lambda \in P_+$ be a dominant weight. Then by $W(\Phi) \cdot \lambda$ we denote the orbit of λ under the action of $W(\Phi)$. Let $\mu \in W(\Phi) \cdot \lambda$ and let $g \in W(\Phi)$ be an element of minimal length such that $g(\lambda) = \mu$. Let r be the length of g. Then we also say that the *length* of μ is r, and we write $\ell(\mu) = r$.

Lemma 8.6.5 *Let* $\lambda \in P_+$ *be a dominant weight and let* $\mu \in W(\Phi) \cdot \lambda$. *Then the length of* μ *is equal to the number of positive roots* α *such that* $(\alpha, \mu) < 0$.

Proof. Let $g \in W(\Phi)$ be an element of minimal length such that $g(\lambda) = \mu$. Let $\alpha \in \Phi^+$ be such that $(\lambda, g^{-1}(\alpha)) = 0$. We claim that this implies that $g^{-1}(\alpha) > 0$. We write α as a linear combination of simple roots with non-negative integral coefficients. If $g^{-1}(\alpha)$ is negative, then it is a linear combination of simple roots with non-positive integral coefficients. As λ is dominant, $(\lambda, g^{-1}(\alpha)) = 0$ implies that $(\lambda, \alpha_i) = 0$ for all α_i appearing in this expression for $g^{-1}(\alpha)$ with non-zero coefficient. Furthermore, $g(\alpha_i) < 0$ for at least one such α_i. By Lemma 8.3.2 this means that $\ell(gr_{\alpha_i}) \leq \ell(g) - 1$. But since $(\lambda, \alpha_i) = 0$ we have that $r_{\alpha_i}(\lambda) = \lambda$, and therefore $gr_{\alpha_i}(\lambda) = \mu$. And as g was chosen to be of minimal length, we have reached a contradiction.

This implies that for any positive root α we have that $g^{-1}(\alpha) < 0$ if and only if $(\lambda, g^{-1}(\alpha)) < 0$. But this is equivalent to $(\mu, \alpha) < 0$. It follows that the number of positive roots α such that $(\alpha, \mu) < 0$ is equal to $n(g^{-1})$; and by Proposition 8.6.2, this is equal to $\ell(g^{-1}) = \ell(g)$. $\square$

Corollary 8.6.6 *Let* λ, μ *be as in Lemma 8.6.5. Write* $\mu = n_1 \lambda_1 + \cdots n_l \lambda_l$ *(linear combination of the fundamental weights). Let* r_{α_i} *be a simple reflection. Then*

$$\ell(r_{\alpha_i}(\mu)) = \ell(\mu) + 1 \ \textit{if} \ n_i > 0,$$
$$\ell(r_{\alpha_i}(\mu)) = \ell(\mu) \ \textit{if} \ n_i = 0,$$
$$\ell(r_{\alpha_i}(\mu)) = \ell(\mu) - 1 \ \textit{if} \ n_i < 0.$$

Proof. First of all, if $n_i = 0$ then $r_{\alpha_i}(\mu) = \mu$. If $n_i > 0$ then we count the number of positive roots α such that $(r_{\alpha_i}(\mu), \alpha) = (\mu, r_{\alpha_i}(\alpha)) < 0$. By Lemma 5.7.1, r_{α_i} permutes the positive roots other than α_i. So the number of $\alpha > 0$ such that $\alpha \neq \alpha_i$ and $(r_{\alpha_i}(\mu), \alpha) < 0$ is the same as the number of $\alpha > 0$ such that $\alpha \neq \alpha_i$ and $(\mu, \alpha) < 0$. Furthermore, $n_i > 0$ means that $(\mu, \alpha_i) > 0$; but then $(\mu, r_{\alpha_i}(\alpha_i)) < 0$. So $\ell(r_{\alpha_i}(\mu)) = \ell(\mu) + 1$. For the remaining case we use a similar argument. $\square$

Let $\lambda \in P_+$. It is straightforward to see that the number of roots $\alpha > 0$ such that $(\alpha, \mu) = 0$ is equal for all $\mu \in W(\Phi) \cdot \lambda$. We denote this number by $n_0(\lambda)$. Let g_0 be the longest element of $W(\Phi)$. Then using Lemma 8.6.5 we see that

$$\ell(g_0(\mu)) = \ell(g_0) - n_0(\lambda) - \ell(\mu). \tag{8.5}$$

Since $\ell(g_0) \geq 0$ we conclude that the highest length obtained in the orbit of λ is $\ell(g_0) - n_0(\lambda)$ (which is the length of $g_0(\lambda)$). Furthermore, the weights of length k are the images under g_0 of the weights of length $\ell(g_0) - n_0(\lambda) - k$. So using the action of the single element g_0, we only have to calculate half the orbit of λ. We note that the action of g_0 is easy to calculate. Since $g_0(\Delta) = -\Delta$ we have that $g_0(\alpha_j) = -\alpha_{\sigma(j)}$, where σ is a permutation of $(1, \dots, l)$. Let $\mu = n_1\lambda_1 + \cdots + n_l\lambda_l$ be a weight and write $g_0(\mu) = m_1\lambda_1 + \cdots + m_l\lambda_l$. Then using Lemma 8.3.1 and the fact that $g_0^2 = 1$, we calculate

$$m_j = \langle g_0(\mu), \alpha_j \rangle = \langle \mu, g_0(\alpha_j) \rangle = -\langle \mu, \alpha_{\sigma(j)} \rangle = -n_{\sigma(j)}.$$

So we only need to calculate the permutation σ once, and then the action of g_0 is easily written down.

The idea of the algorithm for calculating the orbit $\lambda \cdot W(\Phi)$, where $\lambda \in P_+$, is now straightforward. We start with the weight λ, which is the only weight in the orbit of length 0. Then supposing we know all weights of length k, we calculate the weights of length $k+1$ using Corollary 8.6.6: if $\mu = m_1\lambda_1 + \cdots + m_l\lambda_l$ is a weight of length k and $m_i > 0$ for a certain index i, then $r_{\alpha_i}(\mu)$ is a weight of length $k+1$. However, it may happen that $r_{\alpha_i}(\mu) = r_{\alpha_j}(\nu)$ for two weights μ, ν of length k. So when building the set of weights of length $k+1$ we have to discard those weights that we already have constructed. As the number of weights of a certain length can be high, deciding whether or not we already constructed a particular weight, may become a lengthy task. By the next result however we can dispense with this search operation.

Proposition 8.6.7 *Let $\lambda \in P_+$ be a dominant weight. By $\mathcal{L}_k$ we denote the weights of the orbit $W(\Phi) \cdot \lambda$ of length k. Then for every $\mu = m_1\lambda_1 + \cdots + m_l\lambda_l \in \mathcal{L}_{k+1}$ there is a unique weight $\nu \in \mathcal{L}_k$ such that*

1. $r_{\alpha_i}(\nu) = \mu$ for a simple reflection r_{α_i},

2. $m_j \geq 0$ for $i < j \leq l$.

Proof. Let i be the index such that $m_i < 0$ and $m_j \geq 0$ for $j > i$ (note that such an index exists because λ is the only dominant weight of the orbit). Set $\nu = r_{\alpha_i}(\mu)$; then $\nu \in \mathcal{L}_k$ by Corollary 8.6.6. So ν is a weight satisfying the listed requirements. Now suppose that there is a second weight η such that $r_{\alpha_t}(\eta) = \mu$ and $m_j \geq 0$ for $j > t$. Write $\eta = p_1\lambda_1 + \cdots + p_l\lambda_l$. Then by Corollary 8.6.6, $p_t > 0$. But then $m_t = p_t - 2p_t = -p_t < 0$. It follows that $t = i$ and $\eta = \nu$. $\qquad\square$

So by Proposition 8.6.7 each weight ν of length k has a uniquely defined set of *successors* of length $k + 1$. This is used in the following algorithm.

Algorithm WeylOrbit
Input: a weight $\eta \in P$ and the Weyl group $W(\Phi)$ generated by the simple reflections r_{α_i} and with longest element g_0.
Output: the orbit $W(\Phi) \cdot \eta$.

Step 1 (Trivial case.) If η is zero then return $\{\eta\}$.

Step 2 Set $\lambda :=$ ConjugateDominantWeight$(W(\Phi), \eta)$. Set $\mathcal{L}_0 := \{\lambda\}$ and $\mathcal{T}_0 := \{g_0(\lambda)\}$. Set $k := 0$.

Step 3 Set $\mathcal{L}_{k+1} := \emptyset$. For each $\nu = m_1\lambda_1 + \cdots m_l\lambda_l \in \mathcal{L}_k$ and each index i such that $m_i > 0$ do the following:

Step 3a Set $\mu := r_{\alpha_i}(\nu)$, and write $\mu = \sum n_j\lambda_j$.

Step 3b If $n_j \geq 0$ for $j > i$, then add μ to $\mathcal{L}_{k+1}$.

Step 4 Set $\mathcal{T}_{k+1} := g_0(\mathcal{L}_{k+1})$.

Step 5 If $\mathcal{T}_{k+1} = \mathcal{L}_{k+1}$, or the length of the weights in $\mathcal{T}_{k+1}$ is $k + 2$, then return the union of the sets $\mathcal{L}_j$ and $\mathcal{T}_j$ for $1 \leq j \leq k + 1$. Otherwise set $k := k + 1$ and return to Step 3.

Comments: The sets $\mathcal{L}_k$, $\mathcal{T}_k$ contain the weights of length k and $\ell(g_0) - n_0(\lambda) - k$ respectively. Steps 3a. and 3b. are justified by Proposition 8.6.7. If at a certain point the sets $\mathcal{L}_k$, $\mathcal{T}_k$ "meet" (in the sense that the length of an element of $\mathcal{T}_k$ is at most one bigger than the length of an element of $\mathcal{L}_k$), then we have found all elements of the orbit and we return the result.
Remark. As the Weyl group can be very large also the number of weights of a certain length can be large. However, for some applications it is only necessary to loop through the weights in a certain orbit. In these cases it is not necessary to store them all. Using Proposition 8.6.7 an algorithm that does not use much memory can be devised for this. For each weight of length k the proposition defines a unique set of successors of length $k + 1$. So the orbit can be viewed as a tree, with the dominant weight λ at its root. Now using only a small stack of weights we can loop through this tree, for example using the technique known as depth-first search (see, e.g., [75]).

The algorithm WeylOrbit has proved to be very efficient in practice. We refer to [81], [82] for an account of practical experiences with the algorithm.

8.7 Calculating the weights

In this section we describe algorithms for calculating the set of all weights that occur in a given highest-weight module V over L, with highest weight $\lambda \in P_+$.

Lemma 8.7.1 *Let V be an irreducible highest-weight module over L with highest weight λ. Let $\mu \in P(V)$ and $\alpha \in \Phi$. Let r, q be the largest integers such that $\mu - r\alpha, \mu + q\alpha \in P(V)$. Then all $\mu + i\alpha \in P(V)$ for $-r \le i \le q$. Furthermore, $r - q = \langle \mu, \alpha \rangle$.*

Proof. Let W be the space spanned by all $V_{\mu+i\alpha}$ for $-r \le i \le q$. By Root facts 12, 13 there are $x \in L_\alpha$, $y \in L_{-\alpha}$, $h \in H$ spanning a subalgebra K_α of L isomorphic to $\mathfrak{sl}_2(F)$. By Lemma 8.1.2, W is a K_α-submodule of V. On W the element h has eigenvalues $(\mu + i\alpha)(h) = \mu(h) + 2i$. Now by the representation theory of $\mathfrak{sl}_2(F)$ (Theorem 5.1.4) the eigenvalues of h occur in strings with difference 2. So all of $\mu + i\alpha$ must be weights of V for $-r \le i \le q$.

Let r_α denote the reflection corresponding to α. By Theorem 8.4.2 we know that the image of a weight of V under r_α is also a weight. Now $r_\alpha(\mu + i\alpha) = r_\alpha(\mu) - i\alpha$. So the image of $\mu + q\alpha$ must be $\mu - r\alpha$. But this implies $r - q = \langle \mu, \alpha \rangle$. $\square$

Let V be an irreducible highest-weight module over L of highest weight λ. Then by Lemma 8.1.4, the weights of V are of the form $\mu = \lambda - \sum_{i=1}^l k_i \alpha_i$. The number $\sum_{i=1}^l k_i$ is called the *level* of the weight μ. We have an algorithm for computing the weights of a highest-weight module with highest weight λ that is completely analogous to the algorithm CartanMatrixToRootSystem for computing the root system given the simple roots and a Cartan matrix. We start with the highest weight λ (which is the only weight of level 0). Supposing that the weights of level n are computed we compute the weights of level $n+1$ in the following way. Let μ be a weight of level n and $\alpha_i \in \Delta$. Let q be the largest integer such that $\mu + q\alpha_i$ is a weight of V (we can determine this since the weights of level $\le n$ are known). Set $r = q + \langle \mu, \alpha_i \rangle$. Then if $r > 0$ we add $\mu - \alpha_i$ to the set of weights. If we do not find any weights of level $n+1$, then we stop and output the result. Otherwise we set $n := n+1$ and continue. We call this algorithm WeightsOfHWModule.

Lemma 8.7.2 *On input $\lambda \in P_+$ the algorithm WeightsOfHWModule returns the set $P(V)$, where V is the highest-weight module over L with highest weight λ.*

Representations of semisimple Lie algebras

Proof. It is clear that the output consists of elements of $P(V)$ (cf. Lemma 8.7.1). We have to prove that every element of $P(V)$ is constructed. Let $\lambda \neq \mu \in P(V)$, then we claim that there is a simple root $\alpha_i \in \Delta$ such that $\mu + \alpha_i \in P(V)$. Indeed, otherwise $N^+ \cdot V_\mu = 0$ as the root-spaces L_{α_i} generate N^+. This means that μ is a highest weight of V, which contradicts the choice of μ. Now by induction on the level of a weight, $\mu + \alpha_i$ already has been computed. Therefore also μ is constructed during the algorithm. $\square$

If we are only interested in the dominant weights of a highest-weight module, then we can use a different strategy based on the following proposition.

Proposition 8.7.3 *Let V be an irreducible highest-weight module over L with highest weight λ. Let $\mu \in P_+$ be a dominant integral linear function different from λ. Then $\mu \in P(V)$ if and only if there is a positive root $\alpha \in \Phi^+$ such that $\mu + \alpha \in P(V) \cap P_+$.*

Proof. First suppose that $\mu \in P(V)$. Then there is a root $\alpha > 0$ such that $\mu + \alpha \in P(V)$ (otherwise μ is a highest weight of V, which is excluded by Theorem 8.1.5). If $\mu + \alpha \in P_+$ then we are done. But if $\mu + \alpha \notin P_+$, then there is a $\alpha_i \in \Delta$ such that $(\mu + \alpha, \alpha_i) < 0$ (this follows immediately from Lemma 8.3.1). Now since also $r_{\alpha_i}(\mu + \alpha) \in P(V)$ we see by Lemma 8.7.1 that $\mu + \alpha + \alpha_i \in P(V)$. Because $\mu \in P_+$ we have $(\mu, \alpha_i) \geq 0$ and hence $(\alpha, \alpha_i) < 0$. Therefore $\alpha + \alpha_i$ is a root (Proposition 5.4.1). Now we replace α by $\alpha + \alpha_i$ and repeat. Since $P(V)$ is finite after a finite number of steps we find a $\beta \in \Phi^+$ such that $\mu + \beta \in P(V) \cap P_+$.

Now suppose that there is an $\alpha > 0$ such that $\mu + \alpha \in P(V) \cap P_+$. We show that $\mu \in P(V)$. Set $\bar{\mu} = \mu + \alpha$. Then $(\bar{\mu}, \alpha) = (\mu, \alpha) + (\alpha, \alpha) > 0$ since $\mu \in P_+$. Hence $\langle \bar{\mu}, \alpha \rangle > 0$. Furthermore, $r_\alpha(\bar{\mu}) = \bar{\mu} - \langle \bar{\mu}, \alpha \rangle \alpha$ also lies in $P(V)$ so in particular, by Lemma 8.7.1, $\mu = \bar{\mu} - \alpha \in P(V)$. $\square$

The procedure based on this proposition works as follows. Again we start with the highest weight λ. Then in the iteration for every dominant weight μ computed we compute the differences $\mu - \alpha$ for $\alpha > 0$. If such a difference is a dominant weight, then we add it to the set. If no more weights are found this way, then we stop. We call this procedure DominantWeightsOfHWModule. Of course, after having found the dominant weights we can calculate all weights by computing the orbits of the dominant ones under the Weyl group.

Example 8.7.4 Let L be the simple Lie algebra of type G_2, with root system Φ. Suppose that the Cartan matrix of Φ is

$$\begin{pmatrix} 2 & -1 \\ -3 & 2 \end{pmatrix}.$$

Let α_1, α_2 be the simple roots. The other positive roots are $\alpha_1 + \alpha_2$, $2\alpha_1 + \alpha_2$, $3\alpha_1 + \alpha_2$, $3\alpha_1 + 2\alpha_2$. The fundamental weights are given by solving the systems of linear equations $\langle \lambda_i, \alpha_j \rangle = \delta_{ij}$. This yields $\lambda_1 = 2\alpha_1 + \alpha_2$ and $\lambda_2 = 3\alpha_1 + 2\alpha_2$. In our calculations we express every vector as a linear combination of the fundamental weights. We have that $\alpha_1 = 2\lambda_1 - \lambda_2$ and $\alpha_2 = -3\lambda_1 + 2\lambda_2$.

Let V be the highest-weight module over L of highest weight $\lambda = \lambda_1 + \lambda_2$. We calculate the weights of level 1, using the algorithm WeightsOfHWModule. We have that $\lambda - \alpha_1$ and $\lambda - \alpha_2$ are the only possible weights of level 1. In both cases $q = 0$ and $r = q + \langle \lambda, \alpha_i \rangle = 1$; so both are weights of V. Now we determine the weights of level 2. First we try $\lambda - 2\alpha_1$. We have $\lambda - \alpha_1 = -\lambda_1 + 2\lambda_2$, so $\langle \lambda - \alpha_1, \alpha_1 \rangle = -1$. Furthermore $\lambda - \alpha_1 + \alpha_1 = \lambda$ is a weight of V. So in the algorithm WeightsOfHWModule we have $q = 1$ and $r = 0$ and therefore $\lambda - 2\alpha_2$ is not a weight. In the same way it can be seen that $\lambda - 2\alpha_2$ is not a weight. But $\langle \lambda - \alpha_1, \alpha_2 \rangle = 2$. In this case $q = 0$ and hence $r = 2$ and we see that $\lambda - \alpha_1 - \alpha_2$ is a weight of V. We have now determined all weights of levels $0, 1, 2$. Continuing like this we can find all weights of V.

We determine the dominant weights of V using the procedure DominantWeightsOfHWModule. First $\lambda - \alpha_1$ and $\lambda - \alpha_2$ are not dominant, but $\lambda - (\alpha_1 + \alpha_2) = 2\lambda_1$, $\lambda - (2\alpha_1 + \alpha_2) = \lambda_2$, $\lambda - (3\alpha_1 + 2\alpha_2) = \lambda_1$ are. In the next step we find one more dominant weight, namely $\lambda - (3\alpha_1 + 2\alpha_2) - (2\alpha_1 + \alpha_2) = 0$. After this one we find no more and we have determined all dominant weights of V.

8.8 The multiplicity formula of Freudenthal

Let V be an irreducible highest-weight module over L with highest weight λ. Then for an element $\mu \in P$ we set $m_\mu = \dim V_\mu$, where this number is defined to be zero if $\mu \notin P(V)$. The number m_μ is called the *multiplicity* of μ. By Lemma 8.1.4 we know that $m_\lambda = 1$. The following theorem expresses m_μ in terms of $m_{\mu'}$ where $\mu' > \mu$, enabling us to compute m_μ recursively.

Theorem 8.8.1 (Freudenthal's multiplicity formula) *Let V be an irreducible highest-weight module over L with highest weight λ. Then for*

$\mu \in P$ we have

$$\left((\lambda + \rho, \lambda + \rho) - (\mu + \rho, \mu + \rho) \right) m_\mu = 2 \sum_{\alpha \in \Phi^+} \sum_{j=1}^{\infty} (\mu + j\alpha, \alpha) m_{\mu + j\alpha}.$$

In the proof we use a particular element of the universal enveloping algebra $U(L)$. Let $x_1, \ldots, x_n$ and $y_1, \ldots, y_n$ be bases of L that are dual with respect to the Killing form, i.e., $\kappa_L(x_i, y_j) = \delta_{ij}$. Set $c = \sum_{i=1}^{n} x_i y_i$ which is an element $U(L)$. Then exactly in the same way as in the proof of Proposition 4.4.3 we see that $[c, L] = 0$ in $U(L)$. Since L generates $U(L)$ this means that c lies in the centre of $U(L)$. The element c is called a *universal Casimir element*.

Proof. Let $g_1, \ldots, g_l$ be an orthonormal basis of H with respect to the Killing form κ_L (i.e., $\kappa_L(g_i, g_j) = \delta_{ij}$). Furthermore, Root fact 3 implies that there are root vectors $x_\alpha \in L_\alpha$ for $\alpha \in \Phi$ such that $\kappa_L(x_\alpha, x_{-\alpha}) = 1$. So by Root fact 11, $[x_\alpha, x_{-\alpha}] = h_\alpha$ (where h_α is defined by (4.10)). Let ϕ denote the representation afforded by ϕ. Then

$$\phi(x_\alpha)\phi(x_{-\alpha}) = \phi(x_{-\alpha})\phi(x_\alpha) + \phi(h_\alpha). \tag{8.6}$$

Now the bases

$$\{g_1, \ldots, g_l, x_\alpha \text{ for } \alpha \in \Phi\}$$

and

$$\{g_1, \ldots, g_l, x_{-\alpha} \text{ for } \alpha \in \Phi\}$$

are dual with respect to κ_L. So

$$c = \sum_{i=1}^{l} g_i^2 + \sum_{\alpha \in \Phi} x_\alpha x_{-\alpha},$$

is a universal Casimir element. As seen above, $[c, L] = 0$ in $U(L)$, so since the representation ϕ extends to a representation of $U(L)$ we have that $[\phi(c), \phi(L)] = 0$. We calculate,

$$\phi(c) = \sum_{i=1}^{l} \phi(g_i)^2 + \sum_{\alpha \in \Phi} \phi(x_\alpha)\phi(x_{-\alpha})$$

$$= \sum_{i=1}^{l} \phi(g_i)^2 + \sum_{\alpha \in \Phi^+} \phi(x_\alpha)\phi(x_{-\alpha}) + \sum_{\alpha \in \Phi^+} \phi(x_{-\alpha})\phi(x_\alpha)$$

$$= \sum_{i=1}^{l} \phi(g_i)^2 + \sum_{\alpha \in \Phi^+} \phi(h_\alpha) + 2\sum_{\alpha \in \Phi^+} \phi(x_{-\alpha})\phi(x_\alpha) \quad \text{(by (8.6))}$$

$$= \sum_{i=1}^{l} \phi(g_i)^2 + 2\phi(h_\rho) + 2\sum_{\alpha \in \Phi^+} \phi(x_{-\alpha})\phi(x_\alpha). \tag{8.7}$$

Let v_λ be a highest-weight vector. Then

$$\phi(c) \cdot v_\lambda = \left(\sum_{i=1}^{l} \lambda(g_i)^2 + 2\lambda(h_\rho) \right) v_\lambda$$

$$= \big((\lambda, \lambda) + 2(\lambda, \rho) \big) v_\lambda$$

$$= (\lambda, \lambda + 2\rho) v_\lambda.$$

So v_λ is an eigenvector of $\phi(c)$ with eigenvalue $(\lambda, \lambda + 2\rho)$. Now since $[\phi(c), \phi(L)] = 0$ we have that the eigenspace of $\phi(c)$ corresponding to this eigenvalue is a non-zero L-submodule of V. So as V is irreducible we see that $\phi(c)$ acts on V as multiplication by $(\lambda, \lambda + 2\rho)$.

If a is a linear transformation of V, and U is a subspace stable under a then $\text{Tr}_U a$ will denote the trace of the restriction of a to U. So

$$\text{Tr}_{V_\mu} \phi(c) = (\lambda, \lambda + 2\rho) m_\mu.$$

But by using (8.7) we arrive at

$$\text{Tr}_{V_\mu} \phi(c) = (\mu, \mu + 2\rho) m_\mu + 2\sum_{\alpha \in \Phi^+} \text{Tr}_{V_\mu} \phi(x_{-\alpha})\phi(x_\alpha).$$

And hence

$$\big((\lambda, \lambda + 2\rho) - (\mu, \mu + 2\rho) \big) m_\mu = 2\sum_{\alpha \in \Phi^+} \text{Tr}_{V_\mu} \phi(x_{-\alpha})\phi(x_\alpha).$$

Now we use the following fact from linear algebra: if $a : U \to W$ and $b : W \to U$ are two linear maps, then $\mathrm{Tr}_U(ba) = \mathrm{Tr}_W(ab)$. So

$$\begin{aligned}
\mathrm{Tr}_{V_\mu}\phi(x_{-\alpha})\phi(x_\alpha) &= \mathrm{Tr}_{V_{\mu+\alpha}}\phi(x_\alpha)\phi(x_{-\alpha}) \\
&= \mathrm{Tr}_{V_{\mu+\alpha}}(\phi(h_\alpha) + \phi(x_{-\alpha})\phi(x_\alpha)) \quad \text{(by (8.6))} \\
&= (\mu+\alpha)(h_\alpha)m_{\mu+\alpha} + \mathrm{Tr}_{V_{\mu+\alpha}}\phi(x_{-\alpha})\phi(x_\alpha).
\end{aligned}$$

We repeat this and use the fact that $V_{\mu+j\alpha} = 0$ for big j to arrive at

$$\mathrm{Tr}_{V_\mu}\phi(x_{-\alpha})\phi(x_\alpha) = \sum_{j=1}^{\infty}(\mu+j\alpha,\alpha)m_{\mu+j\alpha}.$$

And this implies the formula.   □

In order to be able to use Freudenthal's formula we must show that the factor on the left-hand side is non-zero. This is achieved by the next two lemmas.

Lemma 8.8.2 *Let $\mu \in P_+$ and let $\nu = g^{-1}(\mu)$ for some $g \in W(\Phi)$. Then $(\nu+\rho,\nu+\rho) \le (\mu+\rho,\mu+\rho)$.*

Proof. We calculate

$$\begin{aligned}
(\nu+\rho,\nu+\rho) = (g(\nu+\rho),g(\nu+\rho)) &= (\mu+g(\rho),\mu+g(\rho)) \\
&= (\mu+\rho,\mu+\rho) - 2(\mu,\rho-g(\rho)).
\end{aligned}$$

Now by Lemma 8.3.8 and Proposition 8.3.9 we see that $\rho - g(\rho) \ge 0$. Hence $(\mu,\rho-g(\rho)) \ge 0$.   □

Lemma 8.8.3 *Let V be an irreducible highest-weight module over L with highest weight λ. Then we have $(\lambda+\rho,\lambda+\rho) > (\mu+\rho,\mu+\rho)$ for $\mu \in P(V)$ such that $\lambda \ne \mu$.*

Proof. Since every weight is conjugate to a dominant one (Theorem 8.3.5), we may, by Lemma 8.8.2, assume that $\mu \in P_+$. Since μ is not the highest weight of V we have that $\mu+\alpha \in P(V)$ for some $\alpha \in \Delta$. Fix such an $\alpha \in \Delta$ and set $\nu = \mu+\alpha$; then

$$(\nu+\rho,\nu+\rho) - (\mu+\rho,\mu+\rho) = 2(\mu+\rho,\alpha) + (\alpha,\alpha).$$

The right hand side of this expression is strictly positive since $\mu + \rho \in P_+$ by Lemma 8.3.8. Now let η be an element of P_+ conjugate under $W(\Phi)$ to ν, then by Lemma 8.8.2 we get

$$(\mu + \rho, \mu + \rho) < (\nu + \rho, \nu + \rho) \le (\eta + \rho, \eta + \rho).$$

If $\eta = \lambda$ we are done. Otherwise we replace μ by η and continue. Since $P(V)$ is finite we arrive at a situation where $\eta = \lambda$ after a finite number of steps. $\square$

From Lemma 8.8.3 it follows that the left-hand side of Freudenthal's formula vanishes only if $m_\mu = 0$. So since we know the multiplicity of the highest weight λ we can use this formula to calculate the weights and their multiplicities of an irreducible highest-weight module for L.

Example 8.8.4 Let L, Φ be as in Example 8.7.4. After (maybe) modifying the bilinear form by a scalar (note that as the form appears on both the left-hand and the right-hand side of Freudenthal's formula, this does not affect the result) we may assume that $(\alpha_1, \alpha_1) = \frac{2}{3}$. Now since $\langle \alpha_2, \alpha_1 \rangle = -3$, we have that $2(\alpha_2, \alpha_1) = -2$, i.e., $(\alpha_1, \alpha_2) = -1$. Finally from $\langle \alpha_1, \alpha_2 \rangle = -1$ it follows that $(\alpha_2, \alpha_2) = 2$. This means that $(\lambda_1, \lambda_1) = \frac{2}{3}$, $(\lambda_1, \lambda_2) = 1$, $(\lambda_2, \lambda_2) = 2$ (where the λ_i are the fundamental weights, as calculated in Example 8.7.4).

Now let V be the irreducible highest-weight module over L with highest weight $\lambda = \lambda_1 + \lambda_2$. Since $\rho = \lambda_1 + \lambda_1$ (Lemma 8.3.8) we have $\lambda + \rho = 2(\lambda_1 + \lambda_2)$ and $(\lambda + \rho, \lambda + \rho) = \frac{56}{3}$. We know that the weights of V are of the form $\lambda - k_1 \alpha_1 - k_2 \alpha_2$ for $k_1, k_2 \ge 0$. We calculate the multiplicity of the weight $\mu = \lambda - \alpha_1 = -\lambda_1 + 2\lambda_2$. First of all, $(\mu + \rho, \mu + \rho) = 18$. Furthermore, the sum on the right-hand side of Freudenthal's formula consists of only one term, namely $(\mu + \alpha_1, \alpha_1) m_{\mu + \alpha_1}$ and this equals $\frac{1}{3}$. Consequently $m_{\lambda - \alpha_1} = 1$.

We can of course continue calculating weights and multiplicities using Freudenthal's formula. However, by Theorem 8.4.2 we have that $g(\mu)$ has the same multiplicity as μ for all $g \in W(\Phi)$. So in general it is a good idea after having calculated the multiplicity of a weight, to calculate the orbit of the weight under $W(\Phi)$ (using the techniques of Section 8.6). This way we only have to calculate the multiplicities of the dominant weights with Freudenthal's formula; the other weights all lie in the orbit of some dominant weight.

8.9   Modifying Freudenthal's formula

Let V be the irreducible highest-weight module over L of highest weight λ. In this section we modify Freudenthal's formula so as to reduce the work needed for computing the weight multiplicities of V. We start with a lemma.

Lemma 8.9.1 *Let V be an irreducible highest-weight module of L. Let $\mu \in P(V)$ and $\alpha \in \Phi$, then*

$$-\sum_{j=1}^{\infty}(\mu - j\alpha, \alpha)m_{\mu-j\alpha} = \sum_{j=1}^{\infty}(\mu + j\alpha, \alpha)m_{\mu+j\alpha} + (\mu, \alpha)m_\mu.$$

Proof. We let x_α be as in the proof of Theorem 8.8.1. Again we let ϕ denote the representation of L afforded by V and we use $\mathrm{Tr}_U(ba) = \mathrm{Tr}_W(ab)$ for linear maps $a : U \to W$ and $b : W \to U$. Hence

$$\begin{aligned}
\mathrm{Tr}_{V_\mu}\phi(x_\alpha)\phi(x_{-\alpha}) &= \mathrm{Tr}_{V_{\mu-\alpha}}\phi(x_{-\alpha})\phi(x_\alpha) \\
&= \mathrm{Tr}_{V_{\mu-\alpha}}(\phi(x_\alpha)\phi(x_{-\alpha}) - \phi(h_\alpha)) \quad \text{(by (8.6))} \\
&= -(\mu - \alpha, \alpha)m_{\mu-\alpha} + \mathrm{Tr}_{V_{\mu-\alpha}}\phi(x_\alpha)\phi(x_{-\alpha}).
\end{aligned}$$

From this we have that

$$\mathrm{Tr}_{V_\mu}\phi(x_\alpha)\phi(x_{-\alpha}) = -\sum_{j=1}^{\infty}(\mu - j\alpha, \alpha)m_{\mu-j\alpha}.$$

By (8.6) however, $\mathrm{Tr}_{V_\mu}\phi(x_\alpha)\phi(x_{-\alpha}) = \mathrm{Tr}_{V_\mu}\phi(x_{-\alpha})\phi(x_\alpha) + \mathrm{Tr}_{V_\mu}\phi(h_\alpha)$. Now using the expression for $\mathrm{Tr}_{V_\mu}\phi(x_{-\alpha})\phi(x_\alpha)$ derived in the proof of Theorem 8.8.1, we arrive at the statement of the lemma. $\square$

Let μ be a weight of V and consider the term $(\mu + j\alpha, \alpha)m_{\mu+j\alpha}$ appearing in Freudenthal's formula. Let g be an element of the stabilizer of μ in $\mathrm{W}(\Phi)$. Then by Theorem 8.4.2, $m_{\mu+j\alpha} = m_{\mu+jg(\alpha)}$. So since g leaves the inner product $(\ ,\ )$ invariant, we have $(\mu + jg(\alpha), g(\alpha))m_{\mu+jg(\alpha)} = (\mu + j\alpha, \alpha)m_{\mu+j\alpha}$. As a consequence we only have to calculate the terms $(\mu + j\alpha, \alpha)m_{\mu+j\alpha}$ for representatives α of each orbit in Φ of the stabilizer of μ. We note that by Proposition 8.3.9, the stabilizer of μ in $\mathrm{W}(\Phi)$ is generated by r_{α_i} where $\alpha_i \in \Delta$ is such that $(\alpha_i, \mu) = 0$.

Let I be a subset of $\{1, \ldots, l\}$. Let Φ_I be the subset of Φ consisting of all roots that are linear combinations of the α_i with $i \in I$. Then Φ_I is a root system in its own right, with simple system $\{\alpha_i \mid i \in I\}$. By W_I we denote the subgroup of $\mathrm{W}(\Phi)$ generated by r_{α_i} for $i \in I$. Then W_I is the Weyl group of Φ_I.

For $I \subset \{1, \ldots, l\}$ let W_I^- be the group generated by W_I along with -1. By $W_I^- \cdot \alpha$ we denote the orbit of W_I^- containing α (and similarly for $W_I \cdot \alpha$). Let $\alpha \in \Phi$ and set $O = W_I^- \cdot \alpha$. If $\alpha \in \Phi_I$, then $-\alpha \in W_I \cdot \alpha$ and hence $O = W_I \cdot \alpha$. Now suppose that $\alpha \notin \Phi_I$ and write $\alpha = \sum_{i=1}^l k_i \alpha_i$. Then there is an $i_0 \notin I$ such that $k_{i_0} \neq 0$. Let $\beta \in W_I \cdot \alpha$ and write $\beta = \sum_{i=1}^l m_i \alpha_i$. Then $m_{i_0} = k_{i_0}$ and consequently $W_I \cdot \alpha$ is entirely contained in Φ^+ (if $\alpha > 0$) or entirely contained in Φ^- (if $\alpha < 0$). So $O = W_I \cdot \alpha \bigcup -W_I \cdot \alpha$, where the union is disjoint.

Proposition 8.9.2 *Let V be an irreducible highest-weight module over L with highest weight λ. Let $\mu \in P(V)$, and let $I \subset \{1, \ldots, l\}$ be such that r_{α_i} for $i \in I$ generate the stabilizer of μ in $\mathrm{W}(\Phi)$ (cf. Proposition 8.3.9). Let $O_1, \ldots, O_r$ be the orbits of W_I^- inside Φ. For $1 \leq i \leq r$ let γ_i be a positive root lying in O_i. Then*

$$\left((\lambda + \rho, \lambda + \rho) - (\mu + \rho, \mu + \rho) \right) m_\mu = \sum_{i=1}^r |O_i| \sum_{j=1}^\infty (\mu + j\gamma_i, \gamma_i) m_{\mu + j\gamma_i}.$$

Proof. We suppose that the first s orbits are contained in Φ_I and the other orbits not. Then $O_1 \cup \cdots \cup O_s = \Phi_I$. For $i \leq s$ we have $O_i = W_I \cdot \gamma_i$ and for $i > s$, $O_i = W_I \cdot \gamma_i \cup -W_I \cdot \gamma_i$ (disjoint union). Using Lemma 8.9.1 it is straightforward to see that Freudenthal's formula is equivalent to

$$(\lambda, \lambda + 2\rho) m_\mu = \sum_{\alpha \in \Phi} \sum_{j=1}^\infty (\mu + j\alpha, \alpha) m_{\mu + j\alpha} + (\mu, \mu) m_\mu. \qquad (8.8)$$

For $g \in W_I$ we have $(\mu + jg(\alpha), g(\alpha)) = (\mu + j\alpha, \alpha)$ and $m_{\mu + jg(\alpha)} = m_{\mu + j\alpha}$. So dividing Φ into disjoint W_I-orbits we see that the double sum of (8.8) is equal to

$$\sum_{i=1}^s |W_I \cdot \gamma_i| \sum_{j=1}^\infty (\mu + j\gamma_i, \gamma_i) m_{\mu + j\gamma_i} +$$

$$\sum_{i=s+1}^r |W_I \cdot \gamma_i| \sum_{j=1}^\infty \left((\mu + j\gamma_i, \gamma_i) m_{\mu + j\gamma_i} + (\mu - j\gamma_i, -\gamma_i) m_{\mu - j\gamma_i} \right).$$

Now by Lemma 8.9.1 this is equal to

$$\sum_{i=1}^s |W_I \cdot \gamma_i| \sum_{j=1}^\infty (\mu + j\gamma_i, \gamma_i) m_{\mu + j\gamma_i} +$$

$$\sum_{i=s+1}^r |W_I \cdot \gamma_i| \left(2 \sum_{j=1}^\infty (\mu + j\gamma_i, \gamma_i) m_{\mu + j\gamma_i} + (\mu, \gamma_i) m_\mu \right).$$

Now we collect the coefficients of m_μ together. In the first summand there
are none, and in the second sum the coefficient of m_μ is

$$\sum_{i=s+1}^{r} |W_I \cdot \gamma_i|(\mu, \gamma_i) = \sum_{\alpha \in \Phi^+ \setminus \Phi_I^+} (\mu, \alpha) = \sum_{\alpha \in \Phi^+} (\mu, \alpha) = 2(\mu, \rho).$$

(In the first equality we have used the fact that $W_I \cdot \gamma_i \subset \Phi^+$, and that these
orbits exhaust $\Phi^+ \setminus \Phi_I^+$. In the second equality we have used $(\mu, \alpha_j) = 0$ for
$j \in I$.) Now since $2|W_I \cdot \gamma_i| = |O_i|$ for $i \geq s+1$ we arrive at the formula of
the proposition. $\square$

Remark. If I is empty, then the group W_I^- only consists of two elements
and each orbit is of the form $\{\pm\alpha\}$. In that case the formula of Proposition
8.9.2 reduces to Freudenthal's formula as given in Theorem 8.8.1. So we
benefit most of Proposition 8.9.2 if the group W_I^- is large, i.e, if there are
many simple roots α_i such that $(\alpha_i, \mu) = 0$.

Proposition 8.9.2 gives us an algorithm for calculating the multiplicities
of the dominant weights of the highest-weight module V with highest weight
λ. Using the algorithm DominantWeightsOfHWModule we calculate the set
D of dominant weights of V. We order the set D according to increasing
level. Then dealing with the weights in D in the order in which they appear
in D, we calculate m_μ for $\mu \in D$. First we determine the set I of indices
i such that $(\mu, \alpha_i) = 0$. Furthermore, we calculate the orbits of W_I^- in Φ.
Then we use the formula of Proposition 8.9.2. Let $(\mu + j\gamma_i, \gamma_i)m_{\mu+j\gamma_i}$ be
a term occurring in this formula. If $\mu + j\gamma_i$ is dominant, then we already
know its multiplicity because the level of $\mu + j\gamma_i$ is smaller than the level
of μ. On the other hand, if $\mu + j\gamma_i$ is not dominant, then we calculate
the dominant weight ν conjugate to it under $W(\Phi)$ (using the algorithm
ConjugateDominantWeight). Since by Proposition 8.3.9, $g(\nu) \leq \nu$ for $g \in$
$W(\Phi)$ we see that $\mu < \mu + j\gamma_i \leq \nu$. So the level of ν is smaller than the level
of μ and therefore we already know the multiplicity of ν. Furthermore, by
Theorem 8.4.2, the multiplicity of $\mu + j\gamma_i$ is equal to the multiplicity of ν.
We call this algorithm DominantWeightMultiplicities.

Remark. Once the multiplicities of the dominant weights have been computed, the other multiplicities follow. Indeed, by Theorem 8.3.5, any weight
is conjugate to a dominant weight, and by Theorem 8.4.2 conjugate weights
have equal multiplicity.

Example 8.9.3 Again we consider the Lie algebra L of type G_2 of Example 8.8.4. We let V be the highest-weight module with highest weight $\lambda = \lambda_1 + \lambda_2$. As seen in Example 8.7.4, the dominant weights of V are $\lambda = \lambda_1 + \lambda_2$, $\lambda - \alpha_1 - \alpha_2 = 2\lambda_1$, $\lambda - 2\alpha_1 - \alpha_2 = \lambda_2$, $\lambda - 3\alpha_1 - 2\alpha_2 = \lambda_1$, $\lambda - 5\alpha_1 - 3\alpha_2 = 0$. We have that $r_{\alpha_1}(\lambda_1) = -\lambda_1 + \lambda_2$ and $r_{\alpha_2}(\lambda_2) = 3\lambda_1 - \lambda_2$ ($r_{\alpha_i}(\lambda_j) = \lambda_j$ if $j \neq i$).

Supposing that we know that the multiplicities of the weights $2\lambda_1$ and λ_2 are both 2 we calculate the multiplicity of λ_1. The stabilizer of λ_1 in $W(\Phi)$ is generated by r_{α_2}, so we set $I = \{2\}$. The orbits of W_I^- in Φ are $\pm\{\alpha_1, \alpha_1 + \alpha_2\}$, $\pm\{\alpha_2\}$, $\pm\{2\alpha_1 + \alpha_2\}$ and $\pm\{3\alpha_1 + \alpha_2, 3\alpha_1 + 2\alpha_2\}$. As orbit representatives we choose α_1, α_2, $2\alpha_1 + \alpha_2$ and $3\alpha_1 + \alpha_2$. We calculate the contribution of the term of α_1 to the sum on the right hand side of the formula of Proposition 8.9.2. We have that $\alpha_1 = 2\lambda_1 - \lambda_2$. So the term for $j = 1$ becomes $(3\lambda_1 - \lambda_2, 2\lambda_1 - \lambda_2)m_{3\lambda_1 - \lambda_2}$. Now $\nu = 3\lambda_1 - \lambda_2$ is not dominant, so we calculate the dominant weight to which it is conjugate by the algorithm ConjugateDominantWeight. We have $\langle \nu, \alpha_2 \rangle = -1 < 0$ so we set $\nu_1 = r_{\alpha_2}(\nu) = \lambda_2$ which is dominant. Therefore $m_{3\lambda_1 - \lambda_2} = m_{\lambda_2} = 2$. The inner product evaluates to 1 so the term for $j = 1$ contributes 2 to the sum. For $j = 2$ we get $(5\lambda_1 - 2\lambda_2, 2\lambda_1 - \lambda_2)m_{5\lambda_1 - 2\lambda_2} = \frac{5}{3}m_{5\lambda_1 - 2\lambda_2}$. Now $5\lambda_1 - 2\lambda_1$ is conjugate to $\lambda_1 + \lambda_2$ so that $m_{5\lambda_1 - 2\lambda_2} = 1$. For $j = 3$ we have $\lambda_1 + 3\alpha_1 = 7\lambda_1 - 3\lambda_2$. This weight is conjugate to the dominant weight $2\lambda_1 + \lambda_2$, which is not a weight of V. So we need not go further with the term of α_1. The W_I^--orbit containing α_1 has size 4, so we have to multiply the contribution of α_1 by 4. We leave calculating the contributions of α_2, $2\alpha_1 + \alpha_2$, $3\alpha_1 + \alpha_2$ to the reader. Also we have $(\lambda_1 + \rho, \lambda_1 + \rho) = \frac{26}{3}$. Using this we arrive at

$$10m_{\lambda_1} = 4(2 + \frac{5}{3}) + 2 \cdot 4 + 2 \cdot \frac{8}{3} + 4 \cdot 3 = 40.$$

The conclusion is that $m_{\lambda_1} = 4$. In a similar way we can determine the multiplicity of the weight 0; we get $m_0 = 4$. So we have calculated the multiplicities of all dominant weights of $P(V)$, and hence by taking their orbits under $W(\Phi)$ we get the multiplicities of all weights of V. Multiplying the multiplicities of the dominant weights by the sizes of their orbits we see that the dimension of V is 64.

8.10   Weyl's formulas

We recall that P is the set of weights; it consists of all elements of the form $\sum_{i=1}^{l} n_i \lambda_i$ where $n_i \in \mathbb{Z}$, and $\lambda_1, \ldots, \lambda_l$ are the fundamental weights.

We note that the set P is a group under addition. In this section $\mathbb{R}P$ will denote the group algebra of P over the real numbers, i.e., $\mathbb{R}P$ is the set of all elements $\sum_{\mu \in P} c_\mu e^\mu$, where $c_\mu \in \mathbb{R}$ and $c_\mu = 0$ for all but finitely many $\mu \in P$. On $\mathbb{R}P$ we have a multiplication given by $e^\lambda \cdot e^\mu = e^{\lambda+\mu}$. This multiplication makes $\mathbb{R}P$ into a commutative associative algebra with one (the identity element being e^0). Set $e_i = e^{\lambda_i}$ for $1 \le i \le l$. Let $\mu \in P_+$ (the set of dominant weights), then $\mu = \sum_{i=1}^l n_i \lambda_i$ where $n_i \ge 0$. Hence $e^\mu = e_1^{n_1} \cdots e_l^{n_l}$. It follows that the linear span of the e^μ for $\mu \in P_+$ is a subalgebra of $\mathbb{R}P$ isomorphic to the polynomial ring $\mathbb{R}[X_1, \dots, X_l]$. We denote this subalgebra by $\mathbb{R}P_+$. If $a \in \mathbb{R}P$ is an arbitrary element, then we can write $a = (e_1^{r_1} \cdots e_l^{r_l})^{-1} b$, where $r_i \ge 0$ and b lies in $\mathbb{R}P_+$. From this it follows that $\mathbb{R}P$ does not have zero-divisors (i.e., there are no non-zero $a, b \in \mathbb{R}P$ such that $ab = 0$).

Now let V be a finite-dimensional module over the semisimple Lie algebra L. Then V is the direct sum of weight spaces V_μ for $\mu \in P$. We set $m_\mu = \dim V_\mu$. Then

$$\chi_V = \sum_{\mu \in P} m_\mu e^\mu$$

is called the *formal character* of V. In this section we will obtain a formula for χ_V if V is an irreducible highest-weight module over L with highest weight λ.

The Weyl group $\mathrm{W}(\Phi)$ acts on $\mathbb{R}P$ by $g(e^\mu) = e^{g(\mu)}$. It is straightforward to see that $g(e^\lambda e^\mu) = g(e^\lambda) g(e^\mu)$ so that g is an automorphism of $\mathbb{R}P$ for all $g \in \mathrm{W}(\Phi)$. For $g \in \mathrm{W}(\Phi)$ let the *sign* of g be the determinant of the action of g on the space spanned by the roots. Since the eigenvalues of a reflection are ± 1, and $\mathrm{W}(\Phi)$ is generated by reflections, the sign of $g \in \mathrm{W}(\Phi)$ is ± 1. We denote the sign of g by $\mathrm{sn}(g)$. Note that the sign of a reflection is -1. An element $a \in \mathbb{R}P$ is called *symmetric* if $g(a) = a$ for all $g \in \mathrm{W}(\Phi)$, and it is called *alternating* if $g(a) = \mathrm{sn}(g)a$ for all $g \in \mathrm{W}(\Phi)$.

Example 8.10.1 The character χ_V of an L-module V is symmetric by Theorem 8.4.2. Set

$$Q = e^{-\rho} \prod_{\alpha>0}(e^\alpha - 1) = e^\rho \prod_{\alpha>0}(1 - e^{-\alpha}).$$

Let $\alpha_i \in \Delta$ be a simple root. Then by Corollary 5.7.2, $r_{\alpha_i}(\rho) = \rho - \alpha_i$. Furthermore, by Lemma 5.7.1, r_{α_i} permutes the positive roots other than α_i and it sends α_i to $-\alpha_i$. So

$$r_{\alpha_i}(Q) = \left(\prod_{\alpha_i \ne \alpha > 0}(e^\alpha - 1) \right)(e^{-\alpha_i} - 1)e^{-\rho}e^{\alpha_i} = -Q.$$

Consequently $r_{\alpha_i}(Q) = \mathrm{sn}(r_{\alpha_i})(Q)$, and since the r_{α_i} generate $\mathrm{W}(\Phi)$ this means that Q is alternating.

Set

$$\sigma = \sum_{g \in \mathrm{W}(\Phi)} \mathrm{sn}(g)g$$

which is a linear transformation of $\mathbb{R}P$. For $h \in \mathrm{W}(\Phi)$ we calculate

$$\sigma h = \sum_{g \in \mathrm{W}(\Phi)} \mathrm{sn}(g)gh = (\mathrm{sn}(h))^{-1} \sum_{g \in \mathrm{W}(\Phi)} \mathrm{sn}(gh)gh = \mathrm{sn}(h)\sigma.$$

And in the same way we see that $h\sigma = \mathrm{sn}(h)\sigma$. Hence for $a \in \mathbb{R}P$ and $h \in \mathrm{W}(\Phi)$ we have $h(\sigma(a)) = \mathrm{sn}(h)\sigma(a)$ and therefore $\sigma(a)$ is alternating. Furthermore, if $a \in \mathbb{R}P$ is alternating, then $\sigma(a) = \sum_{g \in \mathrm{W}(\Phi)} \mathrm{sn}(g)^2 a = |W(\Phi)|a$. As a consequence any alternating element is of the form $\sigma(a)$ for an $a \in \mathbb{R}P$. This means that any alternating element a can be expressed as a linear combination of the form

$$\sum_{\mu \in P} c_\mu \sigma(e^\mu). \tag{8.9}$$

But $\sigma h = \pm \sigma$, so the e^μ in (8.9) can be replaced by a $\mathrm{W}(\Phi)$-conjugate. It follows that any alternating element can be expressed as a linear combination of the form (8.9), where the μ are from P_+ (cf. Theorem 8.3.5). Suppose that for such a μ we have $\langle \mu, \alpha_i \rangle = 0$ for some simple root α_i. Then $r_{\alpha_i}(\mu) = \mu$ and hence $\sigma(e^\mu) = -\sigma r_{\alpha_i}(e^\mu) = -\sigma(e^\mu)$, and therefore $\sigma(e^\mu) = 0$. So if $a \in \mathbb{R}P$ is alternating, then by applying the operator $\frac{1}{|W(\Phi)|}\sigma$ to a we obtain an expression for a of the form (8.9). By replacing the e^μ by a conjugate under the Weyl group (if necessary) we obtain an expression for a of the form (8.9) where $\langle \mu, \alpha_i \rangle > 0$ for all μ occurring in the expression, and all simple roots α_i.

Proposition 8.10.2 $Q = \sigma(e^\rho)$.

Proof. We have that Q is a linear combination of terms of the form e^μ where $\mu = \rho - \sum_{\alpha > 0} \varepsilon_\alpha \alpha$ where ε_α is 0 or 1. Now any conjugate of such a μ under $\mathrm{W}(\Phi)$ has the same form. Indeed, since the r_{α_i} generate $\mathrm{W}(\Phi)$ it is enough to show that $r_{\alpha_i}(\mu)$ is again of the same form. But because r_{α_i} permutes the positive roots other than α_i (Lemma 5.7.1) we have

$$r_{\alpha_i}\left(\rho - \sum_{\alpha > 0} \varepsilon_\alpha \alpha\right) = \rho - \alpha_i - \sum_{\alpha > 0} \eta_\alpha \alpha = \rho - \sum_{\alpha > 0} \delta_\alpha \alpha.$$

If $\varepsilon_{\alpha_i} = 1$, then $\eta_{\alpha_i} = -1$ and $\delta_{\alpha_i} = 0$. And if $\varepsilon_{\alpha_i} = 0$, then $\eta_{\alpha_i} = 0$ and $\delta_{\alpha_i} = 1$.

Since Q is alternating, after applying the operator $\frac{1}{|W(\Phi)|}\sigma$ to Q, we get an expression for Q of the form (8.9) where all μ are of the form $\mu = \rho - \sum_{\alpha>0} \varepsilon_\alpha \alpha$. Since any $W(\Phi)$-conjugate of such a μ is of the same form, we may assume that $\langle \mu, \alpha_i \rangle > 0$ for $1 \leq i \leq l$. Write $\mu = \rho - \eta$ where $\eta = \sum_{\alpha>0} \varepsilon_\alpha \alpha = \sum_{i=1}^{l} k_i \alpha_i$, where the k_i are non-negative integers. Then $0 < \langle \mu, \alpha_i \rangle = \langle \rho, \alpha_i \rangle - \langle \eta, \alpha_i \rangle = 1 - \langle \eta, \alpha_i \rangle$ (cf. Lemma 8.3.8) and since $\langle \eta, \alpha_i \rangle \in \mathbb{Z}$ we have that $\langle \eta, \alpha_i \rangle \leq 0$. Therefore,

$$0 \leq (\eta, \eta) = \sum_{i=1}^{l} k_i(\eta, \alpha_i) = \frac{1}{2} \sum_{i=1}^{l} k_i(\alpha_i, \alpha_i)\langle \eta, \alpha_i \rangle \leq 0.$$

And it follows that $\eta = 0$ and $Q = \xi\sigma(e^\rho)$ for some $\xi \in \mathbb{R}$. But

$$\sigma(e^\rho) = \sum_{g \in W(\Phi)} \mathrm{sn}(g)e^{g(\rho)}.$$

By Proposition 8.3.9 together with Lemma 8.3.8 it follows that $g(\rho) \neq \rho$ unless $g = 1$, and if $g \neq 1$, then $g(\rho) < \rho$. So $\sigma(e^\rho) = e^\rho + \sum_{\mu<\rho} c_\mu e^\mu$. But from the definition of Q it follows that $Q = e^\rho + \sum_{\mu<\rho} d_\mu e^\mu$. Therefore ξ must be 1 and $Q = \sigma(e^\rho)$. $\qquad\qquad\square$

Let W be the real Euclidean space spanned by the roots. We consider the tensor product $W \otimes_\mathbb{R} \mathbb{R}P$. The elements of $\mathbb{R}P$ act on this space by

$$a \cdot \left(\sum_i v_i \otimes b_i \right) = \sum_i v_i \otimes ab_i$$

for $a, b_i \in \mathbb{R}P$ and $v_i \in W$. Also we use a bilinear map $(\ ,\ ) : W \otimes_\mathbb{R} \mathbb{R}P \times W \otimes_\mathbb{R} \mathbb{R}P \to \mathbb{R}P$ given by

$$(v \otimes a, w \otimes b) = (v, w)ab$$

where the brackets on the right-hand side denote the bilinear form on W. We have the relations $(c(v \otimes a), w \otimes b) = c(v \otimes a, w \otimes b) = (v \otimes a, c(w \otimes b))$ for $a, b, c \in \mathbb{R}P$ and $v, w \in W$.

The *gradient* of $\mathbb{R}P$ is the linear map $G : \mathbb{R}P \to W \otimes_\mathbb{R} \mathbb{R}P$ given by $G(e^\mu) = \mu \otimes e^\mu$. Furthermore, the *Laplacian* of $\mathbb{R}P$ is the linear map $\Delta : \mathbb{R}P \to \mathbb{R}P$ defined by $\Delta(e^\mu) = (\mu, \mu)e^\mu$. We calculate

$$G(e^\lambda e^\mu) = G(e^{\lambda+\mu}) = (\lambda + \mu) \otimes e^\lambda e^\mu$$
$$= \lambda \otimes e^\lambda e^\mu + \mu \otimes e^\lambda e^\mu$$
$$= e^\mu G(e^\lambda) + e^\lambda G(e^\mu).$$

So since G is linear we have $G(ab) = aG(b) + bG(a)$ for all $a, b \in \mathbb{R}P$. Also

$$\Delta(e^\lambda e^\mu) = (\lambda + \mu, \lambda + \mu)e^\lambda e^\mu$$
$$= (\lambda, \lambda)e^\lambda e^\mu + 2(\lambda, \mu)e^\lambda e^\mu + (\mu, \mu)e^\lambda e^\mu$$
$$= e^\mu \Delta(e^\lambda) + 2(\lambda \otimes e^\lambda, \mu \otimes e^\mu) + e^\lambda \Delta(e^\mu).$$

And because $(\lambda \otimes e^\lambda, \mu \otimes e^\mu) = (G(e^\lambda), G(e^\mu))$ we see, by linearity, that

$$\Delta(ab) = a\Delta(b) + b\Delta(a) + 2(G(a), G(b)) \quad \text{for all } a, b \in \mathbb{R}P. \qquad (8.10)$$

Now we let V be an irreducible highest-weight module over L with highest weight λ. Since $\sum_{\alpha \in \Phi} \alpha = 0$ we have also $\sum_{\alpha \in \Phi} (\mu, \alpha)m_\mu = 0$. Hence, using (8.8) we see that Freudenthal's formula is equivalent to

$$(\lambda, \lambda + 2\rho)m_\mu = \sum_{\alpha \in \Phi} \sum_{j=0}^{\infty} (\mu + j\alpha, \alpha)m_{\mu+j\alpha} + (\mu, \mu)m_\mu.$$

We multiply the left-hand side and the right-hand side by e^μ and sum over $\mu \in P$, to obtain

$$(\lambda, \lambda + 2\rho)\chi_V = \Delta(\chi_V) + \sum_{\mu \in P} \sum_{\alpha \in \Phi} \sum_{j=0}^{\infty} (\mu + j\alpha, \alpha)m_{\mu+j\alpha}e^\mu. \qquad (8.11)$$

Now

$$\prod_{\alpha \in \Phi} (e^\alpha - 1) = \prod_{\alpha > 0} (e^\alpha - 1) \prod_{\alpha > 0} (e^{-\alpha} - 1) = \varepsilon Q^2,$$

where $\varepsilon = 1$, or $\varepsilon = -1$. We multiply (8.11) by εQ^2 and obtain

$$\varepsilon(\lambda, \lambda + 2\rho)\chi_V Q^2 - \varepsilon\Delta(\chi_V)Q^2 =$$
$$\sum_{\alpha \in \Phi} \sum_{\mu \in P} \sum_{j=0}^{\infty} (\mu + j\alpha, \alpha)m_{\mu+j\alpha}(e^{\mu+\alpha} - e^\mu) \prod_{\beta \neq \alpha} (e^\beta - 1). \quad (8.12)$$

Now fix an $\alpha \in \Phi$. For a moment forgetting about the factor $\prod_{\beta \neq \alpha}(e^\beta - 1)$, the coefficient of $e^{\mu+\alpha}$ in the expression on the right of (8.12) equals

$$\sum_{j=0}^{\infty} (\mu + j\alpha, \alpha)m_{\mu+j\alpha} - \sum_{j=0}^{\infty} (\mu + (j+1)\alpha, \alpha)m_{\mu+(j+1)\alpha} = (\mu, \alpha)m_\mu.$$

Consequently, the right-hand side of (8.12) is equal to

$$\sum_{\alpha\in\Phi}\prod_{\beta\neq\alpha}(e^\beta-1)e^\alpha\sum_{\mu\in P}m_\mu(\mu,\alpha)e^\mu =$$

$$\left(\sum_{\alpha\in\Phi}(\alpha\otimes e^\alpha\prod_{\beta\neq\alpha}(e^\beta-1)),\sum_{\mu\in P}\mu\otimes m_\mu e^\mu\right)=$$

$$(\varepsilon G(Q^2),G(\chi_V))=2\varepsilon(QG(Q),G(\chi_V)).$$

So (8.12) is equivalent to $(\lambda,\lambda+2\rho)\chi_V Q^2-Q^2\Delta(\chi_V)=2Q(G(Q),G(\chi_V))$. Since $Q\neq 0$ and $\mathbb{R}P$ does not contain zero-divisors we may divide this relation by Q. Using (8.10) we get

$$(\lambda,\lambda+2\rho)\chi_V Q-Q\Delta(\chi_V)=\Delta(Q\chi_V)-Q\Delta(\chi_V)-\chi_V\Delta(Q).$$

If we set $f=Q\chi_V$, then this means that $(\lambda,\lambda+2\rho)f=\Delta(f)-\chi_V\Delta(Q)$. But by Proposition 8.10.2, $Q=\sum_{g\in W(\Phi)}\mathrm{sn}(g)e^{g(\rho)}$ and since $(g(\rho),g(\rho))=(\rho,\rho)$ this implies that $\Delta(Q)=(\rho,\rho)Q$. Therefore

$$(\lambda+\rho,\lambda+\rho)f=\Delta(f).$$

Now $\chi_V=\sum_{\mu\in P}m_\mu e^\mu$ and $Q=\sum_{g\in W(\Phi)}\mathrm{sn}(g)e^{g(\rho)}$. By multiplying these expressions we see that $\chi_V Q$ is a linear combination of terms of the form $e^{\mu+g(\rho)}$. We calculate

$$\Delta(e^{\mu+g(\rho)})=(\mu+g(\rho),\mu+g(\rho))e^{\mu+g(\rho)}$$

$$=(g^{-1}(\mu)+\rho,g^{-1}(\mu)+\rho)e^{\mu+g(\rho)}.$$

So $e^{\mu+g(\rho)}$ is an eigenvector of Δ with eigenvalue $(g^{-1}(\mu)+\rho,g^{-1}(\mu)+\rho)$. Above we saw that f is an eigenvector of Δ with eigenvalue $(\lambda+\rho,\lambda+\rho)$. So since eigenvectors belonging to different eigenvalues are linearly independent, we have that f is a linear combination of elements $e^{\mu+g(\rho)}$ such that $(g^{-1}(\mu)+\rho,g^{-1}(\mu)+\rho)=(\lambda+\rho,\lambda+\rho)$. But by Lemma 8.8.3 this implies that $g^{-1}(\mu)=\lambda$ and hence f is a linear combination of elements $e^{g(\lambda+\rho)}$.

Since the elements of $W(\Phi)$ act as automorphisms of $\mathbb{R}P$ we have that a product of a symmetric element and an alternating element is alternating. In particular, f is alternating. As a consequence $f=\frac{1}{|W(\Phi)|}\sigma(f)$. So because $\sigma(e^{g(\mu)})=\sigma(e^\mu)$ for all $g\in W(\Phi)$, we see that

$$f=\eta\sigma(e^{\lambda+\rho})=\eta\sum_{g\in W(\Phi)}\mathrm{sn}(g)e^{g(\lambda+\rho)}$$

for some $\eta \in \mathbb{R}$. Now by Proposition 8.3.9 together with Lemma 8.3.8 we have that $g(\lambda + \rho) < \lambda + \rho$ if $g \neq 1$. Hence the coefficient of $e^{\lambda + \rho}$ in f is η. But the coefficient of $e^{\lambda + \rho}$ in $\chi_V Q$ is 1. Therefore $\eta = 1$ and we have proved the following theorem.

Theorem 8.10.3 (Weyl's character formula) *Let V be an irreducible highest-weight module over L with highest weight λ. Then*

$$\chi_V \sum_{g \in W(\Phi)} \mathrm{sn}(g) e^{g(\rho)} = \sum_{g \in W(\Phi)} \mathrm{sn}(g) e^{g(\lambda + \rho)}.$$

Theorem 8.10.4 (Weyl's dimension formula) *Let V be as in Theorem 8.10.3. Then*

$$\dim V = \prod_{\alpha > 0} \frac{(\alpha, \lambda + \rho)}{(\alpha, \rho)}.$$

Proof. By $\mathbb{R}[[t]]$ we denote the ring of formal power series over $\mathbb{R}$ in the indeterminate t. It is the set of all formal expressions $\sum_{i \geq 0} a_i t^i$ where $a_i \in \mathbb{R}$. We consider the homomorphism $\zeta : \mathbb{R}[[t]] \to \mathbb{R}$ given by $\zeta(\sum_{i \geq 0} a_i t^i) = a_0$. Also for $\mu \in \mathbb{R}P$ we have a linear map $\zeta_\mu : \mathbb{R}P \to \mathbb{R}[[t]]$ defined by

$$\zeta_\mu(e^\lambda) = \exp(\lambda, \mu) t = \sum_{k=0}^{\infty} \frac{1}{k!} ((\lambda, \mu) t)^k.$$

Since $\exp(\eta_1 t + \eta_2 t) = \exp(\eta_1 t) \exp(\eta_2 t)$ we have that ζ_μ is a homomorphism, i.e., $\zeta_\mu(e^\lambda e^\nu) = \zeta_\mu(e^\lambda) \zeta_\mu(e^\nu)$. Now because $\chi_V = \sum_{\mu \in P} m_\mu e^\mu$ we see that $\zeta_\nu(\chi_V) = \sum_{\mu \in P} m_\mu \exp(\mu, \nu) t$. Therefore $\zeta(\zeta_\nu(\chi_V)) = \sum_{\mu \in P} m_\mu = \dim V$.

Now we apply $\zeta \zeta_\nu$ to Weyl's character formula. First we calculate

$$\zeta_\nu \sigma(e^\mu) = \sum_{g \in W(\Phi)} \mathrm{sn}(g) \exp(g(\mu), \nu) t$$

$$= \sum_{g \in W(\Phi)} \mathrm{sn}(g) \exp(g^{-1}(\nu), \mu) t$$

$$= \zeta_\mu \sigma(e^\nu).$$

Using this and Proposition 8.10.2 we calculate

$$\zeta_\rho \sigma(e^\mu) = \zeta_\mu \sigma(e^\rho)$$
$$= \zeta_\mu(e^{-\rho}) \prod_{\alpha>0} \zeta_\mu(e^\alpha - 1)$$
$$= \exp(-\rho, \mu)t \prod_{\alpha>0} (\exp(\alpha, \mu)t - 1)$$
$$= \prod_{\alpha>0} (\exp \frac{1}{2}(\alpha, \mu)t - \exp -\frac{1}{2}(\alpha, \mu)t).$$

So if we apply ζ_ρ to Weyl's character formula we get

$$\zeta_\rho(\chi_V) \prod_{\alpha>0} (\exp \frac{1}{2}(\alpha, \rho)t - \exp -\frac{1}{2}(\alpha, \rho)t) =$$
$$\prod_{\alpha>0} (\exp \frac{1}{2}(\alpha, \lambda + \rho)t - \exp -\frac{1}{2}(\alpha, \lambda + \rho)t). \quad (8.13)$$

But

$$\prod_{\alpha>0} (\exp \frac{1}{2}(\alpha, \mu)t - \exp -\frac{1}{2}(\alpha, \mu)t) = (\prod_{\alpha>0}(\alpha, \mu))t^k + O(t^{k+1}),$$

where k is the number of positive roots. So if we divide (8.13) by t^k we get

$$\zeta_\rho(\chi_V) \prod_{\alpha>0} (\alpha, \rho) + O(t) = \prod_{\alpha>0} (\alpha, \lambda + \rho) + O(t).$$

To this expression we apply the homomorphism ζ, which gives us Weyl's dimension formula. $\qquad \square$

Weyl's character formula could in principle be used to obtain the character of an irreducible highest-weight module. However, the sum over the Weyl group and the need to divide two elements of $\mathbb{R}P$ make this formula rather difficult to use. To obtain the character of an irreducible highest-weight module, Freudenthal's formula is much more practical. However, Weyl's dimension formula provides us with an efficient algorithm for obtaining a polynomial $p(X_1, \ldots, X_l)$ such that $p(n_1, \ldots, n_l)$ is the dimension of the highest-weight module with highest weight $\lambda = n_1\lambda_1 + \cdots + n_l\lambda_l$.

Algorithm DimensionPolynomial
Input: a semisimple Lie algebra L with root system Φ of rank l.

Output: a polynomial $p(X_1, \ldots, X_l)$ such that $p(n_1, \ldots, n_l)$ is the dimension of the irreducible highest-weight module over L with highest weight $\lambda = n_1\lambda_1 + \cdots + n_l\lambda_l$.

Step 1 Set $p := 1$, and for each positive root α do the following:

Step 1a Write $\alpha = \sum_{i=1}^{l} k_i\alpha_i$.

Step 1b Multiply p by the polynomial

$$\frac{\sum_{i=1}^{l} k_i(X_i + 1)(\alpha_i, \alpha_i)}{\sum_{i=1}^{l} k_i(\alpha_i, \alpha_i)}.$$

Comments: Set $\lambda = X_1\lambda_1 + \cdots + X_l\lambda_l$, which is an element of the vector space $\mathbb{R}(X_1, \ldots, X_l) \otimes_{\mathbb{R}} \mathbb{R}P$. Let $\alpha = \sum_{i=1}^{l} k_i\alpha_i$ be a positive root. Since $\rho = \sum_{i=1}^{l} \lambda_i$ we have that

$$\frac{(\lambda + \rho, \alpha)}{(\rho, \alpha)} = \frac{\sum_{i=1}^{l} k_i(X_i + 1)(\alpha_i, \alpha_i)}{\sum_{i=1}^{l} k_i(\alpha_i, \alpha_i)}.$$

Example 8.10.5 Let L be the simple Lie algebra of type G_2 of Example 8.8.4. Since in Weyl's dimension formula the form $(\ ,\ )$ appears both in the numerator and in the denominator, nothing changes if we multiply it by a positive scalar. So we may use the values of $(\ ,\ )$ as given in Example 8.8.4. Then the polynomial we get is

$$(X_1 + 1)(X_2 + 1)\frac{X_1 + 1 + 3(X_2 + 1)}{4}\frac{2(X_1 + 1) + 3(X_2 + 1)}{5}$$
$$\frac{2(X_1 + 1) + 2(X_2 + 1)}{4}\frac{2(x_1 + 1) + 4(X_2 + 1)}{6}.$$

If we take $X_1 = 1$ and $X_2 = 1$, then we get 64, in accordance with Example 8.9.3.

8.11 The formulas of Kostant and Racah

Let $\lambda \in P_+$ be a dominant weight, and let V be the irreducible highest-weight module over L with highest weight λ. In this section we derive two more formulas for the multiplicity $m_\mu = \dim V_\mu$ of a weight $\mu \in P$.

Let $\mu \in P$, then by $p(\mu)$ we denote the number of ways of writing $\mu = \sum_{\alpha > 0} k_\alpha\alpha$, where the k_α are non-negative integers. So $p(0) = 1$, and

$p(\mu) = 0$ if μ cannot be written as a linear combination of positive roots with non-negative integer coefficients.

We consider the series $\sum_{\mu \in P} p(\mu)e^{\mu}$, which satisfies the identity

$$\sum_{\mu \in P} p(\mu)e^{\mu} = \prod_{\alpha > 0}(1 + e^{\alpha} + e^{2\alpha} + \cdots).$$

But $(1 - e^{\alpha})^{-1} = 1 + e^{\alpha} + e^{2\alpha} + \cdots$, and hence

$$\left(\sum_{\mu \in P} p(\mu)e^{\mu}\right)\left(\prod_{\alpha > 0}(1 - e^{\alpha})\right) = 1. \tag{8.14}$$

Let $\xi : \mathbb{R}P \to \mathbb{R}P$ be the automorphism given by $\xi(e^{\mu}) = e^{-\mu}$. We apply ξ to the left-hand side and the right-hand side of Weyl's character formula and multiply by e^{ρ} to get

$$\left(\sum_{\mu \in P} m_{\mu}e^{-\mu}\right)\left(\sum_{g \in W(\Phi)} \operatorname{sn}(g)e^{\rho - g(\rho)}\right) = \sum_{g \in W(\Phi)} \operatorname{sn}(g)e^{\rho - g(\lambda + \rho)}. \tag{8.15}$$

By Proposition 8.10.2, $\sum_{g \in W(\Phi)} \operatorname{sn}(g)e^{g(\rho)} = e^{\rho}\prod_{\alpha > 0}(1 - e^{-\alpha})$. To this relation we apply the automorphism ξ and obtain

$$\sum_{g \in W(\Phi)} \operatorname{sn}(g)e^{\rho - g(\rho)} = \prod_{\alpha > 0}(1 - e^{\alpha}).$$

We substitute this in (8.15), multiply on the left and on the right by $\sum_{\mu \in P} p(\mu)e^{\mu}$, and use (8.14) to obtain

$$\sum_{\mu \in P} m_{\mu}e^{-\mu} = \left(\sum_{g \in W(\Phi)} \operatorname{sn}(g)e^{\rho - g(\lambda + \rho)}\right)\left(\sum_{\mu \in P} p(\mu)e^{\mu}\right)$$

$$= \sum_{\mu \in P}\sum_{g \in W(\Phi)} \operatorname{sn}(g)p(\mu)e^{\mu + \rho - g(\lambda + \rho)}.$$

Now by comparing the coefficient of $e^{-\mu}$ on the left and the right we obtain the following theorem.

Theorem 8.11.1 (Kostant) *Let $\lambda \in P_+$ be a dominant weight and let V be the irreducible highest-weight module over L with highest weight λ. Then the multiplicity of the weight $\mu \in P$ is given by*

$$m_{\mu} = \sum_{g \in W(\Phi)} \operatorname{sn}(g)p(g(\lambda + \rho) - \mu - \rho).$$

The next lemma contains a useful recursion formula for the numbers $p(\mu)$.

Lemma 8.11.2 *Let $0 \neq \mu \in P$ be a non-zero weight. Then*

$$p(\mu) = - \sum_{1 \neq g \in W(\Phi)} \text{sn}(g) p(\mu - \rho + g(\rho)).$$

Proof. Set $\lambda = 0$, and let V be the irreducible highest-weight module over L with highest weight λ. Then $m_{-\mu} = 0$ and hence by Theorem 8.11.1,

$$\sum_{g \in W(\Phi)} \text{sn}(g) p(g(\rho) + \mu - \rho) = 0.$$

And this implies the lemma.   □

Proposition 8.11.3 (Racah) *Let $\lambda \in P_+$ and V be as in Theorem 8.11.1. Then the multiplicity of the weight $\mu \in P$ is given by*

$$m_\mu = - \sum_{1 \neq h \in W(\Phi)} \text{sn}(h) m_{\mu + \rho - h(\rho)}.$$

Proof. By combining Theorem 8.11.1 and Lemma 8.11.2 we get

$$m_\mu = - \sum_{1 \neq h \in W(\Phi)} \sum_{g \in W(\Phi)} \text{sn}(g) \text{sn}(h) p(g(\lambda + \rho) - \mu - \rho - \rho + h(\rho)).$$

However, by using Kostant's formula again we have that the right-hand side is equal to

$$- \sum_{1 \neq h \in W(\Phi)} \text{sn}(h) m_{\mu + \rho - h(\rho)}.$$

                                                                              □

We can use Racah's formula to calculate weight multiplicities in a manner similar to Freudenthal's formula. That is we start with the multiplicity of the highest weight λ, and when computing the multiplicity of $\mu \neq \lambda$ of level n we assume that the multiplicities of the weights of levels $0, \dots, n-1$ are known. We note that by Proposition 8.3.9 together with Lemma 8.3.8 we have that $\rho > h(\rho)$ for all $1 \neq h \in W(\Phi)$. So $\mu + \rho - h(\rho) > \mu$, and by assumption we already know $m_{\mu + \rho - h(\rho)}$.

Both Kostant's and Racah's formula can be used to calculate the weight multiplicities of an irreducible highest-weight module. We refer to [5] for a

discussion of practical experiences with these formulas. In this paper the
authors conclude that the formulas by Freudenthal and Racah are more
efficient than Kostant's formula because of the need to calculate the values
of the function p for the latter formula. Furthermore, the size of the Weyl
group increases rapidly when the rank of the root system increases. Because
of this Freudenthal's formula is usually more efficient than Racah's formula
when the rank of the root system is bigger than 3.

8.12 Decomposing a tensor product

Let $\lambda \in P_+$ and let V be the irreducible highest-weight module over L with
highest weight λ. The formal character of V, χ_V is also denoted by χ_λ.
Because χ_λ is the character of an irreducible L-module, we say that χ_λ is
an *irreducible character*.

Let $\lambda_1, \lambda_2 \in P_+$, and let V_1, V_2 the highest-weight modules over L with
highest weights λ_1, λ_2 respectively. The character of $V_1 \oplus V_2$ is $\chi_{\lambda_1} + \chi_{\lambda_2}$.
So taking the direct direct sum of L-modules corresponds to taking the sum
of their characters. Let V be any finite-dimensional L-module. Then V is
completely reducible (Theorem 4.4.6). This means that V decomposes as a
direct sum of irreducible L-modules

$$V = V_1^{\oplus n_1} \oplus \cdots \oplus V_s^{\oplus n_s},$$

where $V_i^{\oplus n_i}$ denotes the direct sum of n_i copies of the irreducible module
V_i. Hence the character χ_V of V can be written as a sum

$$\chi_V = n_1 \chi_{\lambda_1} + \cdots + n_s \chi_{\lambda_s}$$

where the λ_i is the highest weight of V_i. Also if $\lambda \in P_+$, then $\chi_\lambda = \lambda + \sum_{\mu < \lambda} m_\mu e^\mu$. Hence characters belonging to different irreducible highest-
weight modules are linearly independent. So the expression of χ_V as a sum
of irreducible characters is unique. It follows that the decomposition of
χ_V as a sum of irreducible characters gives us the number of irreducible
components of V together with their highest weights.

There is a straightforward method for decomposing the character of V
as a sum of irreducible characters. As before, by $P(V)$ we denote the set
of weights of V. From the knowledge of the character of V we can obtain
this set. Let $\mu \in P(V)$ be a weight that is maximal with respect to the
order $\leq$ defined in Section 8.1. Let $v_\mu \in V$ be a weight vector of weight
μ. Then $x_\alpha \cdot v_\mu = 0$ for all roots $\alpha > 0$. Hence μ is the highest weight of
an irreducible submodule W of V. So we have obtained the highest weight

of one irreducible component of V. By Proposition 4.4.5 there is an L-submodule $U \subset V$ such that $V = U \oplus W$. Now the character χ_μ of W can be found by Freudenthal's formula. Then $\chi_U = \chi_V - \chi_\mu$, and we continue the process with χ_U in place of χ_V. After a finite number of steps we will have found the decomposition of χ_V into irreducible characters. We call this algorithm DecompositionBySuccessiveSubtractions.

Remark. It is possible to make the algorithm DecompositionBySuccessiveSubtractions somewhat more efficient by starting with the dominant part of the character χ_V (i.e., $\sum_{\mu \in P_+} m_\mu e^\mu$). Then by a call to DominantWeightsOfHWModule we only compute the dominant part of the characters χ_μ which we subtract from χ_V.

Let V_1, V_2 be irreducible highest-weight modules of L, then the tensor product $V_1 \otimes V_2$ is also an L-module (cf. Section 1.12). However $V_1 \otimes V_2$ is generally not irreducible. Let $v_{\mu_i}^i$ be a weight vector of weight μ_i in V_i for $i = 1, 2$. Let $h \in H$, then $h \cdot (v_{\mu_1}^1 \otimes v_{\mu_2}^2) = (h \cdot v_{\mu_1}^1) \otimes v_{\mu_2}^2 + v_{\mu_1}^1 \otimes (h \cdot v_{\mu_2}^2) = (\mu_1 + \mu_2)(h)(v_{\mu_1}^1 \otimes v_{\mu_2}^2)$. It follows that $v_{\mu_1}^1 \otimes v_{\mu_2}^2$ is a weight vector in $V_1 \otimes V_2$ of weight $\mu_1 + \mu_2$. So the weights of $V_1 \otimes V_2$ are exactly the sums of the weights of V_1 and V_2. Therefore, if we set $V = V_1 \otimes V_2$, then the character of V is given by

$$\chi_V = \sum_{\mu_1, \mu_2 \in P} m_{\mu_1}^1 m_{\mu_2}^2 e^{\mu_1 + \mu_2},$$

where by $m_{\mu_i}^i$ we denote the multiplicity of μ_i in V_i for $i = 1, 2$. Hence $\chi_V = \chi_1 \chi_2$, where χ_i is the character of V_i for $i = 1, 2$. So knowing the characters of V_1 and V_2 we can easily obtain the character of $V_1 \otimes V_2$. Then by applying the algorithm to decompose χ_V by successive subtractions, we can obtain the highest weights of the modules occurring in the decomposition of $V_1 \otimes V_2$ as a direct sum of irreducible L-modules. However, this can be a lengthy process as the dimension of $V_1 \otimes V_2$ (which is equal to $\dim V_1 \dim V_2$) is generally much bigger than the dimensions of V_1, V_2. It is the objective of this section to describe several other methods by which we can obtain the decomposition of the character of $V_1 \otimes V_2$ into irreducible characters.

Let λ_1, λ_2 and χ_1, χ_2 be the highest weights and the characters of the highest-weight modules V_1, V_2 respectively. Set $V = V_1 \otimes V_2$ and write

$$\chi_V = \chi_1 \chi_2 = \sum_{\lambda \in P_+} n_\lambda \chi_\lambda. \tag{8.16}$$

This is the decomposition of χ_V that we are after. The number n_λ is called the *multiplicity* of χ_λ in χ_V. We multiply (8.16) by $\sum_{g \in W(\Phi)} \mathrm{sn}(g) e^{g(\rho)}$ and apply Weyl's character formula (Theorem 8.10.3) to get

$$\left(\sum_{\mu\in P} m^1_\mu e^\mu\right)\left(\sum_{g\in W(\Phi)} \mathrm{sn}(g)e^{g(\lambda_2+\rho)}\right) = \sum_{\lambda\in P_+} n_\lambda \sum_{g\in W(\Phi)} \mathrm{sn}(g)e^{g(\lambda+\rho)},$$

where m^1_μ denotes the multiplicity of $\mu \in P$ in V_1. We use Kostant's formula (Theorem 8.11.1) for the multiplicities m^1_μ in the first term on the left-hand side. Furthermore we set $n_\lambda = 0$ if $\lambda \notin P_+$ and take the summation on the right to run over all $\lambda \in P$. Doing this we obtain

$$\left(\sum_{\mu\in P}\sum_{g\in W(\Phi)} \mathrm{sn}(g)p(g(\lambda_1+\rho)-\mu-\rho)e^\mu\right)\left(\sum_{g\in W(\Phi)} \mathrm{sn}(g)e^{g(\lambda_2+\rho)}\right) =$$
$$\sum_{\lambda\in P} n_\lambda \sum_{g\in W(\Phi)} \mathrm{sn}(g)e^{g(\lambda+\rho)}.$$

But this is the same as

$$\sum_{\mu\in P}\sum_{g,h\in W(\Phi)} \mathrm{sn}(gh)p(g(\lambda_1+\rho)-\mu-\rho)e^{\mu+h(\lambda_2+\rho)} =$$
$$\sum_{\lambda\in P} n_\lambda \sum_{g\in W(\Phi)} \mathrm{sn}(g)e^{g(\lambda+\rho)}.$$

On the left we substitute $\lambda = \mu + h(\lambda_2+\rho) - \rho$, and on the right $\mu = g(\lambda+\rho)-\rho$. We get

$$\sum_{\lambda\in P}\sum_{g,h\in W(\Phi)} \mathrm{sn}(gh)p(g(\lambda_1+\rho)+h(\lambda_2+\rho)-\lambda-2\rho)e^{\lambda+\rho} =$$
$$\sum_{\mu\in P}\sum_{g\in W(\Phi)} \mathrm{sn}(g)n_{g^{-1}(\mu+\rho)-\rho}e^{\mu+\rho} = \sum_{\mu\in P}\sum_{g\in W(\Phi)} \mathrm{sn}(g)n_{g(\mu+\rho)-\rho}e^{\mu+\rho}.$$

Comparing the coefficients of $e^{\lambda+\rho}$ on both sides we obtain

$$\sum_{g,h\in W(\Phi)} \mathrm{sn}(gh)p(g(\lambda_1+\rho)+h(\lambda_2+\rho)-\lambda-2\rho) = \sum_{g\in W(\Phi)} \mathrm{sn}(g)n_{g(\lambda+\rho)-\rho} =$$
$$n_\lambda + \sum_{1\neq g\in W(\Phi)} \mathrm{sn}(g)n_{g(\lambda+\rho)-\rho}.$$

Now if $\lambda \in P_+$, then also $\lambda + \rho \in P_+$. But then $g(\lambda+\rho) \notin P_+$ if $g \neq 1$ by Theorem 8.3.5. Hence in that case also $g(\lambda+\rho)-\rho \notin P_+$ and $n_{g(\lambda+\rho)-\rho} = 0$. So on the right-hand side we only have the term n_λ, and we have proved the following theorem.

Theorem 8.12.1 (Steinberg) *Let λ_1, λ_2 be the highest weights of the irreducible highest-weight modules V_1 and V_2 respectively. Then the multiplicity of $\lambda \in P_+$ in the character of $V_1 \otimes V_2$ is given by*

$$n_\lambda = \sum_{g,h \in W(\Phi)} \mathrm{sn}(gh) p(g(\lambda_1 + \rho) + h(\lambda_2 + \rho) - \lambda - 2\rho).$$

Proposition 8.12.2 (Racah) *Let λ_1, λ_2, V_1, V_2 and n_λ be as in Theorem 8.12.1. For $\nu \in P$ we denote by m_ν^1 the multiplicity of ν in V_1. Then*

$$n_\lambda = \sum_{h \in W(\Phi)} \mathrm{sn}(h) m^1_{\lambda+\rho-h(\lambda_2+\rho)}.$$

Proof. By Theorem 8.12.1 we have

$$n_\lambda = \sum_{g,h \in W(\Phi)} \mathrm{sn}(gh) p(g(\lambda_1 + \rho) + h(\lambda_2 + \rho) - \lambda - 2\rho)$$

$$= \sum_{h \in W(\Phi)} \mathrm{sn}(h) \sum_{g \in W(\Phi)} \mathrm{sn}(g) p(g(\lambda_1 + \rho) - (\lambda + \rho - h(\lambda_2 + \rho)) - \rho).$$

But by Kostant's formula (Theorem 8.11.1), this is equal to

$$\sum_{h \in W(\Phi)} \mathrm{sn}(h) m^1_{\lambda+\rho-h(\lambda_2+\rho)}.$$

$\square$

Now we describe a third formula for decomposing a tensor product. For $\mu \in P$ we define $s(\mu) \in \{0, \pm 1\}$ as follows. If there is a $1 \neq g \in W(\Phi)$ such that $g(\mu) = \mu$, then $s(\mu) = 0$. Otherwise we let $g \in W(\Phi)$ be such that $g(\mu) \in P_+$ (note that such a g is uniquely determined because, in this case, there are no elements in $W(\Phi)$ that stabilize μ); and we set $s(\mu) = \mathrm{sn}(g)$. Also for $\mu \in P$ we denote the unique element of P_+ to which μ is conjugate by $\{\mu\}$.

Proposition 8.12.3 (Klymik) *Let λ_1, λ_2 and V_1, V_2 be as in Theorem 8.12.1. Let χ_1, χ_2 be the characters of V_1, V_2 respectively. Let m_μ^1 denote the multiplicity of μ in V_1. Then*

$$\chi_1 \chi_2 = \sum_{\mu \in P} m_\mu^1 s(\mu + \lambda_2 + \rho) \chi_{\{\mu+\lambda_2+\rho\}-\rho}.$$

Proof. First we note that if $\nu \in P$ is such that $s(\nu) \neq 0$, then $\{\nu\} - \rho \in P_+$. Indeed, if $s(\nu) \neq 0$ then the stabilizer of ν in $W(\Phi)$ is trivial. But then the stabilizer of $g(\nu)$ in $W(\Phi)$ is trivial for all $g \in W(\Phi)$, and in particular, the stabilizer of $\{\nu\}$ is trivial. Hence for no simple root α_i we have $(\{\nu\}, \alpha_i) = 0$. Therefore $\langle \nu, \alpha_i \rangle > 0$ for all simple roots α_i, and $\{\nu\} - \rho$ is dominant. So all $\{\mu + \lambda_2 + \rho\} - \rho$ appearing on the right are dominant weights.

From Section 8.10 we recall that $\sigma(\nu) = \sum_{g \in W(\Phi)} \mathrm{sn}(g) e^{g(\nu)}$. Multiply $\chi_1 \chi_2$ by $\sigma(\rho)$ and use Weyl's character formula (Theorem 8.10.3) to get

$$\chi_1 \chi_2 \sigma(\rho) = \chi_1 \sigma(\rho + \lambda_2) = \Big(\sum_{\mu \in P} m_\mu^1 e^\mu\Big)\Big(\sum_{g \in W(\Phi)} \mathrm{sn}(g) e^{g(\rho + \lambda_2)}\Big). \quad (8.17)$$

By Theorem 8.4.2 we see that $\sum_{\mu \in P} m_\mu^1 e^\mu = \sum_{\mu \in P} m_\mu^1 e^{g(\mu)}$ for any arbitrary $g \in W(\Phi)$. Furthermore, if for $\nu \in P$ there is a $g \in W(\Phi)$ such that $g(\nu) = \nu$, then there is a reflection r_α that stabilizes ν. To see this we note that there is a $h \in W(\Phi)$ such that $h(\nu) \in P_+$. Furthermore, hgh^{-1} stabilizes $h(\nu)$ so by Proposition 8.3.9 there is a simple root α_i such that $(\alpha_i, h(\nu)) = 0$. Hence the reflection corresponding to $h^{-1}(\alpha_i)$ stabilizes ν. From Section 8.10 we recall that $\sigma h(\nu) = \mathrm{sn}(h)\sigma(\nu)$ for all $h \in W(\Phi)$ and $\nu \in P$. So if we take h to be a reflection stabilizing ν we see that $\sigma(\nu) = 0$. It follows that $\sigma(\nu) = 0$ for all ν such that $s(\nu) = 0$. On the other hand, if $s(\nu) \neq 0$, then let $g \in W(\Phi)$ be such that $g(\nu) \in P_+$. Then $\sigma(\nu) = \mathrm{sn}(g)\sigma(g(\nu)) = s(\nu)\sigma(\{\nu\})$. Using this we see that the right-hand side of (8.17) is equal to

$$\sum_{g \in W(\Phi)} \sum_{\mu \in P} \mathrm{sn}(g) m_\mu^1 e^{g(\rho + \lambda_2 + \mu)} = \sum_{\mu \in P} m_\mu^1 \sigma(\mu + \lambda_2 + \rho)$$

$$= \sum_{\mu \in P} m_\mu^1 s(\mu + \lambda_2 + \rho)\sigma(\{\mu + \lambda_2 + \rho\}).$$

And again using Weyl's character formula, this is equal to

$$\sum_{\mu \in P} m_\mu^1 s(\mu + \lambda_2 + \rho)\chi_{\{\mu + \lambda_2 + \rho\} - \rho}\sigma(\rho).$$

After canceling $\sigma(\rho)$ the desired formula follows. □

Both Steinberg's formula and Racah's formula give a formula for the multiplicity of a weight in the character of $V_1 \otimes V_2$ (whereas Klymik's formula gives the decomposition of the character of the tensor product directly). So to be able to use these formulas we need a finite set of dominant weights that contains all weights that have non-zero multiplicity in the character of $V_1 \otimes V_2$. This is provided by the next lemma.

Lemma 8.12.4 *Let λ_1, λ_2 and V_1, V_2 be as in Theorem 8.12.1. Let λ be a weight of $V_1 \otimes V_2$. Then λ is of the form $\mu + \lambda_2$, for some $\mu \in P(V_1)$.*

Proof. We recall that n_λ denotes the multiplicity of χ_λ in the character of $V_1 \otimes V_2$. From the proof of Proposition 8.12.3 we have the following identity

$$\chi_1 \chi_2 \sigma(\rho) = \sum_{\mu \in P} m_\mu^1 s(\mu + \lambda_2 + \rho) \sigma(\{\mu + \lambda_2 + \rho\}).$$

But using the fact that $\chi_1 \chi_2$ is the character of $V_1 \otimes V_2$ and Weyl's character formula we have $\chi_1 \chi_2 \sigma(\rho) = \sum_{\lambda \in P_+} n_\lambda \chi_\lambda \sigma(\rho) = \sum_{\lambda \in P_+} n_\lambda \sigma(\lambda + \rho)$. Therefore we get

$$\sum_{\lambda \in P_+} n_\lambda \sigma(\lambda + \rho) = \sum_{\mu \in P} m_\mu^1 s(\mu + \lambda_2 + \rho) \sigma(\{\mu + \lambda_2 + \rho\}). \qquad (8.18)$$

We want to compare the coefficients of $e^{\lambda + \rho}$ on both sides of this equation. However the braces on the right-hand side are inconvenient in this respect. We describe how to remove them. First of all, if $\mu + \lambda_2$ is dominant, then so is $\mu + \lambda_2 + \rho$ and hence the braces can be removed. Now suppose that $\mu + \lambda_2$ is not dominant. We may also suppose that the stabilizer of $\mu + \lambda_2 + \rho$ in $W(\Phi)$ is trivial since otherwise $s(\mu + \lambda_2 + \rho) = 0$. Since $\mu + \lambda_2$ is not dominant there is a simple root α_i such that $\langle \mu + \lambda_2, \alpha_i \rangle < 0$. Also since λ_2 is dominant, $\langle \lambda_2, \alpha_i \rangle \geq 0$. Let p, q be such that $r_{\alpha_i}(\mu) = \mu + p\alpha_i$ and $r_{\alpha_i}(\lambda_2) = \lambda_2 - q\alpha_i$. Then because $\langle \mu + \lambda_2, \alpha_i \rangle < 0$ and $\langle \lambda_2, \alpha_i \rangle \geq 0$ we have $p > q \geq 0$. Set $\mu_1 = \mu + (p - q - 1)\alpha_i$, then $r_{\alpha_i}(\mu + \lambda_2 + \rho) = \mu_1 + \lambda_2 + \rho$. We note that $\mu_1 \in P(V_1)$ by Lemma 8.7.1. And $\mu_1 \neq \mu$ because the stabilizer of $\mu + \lambda_2 + \rho$ is trivial. Since $\{\mu_1 + \lambda_2 + \rho\} = \{\mu + \lambda_2 + \rho\}$ we may replace $\{\mu + \lambda_2 + \rho\}$ in (8.18) by $\{\mu_1 + \lambda_2 + \rho\}$. Because $\mu_1 > \mu$ and also the stabilizer of $\mu_1 + \lambda_2 + \rho$ is trivial, we arrive, after a finite number of such replacement steps at a weight μ_k such that $\mu_k + \lambda_2$ is dominant and $\mu_k + \lambda_2 + \rho = h(\mu + \lambda_2 + \rho)$ for some $h \in W(\Phi)$. We denote μ_k by μ'. Then (8.18) is equivalent to

$$\sum_{\lambda \in P_+} n_\lambda \sigma(\lambda + \rho) = \sum_{\substack{\mu \in P \\ \mu + \lambda_2 \text{ dominant}}} m_\mu^1 s(\mu + \lambda_2 + \rho) \sigma(\mu + \lambda_2 + \rho)$$

$$+ \sum_{\substack{\mu \in P \\ \mu + \lambda_2 \text{ not dominant}}} m_\mu^1 s(\mu + \lambda_2 + \rho) \sigma(\mu' + \lambda_2 + \rho).$$

If $\lambda \in P_+$ is such that $n_\lambda > 0$, then $e^{\lambda + \rho}$ appears on the left with coefficient n_λ. (We note that $\sigma(\lambda + \rho)$ has only one term e^μ with μ dominant, namely

$e^{\lambda+\rho}$. This follows from Theorem 8.3.5.) On the right this term appears only if $\lambda = \mu + \lambda_2$ for some $\mu \in P(V_1)$. □

Now both Steinberg's and Racah's formula give an algorithm DecomposeCharacterTensorProduct for decomposing the character of the tensor product $V_1 \otimes V_2$. First we calculate the set $\Lambda = \{\mu + \lambda_2 \mid \mu \in P(V_1) \text{ and } \mu + \lambda_2 \text{ dominant}\}$. Then for each $\lambda \in \Lambda$ we calculate n_λ using the formula. Steinberg's formula is rather difficult to use in practice because of the double sum over the Weyl group. However, Racah's formula has proven to be efficient in practice. We refer to [6] for a discussion of practical experiences with Racah's formula.

Now also Klymik's formula provides a method for DecomposeCharacterTensorProduct. This formula does not involve the Weyl group at all. Furthermore, we can efficiently loop through the weights of V_1 by first calculating the dominant part of the character of V_1 (by the algorithm DominantWeightsOfHWModule), and then looping through the orbit of each dominant weight by the algorithms described in Section 8.6. For this reason Klimyk's formula has proven to be very efficient in practice (see [82]).

Example 8.12.5 Let L be the simple Lie algebra of type G_2 (cf. Example 8.7.4). Let λ_1, λ_2 be as in Example 8.8.4, i.e., the fundamental weights. Let V_1, V_2 be the highest-weight modules over L with highest weights λ_1, λ_2 respectively. We decompose the character of the tensor product $V_1 \otimes V_2$ using Klymik's formula. Using Freudenthal's formula we see that V_1 has two dominant weights, namely $(1,0)$ and $(0,0)$, both of multiplicity 1. So the weights of V_1 are the elements of the orbit of $(1,0)$ together with $(0,0)$. A little calculation shows that the orbit of $(1,0)$ is

$$W(\Phi) \cdot (1,0) = \{(1,0),(-1,1),(2,-1),(-2,1),(1,-1),(-1,0)\}.$$

So V_1 has dimension 7. Now we run through the weights of V_1, for each weight μ calculating $m_\mu^1 s(\mu + \lambda_2 + \rho) \chi_{\{\mu+\lambda_2+\rho\}-\rho}$. First Set $\mu = (1,0)$, then $\mu + \lambda_2 + \rho = (2,2)$, which is dominant. Hence $s(\mu + \lambda_2 + \rho) = 1$. So the character we obtain is the irreducible character belonging to the weight $(1,1)$. Now let $\mu = (-1,1)$. Then $\mu + \lambda_2 + \rho = (0,3)$. But this is stabilized by r_{α_1} and hence $s(\mu + \lambda_2 + \rho) = 0$ so that this weight does not give us a character. Going on like this we see that the only characters we get are those belonging to the weights $(2,0)$ and $(1,0)$. Also if we set $\mu = (-2,1)$, then $\mu + \lambda_2 + \rho = (-1,3)$. Now $r_{\alpha_1}(-1,3) = (1,2)$ so that $s(\mu+\lambda_2+\rho) = -1$. So this contributes the character belonging to $(0,1)$ with multiplicity -1. However, if we set $\mu = (0,0)$ then we get the same character with multiplicity 1. So these characters cancel and we have decomposed that character of $V_1 \otimes V_2$ as

$$\chi_{(1,1)} + \chi_{(2,0)} + \chi_{(1,0)}.$$

8.13 Branching rules

Let K, L be semisimple Lie algebras over an algebraically closed field of characteristic 0. Let H be a fixed Cartan subalgebra of L (which is necessarily split). Furthermore, suppose that there is an injective morphism of Lie algebras $f : K \to L$. Such a map is called an *embedding* of K in L. Let V be the irreducible highest-weight module over L with highest weight λ. Let ϕ denote the representation of L afforded by V. Then $\phi \circ f$ is a representation of K and hence V is also a K-module. Now in general V is reducible as K-module. So the character χ_V of V (viewed as K-module) can be written as a sum

$$\chi_V = n_1\chi_{\lambda_1} + \cdots + n_s\chi_{\lambda_s}, \tag{8.19}$$

where the χ_{λ_i} are irreducible characters (of irreducible K-modules). Now the splitting of V as K-module is called *branching* and the decomposition of its character (8.19) the corresponding *branching rule*. It is the objective of this section to describe an algorithm for computing the branching rule, given L, K, f and V.

Two embeddings $f_1, f_2 : K \to L$ are said to be *equivalent* if for any representation ϕ of L, the representations $\phi \circ f_1$ and $\phi \circ f_2$ of K are equivalent. Equivalent embeddings yield the same branching rule, so we may replace a given embedding by an equivalent one.

Lemma 8.13.1 *Let $f : K \to L$ be an embedding of K into L. Let H be a Cartan subalgebra of L and H_K a Cartan subalgebra of K. Then there is an embedding $f' : K \to L$ such that f' is equivalent to f and $f'(H_K) \subset H$.*

Proof. Let $h \in H_K$, then $\mathrm{ad}_K h$ is a semisimple linear transformation (Root fact 8). Now we consider the representation $\mathrm{ad}_L \circ f : K \to \mathfrak{gl}(L)$, given by $\mathrm{ad}_L \circ f(x)(y) = [f(x), y]$, for $x \in K$ and $y \in L$. Then by Corollary 4.6.4, also $\mathrm{ad}_L f(h)$ is a semisimple linear transformation. Also H is commutative (Root fact 6), so $f(H)$ is a commutative subalgebra of L consisting of semisimple elements. Hence by Lemma 4.14.2, $f(H)$ is contained in a Cartan subalgebra H' of L.

By Theorem 3.5.1 there is a $g \in \mathrm{Int}(L)$ such that $g(H') = H$. We show that the embedding $g \circ f$ is equivalent to f. For that let $\phi : L \to \mathfrak{gl}(V)$

be a representation of L. We have to show that the representations $\phi \circ f$ and $\phi \circ (g \circ f)$ of K are equivalent. But by Lemma 8.5.1 we have that the representations ϕ and $\phi \circ g$ of L are equivalent, so certainly $\phi \circ f$ and $\phi \circ (g \circ f)$ are equivalent.   □

Now by Lemma 8.13.1 we may assume that the embedding f of K into L maps a Cartan subalgebra H_K of K into the given Cartan subalgebra H of L. From now on we omit the map f and think of K as a subalgebra of L, and of H_K as a subalgebra of H.

Let μ be a weight of V (viewed as L-module) with weight space V_μ. Then since $H_K \subset H$, V_μ is also a weight space of V viewed as K-module. Let x_i', y_i', h_i' for $1 \leq i \leq r$ be canonical generators of K, where $h_i' \in H_K$. Also let x_i, y_i, h_i for $1 \leq i \leq l$ be canonical generators of L, where $h_i \in H$. Then we represent a weight μ as a row vector $(\mu(h_1), \ldots, \mu(h_l))$. However, when we view V as a K-module, then we represent μ as the row vector $(\mu(h_1'), \ldots, \mu(h_r'))$. Now because the h_i for a basis of H, there are constants p_{ij} such that $h_i' = \sum_{j=1}^l p_{ij} h_j$ and hence

$$\mu(h_i') = \sum_{j=1}^l p_{ij} \mu(h_j).$$

It follows that by applying the matrix (p_{ij}) to the weights of V (viewed as L-module), we get the weights of V viewed as K-module. Therefore the matrix (p_{ij}) is called the *projection matrix*. These considerations lead to the following algorithm.

Algorithm BranchingRule

Input: semisimple Lie algebras K, L with split Cartan subalgebras H_K, H, an embedding $f : K \to L$ such that $f(H_K) \subset H$, and an L-module V.

Output: the decomposition of the character of V when viewed as a K-module.

Step 1 Calculate canonical generators x_i', y_i', h_i' and x_i, y_i, h_i of K, L respectively, such that $h_i' \in H_K$ and $h_i \in H$.

Step 2 Calculate the projection matrix p.

Step 3 Calculate the set of weights Ω of V (viewed as L-module).

Step 4 By applying p to each element of Ω obtain the character χ_V of the K-module V. Decompose χ_V using the algorithm DecompositionBy-SuccessiveSubtractions, and return the result.

Comments: We note that computing the projection matrix can be done by solving r systems of linear equations, where $r = \dim H_K$. Furthermore it needs to be done once for every embedding $f : K \to L$; it can then be used for every L-module V. Also in the algorithm we only need the character of V (viewed as L-module). So instead of V we can also input its highest weight. Then using the techniques of Section 8.9, we can compute the character of V.

8.14 Notes

Proposition 8.6.2 and the algorithm WeylOrbit are due to D. M. Snow ([81]). Proposition 8.7.3 as well as most material in Section 8.9 is drawn from [63]. Racah's formulas (Propositions 8.11.3 and 8.12.2) are contained in [70], [5], [6]. Klymik's formula appears in [51]; see also [82]. The algorithm for calculating branching rules in Section 8.13 largely follows [66]; see also [61].

Many algorithms discussed in this chapter have been implemented in the computer algebra package LiE (see [21]). In this package many more algorithms for dealing with root systems, weights and Weyl groups have been implemented.

Appendix A

On associative algebras

Associative algebras occur naturally when studying the structure of Lie algebras. Examples can be found in Sections 1.1, 1.15, 2.2, 3.7, 4.11. Apart from being of use for exploring the structure of Lie algebras, associative algebras are also interesting in their own right, and their structure has been an object of intensive study. In this appendix we collect parts of the structure theory of associative algebras. Furthermore we sketch some algorithms for calculating the structure of an associative algebra.

An important example of an associative algebra is the algebra $M_n(F)$ of $n \times n$ matrices over the field F. A subalgebra of an $M_n(F)$ is said to be a *linear* associative algebra. Many associative algebras that occur in our discussions will be linear.

Let A be an associative algebra. An element $e \in A$ is said to be an *identity element* if $ea = ae = a$ for all $a \in A$. It is straightforward to see that such an element (if it exists) is unique. The algebra A is said to be an algebra with one if A has an identity element. In many cases we will restrict our attention to algebras with one. However if the associative algebra A does not have a one, then we can embed it into an algebra with one in the following way. Let B_0 be a 1-dimensional vector space spanned by the element e_0. Set $B = B_0 \oplus A$; where the multiplication on B extends the one of A and furthermore $be_0 = e_0 b = b$ for all $b \in B$. Then B is an associative algebra with one; it is called the *Dorroh extension* of A. Using Dorroh extensions we can often translate results for algebras with one to algebras without a one.

In Section A.1 we show that a finite-dimensional associative algebra has a unique maximal nilpotent ideal, called the radical. We give a characterization of the radical in the case where the ground field is of characteristic 0. This leads to an algorithm for calculating a basis of the radical. We also formulate Wedderburn's structure theorem for semisimple associative algebras. Finally we prove the Wedderburn-Malcev principal theorem (which is an analogue of Levi's theorem) for associative algebras generated by a single element. This proof is used in Section A.2 where we study associative algebras generated by a single element. We show that such an algebra has a system of orthogonal primitive idempotents with sum 1. Also we describe the Jordan decomposition of a linear transformation, and we give an algorithm for calculating this decomposition for a given linear transformation. In Section A.3 we study central idempotents in a semisimple algebra. These central idempotents yield a decomposition of the algebra into a direct sum of simple ideals. We sketch an algorithm for calculating a set of primitive orthogonal and central idempotents with sum 1. We also show how to lift these idempotents modulo the radical, in case the algebra is not semisimple.

A.1   Radical and semisimplicity

A finite-dimensional associative algebra N is said to be *nilpotent* if there is an $m > 0$ such that $a_{i_1} a_{i_2} \cdots a_{i_m} = 0$ for all $a_{i_k} \in N$. Let A be a finite-dimensional associative algebra. In the same way as for Lie algebras (cf. Section 2.2) it can be proved that the sum of two nilpotent ideals of A is a nilpotent ideal of A. From this it follows that A contains a unique maximal nilpotent ideal. It is called the *radical* of A, and denoted by $\mathrm{Rad}(A)$.   If the ground field is of characteristic zero, then we have the following characterization of the radical of linear algebras that contain an identity element.

Proposition A.1.1 *Let A be a finite-dimensional linear associative algebra with one over a field of characteristic 0. Then*

$$\mathrm{Rad}(A) = \{x \in A \mid \mathrm{Tr}(xy) = 0 \text{ for all } y \in A\}.$$

Proof. Set $I = \{x \in A \mid \mathrm{Tr}(xy) = 0 \text{ for all } y \in A\}$. Let $x \in \mathrm{Rad}(A)$. Then $xy \in \mathrm{Rad}(A)$ for all $y \in A$ and therefore xy is nilpotent. Hence $\mathrm{Tr}(xy) = 0$ for all $y \in A$, i.e., $x \in I$. To prove the other inclusion we first show that all elements of I are nilpotent. Let $x \in I$, then since $\mathrm{Tr}(xy) = 0$ for all $y \in A$ we have $\mathrm{Tr}(x^n) = 0$ for $n \geq 1$. Because the ground field is of characteristic 0 this implies that x is nilpotent. Also it is straightforward to see that I is an ideal of A. Now any associative algebra consisting entirely of nilpotent elements must be nilpotent (this can be seen by analogous arguments to those used to prove Engel's theorem in Section 2.1). So I is a nilpotent ideal of A and hence $I = \mathrm{Rad}(A)$.   □

Let A be a finite-dimensional associative algebra. For x in A we define a linear map $\ell(x) : A \to A$ by $\ell(x)(y) = xy$ (left multiplication). If we want to make clear which algebra we mean we also write $\ell_A(x)$ in place of $\ell(x)$. Using Proposition A.1.1 we can now characterize the radical of an arbitrary associative algebra with one of characteristic 0.

Proposition A.1.2 *Let A be a finite-dimensional associative algebra with one defined over a field of characteristic 0. Then*

$$\mathrm{Rad}(A) = \{x \in A \mid \mathrm{Tr}(\ell(xy)) = 0 \text{ for all } y \in A\}.$$

Proof. The map $\ell_A : A \to \mathrm{End}(A)$ is a faithful representation of A because A contains an identity element. So $\ell(A) \cong A$ and therefore the statement follows from Proposition A.1.1.   □

On the basis of Propositions A.1.1 and A.1.2 we can formulate an algorithm Radical for calculating the radical of an associative algebra of characteristic 0. If the algebra is linear then we use Proposition A.1.1 to give us a set of linear equations for the basis elements of the radical. If A is not linear then we use Proposition A.1.2 to the same end. This algorithm does in general not work for associative algebras defined over a field of characteristic $p > 0$. For this case there are other algorithms available. However, discussing these would lead us too far afield and therefore we refer to the literature ([20], [45], [46], [47], [73]).

A finite-dimensional associative algebra A is said to be *semisimple* if $\mathrm{Rad}(A) = 0$. Let A be an associative algebra; then $A/\mathrm{Rad}(A)$ is semisimple. This implies that if I is a nilpotent ideal of A such that A/I is semisimple, then $I = \mathrm{Rad}(A)$. We recall that a field F is said to be perfect if F is of characteristic 0, or F is of characteristic $p > 0$ and $F^p = F$.

Proposition A.1.3 *Let A be a finite-dimensional associative algebra defined over a perfect field F. Let $\tilde{F}$ be an extension field of F, then $\mathrm{Rad}(A \otimes_F \tilde{F}) = \mathrm{Rad}(A) \otimes_F \tilde{F}$.*

Proof. We note that if A is defined over a field of characteristic 0, then this follows from Proposition A.1.2, because this result gives the radical as the solution space of a set of linear equations over F. More generally, let $\{r_1, \dots, r_m\}$ be a basis of $\mathrm{Rad}(A)$. Then $\{r_1 \otimes 1, \dots, r_m \otimes 1\}$ is the basis of a nilpotent ideal I of $A \otimes_F \tilde{F}$. Furthermore, $(A \otimes_F \tilde{F})/I \cong (A/\mathrm{Rad}(A)) \otimes_F \tilde{F}$. Because F is perfect, $A/\mathrm{Rad}(A)$ is a separable algebra (see [69, §10.7]). But this implies that $(A/\mathrm{Rad}(A)) \otimes_F \tilde{F}$ is a semisimple algebra ([69, §10.6]). The conclusion is that $I = \mathrm{Rad}(A \otimes_F \tilde{F})$. □

The algebra A is called *simple* if A has no ideals other than 0 and A. A simple associative algebra with one is necessarily also semisimple. As an example of a simple associative algebra with one we mention the full matrix algebra $M_n(F)$. Now by the next theorem, all semisimple associative algebras are obtained by taking direct sums of full matrix algebras. We recall that a division algebra over F is an associative algebra R over F with one such that for all $x \in R$ there is a $y \in R$ such that $xy = 1$. A commutative division algebra is a field. For a proof of the next theorem we refer to [69].

Theorem A.1.4 (Wedderburn) *Let A be a finite-dimensional associative algebra with one over a field F. Then A decomposes as a direct sum of ideals where every ideal is isomorphic to a full matrix algebra over a division algebra, i.e.,*

$$A \cong M_{n_1}(R_1) \oplus \cdots \oplus M_{n_s}(R_s)$$

where $n_i \geq 1$ and the R_i are division algebras over F.

By the next theorem, the semisimple algebra $A/\mathrm{Rad}(A)$ can be identified as a subalgebra of A. It is analogous to Levi's theorem for Lie algebras. We prove this theorem for the case where A is generated by a single element, because we will need this proof later on. For the proof of the general case we refer to [69].

Theorem A.1.5 (Wedderburn-Malcev principal theorem) *Let F be a perfect field and let A be a finite-dimensional associative algebra with one over F. Then A has a subalgebra S such that $A = S \oplus \mathrm{Rad}(A)$.*

Proof. (For the case where A is generated by a single element.) Suppose that A is generated by the element a. Let $f_a \in F[X]$ be the minimum polynomial of a, and let

$$f_a = p_1^{k_1} \cdots p_r^{k_r}$$

be the factorization of f_a into distinct irreducible polynomials p_i. We suppose that the p_i have leading coefficient 1 and that $\deg(p_i) > 0$ for $1 \leq i \leq r$. First suppose that all $k_i = 1$. Let $b \in \mathrm{Rad}(A)$. Then $b = p(a)$ for some polynomial $p \in F[X]$. Furthermore $b^k = p^k(a) = 0$ for some $k > 0$. Hence f_a divides p^k. But since all $k_i = 1$ this implies that f_a divides p and hence $b = 0$. So in this case $\mathrm{Rad}(A) = 0$ and we are done.

Now we suppose that $k_i > 1$ for some k_i. Set $h = p_1 \cdots p_r$ and $b = h(a)$. Let I be the ideal of A generated by b. We show that $I = \mathrm{Rad}(A)$.

Set $n = \max k_i$, then $b^n = h^n(a) = 0$ so b is nilpotent. Now since A is commutative, also I is nilpotent and hence $I \subset \mathrm{Rad}(A)$. By $\bar{a}$ we denote the image of a in A/I. Let h_1 be the minimum polynomial of $\bar{a}$. We have $h(\bar{a}) = 0$ so that h_1 divides h. But also $h_1(a) \in \mathrm{Rad}(A)$ so that $h_1(a)^N = 0$ for some $N > 0$. As a consequence f_a is a factor of h_1^N and this implies that $h_1 = h$. Hence the minimum polynomial of a generator of A/I is a product of distinct irreducibles. As argued above this implies that A/I is semisimple, so that $\mathrm{Rad}(A) = I$.

Now we construct an element $a' \in A$ such that $h(a') = 0$ and $a = a' \bmod \mathrm{Rad}(A)$. For this we start with the element $a_1 = a$. Then $h(a_1) = 0 \bmod \mathrm{Rad}(A)$ and $a = a_1 \bmod \mathrm{Rad}(A)$. Now inductively we suppose that we have constructed an element $a_k \in A$ such that $h(a_k) = 0 \bmod \mathrm{Rad}(A)^k$ and $a = a_k \bmod \mathrm{Rad}(A)$. We set $a_{k+1} = a_k + r_k$, where $r_k \in \mathrm{Rad}(A)^k$ is to be determined so that $h(a_{k+1}) = 0 \bmod \mathrm{Rad}(A)^{k+1}$. According to Taylor's theorem for polynomials we have

$$h(a_{k+1}) = h(a_k + r_k) = h(a_k) + h'(a_k)r_k + gr_k^2,$$

(where $'$ denotes derivation, and g is a polynomial in a_k and r_k). Now since F is perfect and h is square-free we have that $\gcd(h, h') = 1$. This means that there are polynomials $p, q \in F[X]$ such that $ph + qh' = 1$. We set $r_k = -h(a_k)q(a_k)$, then $r_k \in \mathrm{Rad}(A)^k$. Also

$$h(a_{k+1}) = h(a_k) + h'(a_k)r_k \bmod \mathrm{Rad}(A)^{k+1}$$
$$= h(a_k) - h'(a_k)q(a_k)h(a_k) \bmod \mathrm{Rad}(A)^{k+1}$$
$$= ph^2(a_k) \bmod \mathrm{Rad}(A)^{k+1} = 0 \bmod \mathrm{Rad}(A)^{k+1}.$$

Since $\mathrm{Rad}(A)$ is nilpotent we will find the required element a' after a finite number of steps. Because the minimum polynomial of $\bar{a}$ is h, the minimum polynomial of a' is h as well. Hence the subalgebra of A generated by a' is isomorphic to $A/\mathrm{Rad}(A)$. $\square$

A.2   Algebras generated by a single element

Let V be a finite-dimensional vector space defined over the field F. We recall that $\mathrm{End}(V)$ is the set of all linear maps $a : V \to V$. In this section we study (associative) subalgebras of $\mathrm{End}(V)$, that are generated by a single element.

Let A be an associative algebra. An element $e \in A$ is called an *idempotent* if e is non-zero and $e^2 = e$. Two idempotents e_1, e_2 are called *orthogonal* if $e_1 e_2 = e_2 e_1 = 0$. It is straightforward to see that the sum of orthogonal idempotents is again an idempotent. An idempotent is said to be *primitive* if it is not the sum of orthogonal idempotents.

Lemma A.2.1 *Let $a \in \text{End}(V)$ and let A be the associative algebra (with one) generated by a. Let f_a be the minimum polynomial of a and suppose that*

$$f_a = p_1^{k_1} \cdots p_r^{k_r}$$

is the factorization of f_a into irreducible factors, such that p_i has leading coefficient 1 and $p_i \neq p_j$ for $1 \leq i, j \leq r$. Then A contains r orthogonal primitive idempotents $e_1, \ldots, e_r$ such that $p_i^{k_i}(a)e_i = 0$ and $e_1 + \cdots + e_r = 1$.

Proof. For $1 \leq i \leq r$ let q_i be the product of all factors $p_j^{k_j}$ except $p_i^{k_i}$, i.e.,

$$q_i = p_1^{k_1} \cdots p_{i-1}^{k_{i-1}} p_{i+1}^{k_{i+1}} \cdots p_r^{k_r}.$$

Then since $\gcd(q_1, \ldots, q_r) = 1$ there are polynomials $h_1, \ldots, h_r$ such that $h_1 q_1 + \cdots + h_r q_r = 1$. Set $e_i = h_i q_i(a)$ for $1 \leq i \leq r$. Then $e_1 + \cdots + e_r = 1$. Also for $i \neq j$ we have $e_i e_j = h_i h_j q_i q_j(a) = 0$ since f_a divides $q_i q_j$. Hence

$$e_i = e_i(e_1 + \cdots + e_r) = e_i^2.$$

Furthermore, $p_i^{k_i}(a)e_i = h_i f_a(a) = 0$. Finally we show that the idempotents e_i are primitive. Suppose that $e_i = x + y$, where x and y are orthogonal idempotents in A. Then, since $x, y \in A$, there are polynomials $g, h \in F[X]$ such that $x = g(a)$ and $y = h(a)$. Also $e_i x = (x + y)x = x$ and likewise $e_i y = y$. Now from $xy = 0$ it follows that $gh(a) = 0$ so f_a divides gh. This implies that p_i divides g or p_i divides h. If p_i divides g then there is an $m > 0$ such that $(q_i g)^m(a) = 0$. But $x = g(a) = e_i g(a) = h_i q_i g(a)$ which entails $x^m = 0$, contradicting the assumption that x is an idempotent. If p_i divides h then we get a contradiction in the same way. Therefore e_i is a primitive idempotent. $\qquad\square$

Remark. (Notation as in Lemma A.2.1.) Let X be an indeterminate and suppose that $f_a = p_1 \cdots p_r$, where $p_i = X - \alpha_i$. Suppose further that $\alpha_i \neq \alpha_j$ if $i \neq j$. In this case we can get the idempotents via the *Lagrange polynomials*. The i-th Lagrange polynomial is defined to be

$$\Lambda_i(X) = \prod_{k=1, k\neq i}^{r} \frac{X - \alpha_k}{\alpha_i - \alpha_k}.$$

Now let $y_1, \ldots, y_r$ be r arbitrary elements of the ground field. Then the polynomial

$$g = \Lambda_1 y_1 + \cdots + \Lambda_r y_r$$

is the unique polynomial of degree $r - 1$ such that $g(\alpha_i) = y_i$ for $1 \leq i \leq r$. Taking all y_i equal to 1, this implies that $\sum_i \Lambda_i = 1$. So if we set $e_i = \Lambda_i(a)$, then $e_1 + \cdots + e_r = 1$. Also $e_i e_j = 0$ for $i \neq j$, and hence the e_i are orthogonal idempotents.

We describe how the factors of the minimum polynomial of $a \in \text{End}(V)$ are related to a decomposition of V into a direct sum of subspaces invariant under a. For a univariate polynomial $p \in F[X]$ put

$$V_0(p(a)) = \{v \in V \mid p(a)^m v = 0 \text{ for an } m > 0\}.$$

Lemma A.2.2 *Let a, f_a, p_i and k_i for $1 \leq i \leq r$ be as in Lemma A.2.1. Then V decomposes as a direct sum of subspaces that are invariant under a,*

$$V = V_0(p_1(a)) \oplus \cdots \oplus V_0(p_r(a)).$$

Furthermore, the minimum polynomial of the restriction of a to $V_0(p_i(a))$ is $p_i^{k_i}$.

Proof. Let $v \in V_0(p_i(a))$. Then there is an $m > 0$ such that $p_i(a)^m v = 0$. So $p_i(a)^m (av) = a p_i(a)^m v = 0$ and consequently $V_0(p_i(a))$ is invariant under a.

Let $e_1, \ldots, e_r$ be the primitive orthogonal idempotents provided by Lemma A.2.1. Then since $e_1 + \cdots + e_r = 1$,

$$V = (e_1 + \cdots + e_r)V = e_1 V + \cdots + e_r V.$$

This sum is direct because for $y \in e_i V \cap e_j V$ we have $y = e_i u = e_j v$ for some $u, v \in V$; and as e_i is an idempotent, we obtain $y = e_i u = e_i^2 u = e_i e_j v = 0$. Also, because e_i is a polynomial in a the space $e_i V$ is invariant under a. Now we prove that $V_0(p_i(a)) = e_i V$. Let $y \in e_i V$, i.e, $y = e_i z$. Then by Lemma A.2.1, $p_i^{k_i}(a)y = p_i^{k_i}(a)e_i z = 0$ and $y \in V_0(p_i(a))$. On the other hand suppose that $y \in V_0(p_i(a))$. Then $y = y_1 + \cdots + y_r$ where $y_j \in e_j V$. Furthermore, there is an $m > 0$ such that

$$0 = p_i^m(a)y = p_i^m(a)y_1 + \cdots + p_i^m(a)y_r$$

and $p_i^m(a)y_j \in e_jV$ since e_jV is invariant under a. Because V is the direct sum of these spaces every $p_i^m(a)y_j = 0$. But since, as we just proved, $e_jV \subset V_0(p_j(a))$, also $p_j^{n_j}(a)y_j = 0$ for some $n_j > 0$. Because p_i^m and $p_j^{n_j}$ are relatively prime this implies $y_j = 0$ for $j \neq i$ and we are done.

Let g_i be the minimum polynomial of the restriction of a to $V_0(p_i(a))$, for $1 \leq i \leq r$. Then since $V_0(p_i(a)) = e_iV$ and $p_i^{k_i}(a)e_iz = 0$ for all $z \in V$, we conclude that $p_i^{k_i}(a)y = 0$ for all $y \in V_0(p_i(a))$. So g_i divides $p_i^{k_i}$, and consequently $g_i = p_i^{l_i}$ where $l_i \leq k_i$. Now let g be the product of the g_i for $1 \leq i \leq r$. For $x \in V$ write $x = x_1 + \cdots + x_r$ where $x_i \in V_0(p_i(a))$. Then $g(a)x = g(a)x_1 + \cdots + g(a)x_r = 0$, so that $g(a) = 0$. Hence f_a divides g, forcing $l_i \geq k_i$. $\qquad\square$

Definition A.2.3 *Let V be a vector space defined over a perfect field F. A linear transformation $a \in \mathrm{End}(V)$ is called* semisimple *if the associative subalgebra of $\mathrm{End}(V)$ generated by a is a semisimple algebra.*

Proposition A.2.4 *Let V be a vector space defined over a perfect field F. Let $a \in \mathrm{End}(V)$. Then a is semisimple if and only if the minimum polynomial f_a of a is square-free. If f_a splits into linear factors over F, then a is semisimple if and only if there is a basis of V relative to which the matrix of a is diagonal.*

Proof. Let A be the associative algebra generated by a. Suppose that A is semisimple. If f_a is not square-free, then $f_a = g^2 h$ where $g, h \in F[X]$ are polynomials with leading coefficients 1 and $g \neq 1$. Set $p = gh$, then $p(a)^2 = 0$. Let I be the ideal of A generated by $p(a)$, then every element of I is of the form $q(a)p(a)$, where $q \in F[X]$. Hence $I^2 = 0$ and I is a nonzero nilpotent ideal. But this contradicts the assumption on A.

On the other hand, suppose that f_a is square-free. Let $b \in \mathrm{Rad}(A)$, then $b = p(a)$ for a $p \in F[X]$ and $b^k = 0$. Hence $p^k(a) = 0$, so f_a divides p^k. Since f_a is square-free this can only happen if f_a divides p. But then $b = p(a) = 0$. As a consequence, $\mathrm{Rad}(A) = 0$ and A is a semisimple algebra.

Suppose that $f_a(X) = (X - \lambda_1) \cdots (X - \lambda_t)$ is the factorization of f_a into linear factors. Suppose that a is semisimple so that all these factors are distinct. Then by Lemma A.2.2,

$$V = V_0(a - \lambda_1 \cdot 1) \oplus \cdots \oplus V_0(a - \lambda_t \cdot 1),$$

and the minimum polynomial of the restriction of a to $V_0(a - \lambda_i \cdot 1)$ equals $X - \lambda_i$. Hence a acts as $\lambda_i 1$ on $V_0(a - \lambda_i \cdot 1)$. It follows that a acts diagonally on V. On the other hand, suppose that there is a basis of V with respect to which the matrix of a is diagonal. Then the minimum polynomial is the product of $X - \lambda_i$, where λ_i runs over the set of (distinct) diagonal elements of a. So f_a is square-free. □

Corollary A.2.5 *Let V, F be as in Proposition A.2.4. Let $a, b \in \mathrm{End}(V)$ be two semisimple linear transformations. Furthermore, assume that a and b commute (i.e., $ab = ba$). Then $a + b$ is semisimple.*

Proof. For this we may suppose that the ground field is algebraically closed. Indeed, a, b are also semisimple elements of $\mathrm{End}(V \otimes_F \overline{F})$, where $\overline{F}$ is the algebraic closure of F. Then there is a basis of V with respect to which a acts diagonally. Let λ be an eigenvalue of a, and $V_\lambda = \{v \in V \mid a \cdot v = \lambda v\}$ the corresponding eigenspace. Then V_λ is stable under b since a and b commute. Furthermore, the restriction of b to V_λ is also semisimple. Hence there is a basis of V_λ with respect to which b acts diagonally. Adding these bases together for all eigenspaces V_λ of a, we obtain a basis of V relative to which both a and b act diagonally. Hence also $a + b$ acts diagonally, and therefore $a + b$ is semisimple. □

Proposition A.2.6 *Let V be a finite-dimensional vector space over the perfect field F. Let $a \in \mathrm{End}(V)$ and let A be the associative algebra with one generated by a. Then there are $s, n \in A$ such that*

1. s and n are polynomials in a without constant term,

2. s is semisimple,

3. n is nilpotent,

4. $a = s + n$.

Also, if $f_a = p_1^{k_1} \cdots p_r^{k_r}$ is the minimum polynomial of a (where p_i is irreducible for $1 \le i \le r$). Then $f_s = p_1 \cdots p_r$ is the minimum polynomial of s. Furthermore, if $a = s_1 + n_1$, where s_1 is semisimple and n_1 is nilpotent and $[s_1, n_1] = 0$, then $s_1 = s$ and $n_1 = n$.

Proof. We let a', h and a_k for $k \ge 1$ be as in the (partial) proof of Theorem A.1.5. Put $s = a'$ and $n = a - s$. The statement about the minimum polynomial of s is contained in the (partial) proof of Theorem A.1.5. Hence by Proposition A.2.4 it follows that s is semisimple. Furthermore, n is nilpotent because $a = s \mod \mathrm{Rad}(A)$, and therefore $n \in \mathrm{Rad}(A)$. We show that there are polynomials f, g without constant term such that $s = f(a)$ and $n = g(a)$. First of all, if the minimum polynomial of a has a constant term, then we are done because in that case we can write 1 as a polynomial in a without constant term. If the minimum polynomial of a does not have a constant term, then also the polynomial h has no constant term. From this it follows that all a_k are polynomials in a without constant term. Hence the same holds for the element a'.

As $a = s_1 + n_1 = s + n$ we have $s_1 - s = n - n_1$. Since s and n are polynomials in a and s_1, n_1 commute with a, they also commute with s and n. In particular n and n_1 commute. Hence $n - n_1$ is nilpotent. Also s and s_1 commute and since sums of commuting semisimple transformations are semisimple (Corollary A.2.5), we have that $s_1 - s$ is semisimple. So $s_1 - s$ is both semisimple and nilpotent and hence must be zero.   □

Definition A.2.7 *Let V and a be as in Proposition A.2.6. The decomposition $a = s + n$ provided by Proposition A.2.6 is called the* Jordan *decomposition of a.*

The (partial) proof of Theorem A.1.5 gives us an algorithm for calculating the Jordan decomposition of a linear transformation

Algorithm JordanDecomposition
Input: an endomorphism a of a finite-dimensional vector space, defined over a perfect field.
Output: a tuple (s, n) such that $a = s + n$ is the Jordan decomposition of a.

Step 1 Let f_a be the minimum polynomial of a and let h be the product of the irreducible factors of f_a.

Step 2 Compute polynomials p, q such that $ph + qh' = 1$. Set $a_k := a$.

Step 3 If $h(a_k) = 0$ then return $(a_k, a - a_k)$.

Step 4 Set $a_k := a_k - hq(a_k)$ and return to Step 3.

Comments: The polynomial h in Step 1 can be computed by dividing f_a by $\gcd(f_a, f_a')$. The proofs of Theorem A.1.5 and Proposition A.2.6 show that the output is the Jordan decomposition of a.

A.3   Central idempotents

Let A be an associative algebra with one. We say that an idempotent $e \in A$ is *central* if e lies in the centre of A. Now let A be a semisimple associative algebra. By Wedderburn's structure theorem A decomposes as $A = A_1 \oplus \cdots \oplus A_s$, where the A_i are simple ideals of A. So the identity element uniquely decomposes as $1 = e_1 + \cdots + e_s$, where $e_i \in A_i$. This implies that $e_i e_j = 0$ and hence $e_i = e_i(e_1 + \cdots + e_s) = e_i^2$, so that the e_i are orthogonal idempotents. Furthermore,

$$A = (e_1 + \cdots + e_s)A = e_1 A + \cdots + e_s A$$

and since $e_i A \subset A_i$ we have that $e_i A = A_i$. This also implies that $e_i a = a$ for all $a \in A_i$. Now for $a \in A$ we write $a = a_1 + \cdots + a_s$ where $a_i \in A_i$, and calculate $e_i a = e_i a_i = a_i$ and similarly $a e_i = a_i$. Therefore e_i is central. Also it is clear that e_i is not the sum of two other orthogonal idempotents that are central, because otherwise A_i would split as a direct sum of ideals. So the decomposition of A as a direct sum of simple ideals gives us a set of orthogonal primitive and central idempotents with sum 1. Conversely, such a set of idempotents leads to a decomposition of A into a direct sum of simple ideals. We summarize this in the following proposition.

Proposition A.3.1 *Let A be a semisimple associative algebra with one. Then A decomposes as a direct sum of simple ideals*

$$A = A_1 \oplus \cdots \oplus A_s$$

if and only if there is a set of central primitive and orthogonal idempotents $e_1, \ldots, e_s \in A$ such that $e_1 + \cdots + e_s = 1$ and $A_i = e_i A$.

Let A be a semisimple associative algebra with one over the field F. It is the objective of this section to describe an algorithm for calculating a set of central primitive and orthogonal idempotents in A.

An algorithm for calculating the centre of A can be formulated analogously to the algorithm Centre for Lie algebras. So let T be the centre of A. Then T is a commutative associative algebra containing no nilpotent elements. (Indeed, any nilpotent element of T generates a nilpotent ideal of A.) In particular, T is a semisimple algebra itself. So by Wedderburn's structure theorem we have that T decomposes $T = T_1 \oplus \cdots \oplus T_r$, where T_i is isomorphic to a full matrix algebra defined over a division algebra over F. But T_i is commutative, and hence T_i is isomorphic to an algebra of 1×1-matrices over a division algebra, i.e., T_i is a division algebra over F. Now the commutativity of T_i implies that T_i is a field, so T_i is a field extension of F.

The left-multiplication on T provides a faithful representation of T as T has an identity. Hence T is isomorphic to a linear algebra of $m \times m$-matrices over F, where $m = \dim T$. So we may think of T as an algebra consisting of $m \times m$-matrices. Since T is commutative and consists of semisimple elements we have that these matrices are simultaneously diagonalizable over the algebraic closure $\overline{F}$ of F. This means that there is an $m \times m$-matrix with coefficients in $\overline{F}$ such that xtx^{-1} is in diagonal form for all $t \in T$. For $1 \leq i \leq m$ we define the F-linear function $\alpha_i : T \to \overline{F}$ by letting $\alpha_i(t)$ be the coefficient on position (i, i) of xtx^{-1}. We note that $\alpha_i \neq \alpha_j$ if $i \neq j$ because the dimension of T is equal to the number of distinct α_i.

An element $t_0 \in T$ is said to be a *splitting element* if t_0 generates T.

Lemma A.3.2 *Suppose that the ground field F contains at least $\frac{m(m-1)}{2} + 1$ elements, then T has splitting elements.*

Proof. An element $t_0 \in T$ is a splitting element if and only if $\alpha_i(t_0) \neq \alpha_j(t_0)$ for $i \neq j$. Indeed, the dimension of the algebra generated by t_0 is equal to the number of distinct eigenvalues of t_0. And this number is equal to $\dim T$ if and only if $\alpha_i(t_0) \neq \alpha_j(t_0)$ for $i \neq j$. Now let $\{t_1, \dots, t_m\}$ be a basis of T and for $i < j$ define polynomials

$$f_{ij}(X_1, \dots, X_m) = \sum_{k=1}^{m} (\alpha_i - \alpha_j)(t_k) X_k.$$

We let g be the product of all f_{ij}. Then g has degree $\frac{m(m-1)}{2}$. The lemma follows now from the observation that $t_0 = \sum_{k=1}^{m} \lambda_k t_k$ is a splitting element if and only if $g(\lambda_1, \dots, \lambda_m) \neq 0$. $\qquad\qquad\square$

The proof of this lemma yields a straightforward Las Vegas algorithm for finding a splitting element if the ground field is big. Let g be the polynomial from the proof of Lemma A.3.2. Let Ω be a subset of F of size at least $\frac{m(m-1)}{2\varepsilon}$, and set $t_0 = \sum_{i=1}^{m} \lambda_k t_k$, where the λ_k are randomly chosen from Ω (with uniform distribution). As seen in the proof of Lemma A.3.2, t_0 is a splitting element if and only if $g(\lambda_1, \dots, \lambda_m) = 0$. So by Corollary 1.5.2, the probability that t_0 is a splitting element is at least $1 - \varepsilon$. In the algorithm we select a random element t_0 with coefficients from Ω. Then we check whether t_0 generates T. If so, then we output t_0, otherwise we select a second element and continue. By making the set Ω big we can make the probability that a randomly chosen element t_0 is not a splitting element arbitrarily small.

The proof of Lemma A.2.1 shows how to calculate a set of primitive orthogonal idempotents with sum 1 in T, knowing a splitting element t_0. Let f_{t_0} be the minimum polynomial of t_0 and let $f_{t_0} = p_1 \cdots p_r$ be the factorization of f_{t_0} into irreducible factors. Let q_i be the product of all factors p_j except p_i. Then we calculate polynomials $h_1, \ldots, h_r$ such that $h_1 q_1 + \cdots + h_r q_r = 1$. Set $e_i = h_i q_i(a)$, then the proof of Lemma A.2.1 shows that the e_i are primitive orthogonal idempotents with sum 1.

If A is defined over a small field, then the centre T may not contain splitting elements. In this case we rather work with *decomposing* and *good* elements (cf. Definition 4.11.5). An element $t \in T$ is decomposing if the minimum polynomial of t factors as a product of at least two distinct factors (of degree greater than 1). If $S \subset T$ is a subalgebra, then $s \in S$ is called good with respect to S if s generates S and the minimum polynomial of s is irreducible. A decomposing element leads to a decomposition of T as a sum of two or more ideals (by calculating idempotents in the same way as from a splitting element). A good element with respect to a subalgebra S shows that S is simple. Exactly as in the proof of Proposition 4.11.6 we can show that a random element $s \in S$ is either decomposing or good with probability at least $\frac{1}{2}$. Now using an algorithm analogous to PrimaryDecompositionSmallField we can decompose T into the direct sum of its simple ideals, and this gives us a set of primitive orthogonal idempotents with sum 1. The conclusion is that we have an algorithm for calculating a set of central primitive and orthogonal idempotents in any semisimple associative algebra. We call this algorithm CentralIdempotents.

Suppose that A is not semisimple and set $S = A/\mathrm{Rad}(A)$. Suppose that we have found a set $\{\bar{e}_1, \ldots, \bar{e}_r\}$ of primitive central and orthogonal idempotents in S. By the Wedderburn-Malcev principal theorem, S is isomorphic to a subalgebra of A. Hence there are primitive orthogonal idempotents $e_1, \ldots, e_r \in A$ with sum 1, such that $e_i \bmod \mathrm{Rad}(A) = \bar{e}_i$. We describe an algorithm for finding the e_i, given the $\bar{e}_i$.

Lemma A.3.3 *Let $\bar{e}$ be an idempotent in S. Then we can construct an idempotent $e \in A$ such that $e \bmod \mathrm{Rad}(A) = \bar{e}$.*

Proof. Let $e_0 \in A$ be such that $e_0 \bmod \mathrm{Rad}(A) = \bar{e}$. Then $e_0^2 = e_0 \bmod \mathrm{Rad}(A)$ so that $n_0 = e_0^2 - e_0 \in \mathrm{Rad}(A)$ and consequently n_0 is nilpotent.

Hence there is an integer q such that $n_0^{2^q} = 0$. Now recursively set

$$e_{i+1} = e_i + n_i - 2e_i n_i,$$
$$n_{i+1} = e_{i+1} e_{i+1} - e_{i+1}.$$

We note that all e_i, n_i lie in the commutative subalgebra E_0 of A generated by e_0. Let M_i be the ideal of E_0 generated by $n_0^{2^i}$. A short calculation shows that $n_{i+1} = 4n_i^3 - 3n_i^2$. Hence by induction on i we see that $n_i = 0 \bmod M_i$. Therefore $e_{i+1} = e_i \bmod M_i$, and in particular, $e_{i+1} = e_0 \bmod \mathrm{Rad}(A)$. We conclude that $n_q = 0$ and $e_q e_q = e_q$ and the statement follows. $\qquad\square$

Remark. A small calculation shows that $e_{i+1} = -2e_i^3 + 3e_i^2$. Now if we calculate the Jordan decomposition of e_0 (using the algorithm JordanDecomposition), then we get the same recursion. It follows that the algorithm contained in the proof of Lemma A.3.3 constructs the Jordan decomposition of e_0.

Proposition A.3.4 *Let $\bar{e}_1, \ldots, \bar{e}_r$ be primitive orthogonal idempotents in S with sum $1 \in S$. Then we can construct primitive orthogonal idempotents $e_1, \ldots, e_r$ in A such that $e_i \bmod \mathrm{Rad}(A) = \bar{e}_i$ for $1 \leq i \leq r$, and $e_1 + \cdots + e_r = 1 \in A$.*

Proof. For brevity we set $R = \mathrm{Rad}(A)$. The proof is by induction on r. The case where $r = 1$ is covered by Lemma A.3.3. Suppose that $r > 1$ and that we have constructed primitive orthogonal idempotents $e_1, \ldots, e_{r-1}$ in A such that $e_i \bmod R = \bar{e}_i$ for $1 \leq i \leq r-1$. We describe how to construct e_r. Let $\tilde{e}_r$ be a pre-image of $\bar{e}_r$ in A. Set

$$E = e_1 + \cdots + e_{r-1},$$

and

$$e = \tilde{e}_r - E\tilde{e}_r - \tilde{e}_r E + E\tilde{e}_r E.$$

We have $E\tilde{e}_r = 0 \bmod R$ and $\tilde{e}_r E = 0 \bmod R$, so that $e = \bar{e}_r \bmod R$ and hence $e^2 = e \bmod R$. Now let e_r be the idempotent in A provided by the procedure in the proof of Lemma A.3.3 (where we start with $e_0 = e$). So $e_r = \bar{e}_r \bmod R$. Since $E^2 = E$ we have that $Ee = eE = 0$, and because e_r is a polynomial in e without constant term, it follows that $Ee_r = e_r E = 0$. By the induction hypothesis we have $e_i E = Ee_i = e_i$ for $1 \leq i \leq r-1$. Hence $e_r e_i = e_r E e_i = 0$ and similarly $e_i e_r = 0$.

Finally we show that the idempotents e_i must have sum $1 \in A$. Since $e_i = \bar{e}_i \bmod R$ and the $\bar{e}_i$ have sum 1 in S we see that $e_1 + \cdots + e_r = 1 + u$, for some $u \in R$. But the left-hand side of this expression is an idempotent. So $(1 + u)^2 = 1 + u$. But this is equivalent to $u^2 + u = 0$. Which implies $u = 0$ (in view of $u \in R$). □

A.4 Notes

Proposition A.1.1 first appeared in [25]. It was used by Friedl and Rónyai ([29], [73]) to give an algorithm for computing the radical of an associative algebra of characteristic 0. These papers also contain an algorithm for computing the radical of an algebra defined over a finite field. Furthermore, the authors show that the complexity of this algorithm is polynomial. Also [20] contains a polynomial time algorithm for calculating the radical of an algebra defined over a finite field. In [47] algebras are considered that are defined over fields of the form $\mathbb{F}_q(x_1, \ldots, x_m)$. An algorithm for calculating the radical is given that runs in polynomial time, if the number of indeterminates, m, is a constant.

Splitting elements were introduced by W. Eberly ([26], [27], [28]). In these references full matrix algebras $M_n(F)$ are considered, and an element $a \in M_n(F)$ is defined to be a splitting element if its minimal polynomial is squarefree and of degree n (note that our definition is a special case of this). In [27] splitting elements are used to calculate the decomposition of an associative algebra into a direct sum of simple ideals. A second application of splitting elements can be found in [3], where algorithms are given to decompose a module over a finite group into a direct sum of irreducible modules. Our presentation here is different from the ones in the above references, in order to stress the similarities with splitting elements in Cartan subalgebras (Section 4.11). Decomposing elements appear in [28] (where they are called decomposable elements). Other algorithms for computing idempotents and decompositions of algebras can be found in [24] (for commutative Artinian algebras over finite fields), [32] (over finite fields and the field $\mathbb{Q}$ of rational numbers; in the latter case lifting of idempotents over a finite field is used) [29], [73] (over finite fields, using a splitting procedure that has some similarity to the approach with decomposing elements).

The algorithm for finding the Jordan decomposition of a linear transformation is taken from [2]. In this paper the authors also show that the complexity of the algorithm is polynomial. The algorithm for lifting idempotents is taken from [90].

Bibliography

[1] J. Apel and W. Lassner. An extension of Buchberger's algorithm and calculations in enveloping fields of Lie algebras. *J. Symbolic Comput.*, 6(2-3):361–370, 1988.

[2] L. Babai, R. Beals, J. Cai, G. Ivanyos, and E. M. Luks. Multiplicative equations over commuting matrices. In *Proceedings of the Seventh Annual ACM-SIAM Symposium on Discrete Algorithms*, pages 498–507, New York, 1996. ACM.

[3] L. Babai and L. Rónyai. Computing irreducible representations of finite groups. *Math. Comp.*, 55(192):705–722, 1990.

[4] Y. A. Bahturin, A. A. Mikhalev, V. M. Petrogradsky, and M. V. Zaicev. *Infinite-dimensional Lie superalgebras*. Walter de Gruyter & Co., Berlin, 1992.

[5] R. E. Beck and B. Kolman. Computers in Lie algebras. I. Calculation of inner multiplicities. *SIAM J. Appl. Math.*, 25:300–312, 1973.

[6] R. E. Beck and B. Kolman. Computers in Lie algebras. II. Calculation of outer multiplicities. *SIAM J. Appl. Math.*, 25:313–323, 1973.

[7] R. E. Beck, B. Kolman, and I. N. Stewart. Computing the structure of a Lie algebra. In R. E. Beck and B. Kolman, editors, *Non-associative rings and algebras*, pages 167–188. Academic Press, New York, 1977.

[8] J. G. F. Belinfante and B. Kolman. *A Survey of Lie Algebras with Applications and Computational Methods*. SIAM, Philadelphia, 1972.

[9] G. M. Bergman. The diamond lemma for ring theory. *Adv. in Math.*, 29(2):178–218, 1978.

[10] G. W. Bluman and S. Kumei. *Symmetries and differential equations*. Springer-Verlag, Berlin, Heidelberg, New York, 1989.

[11] L. A. Bokut and A. A. Klein. Serre relations and Gröbner-Shirshov bases for simple Lie algebras. I, II. *Int. J. of Algebra and Computation*, 6(4):389–400, 401–412, 1996.

[12] A. Borel. *Linear algebraic groups*. Springer-Verlag, Berlin, Heidelberg, New York, second edition, 1991.

[13] N. Bourbaki. *Groupes et Algèbres de Lie, Chapitre I*. Hermann, Paris, 1971.

[14] N. Bourbaki. *Groupes et Algèbres de Lie, Chapitres VII et VIII*. Hermann, Paris, 1975.

[15] B. Buchberger. Ein algorithmisches Kriterium für die Lösbarkeit eines algebraischen Gleichungssystems. *Aequationes Math.*, 4:374–383, 1970.

[16] B. Buchberger. Gröbner bases: an algorithmic method in polynomial ideal theory. In N. K. Bose, editor, *Multidimensional Systems Theory*, pages 184–232. D. Reidel, Dordrecht, 1985.

[17] B. Buchberger. Introduction to Gröbner bases. In B. Buchberger and F. Winkler, editors, *Gröbner Bases and Applications*, volume 251 of *LMS Lecture Note Series*, pages 3–31. Cambridge University Press, 1998.

[18] R. W. Carter. *Simple groups of Lie type*. John Wiley & Sons, London-New York-Sydney, 1972. Pure and Applied Mathematics, Vol. 28.

[19] C. Chevalley. *Théorie des Groupes de Lie, Tome III*. Hermann, Paris, 1955.

[20] A. M. Cohen, G. Ivanyos, and D. B. Wales. Finding the radical of an algebra of linear transformations. *J. Pure Appl. Algebra*, pages 177–193, 1997.

[21] A. M. Cohen, M. A. A. van Leeuwen, and B. Lisser. *LiE a Package for Lie Group Computations*. CAN, Amsterdam, 1992.

[22] Computational Algebra Group, School of Mathematics and Statistics University of Sydney, Australia. *The Magma System for Algebra, Number Theory and Geometry*.

[23] D. Cox, J. Little, and D. O'Shea. *Ideals, Varieties and Algorithms: An Introduction to Computational Algebraic Geometry and Commutative Algebra*. Springer Verlag, New York, Heidelberg, Berlin, 1992.

[24] R. A. Davis. Idempotent computation over finite fields. *J. Symbolic Comput.*, 17(3):237–258, 1994.

[25] L. E. Dickson. *Algebras and Their Arithmetics.* University of Chicago Press, Chicago, 1923.

[26] W. Eberly. Decomposition of algebras over finite fields and number fields. *Comput. Complexity*, 1(2):183–210, 1991.

[27] W. Eberly. Decomposition of algebras over R and C. *Computational Complexity*, 1:211–234, 1991.

[28] W. Eberly and M. Giesbrecht. Efficient decomposition of associative algebras. In Y. N. Lakshman, editor, *Proceedings of the 1996 International Symposium on Symbolic and Algebraic Computation: ISSAC'96*, pages 170–178, New York, 1996. ACM.

[29] K. Friedl and L. Rónyai. Polynomial time solutions of some problems in computational algebra. In *Proceedings of the 17th ACM STOC*, pages 153–162, Providence RI, 1985.

[30] W. Fulks. *Advanced Calculus: An introduction to analysis.* John Wiley & Sons Inc., New York, 1961.

[31] The GAP Group, Aachen, St Andrews. *GAP – Groups, Algorithms, and Programming, Version 4.1*, 1999. (http://www-gap.dcs.st-and.ac.uk/~gap).

[32] P. Gianni, V. Miller, and B. Trager. Decomposition of algebras. In P. Gianni, editor, *Symbolic and algebraic computation (Rome, 1988)*, volume 358 of *Lecture Notes in Computer Science*, pages 300–308. Springer, Berlin, 1989.

[33] W. A. de Graaf. Calculating the strucure of a semisimple Lie algebra. *J. Pure Appl. Algebra*, 117&118:319–329, 1997.

[34] W. A. de Graaf. Constructing faithful matrix representations of Lie algebras. In W. W. Küchlin, editor, *Proceedings of the 1997 International Symposium on Symbolic and Algebraic Computation: ISSAC'97*, pages 54–59. ACM Press, 1997.

[35] W. A. de Graaf. Using Cartan subalgebras to calculate nilradicals and Levi subalgebras of Lie algebras. *J. Pure Appl. Algebra*, 139:25–39, 1999.

[36] W. A. de Graaf, G. Ivanyos, A. Küronya, and L. Rónyai. Computing Levi decompositions. *Appl. Alg. in Eng., Comm. and Comp.*, 8(4):291–303, 1997.

[37] W. A. de Graaf, G. Ivanyos, and L. Rónyai. Computing Cartan subalgebras of Lie algebras. *Appl. Alg. in Eng., Comm. and Comp.*, 7(5):339–349, 1996.

[38] W. A. de Graaf and J. Wisliceny. Constructing bases of finitely presented Lie algebras using Gröbner bases in free algebras. In S. Dooley, editor, *Proceedings of the 1999 International Symposium on Symbolic and Algebraic Computation: ISSAC'99*, pages 37–43. ACM Press, 1999.

[39] M. Hall, Jr. A basis for free Lie rings and higher commutators in free groups. *Proceedings of the American Mathematical Society*, 1:575–581, 1950.

[40] M. Hall, Jr. *The Theory of Groups*. Macmillan, New York, 1959.

[41] G. Havas, M. F. Newman, and M. R. Vaughan-Lee. A nilpotent quotient algorithm for graded Lie rings. *J. Symbolic Comput.*, 9(5-6):653–664, 1990.

[42] J. E. Humphreys. *Introduction to Lie Algebras and Representation Theory*. Springer Verlag, New York, Heidelberg, Berlin, 1972.

[43] B. Huppert. *Endliche Gruppen I*. Springer Verlag, New York, Heidelberg, Berlin, 1967.

[44] B. Huppert and N. Blackburn. *Finite Groups II*. Springer Verlag, New York, Heidelberg, Berlin, 1982.

[45] G. Ivanyos. Finding the radical of matrix algebras using Fitting decompositions. *J. Pure Appl. Algebra*, 139:159–182, 1999.

[46] G. Ivanyos and L. Rónyai. Computations in associative and Lie algebras. In A. M. Cohen, H. Cuypers, and H. Sterk, editors, *Some tapas of computer algebra*, volume 4 of *Algorithms and Computation in Mathematics*, pages 91–120. Springer, Berlin, 1999.

[47] G. Ivanyos, L. Rónyai, and A. Szántó. Decomposition of algebras over $\mathbb{F}_q(X_1, \cdots, X_m)$. *Appl. Alg. in Eng., Comm. and Comp.*, 5(2):71–90, 1994.

[48] N. Jacobson. *Lie Algebras*. Dover, New York, 1979.

[49] V. G. Kac. *Infinite Dimensional Lie Algebras*. Cambridge University Press, Cambridge, third edition, 1990.

[50] A. Kandri-Rody and V. Weispfenning. Noncommutative Gröbner bases in algebras of solvable type. *J. Symbolic Comput.*, 9(1):1–26, 1990.

[51] A. U. Klimyk. Decomposition of a direct product of irreducible representations of a semisimple Lie algebra into irreducible representations. In *American Mathematical Society Translations. Series 2, Vol. 76*, pages 63–73. American Mathematical Society, Providence, R.I., 1968.

[52] D. E. Knuth. *The Art of Computer Programming*, volume 1, Fundamental Algorithms. Addison-Wesley, 1968.

[53] D. E. Knuth. *The Art of Computer Programming*, volume 2, Seminumerical Algorithms. Addison-Wesley, third edition, 1998.

[54] A. I. Kostrikin. *Introduction to Algebra*. Springer Verlag, New York, Heidelberg, Berlin, 1982.

[55] G. P. Kukin. The word problem for Lie algebras. *Sibirsk. Mat. Ž.*, 18(5):1194–1197, 1208, 1977.

[56] S. Lang. *Algebraic Number Theory*. Springer Verlag, New York, Heidelberg, Berlin, 1994.

[57] M. A. A. van Leeuwen and M. Roelofs. Termination for a class of algorithms for constructing algebras given by generators and relations. *J. Pure Appl. Algebra*, 117/118:431–445, 1997.

[58] A. K. Lenstra. Factorisation of polynomials. In H. W. Lenstra jr. and R. Tijdeman, editors, *Computational methods in number theory*, volume 154 of *Mathematical Centre Tract*, pages 169–198. Mathematisch Centrum, Amsterdam, 1982.

[59] R. Lidl and H. Niederreiter. *Finite Fields*, volume 20 of *Encyclopedia of Mathematics and its Applications*. Addison-Wesley, Reading, 1983.

[60] S. MacLane and G. Birkhoff. *Algebra*. Macmillan, New York, 1965.

[61] W. McKay, J. Patera, and D. Sankoff. The computation of branching rules for representations of semisimple Lie algebras. In R. E. Beck and B. Kolman, editors, *Computers in nonassociative rings and algebras*, pages 235–277. Academic Press, New York, 1977.

[62] G. Melançon. Combinatorics of Hall trees and Hall words. *J. Combin. Theory Ser. A*, 59(2):285–308, 1992.

[63] R. V. Moody and J. Patera. Fast recursion formula for weight multiplicities. *Bull. Amer. Math. Soc. (N.S.)*, 7(1):237–242, 1982.

[64] F. Mora. Groebner bases for noncommutative polynomial rings. In J. Calmet, editor, *Algebraic algorithms and error correcting codes (Grenoble, 1985), proceedings*, volume 229 of *Lecture Notes in Computer Science*, pages 353–362. Springer, Berlin, 1986.

[65] T. Mora. An introduction to commutative and noncommutative Gröbner bases. *Theoretical Computer Science*, 134(1):131–173, 1994.

[66] A. Navon and J. Patera. Embedding of a simple Lie group into a simple Lie group and branching rules. *J. Mathematical Phys.*, 8:489–493, 1967.

[67] P. J. Olver. *Applications of Lie Groups to Differential Equations*. Springer Verlag, New York, Heidelberg, Berlin, 1993.

[68] J. Patera, P. Winternitz, and H. Zassenhaus. Continuous subgroups of the fundamental groups of physics. I. General method and the Poincaré group. *J. Math. Phys.*, 16:1597–1614, 1975.

[69] R. S. Pierce. *Associative Algebras*. Springer-Verlag, New York, Heidelberg, Berlin, 1982.

[70] G. Racah. Lectures on Lie groups. In *Group theoretical concepts and methods in elementary particle physics (Lectures Istanbul Summer School Theoret. Phys., 1962)*, pages 1–36. Gordon and Breach, New York, 1964.

[71] D. Rand, P. Winternitz, and H. Zassenhaus. On the identification of a Lie algebra given by its structure constants. I. Direct decompositions, Levi decompositions and nilradicals. *Linear Algebra Appl.*, 109:197–246, 1988.

[72] C. Reutenauer. *Free Lie Algebras*, volume 7 of *New Series*. Oxford University Press, Oxford, 1993.

[73] L. Rónyai. Computing the structure of finite algebras. *J. Symbolic Comput.*, 9:355–373, 1990.

[74] J. T. Schwartz. Fast probabilistic algorithms for verification of polynomial identities. *J. Assoc. Comput. Mach.*, 27(4):701–717, 1980.

[75] R. Sedgewick. *Algorithms*. Addison-Wesley Publishing Co., Reading, Mass., 1983.

[76] G. B. Seligman. *Modular Lie Algebras*. Springer Verlag, New York, Heidelberg, Berlin, 1967.

[77] J.-P. Serre. *Lie Algebras and Lie Groups*. W. A. Benjamin, New York, 1965.

[78] J.-P. Serre. *Algèbres de Lie semi-simples complexes*. W. A. Benjamin, inc., New York-Amsterdam, 1966.

[79] A. I. Shirshov. Some algorithmic problems for Lie algebras. *Sib. Mat. Zh.*, 3:292–296, 1962. (Russian).

[80] C. C. Sims. *Computation with Finitely Presented Groups*. Cambridge University Press, Cambridge, 1994.

[81] D. M. Snow. Weyl group orbits. *ACM Trans. Math. Software*, 16(1):94–108, 1990.

[82] D. M. Snow. Computing tensor product decompositions. *ACM Trans. Math. Software*, 19(1):95–108, 1993.

[83] J. Tits. Sur les constantes de structure et le théorème d'existence des algèbres de Lie semi-simples. *Publ. Math. IHES*, 31:21–58, 1966.

[84] P. Turkowski. Low-dimensional real Lie algebras. *J. Math. Phys.*, 29:2139–2144, 1988.

[85] V. A. Ufnarovskij. *Combinatorial and Asymptotic Methods in Algebra*, volume 57 of *Encyclopedia of Mathematical Sciences*, chapter I, pages 1–196. Springer Verlag, Berlin, Heidelberg, New York, 1995.

[86] V. S. Varadaradjan. *Lie Groups, Lie Algebras, and Their Representations*. Prentice-Hall, 1974.

[87] M. R. Vaughan-Lee. An algorithm for computing graded algebras. *J. Symbolic Comput.*, 16(4):345–354, 1993.

[88] D. J. Winter. *Abstract Lie Algebras*. M.I.T. Press, Cambridge, Mass., 1972.

[89] O. Zariski and P. Samuel. *Commutative algebra, Volume I*. D. Van Nostrand Company, Inc., Princeton, New Jersey, 1958. With the cooperation of I. S. Cohen, The University Series in Higher Mathematics.

[90] H. Zassenhaus. Über eine Verallgemeinerung des Henselschen Lemmas. *Arch. Math.*, V:317–325, 1954.

[91] H. Zassenhaus. On the Cartan subalgebra of a Lie algebra. *Linear Algebra Appl.*, 52/53:743–761, 1983.

Index of Symbols

Index of Terminology

Index of Algorithms